Network Management and Control

Volume 2

Network Management and Control

Volume 2

Edited by

Ivan T. Frisch
Polytechnic University
Brooklyn, New York

Manu Malek
AT&T Bell Laboratories
Middletown, New Jersey

and

Shivendra S. Panwar
Polytechnic University
Brooklyn, New York

Plenum Press • New York and London

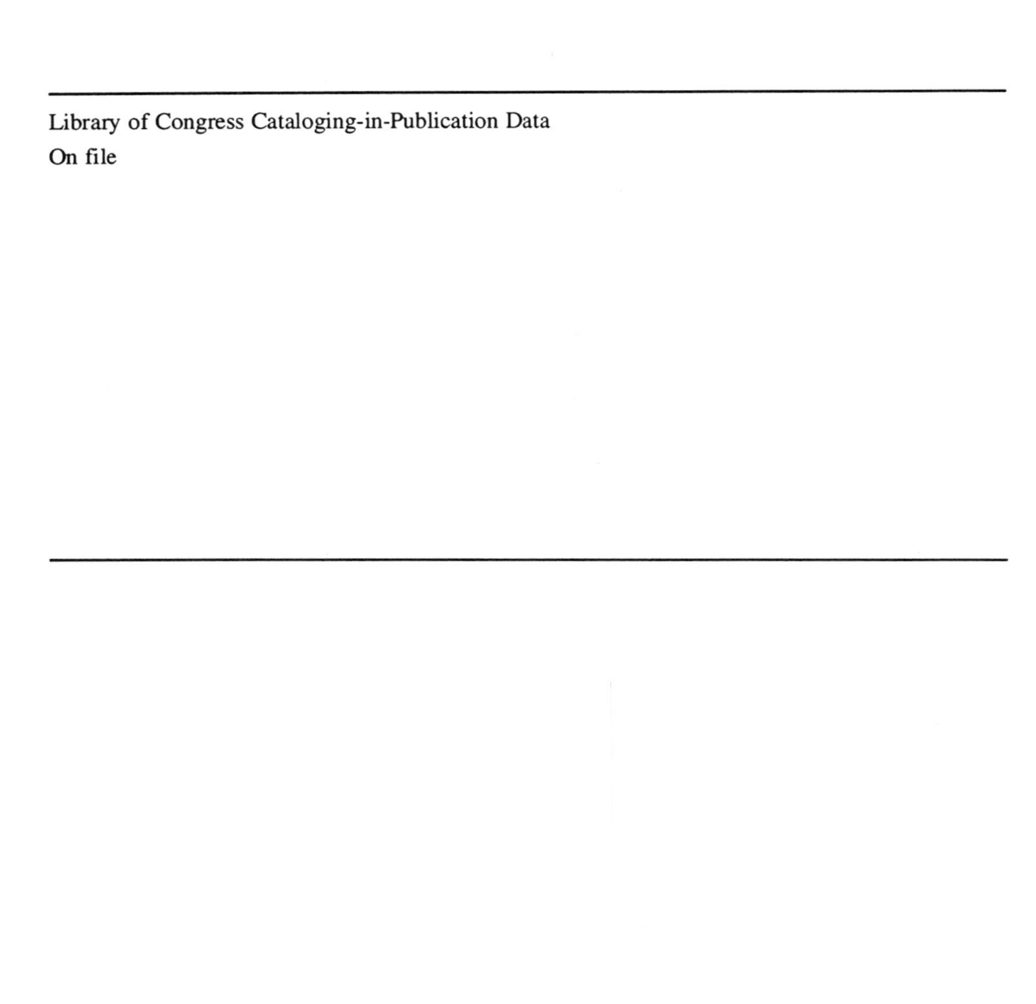

Library of Congress Cataloging-in-Publication Data
On file

Based on the proceedings of the Second IEEE Network Management and Control Workshop,
held September 21–23, 1993, in Tarrytown, New York

ISBN 0-306-44807-6

A Division of Plenum Publishing Corporation
233 Spring Street, New York, N.Y. 10013

Printed in the United States of America

PREFACE

Three speakers at the Second Workshop on Network Management and Control nostalgically remembered the INTEROP Conference at which SNMP was able to interface even to CD players and toasters. We agreed this was indeed a major step forward in standards, but wondered if anyone noticed whether the toast was burned, let alone, would want to eat it.

The assurance of the correct operation of practical systems under difficult environments emerged as the dominant theme of the workshop with growth, interoperability, performance, and scalability as the primary sub-themes.

Perhaps this thrust is unsurprising, since about half the 100 or so attendees were from industry, with a strong contingency of users. Indeed the technical program co-chairs, Shivendra Panwar of Polytechnic and Walter Johnston of NYNEX, took as their assignment the coverage of real problems and opportunities in industry.

Nevertheless we take it as a real indication of progress in the field that the community is beginning to take for granted the availability of standards and even the ability to detect physical, link, and network-level faults and is now expecting diagnostics at higher levels as well as system-wide solutions.

Representatives of two organizations whose business is delivering on-line real-time services, recognized that life is what happens, after you make other plans. They expressed their critical need for early detection of anomalies and faults. In one case, their primary need was to rapidly take the faulty system off-line for diagnosis and switch to a back-up with the briefest restoration time. "I don't want an engineer applying a datascope to an on-line system." For the other, the need is to be able to effectively solve customers' problems without involving them in excessive dialogue. The customer does not appreciate three different people calling him to re-solicit the details of his complaint. Both acknowledged the efficacy of protocols, standards, and network management platforms, but were more concerned with performance and fault monitoring and higher layer protocols, even applications; to solve these problems they had to write a great deal of software themselves.

The criticality of software was highlighted by Larry Bernstein, Chief Technical Officer of Operations Systems at AT&T Network Systems. Many network management functions at the physical, link, and even network layer are built into hardware. However, most systems and management functions are in software, which is difficult to build, test, and diagnose. From the point of view of the producers of major telecommunications switching software, a key to effective systems is "stable software" that can be installed, grown, and used reliably. The key is simple code structures and reusable modules.

However, to the extent that users must write their own software, we are placing the burden on operations less able to effectively deal with these issues, a long way from Bernstein's ideal. Of course the problems will become even tougher. Richard Mandelbaum, Director of the Center for Advanced Technology in Telecommunications at Polytechnic University, highlighted the example of the Internet as a highly utilized system that is growing wildly in its number of users. By December of 1993 there were 2,000,000 hosts on the system. If the present rate of growth were to continue, by the year 2010 the number of hosts would exceed the number of humans on the globe. There will be escalating problems in management, performance, faults, security, and intrusion.

Anna Melamed, Senior Vice President at Technology Solutions Company, works with a number of financial houses. Her view is that users have to solve many problems they would rather have handled by vendors. They have to develop their own systems for the navigation and distribution of alarms, and their own automated knowledge bases for configuration control. A key problem she faces is the lack of scalability of solutions. As systems grow in size and complexity, vendor-provided solutions lose their effectiveness and the customer becomes its own integrator.

One solution discussed was "out-sourcing" network management. The judgment was that this was often a sound business decision. But in other cases it is an approach taken too quickly. In the words of H.L. Mencken, "There is a solution to every problem; simple, quick, and wrong." You cannot out-source a function unless you understand it and can monitor and control it. As one of our tutorial speakers, Kornel Terplan, put it "You cannot out-source chaos."

At the panel discussion we sought to define the model of a system that performs its designated functions and can be successfully managed when there is a problem. In trying to define the paradigm for the future we tried to extrapolate from the present. Although we all know of their failures and shortcomings, the two systems that everyone agreed were most successful were "the telephone network" and the Internet, which are diametrically opposed in most aspects. The telephone network is essentially under central control, is dominated by voice traffic, with controlled access (less so in recent years). The Internet ain't. It is loosely coupled, with both central and distributed control, dominated by bursty data, packet switched, with users easily coming, going, and playing. Of course, the "grade of service," and performance and user requirements for the two are very different. Not suprisingly, some of the most important problems are in medium-scale networks consisting of heterogeneous loosely coupled bridged, tightly coupled LANs and WANs. What does this say for the management in the future of ATM, multimedia systems, hybrid CATV, and telephone networks and personal communication systems?

The workshop was structured to encourage audience participation and interaction. The meeting began with two tutorials. Directions and issues were set in the keynote and plenary session talks. There was also a panel discussion, a luncheon address, and a wrap-up session. The sessions of refereed papers focused on specific developments within the framework of the workshop themes.

This book is based on the presentations given at the workshop and has been organized into sections that encompass most of the presentations on the major themes of the workshop, as well as perspectives on network management in the form of keynote addresses. The sections correspond to workshop sessions.

Session I on Network Management for Security and Reliability covered new techniques for diagnostics and authentication, and examples of practical system implementations. Session II on Management of High-Speed Digital Networks contained papers on SMDS, ATM, shared-media access, and multimedia services. Session III was a panel discussion on Trends in Network Management Capabilities. Session IV on Network Management Protocols contained papers on developments in OSI, semantic-modeling tools, and connection establishment in broadband networks. Session V on Implementation of advanced Technology in Network Management included descriptions of implementations of management for mobile communications, alarm correlation, space telephone networks, and protocol-independent MIBs. Session VI on Decision Support Systems for Planning and Performance discussed network management platforms, adaptive techniques, artificial intelligence, and fuzzy logic approaches to planning.

The editors wish to express their thanks to the authors and to the reviewers of the papers, the organizing and program committees for their help and encouragement in the preparation of the book, for their efforts in making the workshop a success and for bringing together people at the forefront of this technology for an extremely useful interchange of concepts and experiences.

I. Frisch
M. Malek
S. Panwar

ACKNOWLEDGMENTS

The workshop was jointly sponsored by the Polytechnic University and the New York State Science and Technology Foundation and its Center for Advanced Technology in Telecommunications (CATT), NYNEX Corporation, and the IEEE Communications Society's Committee on Network Operations and Management (CNOM).

The General Chairman of the Workshop was Ivan Frisch, Provost of Polytechnic University. Walter Johnston of NYNEX Corporation and Shivendra Panwar of Polytechnic University co-chaired the Technical Program Committee. The Organizing Committee consisted of, in addition to the above, Judy Keller, IEEE Communications Society (Promotion and Publicity Chairperson); Manu Malek, AT&T Bell Laboratories (CNOM Representative); and Hart Rasmussen, Polytechnic University (Local Arrangements and Finance). The Technical Committee consisted of Stephen Brady, IBM T. J. Watson Research Center; Jeffrey D. Case, SNMP Research, Inc.; Roberta Cohen, AT&T Bell Laboratories; Ziny Flikop, NYNEX Corp.; Michael Hluchyi, Motorola-Codex Corp.; Aaron Kershenbaum, IBM T. J. Watson Research Center; Shaygan Kheradpir, GTE Laboratories; Fred Klapproth, Bell Communications Research; Richard Mandelbaum, CATT, Polytechnic University; Prem Mehrotra, AT&T Bell Laboratories; John Paggi, AT&T Bell Laboratories; Steven Rathgaber, NY Switch Corp.; Rajan Rathnasabapathy, NEC America, Inc.; Dror Segal, Securities Industry Automation Corp.; Edward Talley, Grumman Data Systems; and Yechiam Yemini, Columbia University.

Individual workshop sessions were organized and chaired as follows:

Session 1: *Network Management for Security and Reliability* - Session Chairman: Dror Segal, Securities Industry Automation Corp.;

Session 2: *Management of High-Speed Digital Networks* - Session Chairman: Andrew Mayer, Bellcore;

Session 3 (Panel): *Trends in Network Management Capabilities* - Session Chairman: Walter Johnston, NYNEX Corp.;

Session 4: *Network Management Protocols* - Session Chairman: Rajan Rathnasabapathy, NEC America, Inc.;

Session 5: *Implementations of Advanced Technology in Network Management* - Session Chairman: Stephen Brady (IBM);

Session 6: *Decision Support Systems for Planning and Performance* - Session Chairman: John Paggi, AT&T Bell Laboratories.

Dianne Padro's help in preparing the Index is gratefully acknowledged.

The Editors

CONTENTS

I. PERSPECTIVES ON NETWORK MANAGEMENT

II. NETWORK MANAGEMENT FOR SECURITY AND RELIABILITY

III. MANAGEMENT OF HIGH-SPEED DIGITAL NETWORKS

IV. NETWORK MANAGEMENT PROTOCOLS

V. IMPLEMENTATIONS OF ADVANCED TECHNOLOGY IN NETWORK MANAGEMENT

VI. DECISION SUPPORT SYSTEMS FOR PLANNING AND PERFORMANCE

I

PERSPECTIVES ON NETWORK MANAGEMENT

INTRODUCTION AND OVERVIEW

This section consists of papers based on a plenary speech, a tutorial and a luncheon speech. Bernstein and Yuhas present an insightful view of the challenges in telecommunications network management, and illustrate their points with examples. In order to meet the goal of a self-diagnosing, self-healing and intelligent network, many new research areas need to be explored and technologies developed. They present examples of new research work that may help fulfil these goals. Terplan follows with a thorough description of the state-of-the-art in enterprise network management. This article provides an excellent overview of available network management technologies, and provides pointers for network managers as they grapple with the challenge of running complex, mission-critical networks. Finally, Mandelbaum clarifies some of the issues in the management, funding and long-term goals of the planned National Research and Educational Network (NREN). The relationship of the NREN to the current Internet, and how it will fit into the plan for a National Information Infrastructure, the "national information superhighway", are explained. He also describes some of the key issues that are currently unresolved, as well as the payoff in terms of new applications as broadband technologies are implemented in the NREN.

The Editors

THE VISION FOR NETWORKS AND THEIR MANAGEMENT

Lawrence Bernstein and C. M. Yuhas

AT&T Bell Laboratories
184 Liberty Corner Road
Warren, New Jersey 07059

ABSTRACT

The future belongs to those who harness the computer to communication networks. The bringing together of byte people with bit people is essential if this future is to happen. Byte people worry about getting data from networks into computers so that applications software can digest it while bit people worry about getting data from one point to another reliably and quickly. Too often byte and bit people fight about whose work is more important. Both are essential as we move to a telepresence society.

Computers must become so easy to use and data so easy to access that they can solve commuting, education and health problems facing cities around the world. Telecommuting demands will drive the information highway that once built will support teleducation and teledoctoring. Of course, entertainment will come along too.

Too often we do not mention the critical role of software when we discuss the future of computers and communications. This software is key to managing new voice, data, and video networks and even more importantly the wide variety of services they will provide.

Telepresence engineers have an ethical responsibility to make sure that our new systems are safe and work to the betterment of mankind. Commercial success is not enough as lives will depend on the correct and continuous operations of our systems.

This paper describes the technological breakthroughs needed to make the future happen. Special attention is given to software and network reliability issues.

Network Management and Control, Vol.2, Edited by
I.T. Frisch *et al.*, Plenum Press, New York, 1994

INTRODUCTION

Wave after wave of innovation has swept over the globe, and we have changed the lives of people who wouldn't know a bandwidth from a band-aid. Ease of use and total reliability is what the public has been primed to expect from us and our industry. Mark Twain said, "What is it that confers the noblest delight? Discovery! To find a new planet, to invent a new hinge, to find the way to make the lightnings carry your messages." Make the lightnings carry your messages -- that's what our fiber-optic networks do. Today, the newest technology is changing the way we conceive of a network and what constitutes it, and how we operate networks to make them useful and reliable.

Let me share a newspaper clipping about a problem in my home state of New Jersey, one which will surely spread to the rest of the country. It's about getting 1.7 million people to break a basic habit: getting in their cars and driving to work alone. About 5,500 companies which employ 58 per cent of the state's workforce, must change one-fourth of their commuters to "computers" before November 1994. Environmental concerns produced this law. My company, AT&T, as the state's largest private employer, has had car pool and van pool services for many years and has already conducted an informal study of the strategies that will encourage workers to modify their driving habits. Here's the point that is relevant to us today: the big favorites named were telecommuting and a compressed work week. AT&T has determined that such measures not only reduce energy consumption and pollution, but also increase productivity, efficiency, and morale, which in turn reduces stress and turnover. They also reduce the need for office space and parking. Telecommuting worked better than cash incentives in changing people's use of cars. We have a very personal stake in creating efficient information networks and managing them with precision.

TECHNOLOGY

The forces that drive technological innovation are always changing. The move now is towards globalization of companies. Mainframes become data servers in client/server relationships. To move forward with this concept, we needed the breakthrough technologies that could overcome the problems that have prevented embedding management in the network itself. Employing these innovations will require two radical shifts in thinking, one in business and one in science.

In business, the breakthrough technologies mean the difference between just keeping up and jumping out in front. If the current practice of network management was fully implemented, the potential savings in 1994 and beyond would be $23 per line per year. Using the breakthrough technology pushes the savings to over $90 per line. These savings can then be applied to improving customer service and network customization.

In science, the radical shift is the change from dealing with the anatomy of networks to dealing with their physiology, changing the emphasis from form to function. This means instead of switching, think routing; instead of transmission, think transport; instead of operations systems, think network management. Figure 1 shows how we are approaching network management today by using a platform approach. The complexity of interfacing software systems and difficulties in scalability has limited this approach. Now we are moving network management into the networks and services themselves.

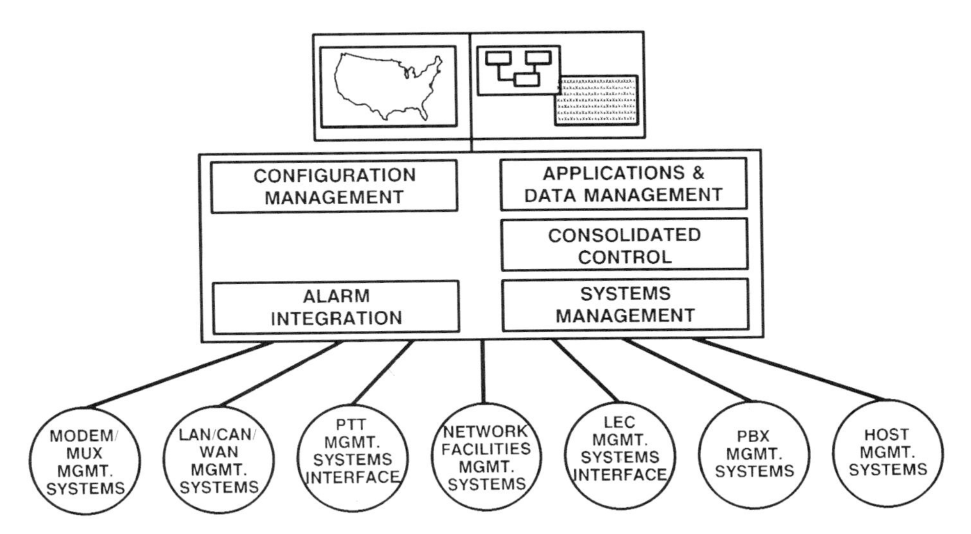

Figure 1. Systems and Network Management for the Nineties.

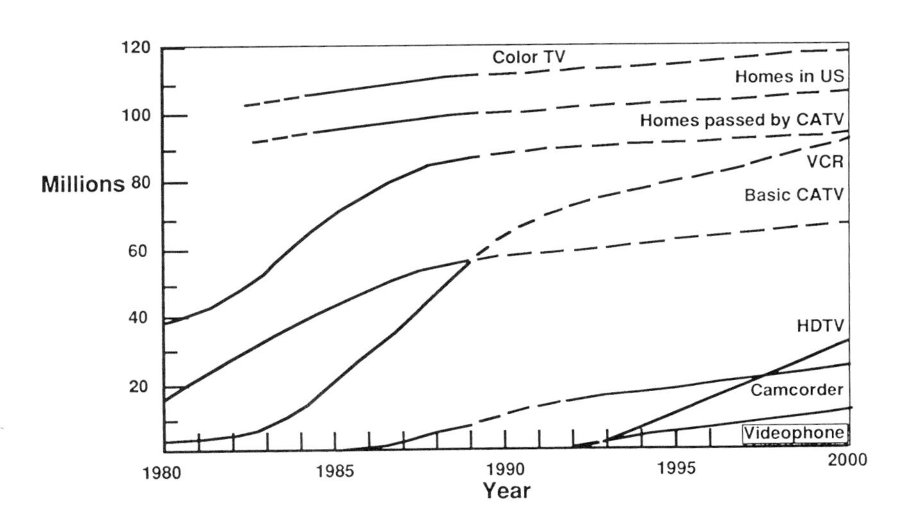

Figure 2. Growth of Video Services in USA.

One of the breakthrough technologies that is so important is Huang's fault-tolerant software library. Useful software solves a problem, but it is only conditionally stable. We have generated thousands of view graphs, millions of words and a hundred million lines of code, but still the stuff hangs and crashes. Robust software (with all the prototyping, building, and testing techniques for getting it) and disciplines for producing systems are current good approaches to the spotty software problem. But they do not yield predictably stable software.

Huang's fault-tolerant software library uses an external "daemon" (the people who name these things have played Dungeons and Dragons far too much) to monitor the process and detect snags. The library provides a pre-ordered set of recovery mechanisms so that the system will continue to work even though the individual transaction may stop. This is not, in itself, news. The beauty of Huang's library is that it is general purpose, can go on a single machine or multiple machines, is a stand-alone library and implements n-version programming (comparing several versions of the same program to take the best match).

Huang's library gets assembled in like any other utility, so the software does not have to be designed with this in mind. Different recovery mechanisms can be chosen. Though it is not yet commercially available, it is in use at AT&T with some success.

The first client/server middleware was established by the Rothschield family in Germany. Papa dispersed his sons through Europe, sending each to a financial capital equipped with only a quick intelligence and a few carrier pigeons. As each learned of a deal in the making, a storm-ruined crop, a ship sunk with valuable cargo, that *information* was sent by carrier pigeon to the son in England, who used the early data to act in the only stock market in Europe before anyone else had a clue. Here was a family that understood that information about money and goods was far more valuable than the money and goods themselves.

Many approaches to networking are hot now. Most are limited by their inability to integrate voice, data, and video. High bandwidth add-drop multiplexing SONET pipes for transport with built-in network management and self-describing features with superfast ATM routers hold the greatest promise for the future. X.25, SNA, frame relay, and TCP/IP protocols will still be used for access to applications at the edge of the network. They will be replaced by an ATM backbone with SONET transport and smart network management systems with distributed functionality will replace today's relatively simple operations support systems. These will be replaced in 20 years by photonic networks, but for now, the SONET/ATM networks offer the greatest hope of embedding network management and handling the diverse needs of an "on screen" society. Figure 2 projects the growth of video services in the U.S.

UNIX systems can support huge data bases. Bandwidth will continue to increase and get cheaper. Sounds great. The problem is that high-speed fiber-optic transport and routers are making networks more fragile, which leads to more complex network management systems. When networks grow like kudzu, we find ourselves with networks and systems that are incompatible. As more and more systems need to talk to each other in logical networks, we get a proliferation of computer communication protocols. Growth of the client/server infrastructure is limited by our inability to follow standards. Contrast that to the ease of using FAX. Here was a capability that became

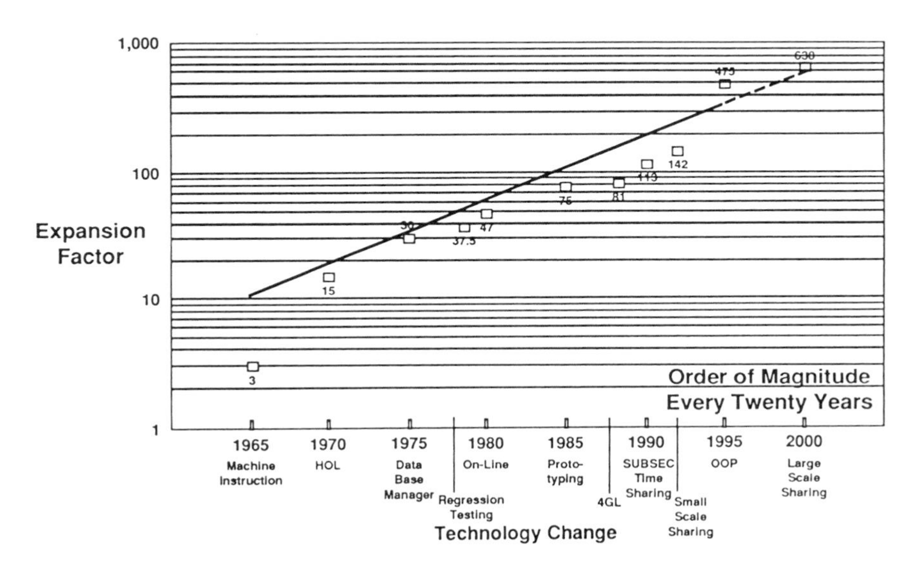

Figure 3. Bernstein's Trends in Software Expansion.

part of everyday life once a standard, well-defined, and well-controlled protocol was used.

At this point, you may feel that I'm just another guy saying software has to be reliable, predictable, reusable, stable, productive, and object-oriented or disaster is around the corner. The message is even more than that. We must change our perspective to THE NETWORK IS THE DATA BASE. This means moving away from the feeling that no trouble means total health, towards more pre-emptive maintenance. With fiber, capacity as a limiting factor is no longer an issue; the creative work will be in developing network capabilities. Industry and universities must work together to develop the theory of software stability and study software. A crazy quilt of multiple equipment vendors cannot produce an integrated, problem-solving network, but if a network is self-diagnosing, self-healing, and intelligent because this management is built in, the available staff can focus on the customer and seize opportunities to introduce new services.

Figure 3 projects the growth in software productivity with a 12% per year annual improvement for those who stay current with software technology.

ELECTRONIC PRESENCE

With digital computers we can begin to match the delivery of information to human senses, and tailor it very specifically. We will need electronic filters and agents to cope with the volumes of electronic mail, voice mail and extensive data bases that will be at our disposal soon.

The state of Iowa started to build a fiber-optic voice-data-video network in 1987 to connect its 90 counties. They are broadcasting college classes, linking state agencies, distributing lottery tickets, updating voter registration lists and exchanging library materials. Vice President Gore envisions a similar nationwide network that could do for communications what the interstate highway system did for driving.

Shoshanna Zuboff, who wrote In the Age of the Smart Machine, addresses a much broader idea. She says the information revolution is changing the very nature of work to the same degree and as radically as the industrial revolution changed the nature of work 100 years ago. Middle managers are finding themselves with new job functions and without the responsibilities they had in the past. The smart machines eliminate the need for them to monitor, gather, and analyze data, their main functions as the industrial revolution matured. Today they need to be facilitators, team builders, and problem solvers. They need to know how to cooperate to arrive at solutions. Zuboff says that these changes are a direct result of the smart machine. Knowledge workers need information, not data, to do their jobs. They need support, not supervision.

ETHICS IN THE INDUSTRY

We do important, sometimes lifesaving, work. Here are some thoughts about the ethics of our industry.

It is difficult to see the large project beyond our own piece. It is difficult to come to industry standards on communication protocols and interfaces. It is difficult to maintain a high level of professional care and attention in programs that will never be reviewed and seem less than critical. It is also absolutely vital that these things be done.

I will close with two final stories: the first is about the London Ambulance Service and the second is about the bombing of the World Trade Center.

The London Ambulance Service has a computer-based dispatching system. The computer system did away with direct human responses to radio and telephone calls, by automatically dispatching the calls.

Calls were disappearing. Some callers were put on hold for up to 30 minutes. Following the cutover to the automated system, there was no backup procedure. The user interface design was not good--the operator had no ability to scroll through the list of pending calls. And the effects of system overload had not been properly anticipated. A government consultant on this system said it was a fundamental requirement to have several layers of defense against failure. He was quoted as saying that the ambulance service was asking a lot of its computer system. "With about a million calls a year, the system has to be more reliable than a nuclear reactor protection system. I would expect to see a detailed safety case for justifying its operations and several different backup systems." As the system went into operation, the only backup it appeared to have was the expectation that people would make their own arrangements if the system failed.

Logged calls were lost. In one particular case, a disabled woman was trapped in her chair by the body of her collapsed husband. She called every 30 minutes, on each subsequent call being told that there was no trace of the earlier call. An ambulance eventually arrived 2-3/4 hours after the initial call, by which time her husband had died.

A competing system development company, whose bid had been rejected, warned that the mapping subsystem which tracked ambulances and dispatched the nearest to a call was not up to the requirements. The rule-based, analytical approach could not deal as well as an experienced operator with the small minority of difficult cases. The system wrongly reduced the influence of operators. RESULT: More than one-fourth of all calls took longer than the 14-minute response time. Up to 20 people died as a result of delays of up to three hours.

Next is an e-mail from my colleague, Rich Wetmore. I can't improve on his own words.

"Team:

It's late Friday night and I've just returned from Manhattan. When the explosion occurred today, we were meeting with New York Telephone at their West Street command center, which is on the 17th floor of the famous "West Street" telephone building and adjacent to the north tower of the World Trade Center.

The explosion was very noticeable -- we heard a loud boom and the building shook. Then the power went out -- but diesel power kicked in immediately. We went to the control room where most of the workstations were black. The only ones running were the ASCII 730s used for the traffic management software system. They were the only ones hooked up to uninterrupted power backup systems since they were deemed "critical."

Next came lots of sirens as the streets below filled with fire trucks, police vehicles, and ambulances. The first report was that a power transformer had exploded and that two separate fires were burning in two of the four WTC buildings. It was about twenty minutes before the first TV news bulletin came on with news of the explosion -- but they had few details and returned to regular programming.

At this point, the NYNEX network was in good shape: no failures and no traffic overloads. But NYNEX has three switching systems on the tenth floor of WTC tower 2, and they were running on batteries. The three switches served a combined 40,000 lines and they were apparently the only technology in the whole WTC complex that was still functioning (no alarms, no public address systems, emergency lighting, elevators, etc.) People in the building could and did use their phones to call out for help. This became a humorous point a bit later because in the excitement, a NYNEX manager was trying to determine how to report this disruption to the FCC (this is required by the FCC within 90 minutes of all major service interruptions). Luckily, he couldn't find the right FCC phone number and when he asked for help, he was reminded that it wasn't yet a reportable event since phone service was working fine -- it was everything else that was broken.

Then, CBS-TV came back on with the headline TERROR IN THE TOWERS and the network overload started. The switches in the WTC went into heavy RADR (receiver attachment delay -- a form of congestion in mf-signaling switches). To combat this, the network manager used the traffic management system to put protective trunk group controls in place in all switching systems in New York with trunk groups into the overloaded WTC switches. This immediately contained and stabilized the congestion.

The next problem was that the E911 systems in Manhattan were reported as overloaded and unreachable. But the data showed that the network was fine -- they simply didn't have enough E911 operators to answer all the calls.

Then the TV began advertising special emergency numbers for various emergency communications. The network manager monitored the network for focused overloads so he would know the digits to put controls on if overloads developed.

Once the situation stabilized and I was sure our systems were doing their job, I turned my attention to recovering my car and getting home. It took 90 minutes of walking and two more hours of driving to get out of Manhattan."

ABOUT THE AUTHORS

Lawrence Bernstein is responsible for the technology supporting software systems development in the Operations Systems Business Unit at AT&T Bell Laboratories. The systems he developed automate operations of telecommunications service providers worldwide.

Since joining Bell Laboratories in 1961, Mr. Bernstein has been involved in computer software and hardware design, including the design of algorithms for parallel processing, intermediate level computer language, and software manufacturing. He was named a Director in 1978 where he managed projects automating the business operations of the then Bell System. Today, the systems he developed are being used throughout the United States. One system handles the wiring records for 100 million telephone users. He led the introduction of expert system technology in network management systems. He is well known for his contribution to software project management, network management, and software technology. At the October 1992 United States Telephone Association meeting, he spoke on technologies which will make networks robust. As plenary speaker to the 1990 Globecom Conference, he spelled out his vision for networked computing which will take us to the 21st century.

He received a Bachelor's degree from Rensselaer Polytechnic Institute in 1961 and Master's degree from New York University in 1963, both in electrical engineering.

Mr. Bernstein is a Fellow of the Institute of Electrical and Electronics Engineers, Inc., an Industrial Fellow of Ball State Center for Information and Communication Sciences, an advisor on undergraduate curricula to Stevens Institute of Technology, and also belongs to Tau Beta Pi and Eta Kappa Nu. He is listed in Who's Who in America.

C. M. Yuhas is a free-lance writer in Short Hills, New Jersey, who has published articles on Network Management and Software Development in IEEE Journal on Selected Areas in Communications, IEEE Network, Datamation, IEEE Communications, Super User, UNIX Review, and ComputerWorld with a Master's degree in Communications from NYU and a Bachelor's degree in English from Douglass. She is a co-author of Thresholds.

REFERENCES

Bozman, Jean S. "Looking forward to office 2001" *Computerworld*, p.28, January 11, 1993.

Feder, Barnaby J. "A Model for a U.S. High-Tech Network? Try Iowa" *New York Times*, March 5, 1993.

Gasman, Lawrence. "The Broadband Jigsaw Puzzle" *Business Communications Review*, pp.35-39, February 1993.

Huang, Yennun and Kintala, Chandra. "Software Implemented Fault Tolerance: Technologies and Experience." *Proceedings of 23rd International Symposium on Fault Tolerant Computing*, Toulouse, France, June 1993.

Neuman, Peter G. "Risks Considered Global(ly)" *Communications of the ACM*, p.154, January 1993.

Randell, Brian, Jenkins, Trevor, Jones, John, Lezard, Tony and Johnson, Paul. "London Ambulance Service" *ACM Sigsoft*, p.28, Jan. 1993.

Wetmore, Richard. Personal Communication.

Zuboff, Shoshanna. *In the Age of the Smart Machine*, Basic Books, New York, 1984.

GLOBAL AREA NETWORK MANAGEMENT

Kornel Terplan
Network Management Consultant
28 Summit Avenue
Hackensack, N.J. 07601

ABSTRACT

Enterprise networks management has become an increasingly complex and challenging task. Hybrid communication architectures, interconnected LANs, MANs and WANs plus the implementation of mission critical applications on departmental servers have exposed the inadequacies of existing network management instruments and techniques. Indeed, some enterprise networks have become practically unmanageable.

The article focuses on effective management of standalone and interconnected LANs, MANs, and WANs with emphasis on management functions, applications, instrumentation, standardization, and platforms. Based on the users' needs, it outlines the trends on network management protocols, it weights the priorities of network management functions, it recommends the optimal mix of instruments, such as monitors, protocol analyzers, software diagnostic tools, element management systems, management integrators, and expert systems. It helps to define mission statements of the corporation for network management integration, centralization, distribution, automation, configuration databases, and outsourcing.

It also gives advice for how to build and to maintain the network management organization.

CHANGES IN NETWORKS

Wide and local area networks are becoming mission critical to many corporations, who need to: provide the end user with computing power, supply bandwidth for integrating communication forms and databases, support better service levels, decrease risks due to fatal outages of focal point resources, and speed up application design cycles.

Networking technology shows a number of dynamic changes. In terms of expectations, users should watch the following developments: total connectivity requires various WAN transport and value-added services, WAN technology will show a number of alternatives,

Network Management and Control, Vol.2, Edited by
I.T. Frisch *et al.*, Plenum Press, New York, 1994

such as fast packet (frame and cell relay), SMDS, B-ISDN, T1/T3 with the ultimate goal to bring all types of WAN-offerings under the umbrella of Synchronous Data Hierarchy (also known as Sonet); Ethernet technology will stay, but market share is expected to dip, and the use of token ring and token bus technologies will rise to share more equally in the market; fast LANs and MANs will penetrate the market rapidly; in terms of media, twisted-pair may penetrate new areas when radiation reduction techniques are more effective; wiring hubs will take the responsibility for being the local management entity, incorporating both physical and logical management; wireless LANs will be used for certain environments and applications, particularly as backup alternatives, but they will remain limited, and a major breakthrough is not expected soon; due to higher flexibility, routers and brouters will take implementation away from bridges, particularly in complex and geographically widespread environments; with bridges, the spanning tree technique or its enhanced versions will most likely be preferred to source routing algorithms. However, new source routing algorithms seem to be more efficient and may delay this process; interconnecting devices will house multiple functions, integrating the capabiltities of a multiplexer, router, packet switch, ATM switch and eventually of a matrix switch. Efficiency may be substantially improved by collapsing many logical network architectures and protocols into one company backbone. From the management point of view, WAN and LAN devices will offer multifunctional, integrated element management systems with or without full enterprise-wide integration capabilities.

CHALLENGES TO ENTERPRISE NETWORK MANAGEMENT

Network management has always been concerned with maintaining costs and improving the efficiency of operators. Various factors will make these goals more challenging in the near future: [3]

- Growth of networks: Enterprise networks continue to expand, both domestically and internationally. In particular, the number of interconnected LANs will grow substantially. In the network management control center, this growth adds to the volume of status and alarm data that an operator must monitor and analyze. Network management instrumentation must allow operators to easily and comprehensively monitor WANs, MANs, and LANs, determine the troubles, and rapidly focus on a magnified portion of the network. At the highest level of monitoring, a several-hundred-segments network must be reduced to a graphic display with well-designed icons and symbols for key network management.

- Continuous operations: Most enterprise networks will operate continuously, around the clock. The challenge for network manangement systems vendors is to maintain operators' attention, and focus their activities on the most relevant actions in emergencies.

- Automation: Network management systems must begin to automate routine functions by improving their capabilities for automated decision making. Thus, network administrators can focus on traffic analysis, trend analysis, and planning.

- Multimedia and multivendor networks: Proprietary designs, separate workstations, dissimilar operator interfaces, and unique command structures can prevent network operators from becoming experienced users in all network management products they operate.

- High cost of reactive network management: The majority of monitors and management systems support reactive techniques instead of proactive ones. Instruments are not yet powerful enough to perform real-time trend analysis on parallel, seemingly unrelated information streams.

CRITICAL SUCCESS FACTORS OF NETWORK MANAGEMENT

The success of network management will depend on three critical success factors; a well-organized set of network management functions, allocated and assigned to instruments and to human skill levels; proper instrumentation with the ability to extract integrated information, to export and import it, to maintain databases, and to provide analysis and performance prediction; personnel who understand their job responsibilities and posess the necessary qualifying skills.

Over a long period of time, it will remain valid to expand network management processes to include configuration, fault, performance, security, accounting management, network documentation, network maintenance, and user administration concerns. Naturally, functions will be constantly added and deleted, depending on actual user needs. However, a core with standard functions and features will remain.

Figure 1 shows the various layers of a generic network management architecture including managed objects, element management systems, management integrators, database entries, and network management functions.

Managed objects are the primary targets of all kinds of network management functions, procedures, applications, and instruments. In order to manage a family of managed objects, element management systems are segmented by the networking objects, they manage. Typical implementation examples are known for modems, bridges, routers, multiplexers, wiring hubs, packet switches, frame switches, and for monitors. Multiple element management systems can be integrated by using master integrators, also known as manager of managers, or by a network management platform. Either way, databases are supporting the management functions.

Typical database segments are: configuration data, performance data, trouble-ticket data, and telemanagement data. Usually, both relational and object-oriented technologies are supported. Network management is subdivided into configuration, fault, performance, security, and accounting management. Each area is supported by a number of functions. Some of these functions are real-time, others are near-real-time oriented. Certain functions, such as accounting and evaluation of various logs may be executed in off-line-modus.

NETWORK MANAGEMENT FUNCTIONS

In terms of network management functions following trends may be summarized: [4], [5]

Configuration management

Recommendations for managed objects and their must-and-can attributes will be wid accepted by users. They will slowly convert existing fragmented files and databases int more integrated structure. More graphics are expected for supporting inventory configuration management by a meaningful combination of a relational database computer aided design.

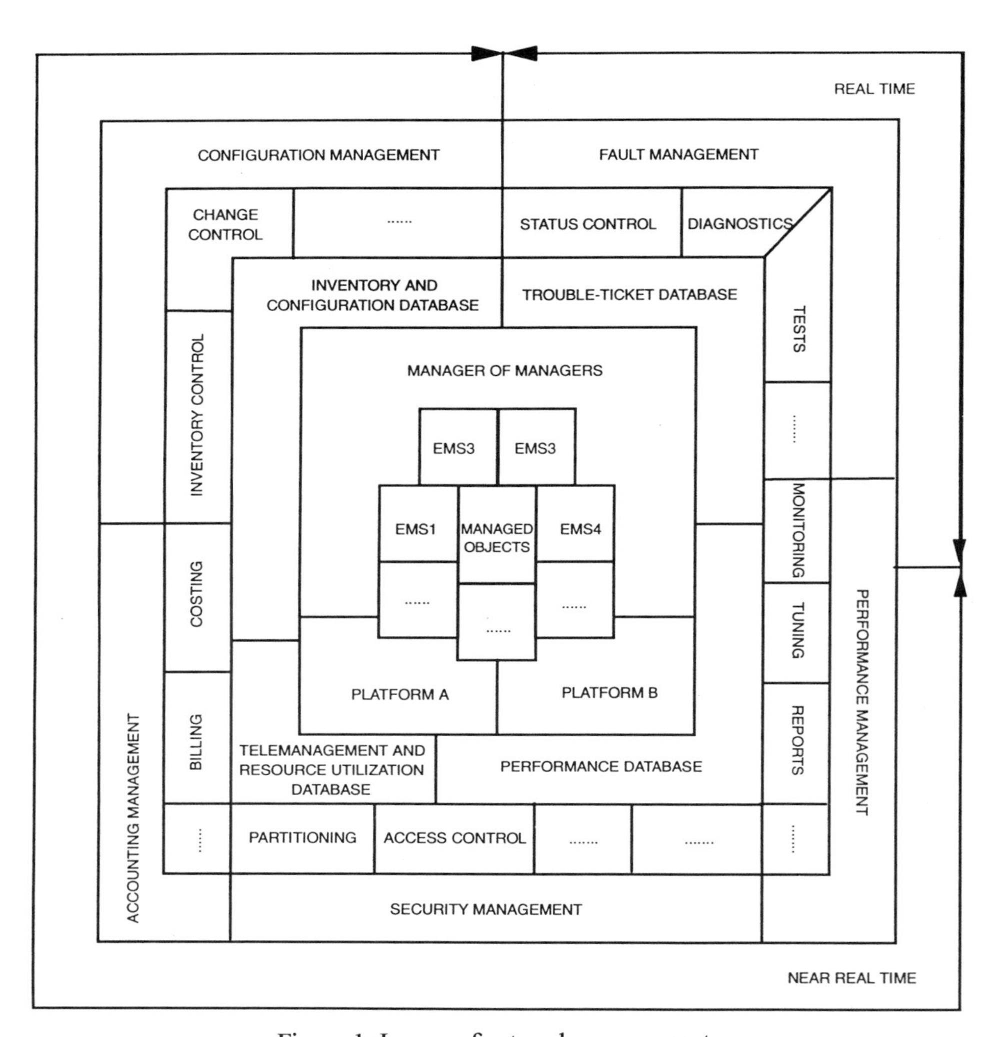

Figure 1. Layers of network management

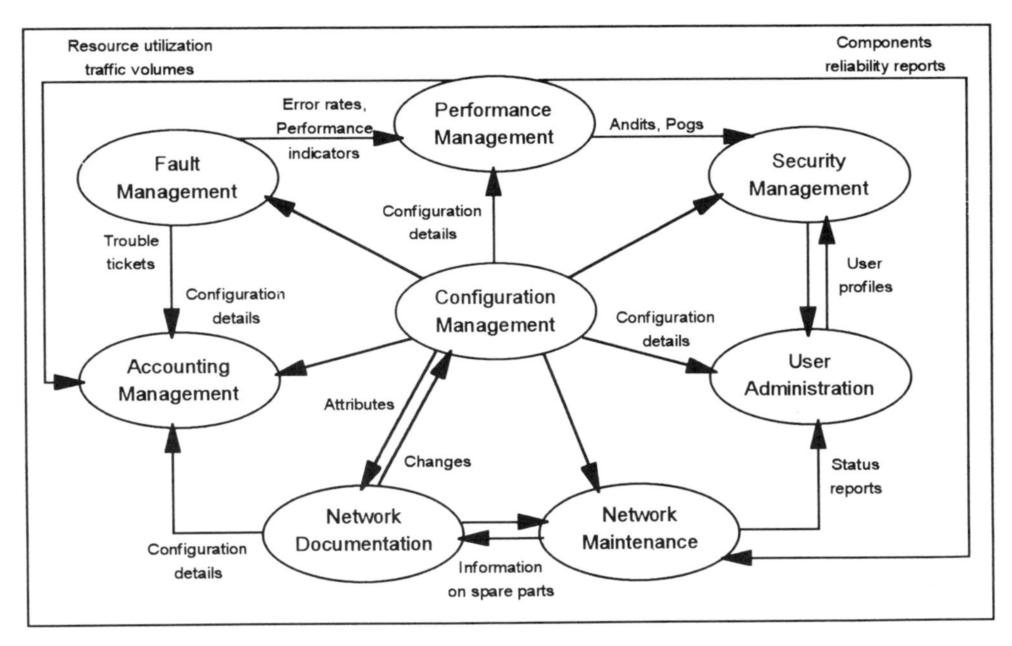

Figure 2. Network management functions - the central role of configuration management

However, the migration to a relational database may be delayed due to the fact of existing products that serve certain architectures in the wide area; examples are Info/System, Info/Master, PNMS, and Netman in the IBM environment. More integrity is expected between configuration and fault management via auto-discovery and auto-mapping features that will be offered by the majority of vendors. Figure 2 shows the central role of this group of functions; up-to-date configuration-related information is distributed to all other groups of management functions.

Fault management

In order to accelerate problem determination and diagnosis, new trouble-ticketing applications are expected. They will provide the flexibility of entering, exporting and importing trouble tickets, using advanced electronic mail features. The same systems will simplify the trouble-ownership questions. More automation is expected, where examples from mainframes will show the directions; NetView with REXX extentions or Net/Master with NCL applications will embed LAN and MAN automation procedures into existing WAN-automation solutions. Existing and new monitors will be equipped with polling and eventing features, or with a combination of both. Fine-tuning polling and eventing will help to reduce the overhead in communication channels.

Performance management

Performance data are expected to be ported from the MIBs of LANs, MANs and WANs management agents into existing performance databases, such as SLR, SAS, and MICS. For a more comprehensive performance evaluation, more help is expected from network operating systems; vendors will offer alternatives - in most cases via NetView - but not necessarily by external and internal monitoring capabilities. Performance evaluation is going to be easier by continuous and distributed monitoring in practically all networking segments.

Security management

Security management will definitively get a face-lift; there will simply be too many mission-critical applications on LANs and MANs. Companies will try to find the optimal combination of organizational, physical, and logical protection; authentication and authorization are becoming equally important. Within a short period of time, a breakthrough is expected with using biometrics as the basis of authentication. Virus detection and removal will remain very high on the priority list.

Accounting management

Accounting management will receive higher priority treatment due to the high cost spent for networks. In addition, as a preparation for evaluating outsourcing, corporate managers will try to estimate the costs of network ownership, breaking the costs down into the following categories: hardware, software, communications, infrastructure, and personnel.

Network documentation

This function will address both the physical and logical levels of networks, and includes the documentation of cabling, physical locations of managed objects, vendors, and occasionally also users. It also includes the administration of backup and escalation procedures that may be utilized by various groups of the network management team. This function is expected to address also the environmental control, such as humidity, temperature, and fire alarm procedures.

Network maintenance

In order to avoid network breakdowns, networking components are expected to be maintained preventively, also including spare components. Aspects of vendor management may be included, as well. In certain cases, software distribution is included here. This function has very tight connections to change management.

User administration

User administration is expected to receive more attention and utilities to simplify the use of networking resources. Also included are the processing of logfiles containing data on resource allocation, troubles and utilization of workstations. Educational plans help to maintain the skill levels of users preventing unnecessary trouble calls at the help-desk. But, most of the features will incorporate some sort of authentication checks in order to avoid security violations. More advanced tools are expected for the help desk, as well.

INNOVATIVE DESIGN PRINCIPLES

When expanding and redesigning processes, several innovative design principles must be considered: [1]

- Information sharing: Information must be made available to anyone who can effectively use it to perform their work. Sharing captured knowledge will avoid having to rediscover problems and changes.

- Responsibility of individuals: The responsibility for the quality of process execution should lay with a single individual. That person should also be responsible for the integrity of the data they create or update. This way, additional integrity checks may be completely eliminated.

- Simplification: Processes should be performed in as simple a fashion as possible, eliminating all steps that don't clearly add value.

- Stratification: After defining levels of process complexity, the lowest level is the primary target for automation. In future steps, process experienceshaveto be used to derive from the highest level to the lowest possible level, and to continue automation.

NETWORK MANAGEMENT STANDARDS

Network management standards will help in accelerating the implementation of functions and in selecting future proof instruments. Specific management application areas that are supported by system management functions and by common management information service elements (CMIP) have been clearly defined by ISO groups. Their implementation, however, depends on how de facto standards perform. OSI network management may include both WAN and LAN management, but the estimated overhead scares both vendors and users away. Similar definitions for the management dialog have been provided for the TCP/IP environment (SNMP). Users have turned to SNMP, and hope to have found a common denominator for at least a number of years. For standardizing the management-agent-dialog of enterprise network management, the following three items must be carefully considered: [2]

How will the management information be formatted and how will the information exchange be regulated? This is actually the protocol definition problem. How will the management information be transported between manager and agent? The OSI standards are using OSI protocol stack, and the TCP/IP standards use TCP/IP protocol stack. In both cases, the management protocol is defined as an application-layer protocol that uses the underlying transport services of the protocol stack. What management information will be exchanged? The collection of management data definitions that a manager or agent know about is called the management information base. MIBs have to be known to both.

In terms of SNMP, the following trends are expected. SNMP agent-level support will be provided by the greatest number of vendors. This support is coming very soon. SNMP manager-level support will be provided by some vendors who most likely will implement on a well-accepted platform, leaving customization and the development of additional applications to vendors and users. Leading manufacturers with network management integrator products, such as NetView, Net/Master, SunNet, Star*Sentry, NetExpert, and OpenView will enable vendors to link their SNMP managers to the integrators. Competition for SNMP manager products and platforms will be significant over the next few years.The MIB private areas are expected to move slowly to the public area and support heterogeneous network management on SNMP basis. The RMON MIB will bridge the gap between the functionality of management systems and analyzers with rich measurement functionality. It defines the next generation of network monitoring with more comprehensive network fault diagnosis, planning, and performance tuning features than any current monitoring solution. It uses SNMP, and its standard MIB design to provide multivendor interoperability between monitoring products and management stations, allowing users to mix and match network monitors and management stations from different vendors.

SNMPv2 will offer answers to some of the shortcomings of SNMPv1. The augmented functionality includes: manager-to-manager communication, support for more underlaying protocol stacks, support of bulk file transfer, and better security features.

The number of SNMP-based managers and objects is increasing. It is not easy to decide which product is best suited to the customers environment. To help to select, the following items will become important. What management functions are supported by the agents and manager(s)? What security features are supported? What objects and services can be managed? What platforms are supported? Is the product helpful for determining any network troubles? Is the product able to determine whether the managed object is operating at its potential? Is the product helpful in assessing whether the network is operating at its best performance capabilitities? Figure 3 outlines the present status of managing agents by various SNMP managers. The private MIB areas are not accessable for the manager of the other product. Manufacturers are not yet willing to disclose the details of their private MIB structure.

As more network management application software gets written, it is important to define standardized APIs so that the software can be easily ported to different platforms, and so that software developed by different vendors can be easily combined on a single platform. The platforms provide a standardized environment for developing and implementing applications, and they also separate management application software from the usual system-level services.

On the basis of platforms, it is expected that independent companies will offer network and system management applications designed to provide real multivendor solutions while taking advantage of the system-level services of platforms. This way, vendors can concentrate on their specific hardware and software, and users can focus on the customization and fine tuning management applications. Figure 4 offers a solution for managing hybrid structures of SNMP and CMIP. Besides homogeneous management, cross-communication is supported by SNMP, CMIP, CMOL and CMOT. If high level integration between the managers is the target, console emulation (e.g., CommandPost from Boole & Babbage) or expert-systems-based integration (e.g., NetExpert from Objective Systems Integrators) may be implemented.

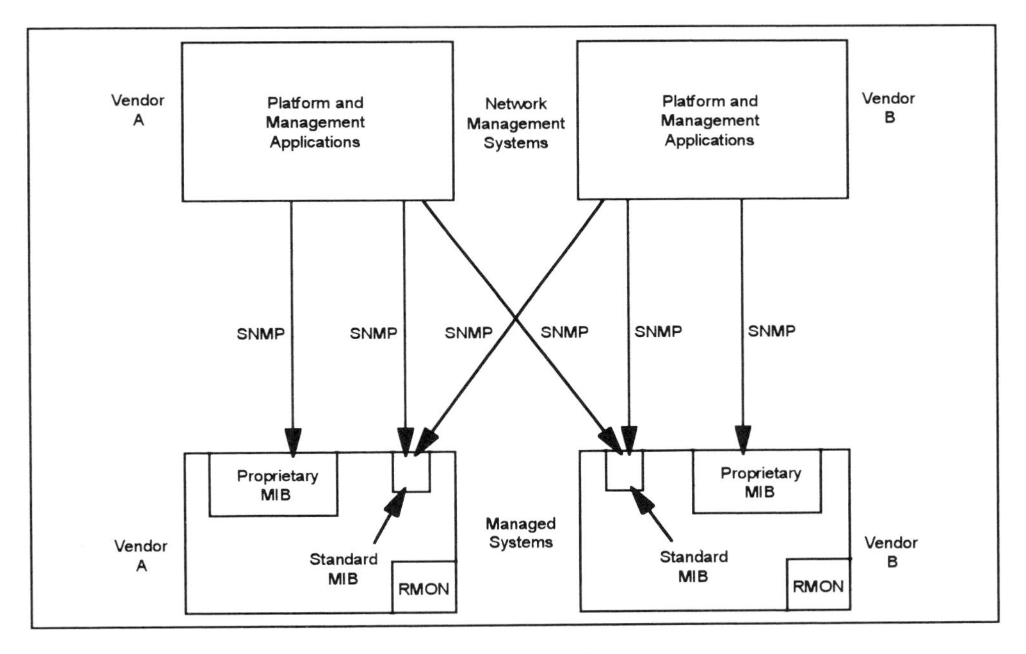

Figure 3. Emerging SNMP network management systems

POSITIONING MONITORING DEVICES

In order to provide supervision of LAN segments, there will be more continuous monitoring supported by inexpensive sensors residing in each segment. At least at the beginning, they will communicatewiththeir master monitor using proprietary protocols. In such environments, both eventing and polling structures and inband and outband transmission options may be implemented. In addition to fault management, structures may also support performance management by distributing analysis capabilities to remote sites.

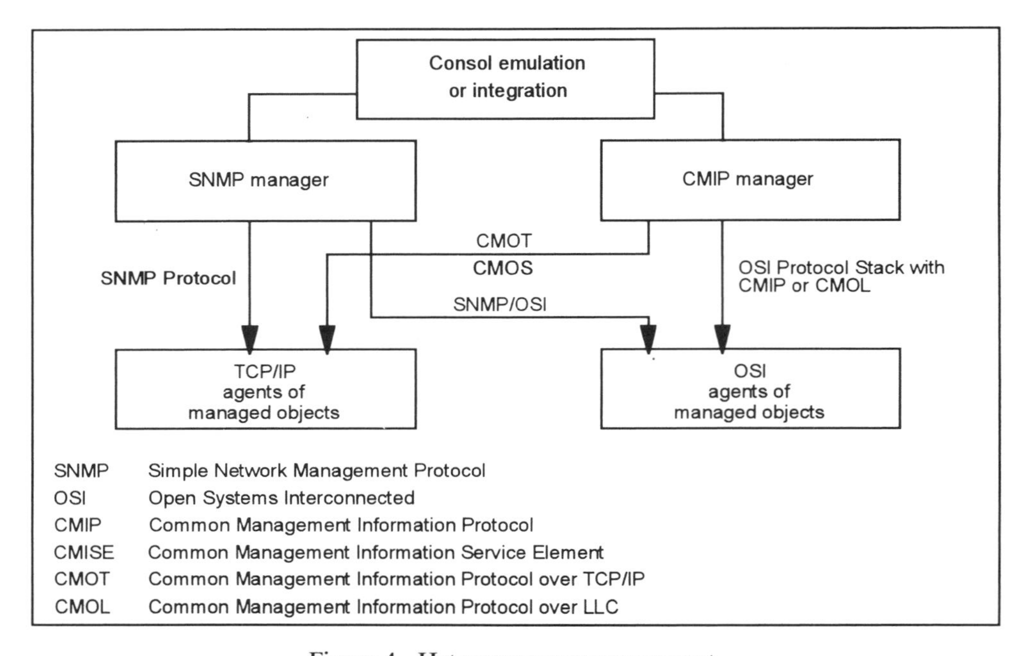

Figure 4. Heterogeneous management

In order to maximize the uptime of LAN segments, outband channnels are preferred as carriers of LAN management information. Depending on budget limits, outband channels may be dedicated or switched. Figure 5 shows a typical example for the distributed architecture that may include both inband and outband management. Not only the independence from communication facilities can be guaranteed, but also the power supply for the sensor is supplied to the sensor residing in the managed object. Health-check recording does not stop when the managed object breaks down.

Monitoring devices may become part of network management structures. These devices expand the capability of managed objects to provide status and performance indicators.

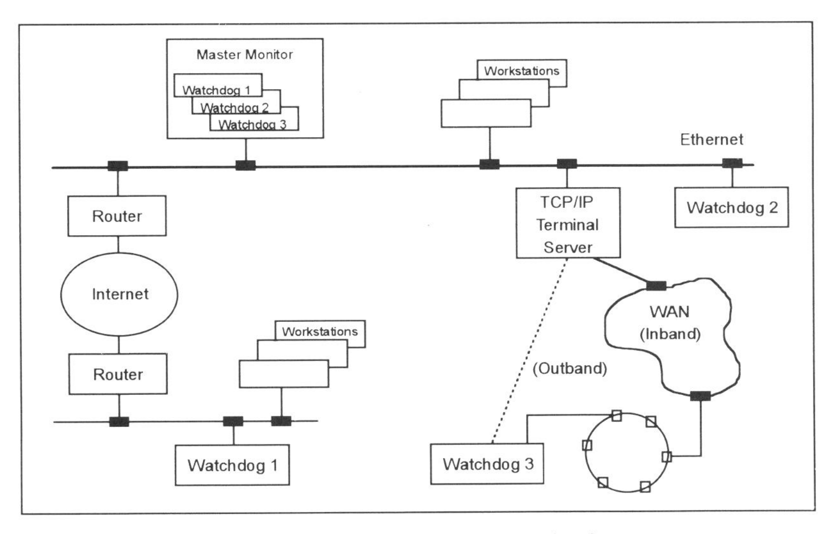

Figure 5. Distributed LAN monitoring

MANAGEMENT CONCEPTS

The structures of the future may follow one of the following basic alternatives: [5]

- Manager of the managers: Hierarchical network management structure with a network management service station in the middle. This station supervises element management systems that are responsible for managing a family of managed objects, such as hubs, segments, routers, bridges, multiplexers, modems, and switches. In this case, investment in an installed base can be preserved. The interfaces are well defined, but the number of managers will probably not be reduced. Typical examples are NetView, Net/Master, NetExpert, MAXM, Spectrum, Lattisnet-NMS, and Allink. Figure 6 shows a typical example.

- Management platform: In this case, system services and clearly defined application programming interfaces are provided by the suppliers, enabling other vendors to develop, implement, and port applications. Typical examples are: HP-OpenView, Network Managers-NMC 1000, 3000 and 6000, Objective Systems Integrators-NetExpert, NetLabs-DiMON, SunConnect-SunNet Manager, and NCR-Start*Sentry. Most of the companies are not eager to offer products for the element management system level. The main goal is to offer integration capabilities. Figure 7 shows a typical example.

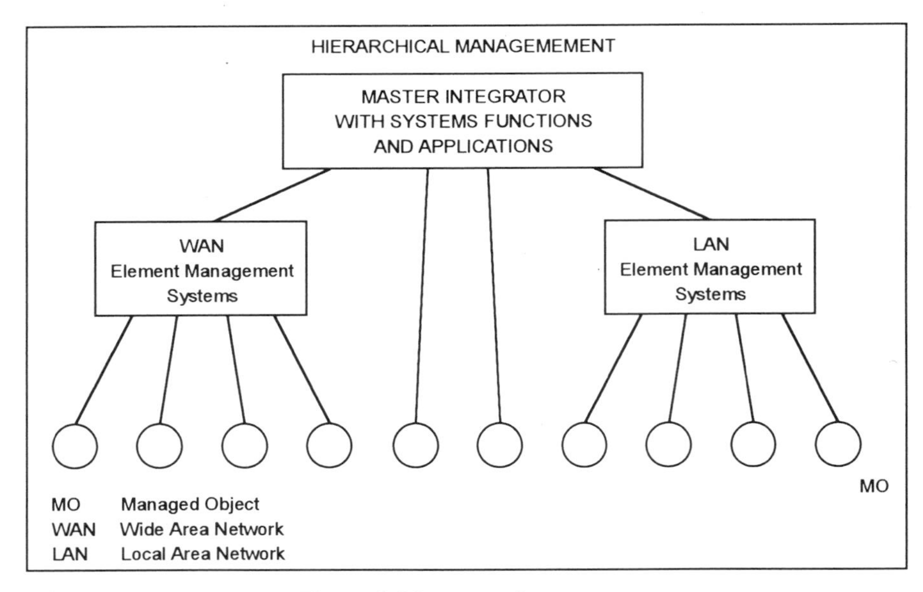

Figure 6. Manager of managers

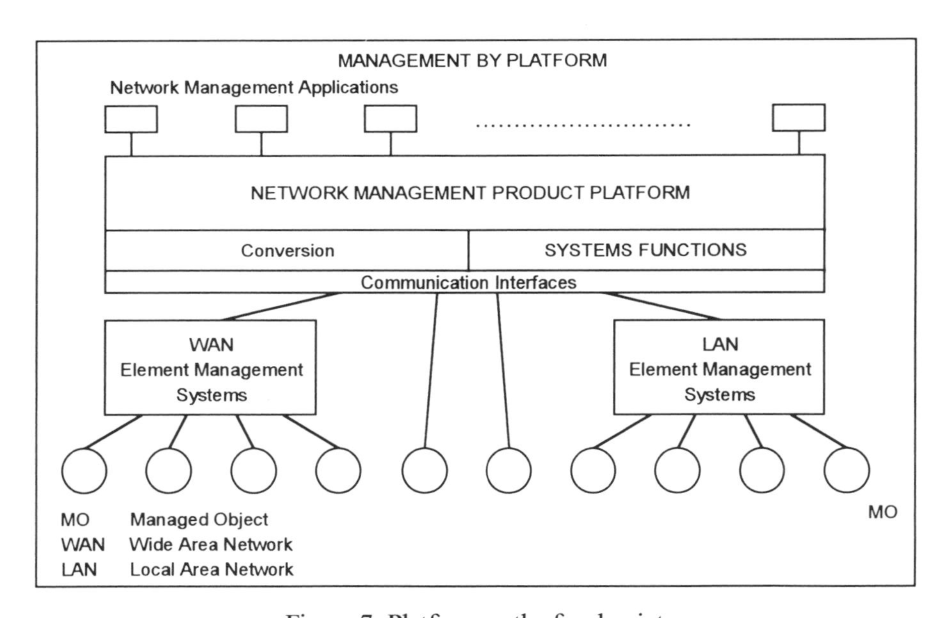

Figure 7. Platfrom as the focal point

In both cases, there are many element management systems that are responsible for a certain class of managed objects. WAN element management system examples are:

- Multiplexer manager
- Modem manager
- DSU/CSU manager
- Packet switch manager
- Cellular manager
- Frame relay manager
- ATM manager

LAN element management system examples are:

- Hub manager
- Bridge manager
- Ethernet segment manager
- Token ring segment manager
- Router manager
- FDDI manager

Integration is not necessary at any price; network management centers can be fully equipped just with element management systems.

DIRECTIONS OF INNOVATIVE DESIGN

In the future, progress is expected in the following areas:

- Graphical user interfaces: Progress has been made with various tools, including X-Windows (low-level window manager running on Unix), Open-Look (high-level graphic tool kit from Sun), Motif (high-level graphic tool kit from DEC and HP), and GMS (Graphic Modeling Systems) for providing graphics routines for all the others. As a result, GUIs (General User Interfaces) can draw complex network diagrams that allow an operator to view the status of hundreds of network nodes and elements simultaneously. Within a short period of time, operators can focus on particular managed objects. In addition, the animation of icons, along with digital speech processing, can attract an operator's attention to a problem.

- Expert systems and neural networks: In order to support proactive network management, many measurements have to be taken at different points in the networking segments and at interconnecting devices. These measurement results must be correlated and analyzed in real time. The neural network is a fundamentally new form of computer processor to collect and correlate high volumes of measurement data and to provide appropriate input to the rule-based expert system. Unlike traditional processors, neural networks are trained to recognize patterns by running simple data through them. They can also process many inputs simultaneously. In managing networks, neural networks can correlate multiple measurement data streams against preprogrammed measurement data ranges that will cause network faults unless corrected proactively. The output from the neural networks is then used by the rule-based expert system to select a corrective action.

- Distributed element management: Managed objects are equipped with enough processing power to manage their own environment, to self-diagnose, and to initiate status reports to the network management center. This trend supports the implementation of robust OSI-based network management standards in the future.

- Database technology: Networking elements are expected to be modelled as objects. Objects interact by sending messages between each other. The called object (e.g., any network element) executes the processes prescribed in the message against its attributes (configuration data and status), and reports the results back to the calling object (e.g., network management station). Object-oriented databases are supporting this type of dialogue with high efficiency. For some applications, e.g., configuration management, they offer considerable better performance than relational databases. At the other hand, fault and performance management are better supported by relational techniques.

As a result of more advanced database and platform technology, client/server-structures may be implemented for supporting distributed network management. Network management clients are workstations or PCs of authorized network management staff; they may be geographically distributed. Network management servers are integrators or element management systems supporting certain physical and logical objects that interact with clients, with managed objects of the network, and one another. Figure 8 shows a simple structure including two servers, only. For the communication between servers and clients both inband and outband techniques may be used.

HUMAN RESOURCES

The distribution of human responsibilities will follow the same path as the functions of network management. Depending on the size of networks, staffing will vary greatly. In terms of network management teams, two subjects have to be remembered: building the team, and keeping the team together. These require considerable managerial skills. The WAN and LAN supervisors will report to the network manager, who may report to the information system manager or to the chief information officer of the corporation. In a

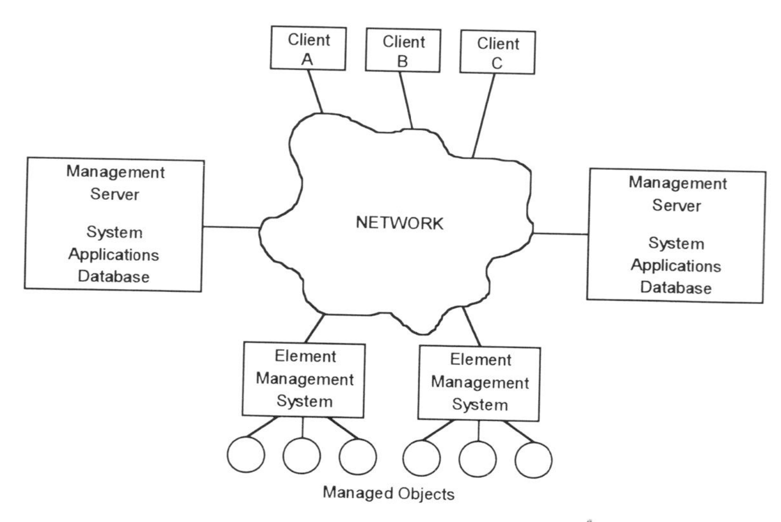

Figure 8. Distributed management architecture

completely decentralized environment, WAN and LAN supervisors will most likely report to business unit management.

SUMMARY

Network management directions may be summarized as follows.There will be integration of WAN and LAN element management systems by a platform product, SNMP manager, or by a manager of managers. In very complex networks, multiple integrators may be implemented. These integrators will use standard protocols for peer-to-peer communication.

There will be integration of WAN and LAN management, first by SNMP, then by CMIP; dual support is expected by integrators and platform providers. This step of integration will take place in multiple steps. We may expect more centralized network management that will centralize control, but distribute certain functions. In particular, monitoring, filtering functions, and reactions to routine messages will be distributed to remote sites. The practical implementation may follow OSI standards or hierarchical SNMP standards.

Due to limited human resources, automation of routine WAN and LAN management functions is absolutely necessary; support is expected by providers of integrated management solutions. Automation packages may migrate to expert systems that can be used offline and then online, offloading network management personnel from routine tasks.

There will be implementation of more powerful databases as support for network management stations, which will consolidate many templates from various LAN and WAN MIBs. Object-orientation is obvious, but relational databases will not lose their importance, in particular, not for fault management.

We may expect a slow move to outsource network management functions. The decision making will depend on the country, industry, and on the importance of networks for critical applications. The expenses of managing networks will increase due to the demand for constantly improved service levels and enhanced management capabilities. The network management market is expected to face a serious shakeout; only products and companies with the best responses to strategic direction demands will survive over the next couple of years.

REFERENCES

[1] Herman, J.: Net Management Directions - Architectures and Standards for Multivendor Net Management, Business Communication Review, June 1991, p. 79-83

[2] Herman, J.: Net Management Directions - Renovating how Network are Managed, Business Communication Review, August 1991, p. 71-73

[3] Swanson, R.H.: Emerging Technologies for Network Management, Business Communication Review, August 1991, p. 53-58

[4] Terplan, K.: Communication Networks Management, Second Edition, Prentice Hall, Englewood Cliffs, USA, 1992

[5] Terplan, K.: Effective Management of Local Area Networks, McGraw-Hill, New York, USA, 1992

ABOUT THE AUTHOR

Dr. Kornel Terplan, is a highly regarded communications expert who has had a fundamental impact on the field of network management. Since establishing his consulting practice in 1983, Dr. Terplan has assisted major corporations with formulation of strategic plans to meet their rapidly changing communication needs.

Kornel Terplan has 20 years of highly successful multi-national consulting experiences. Terplan's books, *Communication Networks Management,* published by Prentice Hall, and *Effective Management of Local Area Networks,* published by McGraw Hill, are viewed as the state-of-the-art compendium throughout the international business community. Dr. Terplan has provided training, consulting and product development services to over *seventy-five* major national and multi-national corporations. He has written over 130 articles, has presented more than 100 papers and has designed 10 networks management, capacity planning and communications performance related courses. Dr. Terplan received his Ph.D. degree at the University of Dresden. He has conducted research, served on the faculty of Polytechnic University of Brooklyn and lectured at numerous universities, including Stanford University, University of California at Berkeley, Los Angeles and Renssalaer Polytechnic Institute. He is Industry Professor at Brooklyn Polytechnic University.

MANAGING THE NREN: ISSUES IN MOVING TOWARDS A GLOBAL DISTRIBUTED ENVIRONMENT

Richard Mandelbaum

Polytechnic University
Brooklyn, NY 11201

One of the nice things about giving luncheon talks is that their function is primarily entertainment. I will therefore try to entertain you, while discussing an issue of critical importance – the evolution of the Internet. The theme of this workshop is that of network management and what I will try to do is emphasize some of the issues involved in managing the NREN.

In 1991 Congress passed the High-Performance Computing Act (HPCA) which became Public Law 102-194 upon its signature by President Bush on Dec 9 of that year. A key component of that legislation was the development of a National Research and Education Network (NREN). The NREN is supposed to be the successor, or at least the conceptual successor, to the Internet.

The basic premise on which the NREN rests is that by establishing a high-capacity national research and education computer network the government will be able to:

- Improve the productivity of researchers and educators, speed the dissemination of scientific and technical knowledge, and shorten product development cycles in industry.
- Advance computer and communications technology and transfer it to private industry.
- Build the knowledge base for the next generation of computer communication technology and services.
- Leverage federal dollars.

In order to better understand what the NREN is supposed to be and what it is supposed to do, we should look at the technical structure of the domestic Internet. The domestic Internet is a mixture of two different types of structures, two tier and three tier structures. By a two tier structure I mean, a network consisting of a national backbone network, covering a significant portion of the United States, and network links directly to end user institution. This has been the model which was generally followed by mission agencies. For example, the Department of Energy (DOE) has a national network which is directly connected to the primary DOE labs in

Network Management and Control, Vol.2, Edited by
I.T. Frisch *et al.*, Plenum Press, New York, 1994

The Internet Today

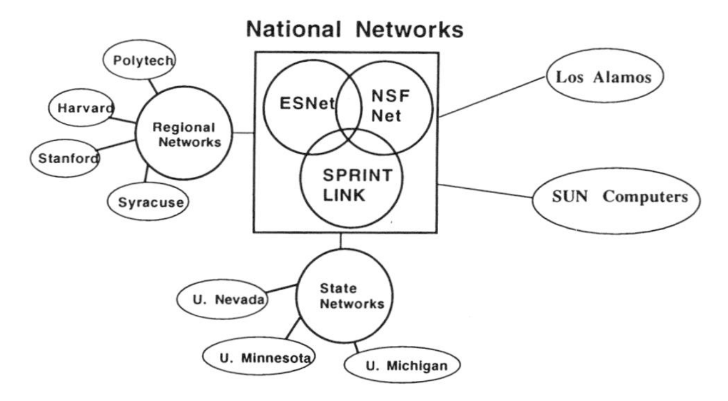

Figure 1. The Two and Three Tier Structure of the Internet.

NEW NSFNET ARCHITECTURE

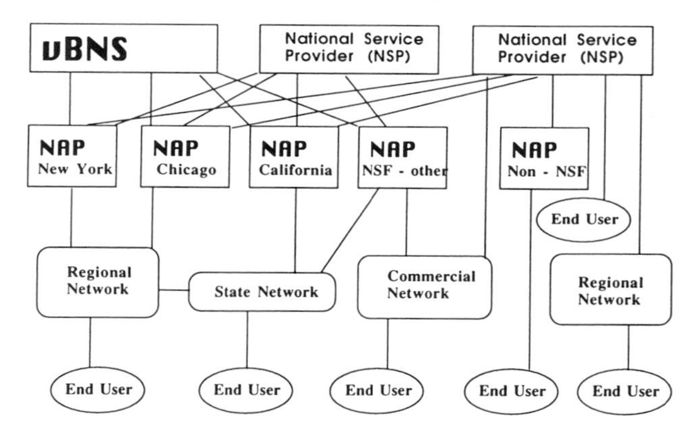

Figure 2. The New NSFNET Architecture.

the country such as Brookhaven, Los Alamos or Lawrence Livermore. Similarly commercial networks, such as PSINET connect their customers direct ly to their national backbone.

The other structure, which was really introduced by the National Science Foundation, is called a three tier structure. **(See Figure 1).** It consists of

1) a national backbone with a relatively small number of nodes. This national backbone is connected to a series of

2) regional networks, each with its own backbone covering a region of the country or a state. These regional backbones are then ultimately connected to

3) end user networks, such as University or Industry LANs or MANS.

The National Science Foundation network national backbone (NSFNET) started out as a 6 node network and at its peak, in terms of its initial incarnation, was never more than about a 20 node network. It wasn't a very complicated network, but was (and continues to be) a relatively high speed clear channel network. **NSFNET** is connected to a collection of regional and commercial networks and these provide services to the bottom tier of institutional LANs.

How do the various component pieces of the Internet interconnect? The Government networks interconnect with each other and with networks from other countries at two primary interconnection points – Federal IntereXchange (FIX) East and FIX West. The FIX's consist of FDDI rings with agency or national network routers attached to them. The commercial backbones, Sprint, ANS, PSI, Alternet etc.do not have a direct peer relationship with the federal backbone, but do have peer relationships with other commercial networks. These networks interconnect at installations which have a technical strucure similar to the FIX's and are called called commercial internet exchanges (CIXs). The backbones of the regional or mid level networks attach either directly to the NSFNET or to a combination of the NSFNET and other agency and commercial networks. In addition several of the mid-level networks connect directly with one another. The NSFNET is currently the primary transit network on the Internet and the NSF takes responsibility for global routing coordination. This is done for the NSF by MERIT Inc. (The Michigan regional network and principal awardee of the 1988 NSFNET backbone agreement.)

On a more strategic level the HPCA established a Federal Networking Council which is charged with the responsibility to provide technical networking policy advice to the government. At a protocol level, although the Internet is primarily a TCP/IP network it is evolving to become a multi-protocol network. Almost from their inception many of the mid-level networks supported the DECNET protocols and most of the European research networks supported X.25. There has been a gradual increase in Internet support of the ISO CLNP protocols and it is anticipated that ISO TP4 will be supported as well.

Before transitioning to a discussion of the future NREN let's pause to give a few figures about the size and growth of the Internet. As of December of 1993, there were about 50000 registered IP networks of which about 20000 could pass traffic to one another and thus constitute the "Internet". Such networks were found in about 100 countries, with E-mail gateways in about another 50. IP networks were growing at a rate of about 12% a month, with Internet connected networks increasing at about 8% a month. (9.5%/month outside the US). The networks on the Internet interconnect a bit more than 2 million hosts and, it is estimated, service about 10-20 million users. If current rates of growth continue these number would reach 10 million hosts and 100 million users by the end of 1997 and about 1 billion users by the year 2000. It is almost impossible to get reasonable statistics on total Internet traffic but NSFNET traffic statistics have been kept since 1987 and show the following pattern:

NSFNET TRAFFIC
OCT 1987 - OCT 1993

Month	G Bytes/Month	Av. Peak Link Volume Kbits/sec
Oct 1987	10	0.01
April 1988	20	0.02
Oct 1988	60	0.07
April 1989	150	0.2
Oct 1989	330	0.4
April 1990	520	0.6
Oct 1990	800	0.9
April 1991	1400	0.6
Oct 1991	1900	1.9
April 1992	2800	3.1
Oct 1992	3900	4.0
April 1993	6200	6.9
Oct 1993	8500	8.4

Before continuing we should clarify some basic assumptions about the future NREN. These are:

- Multiple Federal agency networks will continue to exist.
- The NREN will continue to be provided by more than one provider; be based on multiple Level 2 technologies (Serial Links, SMDS, ATM, Frame Relay); and will be multi-protocol (IP, ISO CLNP,...).
- Federal and Commercial networks will interconnect at analogues of the current CIX's and FIX's. (These are called Network Access Points –NAPS– in the current NSF backbone solicitation) and there will be a combination of new technical and policy mechanisms which will facilitate commercial-federal agency interconnections.

Some of the major requirements for the NREN will be

- Reliable and stable "end-to-end" services with Quality of Service provisions and appropriate mechanisms to safeguard privacy and security.
- Enhanced end user tools and services, including enhanced network information services and network navigation tools.
- The ability to address differing acceptable use policies, including the ability to do source and policy-based routing.
- Automated and enhanced network management tools and procedures.

A more fundamental requirement, reflected in all the legislation pertaining to the NREN, is that the NREN is supposed to be a precursor to the coming National Information Infrastructure (NII). Thus the NREN should lead to increased commercialization and privatization of the technologies and services which the NREN brings about. The NREN is supposed to be distinct from the NII, but a vital component of it. The NREN is supposed to be precisely what its acronym stands for: a vehicle for research and education. The NII is supposed to be something which includes the entirety of commercial telecommunication, computing, entertainment, digital publishing and other efforts. Privatization is, from the government's point of view a key component in setting a foundation for what is supposed to eventually be a global information infrastructure.

A major step in building the NREN is the new NSF solicitation for a successor to the NSFNET backbone. The NSF solicitation appeared in May of 1993 with a due date of August, and the goal of picking awardees by February of 1994. The solicitation has four major components. These are:

- Establishing a very high speed backbone (vBNS) connecting the NSF supercomputer centers and other designated national research resources.
- Managing at least three NSF designated Network Access Points (NAPS) where diverse providers will be able to interconnect their networks.
- Designating a Routing Arbiter.
- Providing for Interregional Connectivity.

In order to understand the solicitation let us begin by describing the new NSFNET architecture which is symbolically represented in **Figure 2.**

The architecture consists of the following elements.

1) A very High Speed Backbone, the vBNS, which is supposed to initially provide OC-3c(155 Mbit/sec) bandwidth and, as needed, be expanded, to OC-12c, OC-48c etc. The vBNS is supposed to initially connect the five NSF financed super computer centers, as well as provide access to individual experimenters, in a fashion as yet undetermined, to do experiments which require very high speed backbone access. The vBNS must be able to switch both IP and CLNP packets. The clear emphasis here is not on aggregation, but on clear channel use of 30 megabits/sec and higher.

2) Network Access Points (NAPs). The network access points are supposed to be the analogues of the current FIXs or CIXs. They are intended to be high speed networks or switches which provide neutral, high speed interconnection points, for various high speed networks, the vBNS, other government agency networks,commercial, mid-lever, state networks etc. The NAPs can operate at a lower protocol level (i.e. level 2) than the vBNS and are not subject to any Acceptable Usage Policy (AUP) restrictions. NAPs can be built around LAN, MAN, SMDS or ATM technologies. There are supposed to be at least three NSF NAPs, in New York, California and Chicago. Any service provider connecting to those three NAPs is labelled a National Service Provider (NSP) by the NSF and is eligible to receive NSF monies for the interregional connectivity portion of the solicitation. The NSF might authorize other NAPs and it has welcomed other entities interested in setting up their NAPs.

3) A Routing Authority, independent of the vBNS provider, empowered to make routing decisions relevant to the operation of the vBNS and the NAPs. The RA is supposed to provide database management information such as network topology, routing policy and interconnection information which can be used by attached networks to build routing tables.

4) Independent mid-level Networks, which are eligible to receive money from the NSF to either connect to a NAP or to purchase connectivity from a NSP.

The vBNS is the closest analog one now has to the NSFNet backbone, but is very different from it. The NSFNet backbone allowed essentially unlimited usage by the university sector as well as usage from the commercial sector as long as such usage adhered to the rules of a very board NSFNet acceptable usage policy (AUP) . Because of this very broad approach to what was acceptable the backbone catalyzed the commercial use and subsequent exponential growth of the Internet. The vBNS, which is almost a duplicate of the 1986 NSFNet will function very differently. The new AUP will be a much more restrictive policy and uses of the vBNS will really have to require very high speed network technologies. Internet mail or File Transfer or remote logins will generally not constitute acceptable use of the vBNS and there is no real plan at this moment to extend the vBNS beyond the supercomputer centers.

The provision of the network access points is really the structure which replaces the old NSFNet backbone. The third tier remains the same, one has regional and state networks which service the end users, but the entry of the national service provider now gives you a mix of the two and three tier structures. NSP's can either be commercial service providers or agency networks which function in the same role. They would connect to NAPS and they might have end users directly connected or the NSPs would have other commercial networks or regional networks as customers and would connect directly them. In turn, the old mid-levels could also opt to connect to the NAPs, rather than buy services from the NSPs.

This new element, which I think is crucial to resolving the debate on the relationship between federally funded networks and commercial entities, is something which literally wasn't even conceived of back in 1986-1987 when the first NSFNet backbone started. There weren't any commercial service providers and it wasn't even clear how to commercialize such a system. The first for-profit commercial Internet provider came into existence in 1989 as a spin-off of NYSERNet, which was the first NSF-funded regional network. NYSERNet was founded in 1985 and initially received most of its funding from NYNEX and the National Science Foundation. In 1989, NYSERNet decided to go out into the market look for a service provider who could offer it Internet connectivity. They quickly found that there were none. As a result of this initial search NYSERNet's previous president and its vice president for technology spawned the first company which offered commercial Internet connectivity, PSI. PSI now has the largest set of customers of any commercial company.

This past May, NYSERNet released an RFI to find out who were the players in the field who might be willing to offer a service. They had a five year contract which was coming to an end and we have gotten twelve replies from entities who are ready to provide Internet connectivity. We now have a whole new set of players who are now in this business and will be the primary providers of Internet service in the future.

What are some of the policy issues in trying to understand where this structure is leading us? I would isolate three general areas:

- Governance
- Funding
- Policy Management Structure

Governance

In the governance arena one has the panoply of questions surrounding the appropriate roles of the Federal and State Governments in determining the evolution of the Internet.

To what extent should the theories which led to the Communications Act of 1934 and the system of regulations under which ATT operated then, and ATT and the RBOCs operate now, be applied to this new industry?

To what extent are there freedom of speech and privacy issues which demand new regulation and legislation?

What is the analogue of universal service and equitable access in the electronic age and must these be guaranteed by the government or can market mechanisms be relied upon?

What should the role of government be in setting the stage for the broader National Information Infrastructure?

In particular, should the government provide for a special role for the research and university sectors orshould the role of these sectors be entirely determined by the marketplace?

Funding issues

The key funding issues are: What Should be Funded and How? What role should "End-User Funding"play in future grant and contract awards? How will Federal funding of networking infrastructure effect competition? How can governements best leverage the funds they allocate to telecommunications infrastructure development? The process by which the federal government funds research is well understand. A faculty member is awarded a grant through his/her institution to conduct research, the research is done, results are published in the scientific literature and one goes on to the next piece of research. Originally, the NSFNet was a purely scientific resource. Funds promoting its expansion and use were a direct benefit to the research community. However that situation has changed drastically. The majority of networks in the Internet are commercial. There are about 10 for-profit commercial entities which are running their own backbones and to the extent that the NSF is funding a regional network or to the extent that the Department of Energy or Defense run their networks and allow external traffic on them they are undercutting the ability of a commercial entity to offer that same service. More worrisome is that it can be argued that such behavior by the Federal Government provides a disincentive for technology infrastructure investments on the part of the private sector. If the private sector has to compete with something which was federally subsidized they will not play the game.

This dichotomy poses a serious equity problem for what is as yet an unregulated domain. If one is interested in reaching community colleges in Plattsburgh, which aren't necessarily one's best telecommunication customers, as opposed to reaching the New York Stock Exchange, which might be a very good customer, how does the government arrange for that to be done.

"End User Funding" is the generic term for the various schemes which try to arrange such connectivity without imposing a regulatory framework. Rather than setting up a network which provides a service to some or all for nothing or for low cost, one should try to funnel Federal money (which isn't directly funding research, but is in support of a research infrastructure) to the end user and then let the end user buy those services they consider appropriate to the extent that services are available. I won't go into all of the issues, but this was really a solution which was proposed by a National Academy of Science (NAS) panel back in 1988. It has been amplified, it has pluses and minuses, it is an issue which is far from closed.

Nevertheless, the recent NSF solicitations are still aimed pretty much at mid-level networks rather than end-users. One of the reasons for this is that the NSF has one of the smallest bureaucracies in Washington to administer any program which is aimed at a large number of potential users. The whole NSF networking program employs under ten people and a large army of reviewers out in the field. One of the end-user funding mechanism which was proposed by the NAS panel was a voucher system. Vouchers were to be given to researchers for networking services. The researchers would then aggregate the vouchers at the university level and the vouchers would be used to support networking services for the university. However such a solution is just not manageable by an agency as small as the NSF.

Policy Management Structure

The issues which come under the policy management rubric include the following. Mechanisms to manage end to end service issues. Processes to air and resolve public policy issues. Entities to monitor congestion and implement rationing-mechanisms such as user fees, "class of service" policies, volume limitations, logically segregated networks and policy based and source based routing.

The last of the crucial policy areas is that of a policy management structure. One needs structure above that of independent autonomous networking entities to be able to actually assure the stability and reliability which we stated earlier were crucial.

Similarly one needs an ability to resolve public policy issues on a global basis rather than based on the policies of individual networks. The trickiest issues is the rationing one. This is closely related to the unresolved question of how to charge for "Internet".

The continuing astonishing growth of the Internet combined with the expected explosion of multimedia applications make this a pressing issue. We note that "charging" is simply one form of a more general set of "rationing" policies. As the Internet grows more commercial the "how" of charging will be implemented by various for-profit vendors, but while there is still a window of opportunity it might be useful for the research community to look at and test a variety of "rationing' policies before having a pure "pay-per-packet" policy imposed on it.

I would like to end with some goals for the NREN. These come out of a white paper issued by the National Telecommunications Task Force, a project of Educom . (Educom is a non profit association that serves as a resource for colleges, universities and research institutions wanting to learn more about information technology. It has about 600+ members in the higher educational community.)

Traditional Goals for the NREN

- Evolve network bandwidth gracefully to SONET bandwidths.
- Extend the NREN to 200 additional higher education institutions per year.
- Extend the NREN to 1000 additional K-12 institutions per year.
- Enhance the reliability and robustness of the network to a level equal to the voice network.
- Develop type of service capabilities.
- Improve network security capability to a level that matches corporate security levels.

" New Payoff " Goals for the NREN

- Develop a policy formulation and accounting framework that encourages delivery of commercial information services over the NREN.
- Facilitate the creation of a standard on network video teleconferencing capability.
- Create a library of scholarly contemporary video materials that can be delivered over the NREN on demand.
- Establish an electronic copyright clearinghouse.
- Implement a single virtual nationwide library of materials useful to education.
- Encourage the publication of electronic journals over the NREN.
- Improve the technology for the creation of networked multimedia scholarly materials.

The traditional goals are simply a continuation of what we have been doing until now. Though admirable, I think, simple extrapolation of current practice is not sufficient to justify the NREN. The "new payoff" goals are, I think, a more appropriate set of criteria by which to judge the success or failure of this new enterprise.

About the Author

Dr. Richard Mandelbaum (rma@poly.edu) is Director of the New York State Center for Advanced Technology in Telecommunications at Polytechnic University and President of NYSERNet, the New York State Education and Research Network. Dr. Mandelbaum is a past president of the Federation of American Research Networks (FARNet) and a former member of the National Science Foundation Network Advisory Panel. He is also on the Policy Committee of the National Telecommunication Task Force (NTTF). Dr. Mandelbaum is a Professor of Mathematics and Electrical Engineering at Polytechnic University and received his Ph.D. in Mathematics from Princeton University in 1970.

II

NETWORK MANAGEMENT FOR SECURITY AND RELIABILITY

INTRODUCTION

The papers and topics selected for this section address some crucial issues facing the industry today: The first paper by Stratman focuses on organizational framework for network and system management in which he described a standard based architected solution. The paper offers a solid and comprehensive approach to the problem of network management that is generic and reusable. The next paper by Veeraraghavan offers a way to model network management procedures using Petri Nets. The paper contributes a new term, "Performability" which could be a very beneficial index for analyzing existing networks and designing new ones. Following that Wang and Schwartz describe a novel approach to diagnose faults, based on probabilistic reasoning, especially in poorly, partially or overinstrumented networks. The last three papers addressed security issues. In the age of viruses, worms, and cyberpunks, security has become a serious threat. Coomaraswamy and Kumar offer an effective authentication and encryption method. Barba and Melus presented a security architecture for the Universal Mobile Telecommunication System (UMTS) which deals with the emerging issues of authentication and secure access in nomadic and mobile networks. Again for the mobile environment, Hager, Hermesmann and Portz determine an authentication scheme suitable for implementation.

We live in revolutionary times. In the last ten years, the complexity, capabilities, and the number of network users expanded exponentially. I would like to further express some views, as seen from the trenches, of someone who participates in planning, installing and operating networks and systems. Unfortunately, the revolution is not user friendly. Networks have become the backbone of our life. Offices, businesses, medical services, and so on, are all networked. Though we are aware of the importance and criticality of networks, there is still a lack of clear understanding of their types, and the modalities of usage, and therefore the required level of management. Most network equipment and technologies available today, were built based on certain design compromises, with control mechanisms that fail to address the role of a component as a part of a whole. When interconnected, these components can create unmanageable complexity. Furthermore, there are no adequate tools to Operate, Administrate and Maintain large scale networks. The grassroots success of the Simple Network Management Protocol (SNMP) offers some relief. However, this very welcome change, which creates a common de facto standard base for management platforms, is still not sufficient. We have become data rich and information poor.

The work presented in this section addresses these issues. I hope that the fruits of these research efforts will find their way into the mainstream of network management tools. The industry has a crucial need for engineered and architected practical solutions to the problems.

Dror Segal
SIAC, Inc.

INTEGRATED STATE AND ALARM MONITORING ACROSS HETEROGENEOUS NETWORKS USING OSI STANDARDS AND OBJECT-ORIENTED TECHNIQUES

Robert H. Stratman

Center for Advanced Aviation System Development
The MITRE Corporation
McLean, Virginia 22102

ABSTRACT

Network management standards for Open System Interconnection (OSI) continue to evolve. Current standards and implementors' agreements being developed by the International Organization for Standardization (ISO) and the Network Management Forum (Forum) are based on an object-oriented approach to network management. These standards concentrate on the overall architecture, the protocols and services that allow a manager system to exchange information with a managed system over an interoperable interface, and the objects necessary to communicate management information over the interoperable interface. The specific manner in which network management is accomplished for functions such as state and alarm monitoring is left up to the implementor. An example of applying OSI standards and object-oriented techniques for state and alarm monitoring in a prototype integrated network management system being developed for the Federal Aviation Administration (FAA) is described.

BACKGROUND

In accomplishing its mission, the FAA is using an increasing number of interconnected, heterogeneous telecommunications networks. The FAA owns some of these networks and leases others. This "network of networks" provides both air-to-ground and ground-to-ground communications services. Increasing air traffic and automation has changed the nature of national ATC from one with fairly autonomous facility operation to one where multiple facilities and telecommunications networks are dependent on one another to ensure efficient operations.

An example of a standards-based open system for network and services management is a prototype system being developed by the FAA and its Center for Advanced Aviation System Development (CAASD) at The MITRE Corporation for managing telecommunications networks and services at Air Route Traffic Control Centers (ARTCCs). An initial version of this prototype, the Telecommunications ARTCC Prototype (TAP), is currently being evaluated in an operational setting. TAP integrates the various communications management capabilities at the ARTCCs into a single management system. This includes making the best use of existing capabilities and defining new capabilities needed for proper operations of communications facilities within the domain of the ARTCCs.

The TAP is based on the OSI definition of network management specified by ISO and the Forum [1,2,3]. The initial version of the TAP is based on Forum Release 1 Specifications. Future versions of the TAP will be compliant with OMNI*Point* specifications

Network Management and Control, Vol.2, Edited by
I.T. Frisch *et al.*, Plenum Press, New York, 1994

[4]. In addition, an object-oriented language (C++) and an object-oriented database (ObjectStore®[1]) are used to implement TAP. The prototype runs on a network of Sun® workstations under SunOS®[2], a UNIX®[3] operating system.

This paper provides an overview of the TAP architecture, and describes how ISO and Forum standards were applied to two application areas: state monitoring and alarm monitoring. It also describes how the standards were extended and interpreted to meet the specific requirements of the TAP integrated manager.

TAP INTEGRATED NETWORK MANAGER ARCHITECTURE

From the perspective of integrating multiple management systems, an OSI architecture consists of at least one manager system and one or more managed systems. The major components of this architecture include managed resources, one or more agents, an interoperable interface, a manager, a set of managed objects, and a set of management applications.

The TAP architecture is based on the OMNI*Point* network management model shown in Figure 1. Communication between management entities (managers and agents) is across an interoperable interface [5,6]. The interoperable interface is a formally-defined set of protocols, procedures, message formats and semantics used to communicate management information. The interoperable interface consists of Communications Services, Management Services, and Information Services. Communications Services specify the protocol required to exchange management information within a standards-based network environment. The management information exchange protocol used by the TAP is the Common Management Information Protocol (CMIP). Management Services specify the functions for transferring management information. The TAP uses the Common Management Information Service (CMIS), which defines a management notification service used to report events to a manager, and operation services used by a manager to direct the agent to retrieve, set, create, and delete information, or take some other action. The System Management Functions (SMFs) supported by the TAP include Configuration Management, Alarm Reporting and Logging, and Event Management. Information Services specify a common naming architecture for managed objects which is laid out in the ISO Guidelines for the Definition of Managed Objects (GDMO). Shared Management Knowledge (SMK) ensures that different management products can interoperate by establishing and sharing a common understanding of management knowledge and capabilities. Within OMNI*Point*, Ensembles are one form of SMK. An Ensemble is a set of requirements, a solution, and detailed specifications that solve a particular management problem. Alarm monitoring in the TAP is based on the Forum *Reconfigurable Circuit Service: Alarm Surveillance Ensemble*.

An overview of the TAP based on the management model is shown in Figure 2. Management interactions between open systems consist of the manager system issuing management operations (e.g., get information, change information) to the managed system and receiving responses and notifications of events (e.g., state changes, alarms) from the managed system. This exchange of operations and responses/notifications is controlled by a manager logically located on the manager system and an agent logically located on the managed system.

Network management systems exchange information modeled in terms of managed objects within an object-oriented paradigm. These objects are stored in the management information base and are abstractions of managed resources representing the "real" communications components, such as equipment and circuits, to be managed. If a resource is not represented by a managed object, it cannot be managed. Groups of these managed resources are controlled and monitored by existing network management systems. A TAP agent uses information from these non-standard management systems to construct OSI standard messages for the manager. In the future, the TAP Integration Manager will

---

1 ObjectStore is a registered trademark of Object Design, Inc.

2 Sun and SunOS are trademarks of Sun Microsystems, Inc.

3 UNIX is a registered trademark of UNIX Systems Laboratories.

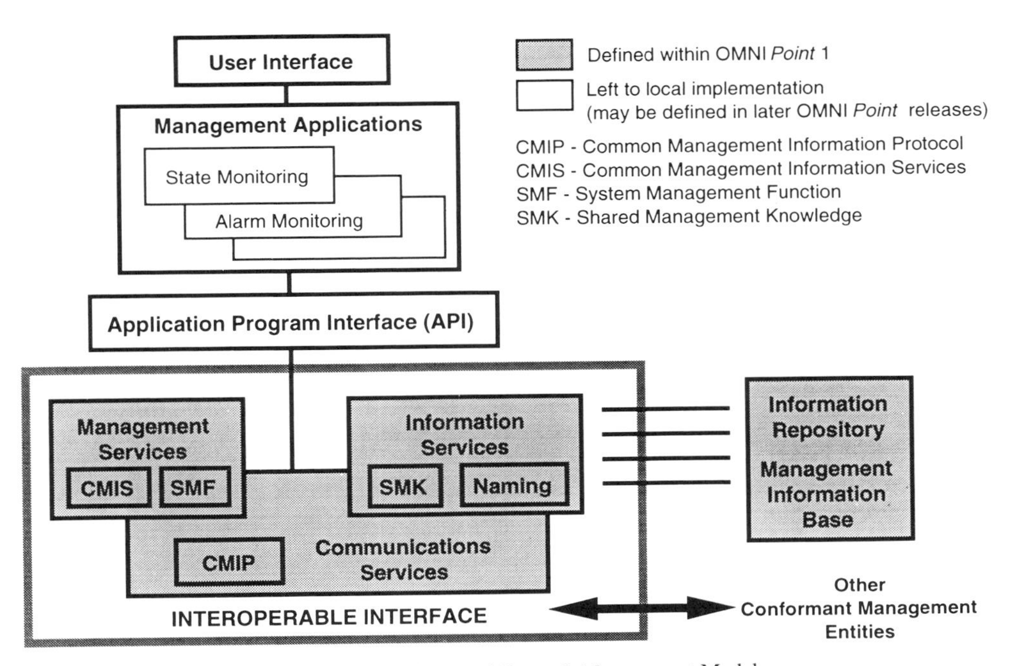

Figure 1. Standards-Based Network Management Model

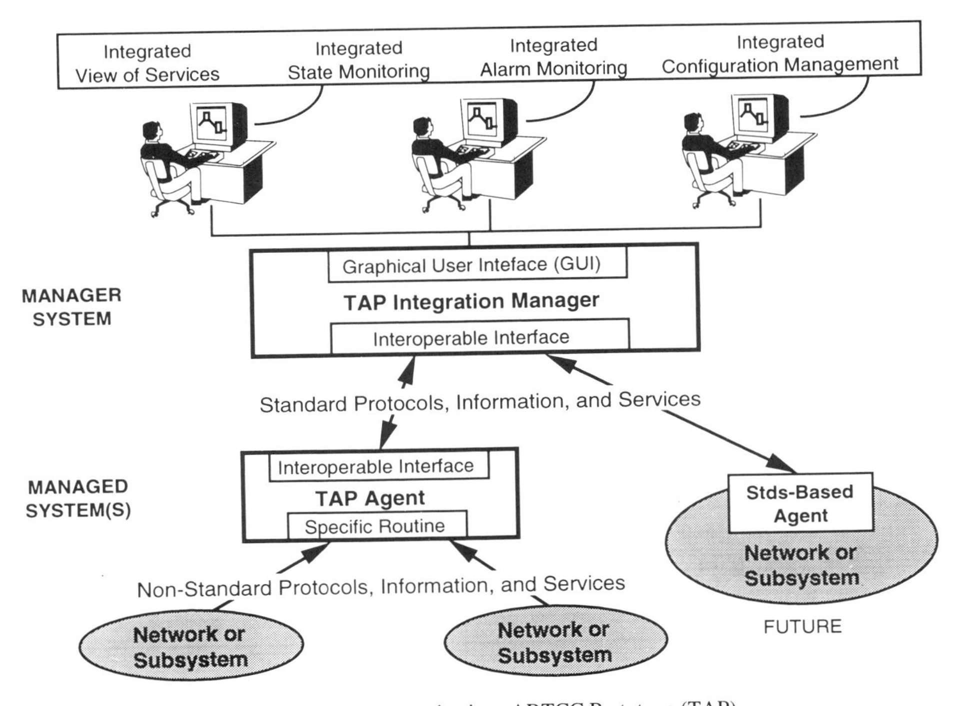

Figure 2. Telecommunications ARTCC Prototype (TAP)

communicate directly with standards-based agents built into the network or subsystem management systems.

The initial version of TAP only monitors managed resources for alarms and state changes; no control operations are performed. Existing network management systems remain in place to perform specific control functions for the individual networks they support. As the TAP evolves, it will include some of these control operations. Access to the TAP Integrated Manager is through a standardized graphical user interface (GUI). The Integrated Manager combines configuration, state, and alarm data, and presents the information to the user through a unified view of the ARTCC networks.

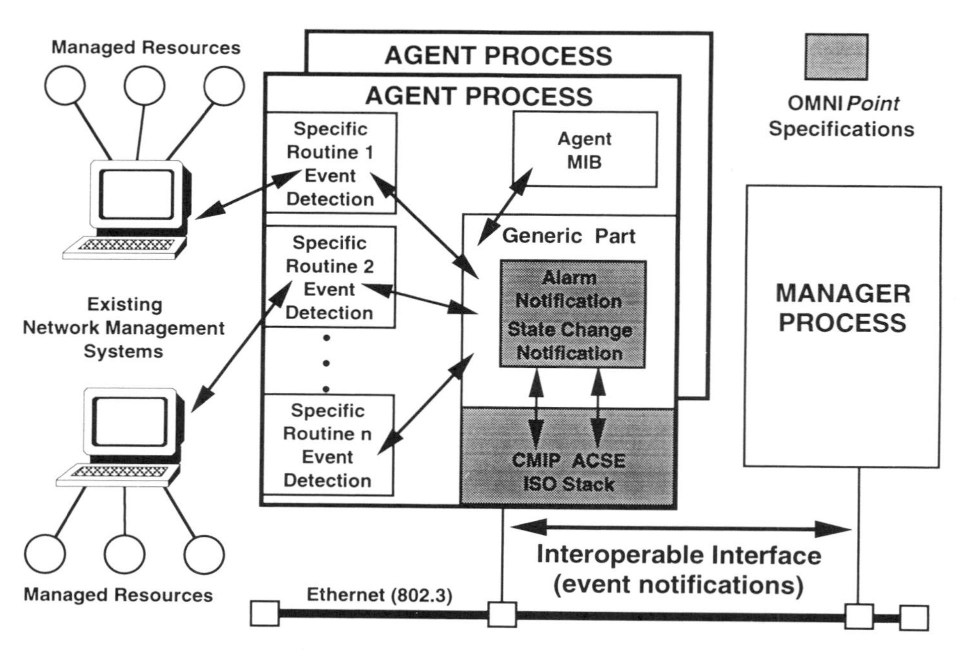

Figure 3. Agent Processing

AGENT AND MANAGER PROCESSING

Agent and manager processing for the TAP is illustrated in Figures 3 and 4. Alarms and state changes for all telecommunications networks and services at the ARTCC are monitored by the TAP. The agent maintains a database of all the TAP equipment and circuit object instances in its monitoring domain. Current alarm state and severity information for all object instances is also maintained by the agent.

Since none of the managed resources or network management systems at the ARTCCs currently conform to ISO or Forum standards for exchanging management information with external systems, the TAP agent includes a set of resource-specific routines to capture monitoring information when managed resources emit notification information. This occurs

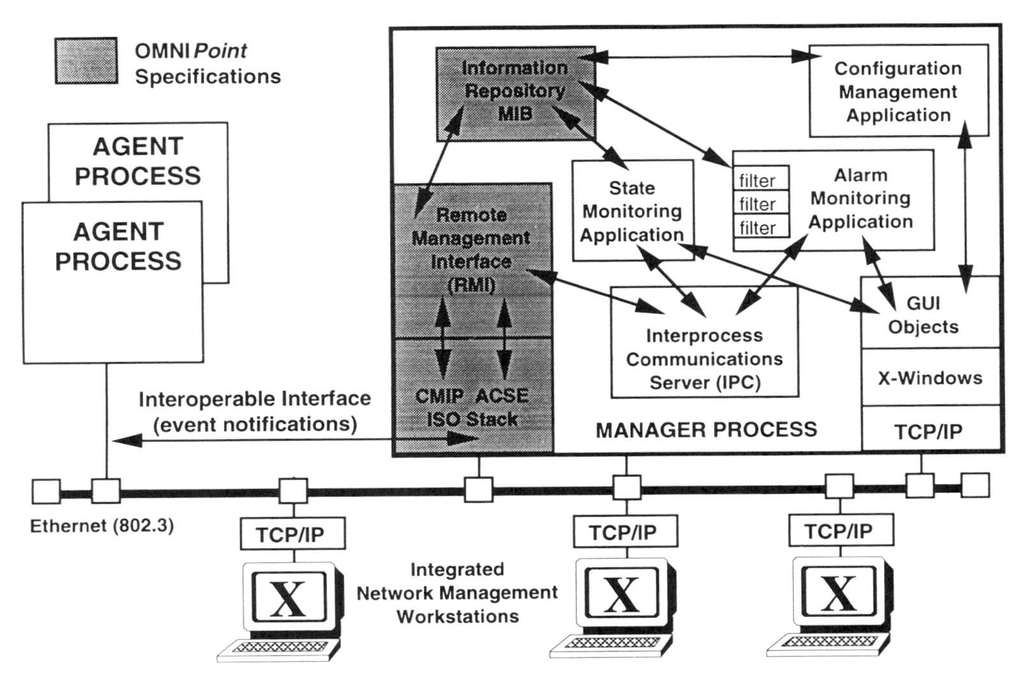

Figure 4. Manager Processing

when some external or internal event occurs. The source of this information can be a particular managed resource, monitoring or control messages exchanged between a managed resource and a network management system, or a network management system itself. Resource-specific information is converted by these routines into OSI management conformant data. For example, a network may include a number of sensors that detect and report alarms for a particular piece of equipment. Each sensor has a predefined alarm severity and identifies a particular problem. The resource-specific routine recognizes each sensor and converts the sensor information into OSI management reporting fields.

State information can be set automatically by a managed resource (a managed resource automatically sets the operational state from ENABLED to DISABLED), or it can be set manually by an operator (e.g., the administrative state is changed from LOCKED to UNLOCKED). Upon receipt of alarm or state information, the agent compares the new state information with the current information. If a change has occurred, the agent constructs an event notification and sends it to the TAP manager over the interoperable interface. The event notification includes an alarm or state change. Management Application Protocol Data Unit (MAPDU) and the CMIS M-EVENT-REPORT function.

Predefined alarm conditions may also trigger state changes. For example, a critical alarm for a power failure on a multiplexer may also indicate a DISABLED operational state. In this case, the agent constructs a separate state change event report along with the alarm event report and sends it to the TAP manager.

Management Services and Information Services for the TAP integrated manager are performed by the Remote Management Interface (RMI). Upon receipt of an event report (alarm or state change), the RMI checks the syntax of the notification. If the syntax check fails, an entry is made in an error log and appropriate messages are sent to the agent and TAP users. If the event report passes the syntax check, the manager creates a State_Change_Record or Alarm_Record object instance with a unique identification number. Using the object class and instance identifier in the event report record, the TAP manager updates (through built-in behaviors) the state and alarm attributes in the corresponding Circuit or Equipment object instance in the management information base.

The Interprocess Communications Server (IPC) is used to notify Status Monitoring and Alarm Monitoring applications when new events have been received by the RMI.

GENERIC NETWORK MODEL

OSI management standards use an object-oriented approach to describe network resources. Each physical or logical network resource that is managed (e.g., a modem, a circuit) is represented in an object class as a managed object instance. The TAP is based on a generic network configuration model derived from objects defined by the ISO and the Forum [7]. An entity-relationship diagram of the generic network model is shown in Figure 5. This network model is sufficiently flexible and extensible to provide physical, logical, and geographical views of the network. The generic model consists of the following object classes:

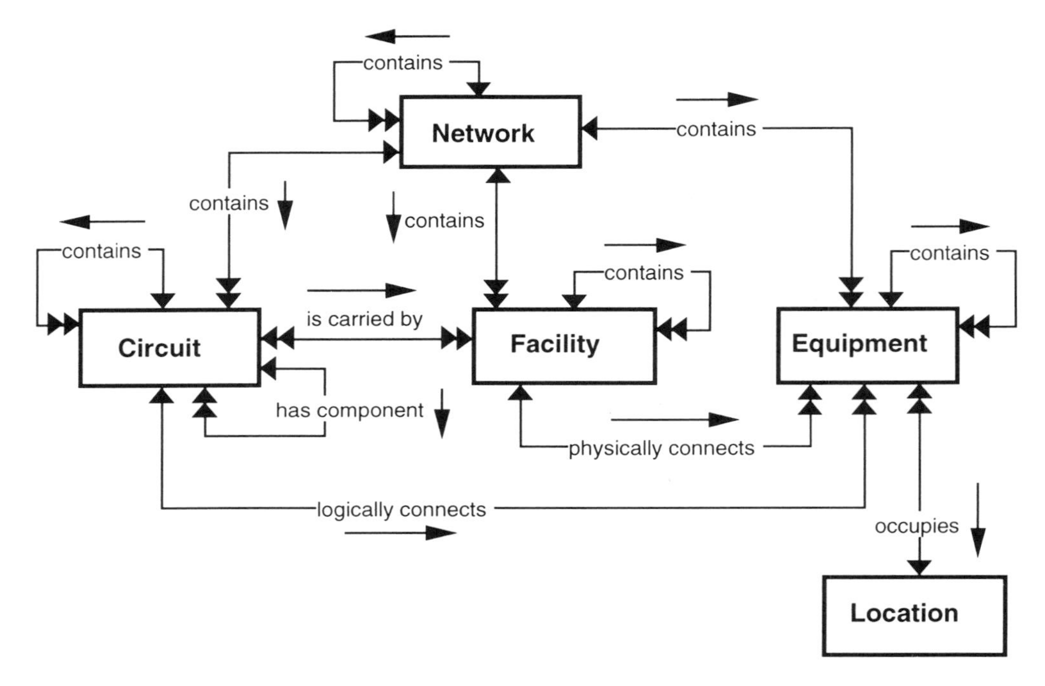

Figure 5. Generic Network Data Model

- The **Network** class refers to a collection of interconnected objects (logical or physical) that together support information exchange. The network class can represent a physical transmission network or a user service network. A network can contain circuit, facility, and equipment managed objects. A network may be contained within another network, thereby forming a superior-subordinate (or containment) relationship.
- The **Circuit** class refers to a connection that is independent of the means of carrying the signal. The circuit provides logical connectivity information. A circuit may consist of an ordered sequence of components. Such a circuit is referred to as a composite circuit.
- The **Facility** class refers to the physical means (e.g., cable) of carrying a signal. Instances of this class carry signals from point-to-point and do not change their contents. Facilities are used to carry circuits.

- The **Equipment** class refers to objects that are physical entities. Equipment may be nested in other equipment, thereby creating a containment relationship (for example, a line card is contained in an equipment shelf that is contained in a switch). Instances of equipment may be the endpoint of a circuit or network. Equipment can include telecommunications systems that provide services to end users, as well as end user host computer systems and terminals. Equipment occupies (is located at) a location.
- The **Location** class refers to a place occupied by one or more objects, organizations, or groups associated with the network. The location is not necessarily part of (i.e., contained by) the network.

INTEGRATED STATE AND ALARM MONITORING DATA MODEL

Additional classes beyond the generic network model were defined to accommodate state and alarm processing for the TAP [8,9,10,11]. An entity-relationship diagram of the data model used for the integrated state and alarm monitoring applications is shown in Figures 6 and 7. Standards-based and CAASD-developed alarm monitoring classes include:

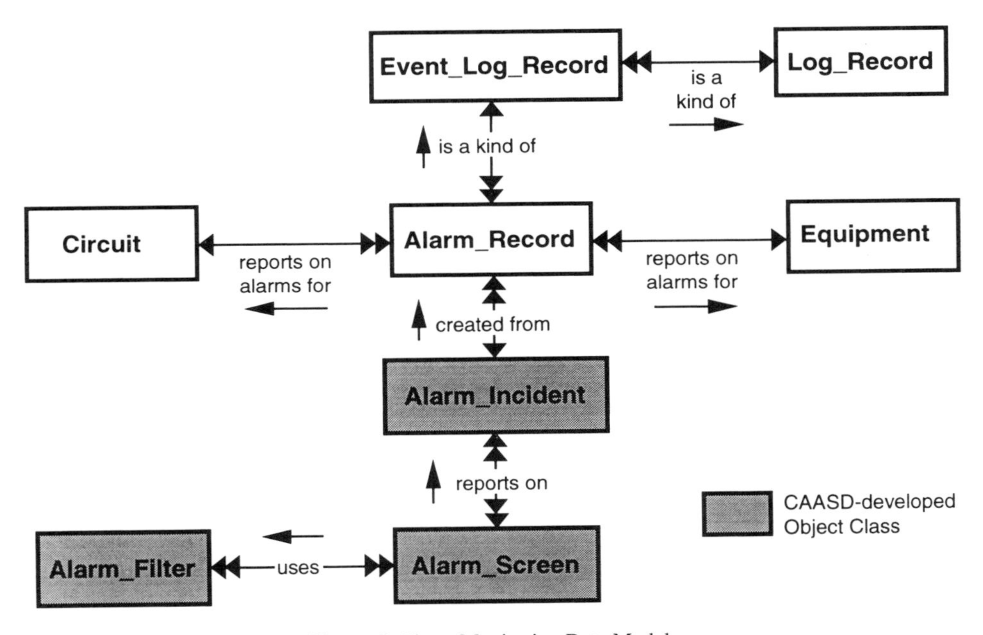

Figure 6. Alarm Monitoring Data Model

- **Log_Record** - Provides common information about records in a log (e.g., an event log).
- **Event_Log_Record** - A subclass of Log_Record that provides common information logged about all events (e.g., alarms and state changes).
- **Alarm_Record** - A subclass of Event_Log_Record that reports on alarm events detected by the agent for managed equipment and circuits.
- **Alarm_Incident** - Records information about each alarm incident. One or more alarm events (recorded in Alarm_Record instances) for a managed resource make up an alarm incident.
- **Alarm_Screen** - Defines how alarm incidents are to be displayed on a monitoring workstation.
- **Alarm_Filter** - Provides a means to suppress certain alarms from appearing on the alarm monitoring workstation.

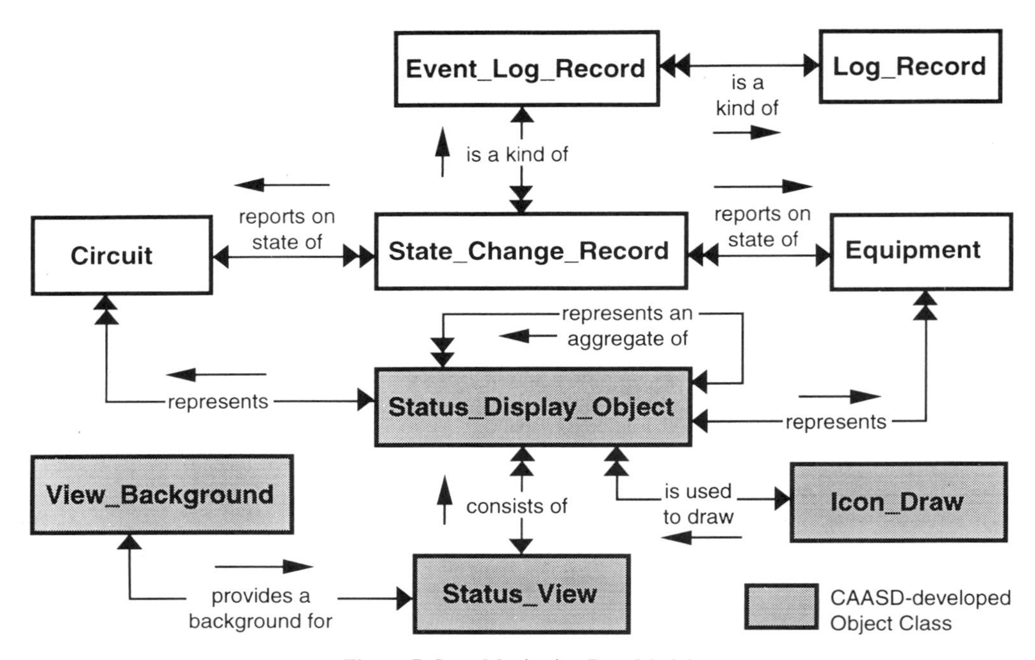

Figure 7. State Monitoring Data Model

Standards-based and CAASD-developed state monitoring classes include:

- **State_Change_Record** - A subclass of Event_Log_Record that reports on the state and status changes for managed equipment and circuits.
- **Status_View** - Specifies customized user-defined views of the network for state monitoring purposes. A view can include any grouping of circuits, equipment or aggregates. An aggregate is a logical grouping of multiple instances of equipment, circuits, or other aggregates for state monitoring purposes.
- **Status_Display_Object** - Identifies specific objects (circuits, equipment, aggregates) to be included in a Status_View and the state monitoring parameters to be displayed for each object.
- **Icon_Draw** - Specifies how an icon (Status_Display_Object) representing a particular managed object is to be drawn and displayed on the monitoring workstation.
- **View_Background** - Defines the characteristics of the background (e.g., state map, regional map, site layout) to be displayed for a particular view.

STATE MONITORING APPLICATION

State monitoring includes observing and recording (making visible) the instantaneous condition of availability of a managed resource. The ISO standards [9] identify three states that can be monitored:

- **Operational** - whether or not the managed resource is physically installed and working (state values are ENABLED and DISABLED).
- **Usage** - whether the managed resource is actively in use, and if so, whether it has spare capacity (state values are IDLE, ACTIVE, and BUSY).

- **Administrative** - permission to use or prohibition against using a managed resource (state values are LOCKED, UNLOCKED, and SHUTTING DOWN).

Status attributes are also defined to qualify one or more state attributes. The ISO standards include status attributes for Alarm, Procedural, Availability, Control, and Standby conditions. For example, the Alarm status attribute indicates whether a particular managed object has an outstanding alarm, is under repair, or has a minor, major, or critical alarm. The Standby status attribute indicates whether a managed resource is a hot standby (will immediately be able to take over), a cold standby, or is currently providing backup service.

The state of one resource can also determine the overall state of other resources. The state of a circuit or equipment object instance depends on the state or condition of resources on which the circuit or equipment is dependent. The Forum [2] defines two types of dependencies:

- The state of a composite circuit is dependent on the state of all its components.
- The state of a subordinate (contained) circuit or equipment is dependent on the superior (containing) circuit or equipment.

These dependency relationships are shown in Figures 8 and 9. The Circuit and Equipment object classes have built-in behaviors to set automatically the state of subordinate circuit or equipment instances, or composite circuits. These behaviors apply when the operational state changes from ENABLED to DISABLED, or when the administrative state changes from UNLOCKED to LOCKED. Based on the composite/component dependencies shown in Figure 8, if circuit *BC* is DISABLED, then circuit *BE* (the composite circuit of circuit *BC*) also becomes DISABLED. Since circuit *BE* is also a component circuit of circuit *AF*, then circuit *AF* also becomes DISABLED. The containment dependency is illustrated by the following example from Figure 9. If equipment *M1* is DISABLED, then subordinate equipment *P1*, *P2*, and *P3* are also set to DISABLED. All circuits whose endpoints are equipment *M1*, *P1*, *P2*, and *P3* must also be set to DISABLED. These would include circuits *C1*, *C2*, *C3*, and *C7*. Since *C7* is a superior circuit containing subordinate circuits. *P1-P4*, *P2-P5*, and *P3-P6*, these subordinate circuits must also be set to DISABLED.

When the operational state or administrative state changes back to a "cleared" or "normal" state (i.e., operational state changes from DISABLED to ENABLED, or administrative state changes from LOCKED to UNLOCKED), behaviors similar to the ones described above are executed, but with additional checks. These additional checks are needed to ensure that other related equipment or circuits are not still in a DISABLED or LOCKED state. This includes checking that other component circuits of a composite circuit are not still in a DISABLED or LOCKED state, and checking that other endpoints of an affected circuit are not still DISABLED or LOCKED. The following example illustrates this point. In the composite/component example above, if circuit *BC* is DISABLED, then composite circuits *BE* and *AF* are also DISABLED. Suppose that circuit *EF* also becomes DISABLED while circuit *BC* is DISABLED. When circuit *BC* becomes ENABLED, the dependency logic should also change composite circuits *BE* and *AF* to an ENABLED state. This would be correct for circuit *BE*, because none of its components are DISABLED. However, circuit *EF*, which is a component of circuit *AF*, is still DISABLED. Therefore, circuit *AF* cannot be set to an ENABLED state until both circuits *BC* and *EF* are ENABLED.

DIFFERENT VIEWS OF THE NETWORK

The TAP supports a wide range of users who want to view the network from different perspectives. These perspectives can range from an end-to-end user service view to individual equipment components. To support state monitoring for different perspectives of the network, a view/aggregate concept was developed. Views represent workstation screens that collect and display state, status, and alarm information according to the needs of a particular user. Specific equipment and circuit object instances may be specified to appear in

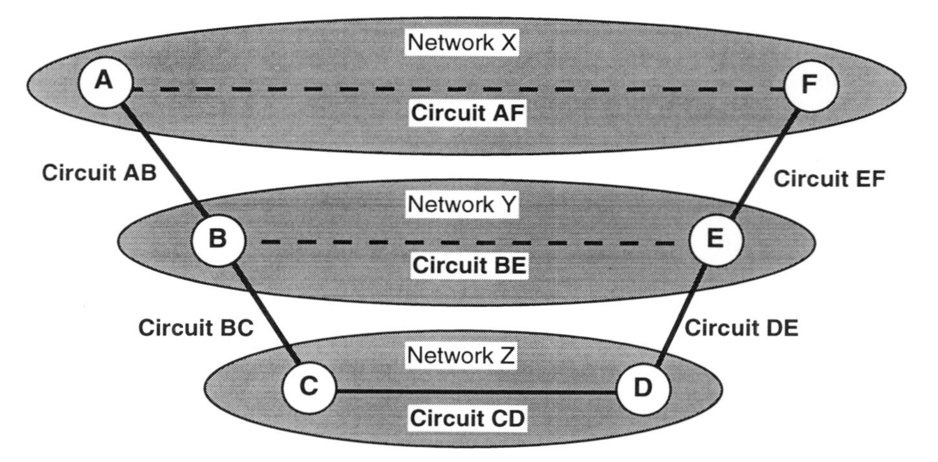

- Circuit BE is the composite circuit of component Circuits BC, CD, and DE
  - Circuit BE is also a component circuit of the composite Circuit AF

Figure 8. Composite/Component Relationship

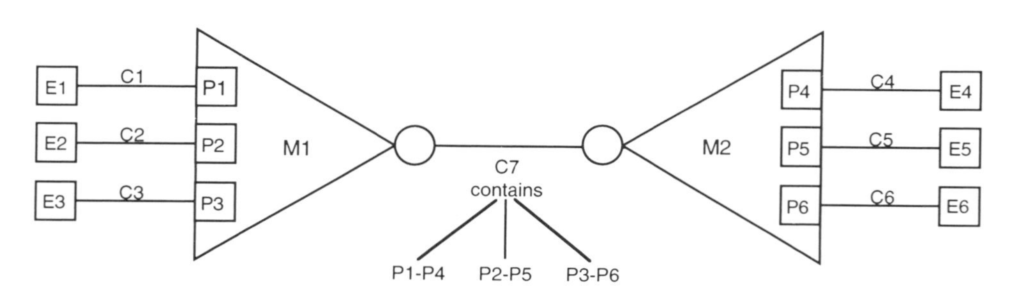

- C7 is a superior circuit that contains subordinate circuits P1-P4, P2-P5, and P3-P6
- M1 is a superior equipment that contains subordinate equipment P1, P2, and P3

Figure 9. Superior/Subordinate Containment Relationship

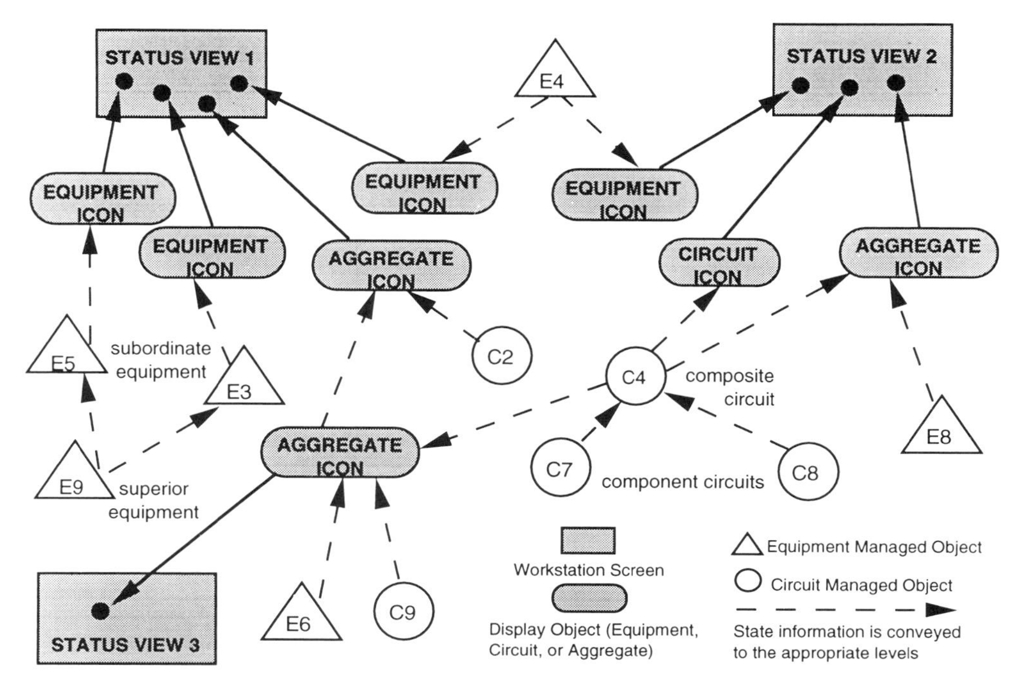

Figure 10. View/Aggregate Concept

a View. Aggregates representing logical groupings of circuit or equipment instances may also be specified. The view/aggregate concept illustrated in Figure 10 shows that:

- Aggregates represent a set of equipment, circuits, or other aggregates.
- Equipment, circuits, and aggregates can appear in multiple views.

The aggregate provides a convenient way to monitor a large group of managed objects without having to display each instance on the screen. Aggregates do not have any inherent state or status information that is detected or reported by the agent. The TAP user would have to create a sub-view in order to see the state and status of specific managed objects in the aggregate. The state and status of the aggregate is determined by the overall state and status of the group of objects that the aggregate represents. For example, if any of the individual managed objects in the aggregate has an operational state of DISABLED, then the operational state of the aggregate is set to DISABLED. For alarm status, the aggregate assumes the highest level alarm status of any managed object in the group. For example, if an aggregate represents three equipment instances, and those instances have alarm status's of MINOR, MAJOR, and CRITICAL, then the alarm status of the aggregate would be set to CRITICAL. State and status information for aggregates is set automatically through built-in behaviors that are triggered when state and alarm changes are set in managed objects that the aggregate represents.

Display functions built into the configuration objects are used to construct a predefined set of network status views the TAP user may choose to see. The Status_View object class identifies a particular view, indicates the background to be used (e.g., superimpose a network view on a regional map), and specifies the managed objects to be displayed. Each display object represents a circuit, equipment, or aggregate.

ALARM MONITORING APPLICATION

Alarms are event notifications concerning detected faults or abnormal conditions [10]. Alarms for all ATC telecommunications networks at the ARTCC are monitored by the TAP.

The manager notifies the alarm monitoring application when additional Alarm_Record instances are created. The alarm monitoring application maintains a record of alarm incidents using the Alarm_Incident class. An alarm incident to the TAP user includes the information from all alarm report events from the time a managed resource changes to an "active" alarm condition (i.e., an outstanding alarm) to the time it reverts back to a "cleared" alarm condition (i.e., no outstanding alarms). Each Alarm_Record (event report) may not necessarily identify a new alarm incident. For example, if the agent reports that a managed object has changed from an alarm severity of MINOR to a severity of CRITICAL, the alarm report is not considered a new alarm incident.

New alarm incidents are displayed on the alarm screen as a blinking line of text information. Alarms with a CRITICAL severity appear in red. The operator can specify whether or not an audible alert is to be sounded when a new alarm is displayed on the screen. New alarms stay blinking until the operator acknowledges the alarm. By acknowledging the alarm, the operator can have the alarm message remain on the screen (but not blinking), move the alarm to an outstanding alarm condition (the alarm requires further action before it can be closed), or close the alarm (the alarm is no longer displayed but remains in the alarm report log until it is archived). An alarm incident cannot be closed by the TAP user until the network has detected a "cleared" (i.e., no outstanding alarms) situation for the managed object. The alarm monitoring application maintains four date/times for each alarm incident: detected by the managed resource, acknowledged by the operator, cleared by the managed resource, and closed by the operator. At any time, the operator can enter a comment for a particular alarm incident or view all previously entered comments. The alarm application records the date, time, and ID, and text of the comment.

To provide each group with the flexibility to see only alarms in their domain, alarms can be filtered (i.e., suppressed) so that they will not appear on the alarm screen. However, the alarm is still recorded in the log. Criteria for filtering alarms can be any combination of date/time (from/to), alarm type, probable cause, location, object class, network, or threshold (i.e., the number of times an alarm must be activated over a given time period before the alarm is displayed on the screen. Multiple filters can be in effect for a given user group at any given time). The user can specify the order in which alarm incidents are displayed on the screen using any combination of date/time, severity, alarm type, location, object class, or network.

CONCLUSIONS

OSI network management standards continue to evolve. Current standards concentrate on the overall architecture, the protocols and services that allow a manager system to exchange information with a managed system over an interoperable interface, and the objects necessary to communicate management information over the interoperable interface. The specific manner in which network management is accomplished for functions such as state and alarm monitoring, as well as the interpretation of some standards, is left to the developer.

Release 1 of the Forum implementation agreements is aligned with the finalized and draft versions of the ISO standards that existed at the time the agreements were published in June 1990. The initial version of TAP reflects ISO standards and Forum implementation agreements as of June 1990. Standards bodies and implementors workshops are continuing to develop additional standards beyond Forum Release 1 and the finalized ISO standards. In early 1992, an industry-wide effort directed at developing more widely accepted implementation profiles resulted in the development of the Open Management Roadmap. The Roadmap specifications will be issued in a series of OMNI*Point* releases. Each OMNI*Point* release is expected to address a defined set of user needs (management functions) that can be met by applying available and/or anticipated international standards. OMNI*Point* 1, released at the end of 1992, specifies changes to the managed object class library documented in Forum Release 1 and the replacement of previous Forum-specified management functions by the corresponding OSI management function specifications. Potential impacts of OMNI*Point* 1 on the TAP were identified, and appropriate change made to the TAP. Implementation of a network management system that adheres to OSI standards requires constant monitoring of standards development.

ABOUT THE AUTHOR

Robert H. Stratman (rstratma@mitre.org) joined the MITRE Corporation in 1981 and is currently a Department Assistant in The MITRE Corporation's Center for Advanced Aviation System Development (CAASD). He is providing system engineering support to the Federal Aviation Administration's Airway facilities (AF) in their efforts to improve the management of the National Airspace System (NAS) communications infrastructure and the services it provides. Areas of interest include communications architecture and network management. Prior to joining MITRE, Mr. Stratman was involved in database and information system design and analysis at Boeing Computer Services and Booz, Allen & Hamilton. He received a B.A. in mathematics from Pennsylvania State University.

REFERENCES

1. *ISO 7498-4, Information Processing Systems - Open Systems Interconnection - Basic Reference Model - Part 4: Management Framework,* 1989.

2. OSI/Network Management Forum, June 1990, *Release 1 Specification,* OSI/Network Management Forum, Bernardsville, NJ.

3. *ISO 10165-1, Information Processing Systems - Open Systems Interconnection - Structure of Management Information - Part 1: Management Information Model,* 1991.

4. Network Management Forum, 1993, *OMNIPoint 1 Specification*, Network Management Forum, Bernardsville, NJ.

5. *ISO/IEC 9596-1, Information Technology - Open Systems Interconnection - Common Management Information Protocol Specification*, November 1990.

6. *ISO/IEC 9595, Information Technology - Open Systems Interconnection - Common Management Information Service Definition*, November 1990.

7. Kennedy, T. W., Riegner, S. E. M., August 1991, *An Object-Oriented Data Model for the Operations, Administration, and Maintenance (OA&M) of the Federal Aviation Administration Telecommunications (Release 1.02)*, WP-91W00134, The MITRE Corporation, McLean, VA.

8. *ISO 10164-1, Information Processing Systems - Open Systems Interconnection - Systems Management - Part 1: Object Management Function*, 1991.

9. *ISO 10164-2, Information Processing Systems - Open Systems Interconnection - Systems Management - Part 2: State Management Function*, 1991.

10. *ISO 10164-4, Information Processing Systems - Open Systems Interconnection - Systems Management - Part 4: Alarm Reporting Function*, 1991.

11. *ISO 10164-5, Information Processing Systems - Open Systems Interconnection - Systems Management - Part 5: Event Report Management Function*, 1991.

COVERAGE MODELING FOR THE DESIGN OF NETWORK MANAGEMENT PROCEDURES

M. Veeraraghavan

AT&T Bell Laboratories
Holmdel, NJ 07733

ABSTRACT

Fault-tolerant networks used in applications with high availability requirements need built-in automatic network management schemes to handle faults and congestion. Different timers and threshold parameters need to be specified for these network management procedures. Besides high availability requirements, fault-tolerant networks often have stringent performance requirements as well. Hence a design of network management procedures should simultaneously consider both availability and performance metrics.

In this paper, we define coverage of network management procedures for a given set of performance measures. We outline a coverage modeling technique using Generalized Stochastic Petri Nets (GSPN) to obtain coverage measures. These measures are then combined with a fault-occurrence and repair model to obtain combined performance and availability measures. A GSPN model of a network management procedure called *changeover* used in Signaling System No. 7 (SS7) protocol based networks, is used to illustrate the coverage and performability modeling techniques.

INTRODUCTION

Networks used in applications with high availability requirements need automatic network management schemes to handle faults and congestion. These schemes are used to detect and confine faults/congestion, reconfigure the network with adaptive routing techniques, and restore the original configuration upon restart of the faulty component or abatement of congestion. Different timers and threshold parameters need to be specified while defining these network management procedures. Often, networks with high availability requirements are used in applications that have high performance objectives as well. Hence a design of network management schemes should simultaneously consider both availability and performance metrics [1].

Network Management and Control, Vol.2, Edited by
I.T. Frisch *et al.*, Plenum Press, New York, 1994

Network management schemes used to handle faults in a network are analogous to fault handling schemes used in fault-tolerant computer systems. The notion of *coverage* has been used to model and characterize fault handling schemes in the design of repairable [2] and non-repairable [3] computer systems. With respect to networks, imperfect coverage has been defined in [4] as the failure of the network to re-route affected traffic, even when spare capacity is available.

In this paper, we extend this definition of coverage, and describe a technique to model network management procedures for coverage measures. We also illustrate the use of these coverage measures in performability models of the network to obtain combined performance and reliability measures.

The definition of coverage is extended in two ways. First, it is extended to be applicable to any network management procedure, and hence imperfect coverage will include not just the failure of a network to re-route affected traffic, but will reflect the failure of any network management procedure. For example, if the network management procedure is being used to restore a node after repair into a functioning network, imperfect coverage of this procedure reflects the failure of the network to successful reintegrate the node. Secondly, coverage of a network management procedure is redefined as a function of a given set of performance measures. For example, when a link fault occurs in a network, a network management procedure may successfully reconfigure the network and recover to a stable state with one less link. The overall network measures such as delay and throughput will be affected due to the unavailability of this faulty link. However, additionally, the network management procedure may lead to a state where not only is the network reconfigured around the faulty link, but due to abnormal (error) conditions in the network management procedure itself, other performance measures, such as the percentage of lost messages, mis-sequenced messages and duplicate messages, are also adversely affected. Coverage models described in this paper allow for these possibilities.

To describe the coverage modeling technique, we use a network based on Signaling System No. 7 (SS7) [5] as an example. The SS7 protocol provides extensive dynamic fault handling and congestion control schemes to automatically detect, isolate, reconfigure, recover and finally restore the network following faults or congestion. A coverage model of the SS7 **changeover procedure** used for the reconfiguration and recovery phases of handling a link fault is used in this paper for illustration purposes.

Furthermore, this paper describes a performability model of a network with Fault/Error Handling Models (FEHM) to model network management procedures and Fault Occurrence and Repair Models (FORM) [3] to capture the fault occurrence and repair behavior of the network. Previous efforts to model SS7-based networks have been carried out for pure availability measures [6] and for pure performance measures [7]. Using the techniques described in this paper, combined measures of performance and availability can be obtained for SS7-based networks and other fault-tolerant networks.

In this paper, a signaling network architecture and the **changeover procedure** are described. Coverage of a network management procedure for a given vector of performance measures is defined. A coverage modeling technique using Generalized Stochastic Petri Nets (GSPN) is described and illustrated using the SS7 changeover procedure. A combined performance/availability model to illustrate how coverage parameters are incorporated into the FORM to obtain network level measures is then outlined.

SIGNALING NETWORK DESCRIPTION

In this section, we describe a fault-tolerant, highly available, long-haul network with stringent performance criteria. The network architecture is outlined in the first sub-section. The next sub-section describes the SS7 changeover procedure in detail. This section is included here as a refresher to enable the reader to match different aspects of the procedure with its GSPN coverage model (FEHM) described later.

Architecture

A signaling mesh network is shown in Figure 1. The network consists of independent

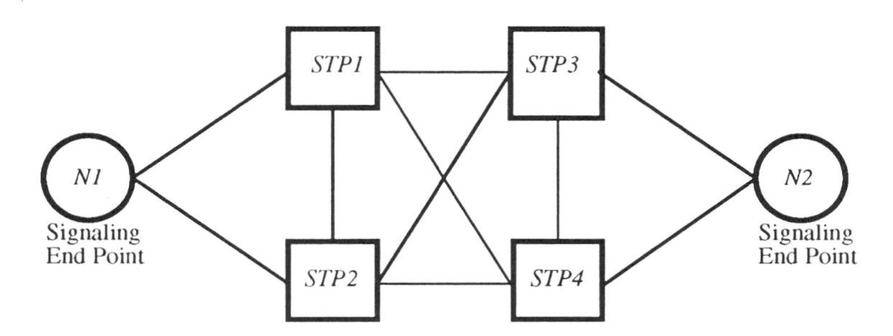

Figure 1. Long-Haul Signaling Mesh Network

nodes that communicate with each other using SS7 messages. There are two types of nodes in the signaling network, signaling end points and Signaling Transfer Points (STPs). A signaling end point is the source or destination of user level SS7 messages (Nodes *N1* and *N2* in Figure 1). Signaling transfer points only transfer user level messages received on one link to another link (Nodes *STP1*, *STP2*, *STP3* and *STP4* in Figure 1). Signaling transfer points have network management functions and are the controlling nodes in the network. The Message Transfer Part (MTP) provides the network, data link and physical layer functions of the OSI seven-layered protocol model for the SS7 protocol. MTP has three levels, Level 1 (L1), Level 2 (L2) and Level 3 (L3) as shown in Figure 2.

MTP L3 (Network Layer Functions)
MTP L2 (Data Link Layer Functions)
MTP L1 (Physical Layer Functions)

Figure 2. Message Transfer Part Levels of the SS7 Protocol

Changeover Procedure

The MTP L3 layer of the SS7 protocol stack provides network management procedures to automatically handle a node/link fault, a node/link restoral, and congestion conditions. In this section, we describe the changeover procedure, which is used in the SS7 network to reconfigure the network around a link fault. We assume that a link, *L1*, between node *N1* and *STP1* in Figure 1 has a fault.

In response to a link out-of-service primitive from MTP L2, MTP L3 network management at nodes *N1* and *STP1* perform a procedure called **changeover** [5]. MTP L2 provides error correction using a scheme based on acknowledgments (one per message, and piggybacked on messages in the opposite direction) and *go-back-N* retransmission in case of errors. The changeover procedure is used to retrieve unacknowledged messages that were transmitted earlier on link *L1* and to retransmit them on an alternative route. MTP L3 first determines if there is an alternative signaling route, and if the changeover procedure is needed and can be done. If there is an alternative signaling route and changeover needs to be done (there are unacknowledged messages queued up on the faulty link) and changeover can be done (both nodes *N1* and *STP1* are functioning), then we assume that MTP L3 at node *N1* initiates the changeover procedure. MTP L2 keeps track of the sequence number of the last message it acknowledged, i.e., the sequence number of the last message it successfully received from *STP1*. So MTP L3 at node *N1* issues a primitive to its MTP L2 to retrieve this sequence number, x. Upon reception of the sequence number x from its MTP L2, MTP L3 at node *N1* issues a *Changeover Order message* to *STP1*, containing the sequence number x. Node *N1* initiates *timer T2* within which it expects to receive a *Changeover Acknowledgment message* from *STP1*. The changeover acknowledgment contains a sequence number, y, which corresponds to the last message *STP1* successfully received from node *N1*. When MTP L3 at node *N1* receives the changeover acknowledgment message from *STP1*, it issues a *Request primitive* to its MTP L2 to retrieve messages starting after the sequence number y. MTP L2 at node *N1* responds with all the message addresses for these retrieved messages to its MTP L3. A similar action occurs at *STP1*. The retrieved messages are sent over the alternative route. MTP L3 at both nodes *N1* and *STP1* update their corresponding routing tables.

If there is no alternative route between nodes *N*1 and *STP1*, then changeover cannot be performed across these nodes. However, an alternative route may exist between the signaling end points *N1* and *N2* that need to exchange user level SS7 messages. New messages destined to node *N2* from *N1* will be routed via this alternative path after *timer T1* expires (*time-controlled changeover*)[1], but unacknowledged messages that were routed through node *STP1* are dropped. This may result in mis-sequenced messages (messages sent first via *STP1* reaching after messages sent later on the alternative route) and lost messages (messages that were sent via *STP1* that never reached node *N2* will be dropped at node *N1* due to incomplete changeover between nodes *N1* and *STP1*). These cases illustrate the possibility that during a network management procedure, there may be a degradation of performance in the form of lost or mis-sequenced messages.

COVERAGE

We first define coverage measures for use in modeling fault-tolerant repairable networks. Next, we describe the behavioral decomposition technique in which these coverage measures are used.

Definition of Coverage in Fault-Tolerant Repairable Networks

For fault-tolerant networks, we define coverage measures to characterize network management procedures.

Let Λ *represent a network management procedure,*

Event E_Λ *represent the completion of procedure* Λ,

Vector $\boldsymbol{P}_\Lambda$ *represent a set of performance measures for procedure* Λ, *and*

$\wp_i$ *represent the* i^{th} *set of numeric values for the elements of the vector* $\boldsymbol{P}_\Lambda$, *then*

$$\text{Coverage, } C_i(E_\Lambda, \boldsymbol{P}_\Lambda, \wp_i) = Pr((E_\Lambda \cap (\boldsymbol{P}_\Lambda = \wp_i)) \mid \Lambda \text{ was initiated}),$$

1. This time delay is introduced to minimize the probability of mis-sequencing.

where $Pr(\bullet)$ represents the Probability function, and $i = 1, 2,... \rho$, where ρ is the total number of the possible outcomes of the procedure Λ.

Network management procedures include different phases of fault and congestion handling. Typically, different stages in handling a fault include fault detection, retry, reconfiguration, fault confinement, recovery, restart, and reintegration. Different stages related to congestion control include detection, confinement, reconfiguration, and recovery. Any network management procedure Λ that perform functions related to one or more of these phases can be characterized by the defined coverage measure.

Based on the application for which a given network and a corresponding network management procedure Λ are used, vector $\mathbf{P}_\Lambda$, could consist of different performance measures. Typical examples of performance measures include number or percentage of lost messages, number or percentage of out-of-sequence messages, number or percentage of duplicated messages, and message delay.

For a given network management procedure Λ, and a corresponding performance vector P_Λ, we only consider a integral number of $\wp_i$ for modeling purposes so that upon completion of the network management procedure, the network is one of a discrete number of states. This assumption is made so that resulting models are tractable. Since times for completion of network management procedures are either deterministic or can be bounded, this assumption leads to fairly realistic models. Thus for a discrete set of $\wp_i$'s, a discrete set of coverage measures C_i's are obtained.

To illustrate the use of this coverage measure $C_i (E_\Lambda, \boldsymbol{P}_\Lambda, \wp_i)$, the network management procedure Λ under consideration is assumed to be the MTP L3 changeover procedure, and the performance measures vector, $\mathbf{P}_\Lambda$ = [number of lost messages, number of out-of-sequence messages]. Let $\wp_1$= [0, 0], then coverage $C_1 (E_\Lambda, \boldsymbol{P}_\Lambda, \wp_1)$ represents the probability that the network reconfigured and recovered from the link fault (i.e., the changeover procedure was completed) with 0 lost and 0 out-of-sequence messages. For $\wp_2$= [10, 4], then $C_2 (E_\Lambda, \boldsymbol{P}_\Lambda, \wp_2)$, is the probability that the network reconfigured and recovered from the link fault with 10 lost and 4 out-of-sequence messages. As described earlier, if there is no alternative route, changeover cannot be performed without the possibility of lost messages, since there is no way to retrieve the sequence number of the last correctly received message. There exists the possibility of out-of-sequence messages even if changeover is successful. Messages successfully received at STP1 before the link fault would be en route to node N2 via other STPs[2] such as STP3 in Figure 1. Subsequent messages sent from node N1 to N2 via an alternative route may reach before these previously sent messages. Thus, there exists a possibility for out-of-sequence messages.

If the network management procedure under consideration is a network node restart and reintegration procedure, the coverage measure is a *restoral coverage*. Thus, coverage measures can be defined for every network management procedure for corresponding sets of performance measures.

Use of Coverage Measures in Reliability Models

Fault/Error Handling Models (FEHM) are used to model network management procedures for coverage measures. These models are used in conjunction with Fault Occurrence and Repair Models (FORM) in a technique called behavioral decomposition [3]. FEHMs are first constructed and solved separately, one FEHM for each component type or one per transition of a Markov model representing the FORM. Coverage parameters from the FEHMs are then incorporated into the FORM and the complete model solved for network level measures that reflect both performance and reliability. The basis for decomposing models along

2. STPs merely act as routing nodes for signaling messages generated by node N1 destined for node N2.

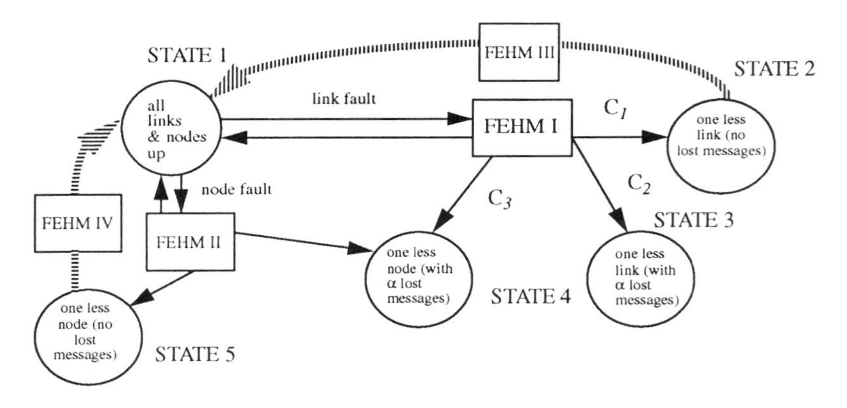

Figure 3. Coverage Models for Repairable Networks

these lines comes from the fact that fault handling rates are typically much higher than fault-occurrence or repair rates.

Part of a combined FORM-FEHM model for a repairable fault-tolerant network is shown in Figure 3. FEHMs I and II represent fault/error handling procedures following a link fault and node fault respectively. Arcs leading from FEHM I places the network in one of four states, state with one fewer link with no lost messages (STATE 2), state with one fewer link/node with α lost messages (STATES 3 or 4), and the original state (STATE 1). FEHMs III and IV represent procedures used to restore a node and link respectively. All the arcs leading from these FEHMs are not shown for simplicity reasons. Coverage parameters C_1, C_2, and C_3 are shown in Figure 3 on the arcs leading off from FEHM I to states 2, 3, and 4, respectively. Similar coverage measures are also obtained by solving FEHMs II, III and IV.

COVERAGE MODELING TECHNIQUE

To describe the modeling technique, a Markov model is used as the FORM and a Generalized Stochastic Petri Net (GSPN) [8] as the FEHM. Assuming that most readers are familiar with Markov models, we present a brief description of GSPNs. The GSPN FEHM used to model the **changeover procedure** is then described.

Overview of Generalized Stochastic Petri Nets

A GSPN consists of *places* (drawn as circles) and *transitions*, *timed* (drawn as bars) and *immediate* (drawn as horizontal lines) and a set of directed *arcs* from places to transitions and from transitions to places. When all input places to a transition have tokens, the transition is enabled. An enabled transition fires, by removing tokens from the input places and depositing a token in each output place. A timed transition fires after a time period that may be deterministic or have a general distribution. An *inhibitor arc* from a place to a transition has a small circle instead of an arrowhead. The firing rule for transitions that have inhibitor arcs as inputs is that all normal places from which there are directed arcs should have a token and all places from which there are inhibitor arcs to the transition should be empty in order for the transition to fire. If multiple immediate transitions are enabled simultaneously, a *random switch* with an associated *switching distribution* specifies which transition fires.

Fault/Error Handling Model of the Changeover Procedure

Based on the definition of coverage given in this paper, we first identify a set of performance measures to characterize the coverage of procedure Λ, set to be the SS7 MTP L3 changeover procedure. The performance vector, P_Λ, and its numeric values for the 4 possible outcomes of procedure Λ are assumed to be:

$$\mathbf{P}_\Lambda \equiv [\text{number of lost messages, number of links removed, number of nodes removed}]$$

$$\wp_1 = [0, 1, 0];\ \wp_2 = [\alpha, 1, 0];\ \wp_3 = [\alpha, 0, 1];\ \wp_4 = [0, 0, 0],$$

where α, the number of lost messages, is based on the time for link fault detection and other parameters. Then the corresponding coverage measures are $C_1(E_\Lambda, \mathbf{P}_\Lambda, \wp_1)$, $C_2(E_\Lambda, \mathbf{P}_\Lambda, \wp_2)$, $C_3(E_\Lambda, \mathbf{P}_\Lambda, \wp_3)$, and $C_4(E_\Lambda, \mathbf{P}_\Lambda, \wp_4)$. For brevity, these coverage measures are henceforth referred to as C_1, C_2, C_3, and C_4, respectively.

Figure 4 shows a fault/error handling model for ChangeOver (CO) following a link fault in the network. Tables 1 and 2 summarize the conditions represented by places in Figure 1 and the events performed by transitions. The initial marking of the GSPN consists of a token

Table 1. Conditions Represented by Places in the GSPN of Figure 4

Place	Condition
P1	Network functional
P2	Link fault
P3	Link available
P4	Alternative route available, CO not possible
P5	Alternative route available, CO possible
P7	Alternative route not available
P8	One less link with no lost messages
P9	One less link with α lost messages
P10	Check if transient

Place	Condition
P11	Permanent fault
P12	Transient fault
P13	Check for alternative routes and if CO is possible
P15	CO Acknowledgment arrived
P16	CO Ack in transition
P17	Waiting for CO Ack
P18	CO Order in transit
P19	Start timer T2 for reception of CO Acknowledgment
P20	Timer T2 expired

in places, P1 and P2. When a link fault occurs, a token is deposited in place P2 and transition TR15 fires. The token is deposited in output place P10. Transitions TR2 and TR5 are both enabled. With probability q, transition TR5 fires and with probability $1-q$, transition TR2 fires. This captures the possibility that the fault may be a transient. If TR5 fires, the token is absorbed in place P3 representing that CO is not carried out since the fault was a transient. If TR2 fires, the token is deposited in place P11, which in turn causes transition TR8 to fire.

Three transitions lead from place P13, the output place of TR8. As these transitions are simultaneously enabled, the random switching distribution is represented with probabilities, s, u, and $1-s-u$, to represent the three cases when an alternative route is available but CO is

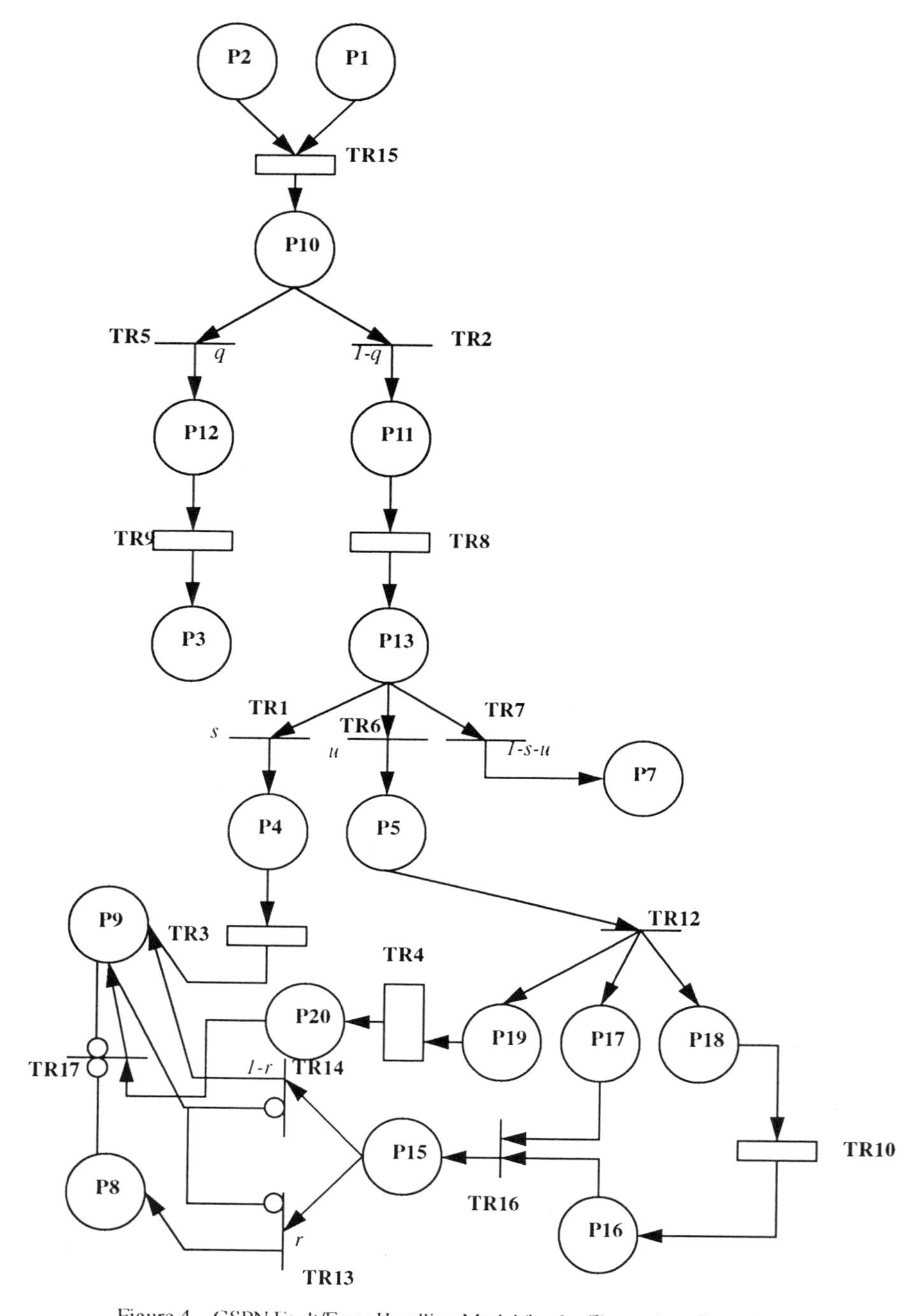

Figure 4. GSPN Fault/Error Handling Model for the Changeover Procedure

not possible, an alternative route is available and CO is possible, and the case where there is no alternative route to the remote node. Based on the state of the network (state in the Markov model FORM) from which this FEHM (GSPN) is entered (in terms of available nodes, links and routes), one of the three transitions has a probability 1 and the other two 0. For example, if all routes to a destination are unavailable, then TR7 is the selected transition which leads to place P7. On the other hand, if an alternative route to node *N2* is available, but changeover is not possible because there is no signaling route to node *STP1*, then probability s in Figure 4 equals 1 and TR1 is the selected transition which leads to place P4. An example state of the network in Figure 1, when this may happen, is if all linksets between *STP1* and every other neighboring node are unavailable, and the last available link from node *N1* to node *STP1* just had a link fault. In this case, while node *N1* may still have a signaling route to node *N2*, there is no way for node *N1* to determine which of the earlier sent messages reached their final destination *N2* successfully, since MTP L2 error correction schemes is carried out on a link-per-link basis (*N1* to *STP1*).

Table 2. Events Performed by Transitions in the GSPN of Figure 4

Transition	Event
TR1	Alternative route available, CO not possible
TR2	Permanent fault
TR3	Timer T1
TR4	Timer T2
TR5	Transient fault
TR6	Alternative route available, CO possible
TR7	Alternative route not available
TR8	Check to see if CO is possible and alternative routes are available

Transition	Event
TR9	Transient disappears
TR10	Process CO order and generate CO Ack
TR12	Initiate CO
TR13	CO Ack correct
TR14	CO Ack in error
TR15	Recognize link fault
TR16	CO Ack received
TR17	Handle timer T2 expiry

If the token is in place P4, time controlled changeover is initiated to minimize duplicated messages. This is represented by TR3, which then leads to an absorbing place P9. Since CO is not possible in this case, messages queued for the faulty link are dropped. Thus place P9 represents CO completion with p lost messages. The MTP L3 protocol standard [5] specifies that a timer T1 be used for time controlled changeover.

If transition TR6 fires instead of TR1 or TR7, the token is deposited in place P5 which enables transition TR12. This corresponds to the case when an alternative route is available and CO is possible. A CO order message is generated (P18), timer T2 is started within which a CO acknowledgment is expected (P19), and a token is deposited in place P17 to represent the condition that the transmitting node is waiting for the CO acknowledgment. Place P18 and P16 represent the signaling link [9] with the CO order and CO acknowledgment messages in transit. Transition T10 is enabled where the CO order is processed and a CO acknowledgment message generated. With the arrival of a token in P16, transition TR16 fires, since place P17 already holds a token.

When TR16 fires, a token is deposited in place P15 which represents that the CO acknowledgment arrived. There is however a small probability that this message is in error. This is represented by the random switch shown in transitions TR13 and TR14. If the message is in error TR14 fires, leading to place P9, where messages queued for the faulty link are dropped. If the CO acknowledgment is not in error, transition TR13 fires and the token is deposited in place P8, which represents successful changeover with no lost messages. Both transitions TR13 and TR14 fire only if place P9 is empty (inhibitor arc). This models the fact that if timer T2 expires before the CO acknowledgment message is received (TR4 has fired and a token is present in place P9) and then a token appears later in place P15, then neither TR13 nor TR14 should fire.

To represent the possibility that timer T2 may expire before CO acknowledgment is received, transition TR4 fires when timer T2 expires, and a token is deposited in place P20. If no tokens are present in places P8 or P9 (represented by inhibitor arcs), i.e., CO acknowledgment was not received, transition TR17 fires causing the token to be deposited in place P9.

Solution of the GSPN FEHM

The GSPN in Figure 4 was solved for the probability of being in absorbing places, P3, P7, P8, and P9 for a set of input parameters as shown in Table 3. All the timed transitions are assumed to have exponentially distributed times to firing. The entries corresponding to these

Table 3. Input Parameters and Output Measures for GSPN in Figure 4

Input Parameter	Value	Input Parameter	Value
r	0.9	TR3	1
q	0.1	TR4	1
s	0.005	TR8	50
u	0.99	TR9	100
TR15	100	TR10	10

Output Measure	Value
P3 (C_4)	0.1
P7 (C_3)	0.0045
P8 (C_1)	0.729
P9 (C_2)	0.1665

transitions in Table 3 represent their rates in units of */sec*. The tool, SHARPE [10], is used to solve this model. The input parameters r and q are arbitrarily selected, where r represents the probability that the CO acknowledgment message is not in error, and q the probability of the fault being a transient. Parameters s and u should have been 0 and 1 respectively, for the case when this GSPN model is entered from a state in the FORM (Markov model) where an alternative route is available and CO is possible. However, the implementation of the GSPN model solution in SHARPE does not allow for a transition to have implicit weight of 0. Hence, an approximate value of 0.99 is assumed for u and 0.005 for s. The impact of this assumption is currently being evaluated through the use of other models, such as Stochastic Reward Nets [11]. Timer values for transitions TR3 and TR4 are selected to be 1000 ms as an approximate figure based on the specifications of the MTP L3 protocol[5]. The firing rate for TR9 is taken from [3]. Rates for the other transitions are selected based on estimates of the time needed to perform the functions they represent. The resulting output measures in Table 3 indicate the steady state probability of the absorbing places, P3, P7, P8, and P9 being non-empty.

COMBINED PERFORMANCE AND RELIABILITY MODELING

Figure 5 shows FEHM 1 incorporated into part of a FORM showing a link fault. The

FEHM I in Figure 5 is the GSPN shown in Figure 4. Place P8 in Figure 4 corresponds to state 2 in Figure 5. Place P7 in Figure 4 corresponds to state 4 in Figure 5. Place P9 in Figure 4 corresponds to state 3 in Figure 5. Place P1 in Figure 4 corresponds to state 1 in Figure 5 which represents the network in its original state. The four arcs leading from FEHM I in Fig-

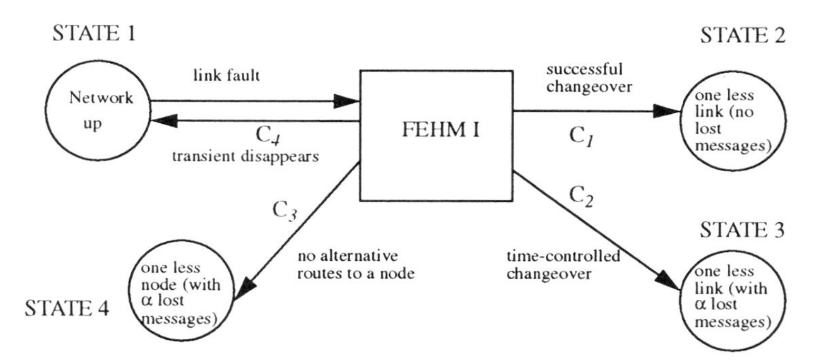

Figure 5. Part of the FORM for the network with FEHM I for a link fault

ure 5 can be traced to the directed arcs from transitions to places, P3, P7, P8 and P9 in Figure 4. As FEHM I only models the changeover procedure, which does not include error detection, we have assumed error detection to have perfect coverage in Figure 5. The probabilities of being in states 3 and 4 in Figure 4 add to the probability of lost messages. The combined model can thus be solved for performability measures, that reflect both performance and availability. Figure 6 shows a part of a Markov model integrated with the coverage parameters obtained by solving the FEHM. If a performance value (reward rate) is associated with each

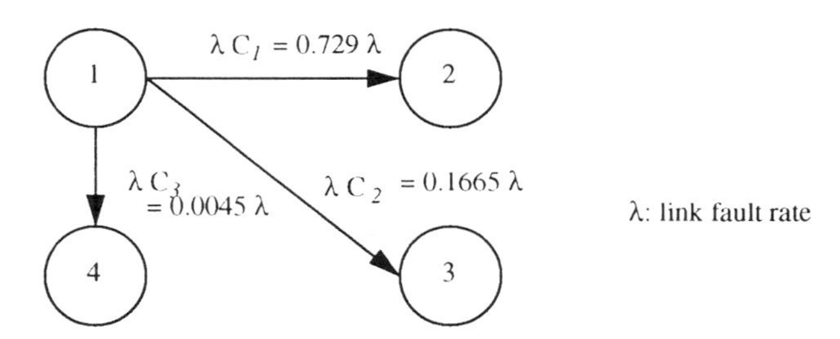

Figure 6. Part of the Combined FORM-FEHM Model

state (Markov reward model[12]), the model can be solved for performability measures. The parameter α representing the number of lost messages is computed as follows:

Assumptions:

Mean link fault detection time = 128 ms (for 64 kbps links: MTP L2 specification)

Average message size = 50 bytes

Link loading factor = 0.4

Result:

Number of messages queued for faulty link $= 64000$ *bits/sec* $\times 0.4 \times 128$ *msec/*(50×8) *bits/msg.*

≈ 8 *messages*

Figure 6 only shows part of the combined FORM-FEHM model for description in this paper. The complete model of the network with coverage probabilities incorporated can be solved for steady state probabilities of its states, average holding time in each state, and average number of times a state is visited. The Markov chain (FORM) will be irreducible since the network nodes and links are assumed to be repairable. Combining the performance measure (such as number of lost messages) with the state steady state probabilities, holding times in each state, and average number of times a state is visited, the network designer can obtain combined performance and reliability measures, such as the percentage of lost messages. Work is currently ongoing to solve FORMs of typical SS7 networks with incorporated FEHMs.

The modeling technique described above allows a modeler to carry out sensitivity analysis of network performability measures, such as percentage of lost messages in both normal operating conditions and under fault conditions, to parameters used in network management procedures such as values of *timers, T1* and *T2*. Detailed analysis of an SS7 or any other fault-tolerant network can be carried out using these techniques. We have not included an elaborate analysis of an SS7 network in this paper, since the intent of this paper is only to illustrate the coverage and performability modeling techniques.

CONCLUSION

The definition of coverage as applied to network management procedures is extended to capture the notion that the outcome of a network management procedure in repairable networks cannot be classified as "failed" or "successful". Instead, a network management procedure could result in the network being in one or more states that represent different values for a given set of performance measures. Thus coverage of a network management procedure is redefined as a function of a given set of performance measures.

A Generalized Stochastic Petri Net model of a network management procedure to reconfigure and recover from a link fault is used to illustrate the coverage modeling technique. Coverage measures obtained from this model are then used in a model of the network that captures its fault occurrence and repair behavior. Such combined models can then be solved for overall network measures that reflect both performance and availability.

ABOUT THE AUTHORS

Malathi Veeraraghavan is a member of technical staff at AT&T Bell Laboratories. Her research interests include signaling and control of networks, fault-tolerant systems design, and reliability and performance modeling. Dr. Veeraraghavan received a BTech degree in Electrical Engineering from Indian Institute of Technology (Madras) and MS and PhD degrees in Electrical Engineering from Duke University, Durham. She is a member of the IEEE and an Associate Editor of the IEEE Transactions on Reliability.

REFERENCES

[1] F. Meyer, "On Evaluating the Performability of Degradable Computer Systems", *IEEE Transactions on Computers*, Vol. 29, No. 8, pp. 720-731, 1980.

[2] T. C. Arnold, "The Concept of Coverage and Its Effect on the Reliability Model of a Repairable System", *IEEE Transactions on Computers*, Vol. C-22, No. 3, March 1973.

[3] K. S. Trivedi, J. Bechta Dugan, R. Geist, M. Smotherman, "Modeling Imperfect Coverage in Fault-Tolerant Systems", *Proc. of the Fourteenth International Conference on Fault-Tolerant Computing*, 1984.

[4] A. Reibman and H. Zaretsky, "Modeling Fault Coverage and Reliability in a Fault-Tolerant Network", *Globecom*, pp. 689-692, 1990.

[5] CCITT Specifications of Signalling System No. 7 Recommendations Q.700-Q.716, Blue Book, Volume VI - Fascicle VI.7, Geneva 1989.

[6] M. Hamilton and N. A. Marlow, "Analyzing Telecommunication Network Availability Performance Using the Downtime Probability Distribution", *Globecom*, pp. 590-596, 1991.

[7] G. Willman and P. J. Kuehn, "Performance Modeling of Signaling System No. 7," *IEEE Communication Magazine*, Vol. 28, No. 7, pp. 44-56, July 1990.

[8] M. Ajmone Marsan, G. Balbo and G. Conte, "Performance Models of Multiprocessor Systems", The MIT Press, 1986.

[9] P. M. Merlin, "Specification and Validation of Protocols", *IEEE Transaction on Communications*, Vol. COM-27, No. 11, November 1979.

[10] R. A. Sahner and K. S. Trivedi, "SHARPE - Symbolic Hierarchical Automated Reliability and Performance Evaluator Introduction and Guide for Users", Duke University, Sept. 1986.

[11] K. Muppala and K. S. Trivedi, "Composite Performance and Availability Analysis Using a Hierarchy of Stochastic Reward Nets", in *Proc. of the Fifth Intl. Conf. on Modeling Techniques and Tools for Computer Performance Evaluation*, Torino, Italy, 1991, Editor G. Balbo.

[12] K. S. Trivedi, J. K. Muppala, S. P. Woolet, B. R. Haverkort, "Composite Performance and Dependability Analysis", *Performance Evaluation*, Vol. 14, 1992.

FAULT DIAGNOSIS OF NETWORK CONNECTIVITY PROBLEMS BY PROBABILISTIC REASONING

Clark Wang[1] and Mischa Schwartz[2]

[1]Eigen Research
Scarsdale, NY 10583
[2]Center for Telecommunications Research and
Department of Electrical Engineering
Columbia University
New York, NY 10027

ABSTRACT

In this paper, we propose a framework for identifying most probable faulty network components from a set of generic alarms in a heterogeneous environment. A designated network entity with management responsibilities determines a fault has occurred due to its inability to communicate with certain other entities. Given this information as well as the information that it can communicate with another specified set of entities, one would like to identify as quickly as possible a ranked list of the most probable failed network resources. A method based on maximum a posteriori (MAP) estimation is presented.

INTRODUCTION

Communication networks have increased dramatically in both size and complexity in the last few years. However, the new power brought with modern communication networks creates greater vulnerability. As networks become more heterogeneous, and more hardware and software from various vendors, with different architectures, are used, the whole picture of the specification becomes bewildering. Though the openness due to multiple vendors allows the system to be used in many innovative ways, it also multiplies the potential for incompatibility and lack of operability between subsystems. This brings out the need for a unified approach or principles to the area of Fault Management. Rather than develop ad hoc fault detecting mechanisms for each device or system in a network, it would be desirable to develop a unified approach on which the detection and identification of a fault in any system could be based.

The current trend of network management research is to design a sophisticated architecture, Management Information Database (MIB), and network message exchange mechanism which inform network entities of the current operational and statistical status of

*This work was supported by ONR Grant N00014-90-J-1289 and NSF Grant # CDR-88-11111.

the network. Based on this sophisticated infrastructure, one can then confidently talk about dynamic facility and traffic restoration. However, it is inevitable that a heterogeneous network has more than a dozen interconnected different vendors' products, with various architectures. This poses serious questions regarding the interchange of network information.

Our goal is to emphasize the importance of probabilistic assessment given that network information provided might be incomprehensible and incomplete particularly in heterogeneous environments. Our previous work was to diagnose faults through model-based analysis [3], [11] using one or a set of simple observers. In this paper, we propose a framework for identifying most probable faulty network components from a set of generic alarms in a heterogeneous environment. We model the network as a collection of a number of *network resources* or *network elements*. The network provides a necessary connection between two *network entities* by utilizing a set of network resources. For example, a physical connection between two network nodes can be a concatenation of several network links. In ATM terminology, a connection between two access points or ATM entities is called a *virtual path*. The ATM entities and/or the virtual path could be treated as network resources.

A designated network entity (an entity with management responsibilities, for example) determines a fault has occurred due to its inability to communicate with certain other entities. Given this information as well as the information that it can communicate with another specified set of entities, one would like to identify as quickly as possible a ranked list of the most probable failed network resources. In other words faults are regarded as a loss of system connectivity due to the malfunction or breakdown of a number of network resources. We present a *maximum a posteriori* (MAP) method that carries out this identification rapidly. This method can be used in conjunction with testing to pinpoint the actual faulty network component(s). Thus, instead of exhaustive testing over all possible network resources, requiring possible substantial network signaling bandwidth for testing traffic and possibly interrupting normal network operations, we only have to test the most likely faulty components which have been identified by this method. As is true with *maximum a posteriori* tests, this method requires prior knowledge of, or estimates of, failure probabilities. The collection of the prior failure statistics can be automated from the past operational statistics or assigned by a network manager according to the confidence level associated with each network resource.

Recent studies on Bayesian theory [4], [9], [10] have focused on a more general framework for the next generation of Expert Systems with probabilistic inference. These appear to be not easily applied to the particular problem we have investigated here unless the network possesses certain topological properties in order to reduce the computational complexity [5].

To develop the method more easily, we use the terms *nodes* and *links* to represent the network entities and their connections in the following analysis. The model and assumptions are outlined in next section.

THE MODEL OF NETWORKS

Consider a network with a set of nodes $N = \{n_1, n_2, \cdots, n_\alpha\}$ and a set of links $L = \{l_1, l_2, \cdots, l_\beta\}$. One node n_i is taken as the designated node. A routing table, specifying routes available from n_i to all other nodes, is assumed available. We assume that n_i can set up a physical connection to n_j if there is a path, consisting of concatenation of links, between n_i and n_j,

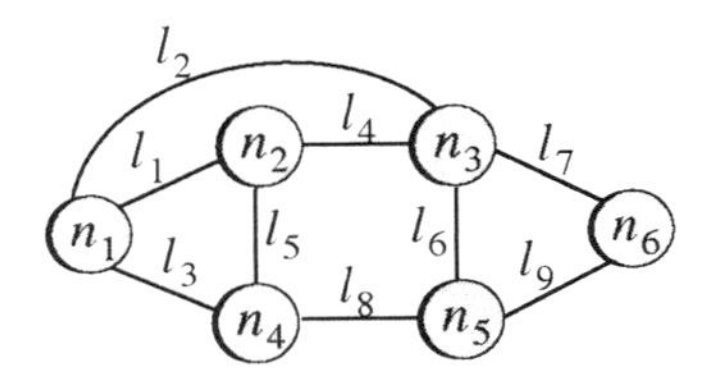

Figure 1. A 6-node example

as specified by the routing table. For the example in Figure 1, say that node n_1 can open a connection to node n_3, but n_1 cannot open any connection to n_5 given in the routing table (this implies at least one link must have failed for each of the routes specified).

We say that link l_3 is more likely to have a fault than link l_1 if $P($link l_3 has failed $|\, n_1$ cannot set up a connection to n_5, but n_1 can set up a connection to n_3) > $P($link l_1 has failed $|\, n_1$ cannot set up a connection to n_5, but n_1 can set up a connection to n_3). Our goal is to find which link l_i is the most likely faulty link over all links that could have failed. Before we go further, let us define some notation first.

$l_i = 0$: link l_i has failed (is faulty)

$l_i = 1$: link l_i is operational (non-faulty)

$n_i \to n_j$: there is at least one non-faulty path from node n_i to node n_j available drawn from the specified set of routes; in other words, node n_i can set up a connection with node n_j

$\overline{n_i \to n_j}$: node n_i cannot set up a connection with node n_j using any of the routes specified

$\overline{\Theta}_i$: the set of nodes n_j such that $\overline{n_i \to n_j}$

Θ_i: the set of nodes n_j such that $n_i \to n_j$

$P\{l_i = 0\} = p_i$: the probability link l_i is faulty, (i.e. $P\{l_i = 1\} = 1 - p_i$)

Our method is applied to a node n_i with partial information available to this node only, such as $\{\overline{n_i \to n_s}\}$ and $\{n_i \to n_t\}$ for some $1 \le s, t \le \alpha$.

Classification of the Events

Assume the information (or *evidence*) available to the network node is in the form $\{\overline{n_i \to n_s}, s \in \overline{\Theta}_i,\ n_i \to n_t, t \in \Theta_i\}$. Let $\tilde{\Theta}_i = \{N - n_i\} - \overline{\Theta}_i - \Theta_i$ be the set of nodes which are not specified in this evidence. As an example in Figure 1, say $\{\overline{n_1 \to n_5}, n_1 \to n_2, n_1 \to n_3, n_1 \to n_4\}$. This means $\overline{\Theta}_i = \{5\}$, $\Theta_1 = \{2,3,4\}$, and $\tilde{\Theta}_1 = \{6\}$. In addition, node n_i has the topological information of the network required to establish its routing table, and the a priori probabilities of failure p_i associated with each link.

Statistical Assumptions and Criteria

We assume that every link will fail independently with link i having a probability p_i of failure: p_i can be measured by the average fraction of time that the link is down. All evidence is assumed to be correct. Link l_j is identified as faulty if

$$j = \arg\max_k \quad P\{l_k = 0 \,|\, \overline{n_i \to n_s}, s \in \overline{\Theta}_i, n_i \to n_t, t \in \Theta_i\} \tag{1}$$

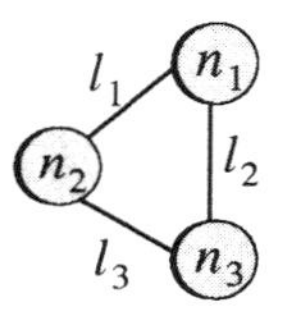

Figure 2. A 3-node example

The maximization is carried out only over the sub-domain of links included in the routes. By the Bayes rule, the above expression can be written as

$$\begin{aligned} & P\{l_k = 0 \mid \overline{n_1 \to n_s, s \in \overline{\Theta}_i}, n_1 \to n_t, t \in \Theta_i\} \\ & = \frac{P\{\overline{n_1 \to n_s, s \in \overline{\Theta}_i}, n_1 \to n_t, t \in \Theta_i \mid l_k = 0\} \cdot p_k}{P\{\overline{n_1 \to n_s, s \in \overline{\Theta}_i}, n_1 \to n_t, t \in \Theta_i\}} \end{aligned} \tag{2}$$

However, the numerator of (2) is difficult to compute in our model since, in general, depending on routes specified, the nodes can be inter-connected in any fashion. The numerator term can be rewritten and calculated by summing all possible combinations of network link states, up or down. Letting β be the total number of links in the network, we rewrite the expression as follows,

$$\begin{aligned} & P\{l_k = 0 \mid \overline{n_i \to n_s, s \in \overline{\Theta}_i}, n_i \to n_t, t \in \Theta_i\} \\ & = \frac{P\{l_k = 0, \overline{n_i \to n_s, s \in \overline{\Theta}_i}, n_i \to n_t, t \in \Theta_i\}}{P\{\overline{n_i \to n_s, s \in \overline{\Theta}_i}, n_i \to n_t, t \in \Theta_i\}} \\ & = \frac{\sum_{i_1=0}^{1} \sum_{i_2=0}^{1} \cdots \sum_{i_\beta=0}^{1} P\{l_k = 0, \overline{n_i \to n_s, s \in \overline{\Theta}_i}, n_i \to n_t, t \in \Theta_i \mid l_1 = i_1, l_2 = i_2, \cdots, l_\beta = i_\beta\} \cdot P\{l_1 = i_1, l_2 = i_2, \cdots, l_\beta = i_\beta\}}{P\{\overline{n_i \to n_s, s \in \overline{\Theta}_i}, n_i \to n_t, t \in \Theta_i\}} \end{aligned}$$

(The summation is carried out only over routes specified.) Since the denominator $P\{\overline{n_i \to n_s, s \in \overline{\Theta}_i}, n_i \to n_t, t \in \Theta_i\}$ is constant over all possible k and $P\{l_1 = i_1, l_2 = i_2, \cdots, l_\beta = i_\beta\} = (1-p_1)^{i_1} p_1^{1-i_1} \cdot (1-p_2)^{i_2} \cdot p_2^{1-i_2} \cdots (1-p_\beta)^{i_\beta} p_\beta^{1-i_\beta}$, it is sufficient to maximize

$$\begin{aligned} j = \operatorname*{argmax}_k \sum_{i_1=0}^{1} \sum_{i_2=0}^{1} \cdots \sum_{i_\beta=0}^{1} & P\{l_k = 0, \overline{n_i \to n_s, s \in \overline{\Theta}_i}, n_i \to n_t, t \in \Theta_i \mid l_1 = i_1, l_2 = i_2, \cdots, l_\beta = i_\beta\} \cdot \\ & (1-p_1)^{i_1} p_1^{1-i_1} \cdot (1-p_2)^{i_2} \cdot p_2^{1-i_2} \cdots (1-p_\beta)^{i_\beta} p_\beta^{1-i_\beta} \end{aligned} \tag{3}$$

Note that $P\{l_k = 0, \overline{n_1 \to n_s, s \in \overline{\Theta}}, n_1 \to n_t, t \in \Theta \mid l_1 = i_1, l_2 = i_2, \cdots, l_\beta = i_\beta\}$ is equal to 1 or 0. It is prohibitive to carry out the summation in (3) for a large and complicated network. For a network with 20 links with all possible routes specified, there are 2^{20-1} terms in (3) in the worst case. In the next section, we describe our approach, the uncertain-cut method, which substantially reduces the number of terms for the summation.

THE UNCERTAIN-CUT METHOD

Working Through Examples

Let us go through some simple examples before investigating a generalized formulation. Consider the 3-node network of Figure 2. Assume that we are at node n_1 and are given the

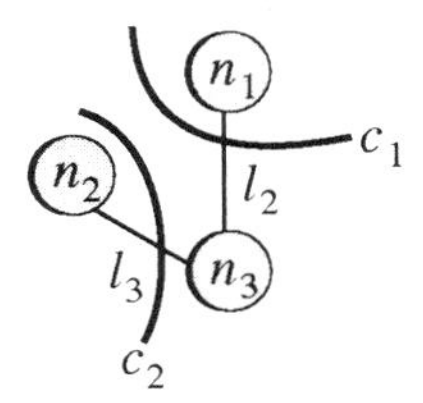

Figure 3. The 3-node example with one link removed

information $\overline{n_1 \to n_2}$ which means n_1 cannot set up a connection to n_2 using link l_1 or links l_2, l_3 as the routes. We would like to identify the faulty link. As noted above in (1), the goal is to compare $P(l_1 = 0 \mid \overline{n_1 \to n_2})$, $P(l_2 = 0 \mid \overline{n_1 \to n_2})$ and $P(l_3 = 0 \mid \overline{n_1 \to n_2})$. The largest value corresponds to the link with the highest probability of being faulty. Since

$$P(l_1 = 0 \mid \overline{n_1 \to n_2}) = \frac{P(\overline{n_1 \to n_2} \mid l_1 = 0) \cdot p_1}{P(\overline{n_1 \to n_2})}$$

It is sufficient to compare $P(\overline{n_1 \to n_2} \mid l_1 = 0) \cdot p_1$, $P(\overline{n_1 \to n_2} \mid l_2 = 0) \cdot p_2$ and $P(\overline{n_1 \to n_2} \mid l_3 = 0) \cdot p_3$. Consider the first term which can be computed as suggested in (3),

$$\begin{aligned}
& P\{l_1 = 0, \overline{n_1 \to n_2} \mid l_1 = 0\} \cdot p_1 \\
= \; & P\{l_1 = 0, \overline{n_1 \to n_2} \mid l_1 = 0, l_2 = 0, l_3 = 0\} \cdot p_1 p_2 p_3 \\
& + P\{l_1 = 0, \overline{n_1 \to n_2} \mid l_1 = 0, l_2 = 0, l_3 = 1\} \cdot p_1 p_2 (1 - p_3) \\
& + P\{l_1 = 0, \overline{n_1 \to n_2} \mid l_1 = 0, l_2 = 1, l_3 = 0\} \cdot p_1 (1 - p_2) p_3 \\
& + P\{l_1 = 0, \overline{n_1 \to n_2} \mid l_1 = 0, l_2 = 1, l_3 = 1\} \cdot p_1 (1 - p_2)(1 - p_3) \\
= \; & 1 \cdot p_1 p_2 p_3 + 1 \cdot p_1 p_2 (1 - p_3) + 1 \cdot p_1 (1 - p_2) p_3 + 0 \cdot p_1 (1 - p_2)(1 - p_3) \\
= \; & p_1 (p_2 + p_3 - p_2 p_3) \qquad (4)
\end{aligned}$$

The expression in (4), given known values for p_1, p_2, p_3, is then compared to similarly determined values of the other two probabilities listed above to see which is largest. As noted earlier, this approach is, in general, prohibitive. We now describe our new method for obtaining the same values of the desired probabilities with considerably less computation. To demonstrate the approach, we indicate how to compute $P(\overline{n_1 \to n_2} \mid l_1 = 0) \cdot p_1$.

Since $l_1 = 0$ is given, we simply remove link l_1. The resulting graph is shown in Figure 3. $P(\overline{n_1 \to n_2}) \cdot p_1$ is equivalently computed for this graph. Note that there are two cuts (the formal definition of a ***cut*** will be given later) shown in Figure 3. Define the random variables $c_1 = l_2$ and $c_2 = l_3$. More generally, these two random variables are defined as the sum of l_i for the links the cut has crossed. In this particular example, only one link is crossed by each cut. Thus only one term shows up in the expressions for c_1 and c_2. Rewrite $P(\overline{n_1 \to n_2}) \cdot p_1$ by Bayes rule, conditioning on event X, as

$$P(\overline{n_1 \to n_2}) \cdot p_1 = \{P(\overline{n_1 \to n_2} \mid X) \cdot P(X) + P(\overline{n_1 \to n_2} \mid \overline{X}) \cdot P(\overline{X})\} \cdot p_1 \qquad (5)$$

We choose $X = \{c_1 = 0 \vee c_2 = 0\}$, $\overline{X} = \{\overline{c_1 = 0 \vee c_2 = 0}\} = \{\overline{c_1 = 0} \wedge \overline{c_2 = 0}\} = \{c_1 > 0 \wedge c_2 > 0\}$. Note that $P(\overline{n_1 \to n_2} \mid \overline{X}) = 0$ with $\overline{X}$ true since there is a path available from n_1 to n_2. $l_2 = l_3 = 1$ which contradicts $\overline{n_1 \to n_2}$. Furthermore, $P(\overline{n_1 \to n_2} \mid X) = 1$. We get

$$\begin{aligned}
P(\overline{n_1 \to n_2}) \cdot p_1 = \; & \{P(X)\} \cdot p_1 \\
= \; & \{P(c_1 = 0) + P(c_2 = 0) - P(c_1 = 0 \wedge c_2 = 0)\} \cdot p_1 \\
= \; & (p_2 + p_3 - p_2 p_3) \cdot p_1
\end{aligned}$$

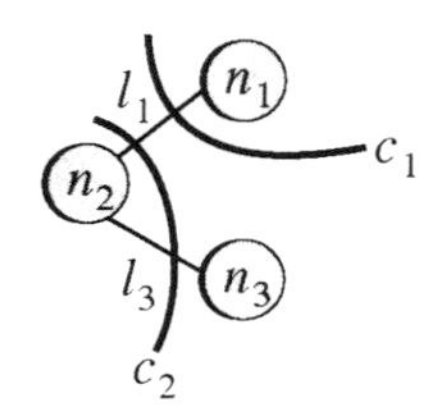

Figure 4. The 3-node example with another link removed

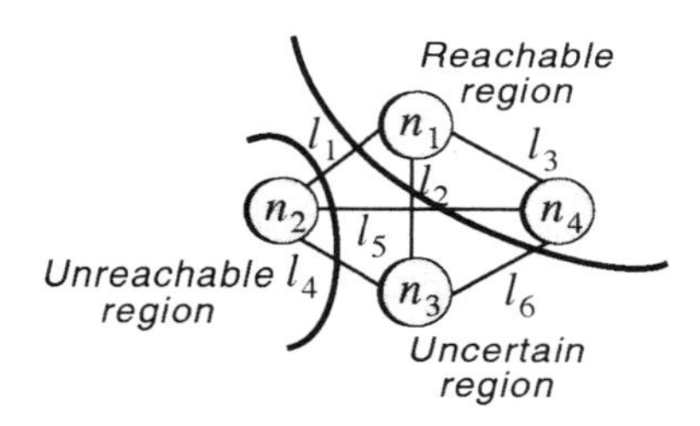

Figure 5. A 4-node example

which agrees with (4). Though we have not shown how to get the two cuts in Figure 3, it is not hard to understand that these two cuts *must* satisfy $P(\overline{n_1 \to n_2} \mid \overline{X})=0$ in order to eliminate the second term on the right hand side of (5).

In a similar manner, the graph for computing $P(\overline{n_1 \to n_2} \mid l_2=0)\cdot p_2$, created by removing link l_2 from Figure 2, is shown in Figure 4. In this case, $c_1=l_1$, $c_2=l_1+l_3$. Recall that we define these two random variables as the sum of l_i for the links the cut has crossed. Rewriting $P(\overline{n_1 \to n_2} \mid l_2=0)\cdot p_2$ as we did in (5), and using $P(\overline{n_1 \to n_2} \mid \overline{X})=0$ with X defined as previously in terms of two cuts, we get

$$P(\overline{n_1 \to n_2})\cdot p_2=\{P(X)\}\cdot p_2=(p_1+p_1p_3-p_1p_3)p_2=p_1p_2$$

Similarly, we could get $P(\overline{n_1 \to n_2} \mid l_3=0)\cdot p_3=p_1p_3$. The remaining job for this example is to compare the numerical values of $P(\overline{n_1 \to n_2} \mid l_1=0)\cdot p_1$, $P(\overline{n_1 \to n_2} \mid l_2=0)\cdot p_2$ and $P(\overline{n_1 \to n_2} \mid l_3=0)\cdot p_3$. We then select the largest value which corresponds to the link having the highest probability of being faulty.

Now consider the example in Figure 5. We will try to formalize the idea this time. Say we are given that $\{\overline{n_1 \to n_2}, n_1 \to n_4\}$ with the routing table shown in Table 1. We would like to find the most likely faulty link l_j which satisfies the following expression

Table 1. Routing table for the 4-node example

Source	Destinatio n	Links in the route
n_1	n_2	l_1
		l_3, l_5
		l_2, l_4
n_1	n_4	l_3
		l_2, l_6

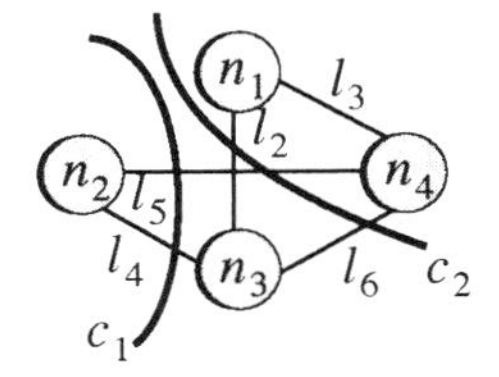

Figure 6. The 4-node example with one link removed

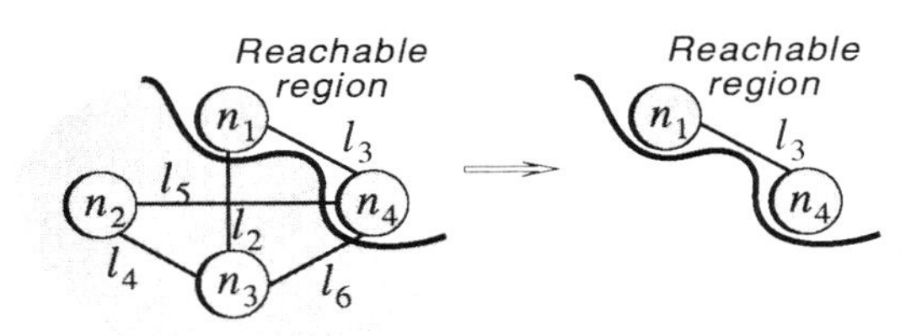

Figure 7. The subgraph of the 4-node example

$$j = \arg\max_k \quad P\{l_k = 0 \mid \overline{n_1 \to n_2}, n_1 \to n_4\} \tag{6}$$

Again, it suffices to maximize the following,

$$j = \arg\max_k P\{\overline{n_1 \to n_2}, n_1 \to n_4 \mid l_k = 0\} \cdot p_k \tag{7}$$

Consider the case $k = 1$ at first. The graph for this case is shown in Figure 6 which was created by removing link l_1 from Figure 5. Note that $\Theta_1 = \{4\}$, $\overline{\Theta}_1 = \{2\}$, and the unspecified set of nodes, in this case, is given by $\tilde{\Theta}_1 = \{3\}$. We call $\{n_1, n_4\}$ the *reachable region*, $\{n_2\}$ the *unreachable region*, and $\{n_3\}$ the *uncertain region*. We will choose two cuts such that the reachable region and unreachable region are not separated up by the cuts. In this example only two cuts are again required. More generally, we will show later that the number of cuts is proportional to the number of nodes in the uncertain region.

To get c_1, let n_3 join the reachable region. The graph reduces to two regions only. c_1 is the boundary between these two regions. To get c_2, let n_3 join the unreachable region where c_2 is the boundary between the reachable and unreachable region. In this example, random variable c_1 is the sum of l_5 and l_4; similarly, $c_2 = l_2 + l_5$ (l_6 is not included in c_2 because link l_6 is not used in any routes from n_1 to n_2). To compute $P(\overline{n_1 \to n_2}, n_1 \to n_4 \mid l_1 = 0) \cdot p_1$ (the case $k = 1$ in (7)) or $P(\overline{n_1 \to n_2}, n_1 \to n_4) \cdot p_1$ (with respect to the graph in Figure 6), we rewrite it, by Bayes rule, conditioning on event $X = \{c_1 = 0 \vee c_2 = 0\}$:

$$\begin{aligned} P(\overline{n_1 \to n_2}, n_1 \to n_4) \cdot p_1 = \{ & P(\overline{n_1 \to n_2}, n_1 \to n_4 \mid X) \cdot P(X) \\ & + P(\overline{n_1 \to n_2}, n_1 \to n_4 \mid \overline{X}) \cdot P(\overline{X}) \ \} \cdot p_1 \end{aligned}$$

Or

$$\begin{aligned} P(\overline{n_1 \to n_2}, n_1 \to n_4) \cdot p_1 = \{ & P(\overline{n_1 \to n_2}, n_1 \to n_4 \mid c_1 = 0 \vee c_2 = 0) \cdot P(c_1 = 0 \vee c_2 = 0) \\ & + P(\overline{n_1 \to n_2}, n_1 \to n_4 \mid c_1 > 0 \wedge c_2 > 0) \cdot P(c_1 > 0 \wedge c_2 > 0) \ \} \cdot p_1 \end{aligned} \tag{8}$$

Note that $\{c_1 > 0 \wedge c_2 > 0\}$ implies at least one path is available from n_1 to n_2 from the reachable region to the unreachable region, which contradicts $\overline{n_1 \to n_2}$. Hence,

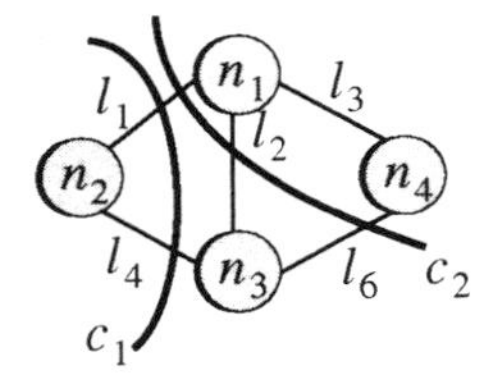

Figure 8. The 4-node example with another link removed

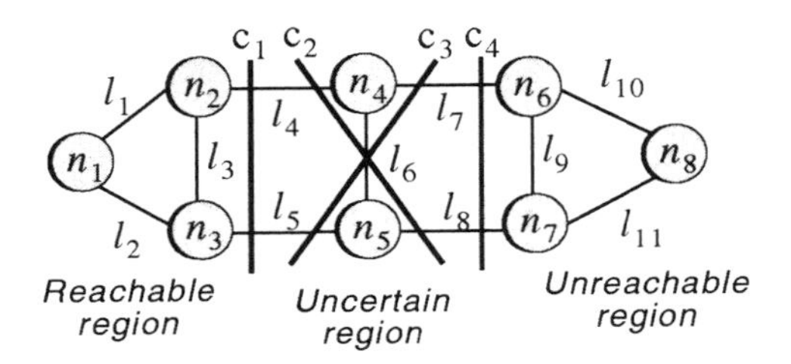

Figure 9. A 8-node example

$P(\overline{n_1 \to n_2}, n_1 \to n_4 \mid c_1 > 0 \wedge c_2 > 0) = 0$. Since we choose a cut that does not separate the unreachable region, it must satisfy $\{\overline{n_1 \to n_2}\}$. We get

$$P(\overline{n_1 \to n_2}, n_1 \to n_4 \mid c_1 = 0 \vee c_2 = 0) = P(n_1 \to n_4 \mid c_1 = 0 \vee c_2 = 0)$$

Equation (8) simplifies to

$$P(\overline{n_1 \to n_2}, n_1 \to n_4) \cdot p_1 = P(n_1 \to n_4 \mid c_1 = 0 \vee c_2 = 0) \cdot P(c_1 = 0 \vee c_2 = 0) \cdot p_1 \tag{9}$$

since the connection $\{n_1 \to n_4\}$ does not traverse nodes n_2 and n_3 in the unreachable and uncertain region in order to reach n_4. Otherwise, we declare $\{\overline{n_1 \to n_2}, \overline{n_1 \to n_3}\}$ instead of $\{\overline{n_1 \to n_2}\}$. To compute $P(n_1 \to n_4 \mid c_1 = 0 \vee c_2 = 0)$ on the right hand side of (9), we only have to consider the sub-graph composed of nodes in the reachable region as shown in Figure 7. This is equivalent to computing $P(n_1 \to n_4)$ for this sub-graph which is $(1 - p_3)$. Thus Equation (9) is reduced to

$$P(\overline{n_1 \to n_2}, n_1 \to n_4) \cdot p_1 = (1 - p_3) \cdot P(c_1 = 0 \vee c_2 = 0) \cdot p_1$$

Recall that l_2, l_4, l_5, l_6 are independent random variables with probabilities p_2, p_4, p_5, p_6, respectively, of zero, while $c_1 = l_5 + l_4$ and $c_2 = l_2 + l_5$. Hence,

$$P(c_1 = 0 \vee c_2 = 0) = p_4 p_5 + p_2 p_5 - p_2 p_4 p_5$$

We finally have the case of $k = 1$ for Equation (7)

$$P(\overline{n_1 \to n_2}, n_1 \to n_4 \mid l_1 = 0) \cdot p_1 = (1 - p_3) \cdot (p_4 p_5 + p_2 p_5 - p_2 p_4 p_5) \cdot p_1$$

We now demonstrate this method for the case $k = 5$ for Equation (7). The graph with link l_5 removed is shown in Figure 8. Note that $c_1 = l_1 + l_4$ and $c_2 = l_1 + l_2$.

$$\begin{aligned} & P(\overline{n_1 \to n_2}, n_1 \to n_4 \mid l_5 = 0) \cdot p_5 \\ &= P(n_1 \to n_4 \mid c_1 = 0 \vee c_2 = 0) \cdot P(c_1 = 0 \vee c_2 = 0) \cdot p \\ &= (1 - p_3) \cdot (p_1 p_4 + p_1 p_2 - p_1 p_2 p_4) \cdot p_5 \end{aligned}$$

In the next paragraph, we will derive a general expression for this. In general, multiple cuts will be required. Consider the network of Figure 9 as an example. The following scenario is assumed with information provided to node n_1: $\{n_1 \to n_2,\ \overline{n_1 \to n_3},\ \overline{n_1 \to n_6},\ \overline{n_1 \to n_7},\ \overline{n_1 \to n_8}\}$. Here all possible routes are assumed specified. We require that event X have the form of $X = \{c_1 = 0 \vee c_2 = 0 \vee \cdots \vee c_r = 0\}$. In this case, there are four cuts such that $P(n_1 \to n_2,\ \overline{n_1 \to n_3},\ \overline{n_1 \to n_6},\ \overline{n_1 \to n_7},\ \overline{n_1 \to n_8} \mid \overline{X}) = 0$ with $\overline{X} = \{c_1 > 0 \wedge c_2 > 0 \wedge c_3 > 0 \wedge c_4 > 0\}$.

Generalized Formulation

As indicated in (1) and (2), link l_j is identified as a faulty link if

$$j = \arg\max_k \Psi(i,k) \cdot p_k$$

where

$$\begin{aligned}\Psi(i,k) &= P\{\overline{n_i \to n_s, s \in \overline{\Theta}_i}, n_i \to n_t, t \in \Theta_i \mid l_k = 0\} \\ &= P_{(l_k=0)}\{\overline{n_i \to n_s, s \in \overline{\Theta}_i}, n_i \to n_t, t \in \Theta_i\}\end{aligned}$$

$P_{(l_k=0)}\{\bullet\}$ represents the probability with respect to the graph with link l_k removed. Hereafter, we will focus on the graph with link l_k removed. We classify the nodes into the following three categories where node n_i is considered the designated node:

- ***Reachable region*** (or sub-graph) $\Re = \{n_t \mid t \in (\Theta_i \cup \{i\})\}$: node n_i and all nodes in Θ_i.
- ***Unreachable region*** $\overline{\Re} = \{n_t \mid t \in \overline{\Theta}_i\}$: all unreachable nodes from node n_i.
- ***Uncertain region*** $\tilde{\Re} = \{n_t \mid t \in \tilde{\Theta}_i\}$: all nodes which are not specified by the given evidence. Let $\tilde{\Re}^s$ denote the set of all subsets of $\tilde{\Re}$ such that $\tilde{\Re}^s_n$ is the n-th subset of $\tilde{\Re}$ or n-th element of $\tilde{\Re}^s$ where there are a total of r $(= 2^{|\tilde{\Re}|})$ subsets $\tilde{\Re}^s_n$. For example, in Figure 5, $\tilde{\Re} = \{n_3\}$ and there are two subsets of $\tilde{\Re}$: $\tilde{\Re}^s_1 = \varnothing$, and $\tilde{\Re}^s_2 = \{n_3\}$ $(r = 2)$.

Consider the graph removing all links not included in the routes specified. For a subset $\tilde{\Re}^s_n$ of $\tilde{\Re}$, define a *cut* c_n to be the sum of random variables l_m. The set l_m corresponds to the remaining links from any nodes in the union $\tilde{\Re}^s_n \cup \Re$ to the nodes not in $\tilde{\Re}^s_n \cup \Re$. Thus, there are r cuts, $c_1, c_2, \cdots, c_r$. We now compute $\Psi(i,k)$ by conditioning on the events $X = \{c_1 = 0 \vee c_2 = 0 \vee \cdots \vee c_r = 0\}$ and $\overline{X}$:

$$\begin{aligned}\Psi(i,k) = {} & P_{(l_k=0)}\{\overline{n_i \to n_s, s \in \overline{\Theta}_i}, n_i \to n_t, t \in \Theta_i \mid c_1 = 0 \vee \cdots \vee c_r = 0\} \cdot P_{(l_k=0)}(c_1 = 0 \vee \cdots \vee c_r = 0) \\ & + P_{(l_k=0)}\{\overline{n_i \to n_s, s \in \overline{\Theta}_i}, n_i \to n_t, t \in \Theta_i \mid c_1 > 0 \wedge \cdots \wedge c_r > 0\} \cdot P_{(l_k=0)}(c_1 > 0 \wedge \cdots \wedge c_r > 0)\end{aligned}$$

Note that $\{c_1 > 0 \wedge c_2 > 0 \wedge \cdots \wedge c_r > 0\}$ implies that at least one non-faulty link exists between $\Re$ and any nodes in $\tilde{\Re}$ and between any nodes in $\tilde{\Re}$ and $\overline{\Re}$. Thus, at least one non-faulty link exists from $\Re$ to $\overline{\Re}$ which contradicts the evidence $\{\overline{n_i \to n_s, s \in \overline{\Theta}_i}\}$. Hence $P_{(l_k=0)}\{\overline{n_i \to n_s, s \in \overline{\Theta}_i}, n_i \to n_t, t \in \Theta_i \mid c_1 > 0 \wedge \cdots \wedge c_r > 0\} = 0$. Event $\overline{X}$ thus contributes 0 to $\Psi(i,k)$. We get

$$\Psi(i,k) = P_{(l_k=0)}\{\overline{n_i \to n_s, s \in \overline{\Theta}_i}, n_i \to n_t, t \in \Theta_i \mid c_1 = 0 \vee \cdots \vee c_r = 0\} \cdot P_{(l_k=0)}(c_1 = 0 \vee \cdots \vee c_r = 0)$$

Consider a cut c_n. If the links along this cut have failed $(c_n = 0)$, then n_i cannot connect to n_s for $s \in \overline{\Theta}_i$. This means that $\{c_1 = 0 \vee \cdots \vee c_r = 0\}$ implies $\{\overline{n_i \to n_s, s \in \overline{\Theta}_i}\}$. $\Psi(i,k)$ thus simplifies to

$$\begin{aligned}\Psi(i,k) &= P_{(l_k=0)}\{n_i \to n_t, t \in \Theta_i \mid c_1 = 0 \vee \cdots \vee c_r = 0\} \cdot P_{(l_k=0)}(c_1 = 0 \vee \cdots \vee c_r = 0) \\ &= P_{(l_k=0)}\{n_i \to n_t, t \in \Theta_i, (c_1 = 0 \vee \cdots \vee c_r = 0)\}\end{aligned}$$

Furthermore, the connections for $\{n_i \rightarrow n_t, t \in \Theta\}$ do not traverse any nodes in the uncertain region in order to reach n_t for $t \in \Theta_i$. Also note that every cut does not cross any link in the sub-graph $\Re$; in other words, c_n is the sum of random variables l_m none of which correspond to any link in $\Re$. Since $\{n_i \rightarrow n_t, t \in \Theta\}$ can be determined only by the sub-graph $\Re$, $\{n_i \rightarrow n_t, t \in \Theta\}$ and $\{c_1 = 0 \vee \cdots \vee c_r = 0\}$ are independent. We finally have

$$\Psi(i,k) = P_{(l_k=0)}\{n_i \rightarrow n_t, t \in \Theta_i\} \cdot P_{(l_k=0)}(c_1 = 0 \vee \cdots \vee c_r = 0)\}$$

We can compute $P_{(l_k=0)}\{n_i \rightarrow n_t, t \in \Theta_i\}$ by examining the sub-graph $\Re$ only. In summary, link l_j is identified as faulty if

$$j = \arg\max_k P_{(l_k=0)}\{n_i \rightarrow n_t, t \in \Theta_i\} \cdot P_{(l_k=0)}\{c_1 = 0 \vee \cdots \vee c_r = 0\} \cdot p_k \tag{10}$$

This generalizes (5) and (9) in the examples above.

Identification of Faulty Links in the Uncertain and Unreachable RegionWe now consider the case of the MAP test restricted to links with at least one end connected to nodes in the uncertain region ($\tilde{\Re}$) or unreachable region ($\overline{\Re}$). For the example in Figure 9 with the following information provided to node n_1: $\{n_1 \rightarrow n_2,\ n_1 \rightarrow n_3,\ \overline{n_1 \rightarrow n_6},\ \overline{n_1 \rightarrow n_7},\ \overline{n_1 \rightarrow n_8}\}$ with all possible routes again allowed, one might be interested in the identification of faulty links among links $l_4, l_5, l_6, \cdots, l_{11}$. Because n_1 is able to communicate with n_2 and n_3, the network manager might not be interested in determining whether one of the links l_1, l_2, l_3 in the reachable region is faulty. Knowing one of the links l_1, l_2, l_3 is faulty does not help to recover other faulty links causing the problem $\{\overline{n_1 \rightarrow n_6},\ \overline{n_1 \rightarrow n_7},\ \overline{n_1 \rightarrow n_8}\}$. To recover the system connectivity, one might focus on finding probable faulty links among $l_4, l_5, l_6, \cdots, l_{11}$. Let ℓ denote the set of links connected to nodes in the uncertain or unreachable region. Equation (10) is rewritten as

$$j = \arg\max_{l_k \in \ell} P_{(l_k=0)}\{n_i \rightarrow n_t, t \in \Theta_i\} \cdot P_{(l_k=0)}\{c_1 = 0 \vee \cdots \vee c_r = 0\} \cdot p_k$$

Note that $P_{(l_k=0)}\{n_i \rightarrow n_t, t \in \Theta_i\}$ is constant over all $l_k \in \ell$. For example, $P_{(l_{10}=0)}\{n_1 \rightarrow n_2, n_1 \rightarrow n_3\} = P_{(l_{11}=0)}\{n_1 \rightarrow n_2, n_1 \rightarrow n_3\}$ in Figure 9. We finally get

$$j = \arg\max_{l_k \in \ell} P_{(l_k=0)}\{c_1 = 0 \vee \cdots \vee c_r = 0\} \cdot p_k \tag{11}$$

COMPUTATIONAL ASPECTS

We will illustrate the computational steps through the example in Figure 9. The following scenario is assumed: the information provided to node n_1 is given by$\{n_1 \rightarrow n_2,\ n_1 \rightarrow n_3,\ \overline{n_1 \rightarrow n_6},\ \overline{n_1 \rightarrow n_7},\ \overline{n_1 \rightarrow n_8}\}$. In this case, $\Re = \{n_1, n_2, n_3\}$, $\overline{\Re} = \{n_6, n_7, n_8\}$ and $\tilde{\Re} = \{n_4, n_5\}$.

The Brute Force Way

To compute $P_{(l_k=0)}\{c_1 = 0 \vee \cdots \vee c_4 = 0\}$, for the example above, we expand it as follows:

$$\begin{aligned} P_{(l_k=0)}\{c_1 = 0 \vee \cdots \vee c_4 = 0\} = {} & P(c_1 = 0) + P(c_2 = 0) + P(c_3 = 0) + P(c_4 = 0) \\ & -\{P(c_1 = 0 \wedge c_2 = 0) + P(c_1 = 0 \wedge c_3 = 0) + P(c_1 = 0 \wedge c_4 = 0) \\ & + P(c_2 = 0 \wedge c_3 = 0) + P(c_2 = 0 \wedge c_4 = 0) + P(c_3 = 0 \wedge c_4 = 0)\} \\ & +\{P(c_1 = 0 \wedge c_2 = 0 \wedge c_3 = 0) + P(c_1 = 0 \wedge c_2 = 0 \wedge c_4 = 0) \\ & + P(c_1 = 0 \wedge c_3 = 0 \wedge c_4 = 0) + P(c_2 = 0 \wedge c_3 = 0 \wedge c_4 = 0)\} \\ & - P(c_1 = 0 \wedge c_2 = 0 \wedge c_3 = 0 \wedge c_4 = 0) \end{aligned} \tag{12}$$

To be specific, consider the case of $k=6$. Here $c_1 = l_4 + l_5$, $c_2 = l_4 + l_8$, $c_3 = l_5 + l_7$, and $c_4 = l_7 + l_8$. Recall that l_i is modeled as a discrete random variable

$$\begin{cases} P(l_i = 0) = p_i \\ P(l_i = 1) = 1 - p_i \end{cases}$$

Thus $P(c_1 = 0) = p_4 p_5$. Similarly, we have $P(c_2 = 0) = p_4 p_8$, $P(c_3 = 0) = p_5 p_7$, and $P(c_4 = 0) = p_7 p_8$. Furthermore, $P(c_1 = 0 \wedge c_2 = 0) = p_4 p_5 p_8$, $P(c_1 = 0 \wedge c_3 = 0) = p_4 p_5 p_7$, etc. Note that every term in (12) can be easily calculated, but, the number of terms grows exponentially with the number of cuts. However, if $p_i << 1$, we limit the above computation to the second order in p_i, producing satisfactory results in polynomial time. In the next paragraph, we will employ dependency among cuts to reduce computational complexity, as is similar to the key idea in [10].

The Computation Procedure

Rewrite $P_{(l_k=0)}\{c_1 = 0 \vee \cdots \vee c_4 = 0\}$ as

$$\begin{aligned} &1 - P\{(c_1 > 0) \wedge (c_2 > 0) \wedge (c_3 > 0) \wedge (c_4 > 0)\} \\ &= 1 - P(c_1 > 0) P(c_2 > 0 | c_1 > 0) P(c_3 > 0 | c_1 > 0, c_2 > 0) P(c_4 > 0 | c_1 > 0, c_2 > 0, c_3 > 0) \end{aligned} \tag{13}$$

We neglect the subscript $l_k = 0$ in the analysis; however we are still assuming that all probabilities are taken with respect to the sub-graph with link l_k removed. Let

$$\begin{aligned} \phi_1 &= P(c_1 > 0) = 1 - p_4 p_5 \\ \phi_2 &= P(c_2 > 0 | c_1 > 0) \\ \phi_3 &= P(c_3 > 0 | c_1 > 0, c_2 > 0) \\ \phi_4 &= P(c_4 > 0 | c_1 > 0, c_2 > 0, c_3 > 0) \end{aligned}$$

Note that since c_2 and c_3 are independent, ϕ_3 is reduced to $P(c_3 > 0 | c_1 > 0)$. Similarly, ϕ_4 is reduced to $P(c_4 | c_2 > 0, c_3 > 0)$ since c_1 and c_4 are independent. Note that Equation (13) can also be expressed as

$$\begin{aligned} &1 - P\{(c_1 > 0) \wedge (c_2 > 0) \wedge (c_3 > 0) \wedge (c_4 > 0)\} \\ &= 1 - P(c_{K_1} > 0) P(c_{K_2} > 0 | c_{K_1} > 0) P(c_{K_3} > 0 | c_{K_1} > 0, c_{K_2} > 0) \cdot P(c_{K_4} > 0 | c_{K_1} > 0, c_{K_2} > 0, c_{K_3} > 0) \end{aligned} \tag{14}$$

where $(c_{K_1}, c_{K_2}, c_{K_3}, c_{K_4})$ can be any permutation of (c_1, c_2, c_3, c_4). We say that c_1 has a degree of dependency 2, since c_1 is dependent on c_2, c_3. In general, it is better to first find the degree of dependency of each cut. Choose the permutation such that the degree of dependency of c_{K_M} is larger than the degree of dependency of c_{K_N} for $M > N$. Then remove the unnecessary term in Equation (14). In other words, minimize the number of terms such as $c_{K_M} > 0$ in Equation (14). For the particular example we have been discussing, we get

$$\begin{aligned} &1 - P\{(c_1 > 0) \wedge (c_2 > 0) \wedge (c_3 > 0) \wedge (c_4 > 0)\} \\ &= 1 - P(c_1 > 0) P(c_2 > 0 | c_1 > 0) P(c_3 > 0 | c_1 > 0) P(c_4 > 0 | c_2 > 0, c_3 > 0) \end{aligned} \tag{15}$$

We now discuss the two cases of computing ϕ_i

- ϕ_i is in the form of $P(c_m > 0 | c_n > 0)$: It can be shown that

$$P(c_m > 0 | c_n > 0) = 1 - \frac{P(c_m = 0) - P(c_m = 0 \wedge c_n = 0)}{1 - P(c_n = 0)} \tag{16}$$

- ϕ_i is in the form of $P(c_m > 0 | c_{n_1} > 0, c_{n_2} > 0, \cdots, c_{n_K} > 0)$: Similarly, we have

$$P(c_m > 0 | c_{n_1} > 0, c_{n_2} > 0, \cdots, c_{n_K} > 0)$$
$$= 1 - \frac{P(c_m = 0) - P\{c_m = 0 \wedge (c_{n_1} = 0 \vee c_{n_2} = 0 \vee \cdots \vee c_{n_K} = 0)\}}{1 - P(c_{n_1} = 0 \vee c_{n_2} = 0 \vee \cdots \vee c_{n_K} = 0)} \quad (17)$$

The only unknown is $P\{c_m = 0 \wedge (c_{n_1} = 0 \vee c_{n_2} = 0 \vee \cdots \vee c_{n_K} = 0)\}$. To calculate this, since $c_m = 0$ we only have to consider the sub-graph with links removed that are crossed by the cut c_m. This is equivalent to computing $P(c_{n_1} = 0 \vee c_{n_2} \vee \cdots \vee c_{n_K} = 0)$ with respect to the reduced sub-graph. Again, we expand this as we did in Equation (14). This expansion has a much lower degree of dependency than the original expansion since we have removed some links in the sub-graph. Let us come back to the example. Equation (15) is reduced to

$$P\{c_1 = 0 \vee \cdots \vee c_4 = 0\}$$
$$= 1 - P(c_1 > 0)P(c_2 > 0 | c_1 > 0)P(c_3 > 0 | c_1 > 0)\{P_{(c_4=0)}(c_2 > 0) \cdot P_{(c_4=0)}(c_3 > 0 | c_2 > 0) \cdot P(c_4 = 0)\}$$

Finally we may apply Equation (16) to the above to get the numerical result. The actual computational complexity strongly depends on the topology of the network. We will show the complexity upper bound next.

Complexity

We provide a conservative analysis of the computational complexity of the Uncertain Cut method by studying a fully connected network. Using exhaustive search, the complexity grows exponentially with the number of links in a network.

Let the number of cuts be r. This is proportional to the number of routes. It follows immediately from (14) that we expand $P_{(l_k=0)}\{c_1 = 0 \vee \cdots \vee c_r = 0\}$ as

$$P_{(l_k=0)}\{c_1 = 0 \vee \cdots \vee c_r = 0\} = 1 - \phi_1 \cdot \phi_2 \cdots \phi_r$$

To compute each ϕ_i, use the expansion of (17) repeatedly (at most i times) until we get the form of (16). Thus the computation of $P_{(l_k=0)}\{c_1 = 0 \vee \cdots \vee c_r = 0\}$ needs no more than $1 + 2 + \cdots + r = \frac{r^2 + r}{2}$ steps. Let L be the number of links. The MAP test shown in (11) requires repeated computation of $P_{(l_k=0)}\{c_1 = 0 \vee \cdots \vee c_r = 0\}$ less than L times. Thus the complexity of this method applied on a fully connected network is $O(r^2 \cdot L)$. Other networks, sparser than a fully connected network, have a lower complexity since the number of links is smaller.

Note that for a given limited number of routes, the exponential complexity required for exhaustive search has been reduced to polynomial complexity using this technique. However, if all possible routes are specified, this could result in exponential complexity as well.

A Numerical Example

In Figure 10, we show the numerical result of the example of Figure 9 we have been discussing. We set p_i equal to 0.1 for all links except link l_4 where p_4 takes values from 0.1 to 0.8. The results for links $l_4, \cdots, l_9$ only are shown. The height of each bar represents the probability of a link being faulty with a corresponding value of p_4 shown on the x-axis. As a check, there are four links (l_4, l_5, l_7, l_8) equally likely to be faulty for $p_4 = 0.1$. This agrees with the symmetry property of the graph. For the case of $p_4 > 0.1$, link l_4 is the most probable faulty link. Though we have changed the value of p_4 dramatically, the ranking among links does not change. In other words, we can tolerate some degree of inaccuracy for the prior probabilities. Our experience from several numerical examples shows that the model here seems to tolerate large deviations in the prior probabilities, however, it still can lead to the

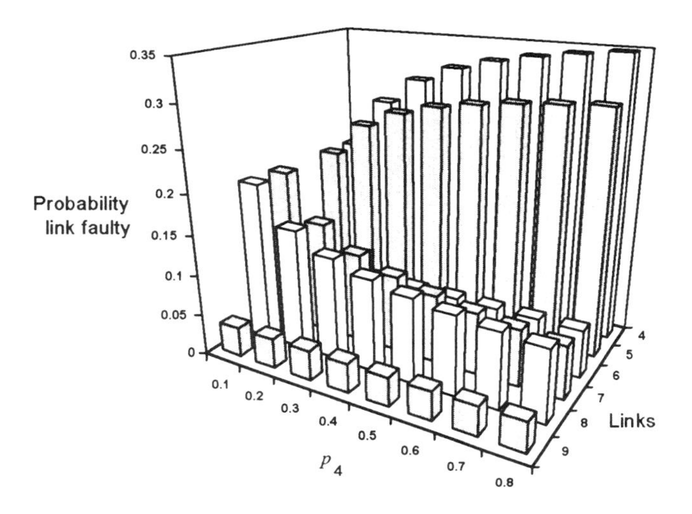

Figure 10. The numerical result of the 8-node example

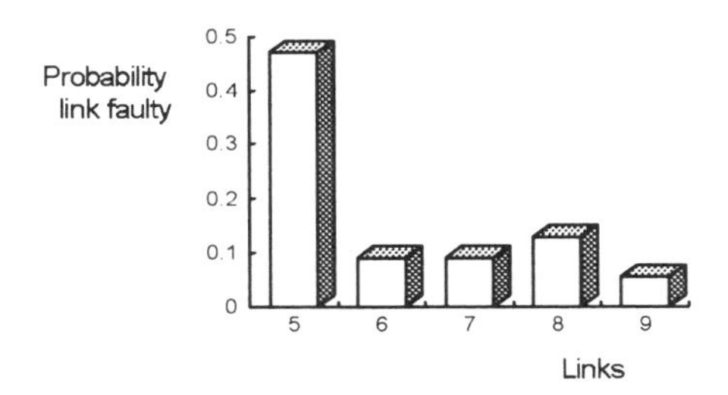

Figure 11. The new result after one link is confirmed faulty

correct decision. This appears to agree with Ben-Bassat's conclusion for Bayesian models [1]. This is helpful when relative frequencies such as average down-time or average number of failures during the last month, for example, are not available. A human expert might simply assign the prior probabilities according to the confidence level.

APPLICATIONS

Multiple Link Failures

Since we have made no assumption about how many faulty links are presented in the analysis, the previous result is still optimal in the situation of multiple link failures. The most probable faulty link remains the same regardless of the number of faulty links. However, as suggested earlier, this method can be used in conjunction with testing to quickly pinpoint actual faulty links. Whenever an actual faulty link l_j is identified, we remove this link from the graph or the network, repeat the computation, and generate a new ranked list of most probable faulty links. We then test for the next most probable faulty links, etc.

For example, we test the most probable faulty link l_4 according to the plot in Figure 10. However, link l_4 is not the only faulty link since n_1 is unable to communicate with n_6,n_7 and

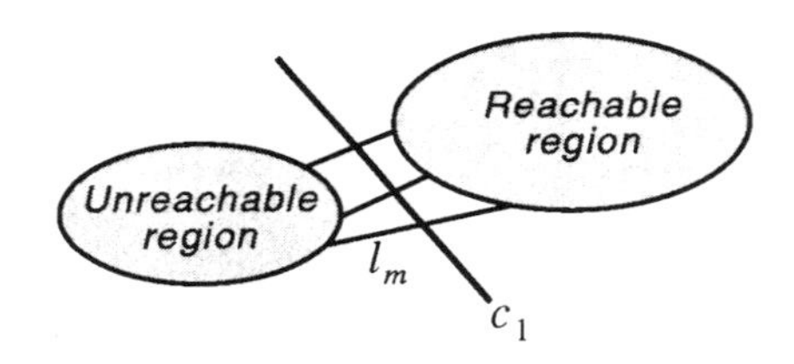

Figure 12. A network with only two regions

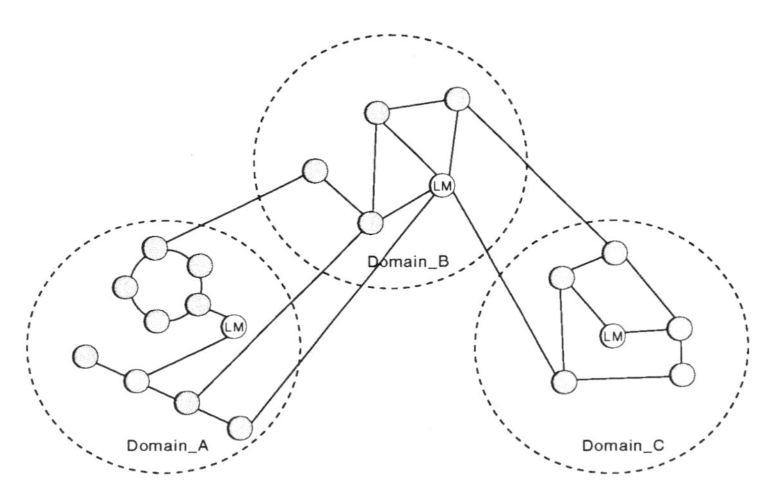

Figure 13. A heterogeneous network

n_8. That implies at least one more faulty link must be presented. Figure 11 shows the new result after link l_4 is confirmed faulty. Link l_5 is thus the next target to be tested.

Complete Information Given

Consider the special case where the information provided to the network node n_i is *complete*. We mean all the nodes are either in a reachable (Θ_i) or unreachable ($\overline{\Theta}_i$) region. This is shown in Figure 12. There is only one cut c_1 which is the summation of l_m corresponding to the links from the reachable to the unreachable region. Equation (11) is reduced to

$$j = \arg\max_{l_k \in \ell} P_{(l_k=0)}\{c_1 = 0\} \cdot p_k$$

We now try each link l_k in c_1, one at a time.

Location of the Network Manager

We assume the network manager can reside at any designated node. This convenience makes virtually any node a possible site for the network manager.

Figure 13 shows a network consisting of several domains of sub-networks. Each proprietary domain is managed by a Local Manager (LM). Consider the following scenario: Domain_A includes a Token-ring and an Ethernet; Domain_B is a public network; and Domain_C is another private network. The Local Manager of Domain_C does not have the privilege of accessing the link level protocol of Domain_A. A problem, such as a host in Domain_C which is unable to communicate with certain nodes of Domain_A, might be first noticed by the Local Manager at Domain_C. This LM could execute the fault-finding algorithm. If it suspects that the problem is caused by faulty links in Domain_A, then it could make a suggestion to the Local Manager at Domain_A for appropriate testing and repair.

The Network Management functions could be built in at the application level. This would work even if the managed entities do not share a common link-level protocol. All that is required is that end-to-end transport service be available. This reduces the protocol complexity of the Network Management functions across different proprietary networks without accessing divergent link protocols.

Fault Diagnosis in the Open System Interconnection (OSI) Model

We can extend the generic representation of Figure 1 using an object-oriented model although we have used the terms *links* and *nodes* throughout the analysis of Section 3. In this case, we model the network as a collection of *network resources* or *network elements*. Let the node in Figure 1 represent a network entity and the link represent a network resource. The network provides a necessary connection between two network entities by utilizing a set of network resources. Faults are treated as a loss of system connectivity due to the malfunction or breakdown of a number of network resources. For example, a physical connection between two network nodes can be formed by several network links. In ATM terminology, a connection between two access points or ATM entities is called a *virtual path*. The ATM entities and/or the virtual path can be treated as network resources. The graph with undirected arcs in Figure 1 then represents the relationship between entities and network resources (network elements).

The International Standards Organization (ISO) has adopted the object-oriented model for the management of OSI networks [7]. This eases inter-operability in a heterogeneous environment. However, to resolve a problem in which a network node is unable to communicate with a set of other nodes might require many network queries to find vendor- and protocol-specific properties before testing components along all possible paths [6]. The method described in this paper could speed up the diagnostic process by testing a ranked list of possible faulty network resources instead of testing exhaustively and thereby reducing network bandwidth and processing. Furthermore, this approach has the advantage of being implemented at the Application Layer, since current standards activities have concentrated on the Application Layer. For example, CMIP (Common Management Information Protocol) [8] could be used by Application Layer entities to exchange management information.

SUMMARY

A method based on Bayesian theory (MAP) is presented for identification of faulty links in a communication network. We stress again that our model, although clearly applicable to physical links and nodes in a network, can be applied to any set of interconnected network entities. The interconnecting resources could be bridges, routers, subnetworks, ATM virtual paths, etc. Our goal is to maintain a relatively abstract view of the network and to show the applicability of the technique described to a variety of scenarios not necessarily at the physical layer.

However, it is clear that the approach presented does not apply to relatively small homogenous networks such as closed-domain LANs. Vendors could easily associate each network resource in this case with a dedicated sensor and corresponding alarm. It is possible to achieve fault diagnosis for a homogenous networking environment in a *deterministic* fashion. Assume first that every network element, which can be a device or a single layer of a protocol stack, generates an alarm when a fault occurs within it. Second, assume that the alarm is always received faithfully by a network manager. The network manager can then allocate the remaining available network resources to users whenever faulty elements are reported. For more complex, hybrid networks, there is a need for non-deterministic reasoning. In general a network manager may encounter different levels of warning indicating whether one component or portion of the network is in trouble, possibly with some degree of uncertainty when failure information is insufficient. To locate and identify actual faults from a set of mixed alarms associated with different levels (richness) of information is an intricate task. There are several deterministic approaches that have been used before. One is complete relational exploration which gives a list of potentially troubled network elements. For example, we could find out the intersection of the set of network elements indicated by each alarm; however, we will not see the difference in a degree of significance among network elements. In addition, for a system based on deterministic diagnosis, the designer attempts to define a set of alarms as large as possible in order to include all fault scenarios. A single fault can trigger several alarms. One alarm can be issued by the faulty entity itself and, in addition, many other alarms can be issued by neighboring entities in failing to interact with the faulty entities. This can cause networks to suffer excessive network management messages during the fault diagnosis process.

Recent CCS (Common Channel Signaling) network incidents [2] indicate that a network manager might not be able to obtain complete status information about the whole network during an outage. Though enough redundancy was implemented in the CCS network, congestion due to equipment failure and the sharp increase in network management messages prevented network operators from obtaining complete and accurate information. This scenario demonstrates the need for non-deterministic reasoning.

Finally we noted two drawbacks in the model; the independence assumption and the lack of timing information need further refinement. Hardware faults may or may not occur independently; software faults may be prone to trigger others, making the independence assumption invalid.

Clark Wang received the B.S. degree in nuclear engineering from National Tsing Hua University, Taiwan, in 1986 and the M.S. degree and the Ph.D. degrees in electrical engineering from Columbia University, New York, in 1990 and 1993, respectively.

In 1990, he received the Edwin Howard Armstrong Memorial Award from Columbia University. He cofounded Eigen Research Inc., a recent start-up company that builds distributed processing systems. His current research interests include network management, protocols, and multimedia applications.

Mischa Schwartz is Charles Bachelor Professor of Electrical Engineering at Columbia University. He is a member of The National Academy of Engineering, a Fellow of the IEEE, and a Fellow of the AAAS. He was a Director of the IEEE in 1978, 1979 and was President of the IEEE Communications Society in 1984, 1985. He received the IEEE Education Medal in 1983, the Cooper Union Gano Dunn Award for achievement in science and technology in 1986, and the IEEE Region Award I in 1990. He served as Director of Columbia's Center for Telecommunications Research, and NSF Engineering Center, from

1985-1988. He is author of 8 books in communication systems, signal processing, and communication networks.

REFERENCES

[1] M. Ben-Bassat, K. L. Klove, and M. H. Weil, Sensitivity Analysis in Bayesia Classification Models: Multiplicative Deviations, IEEE Transactions on Patter Analysis and Machine Intelligence, Vol. PAMI-2, No. 3, 1980.

[2] Bellcore, CCS Network Outages in Bell Atlantic and Pacific Bell - June 10 through July 2, 1991, Special Report, SR-NWT-002149, Issue 1, NJ, 1991.

[3] A. Bouloutas, G. Hart, M. Schwartz, On the Design of Observers for Failure Detectio of Discrete Event Systems, in Network Management and Control, A. Kershenbaum, M Malek, M. Wall, eds, Plenum Press, NY, 1990.

[4] P. Cheeseman, A Method of Computing Generalized Bayesian Probability Values fo Expert Systems, in Proceedings of 8th International Joint Conference on Artificia Intelligence, Los Angels, CA, 1983.

[5] R. H. Deng, A. A. Lazar and W. Wang, A Probabilistic Approach to Fault Diagnosis i Linear Lightwave Networks, IEEE Jounal on Selected Area in Communications, Vol 11, No.9, Dec. 1993.

[6] A. Dupuy, S. Sengupta, O. Wolfson and Y. Yemini, NETMATE: A Networ Management Environment, IEEE Network, Vol. 5, No. 2, Mar. 1991.

[7] International Organization for Standardization, Information Processing Systems - Ope Systems Interconnection - Basic Reference Model Part 4: OSI Managemen Framework, No. 7498-4, 1989.

[8] International Organization for Standardization, Information Processing Systems - Ope Systems Interconnection - Management Information Protocol Specification Part 2 Common Management Information Protocol, No. 9596-2, 1989

[9] S. L. Lautitzen and D. J. Spiegeelhalter, Local Computation with Probabilities o Graphical Structures and their Application to Expert Systems, Journal of Roya Statistical Society, Ser. B, Vol. 50, No. 2, 1988.

[10] J. Pearl, Fusion, Propagation, and Structuring in Belief Networks, Artificia Intelligence, 29, 1986.

[11] C. Wang and M. Schwartz, Fault Detection with Multiple Observers, IEEE/AC Transactions on Networking, Vol. 1, No. 1, Feb. 1993.

KEEPING AUTHENTICATION AND KEY EXCHANGE ALIVE IN COMMUNICATION NETWORKS

Gnanesh Coomaraswamy and Srikanta P. R. Kumar

Northwestern University
Department of Electrical Engineering and Computer Science
Evanston, IL 60208-3118

ABSTRACT

Authentication and exchange of encryption keys are critical to network management. We propose a distributed scheme which, in addition to accomplishing authentication, also produces a session key at the end of the procedure. This is achieved by combining a zero-knowledge authentication scheme and a public key exchange protocol. Due to this combination, both protocols gain additional security against otherwise successful attacks. Performance and application to cellular networks are also discussed.

1. INTRODUCTION

Achieving security through cryptography is the most feasible and cost effective way of protecting information transmitted over a public network. In this approach to security, messages to be encrypted are transformed by a function that is parameterized by a key. The recovery of the messages from the encrypted text requires a decryption key which is known only to the intended receiver. These decryption keys themselves need to be conveyed to the receiver through the public network in a non-readable form, which creates a *key distribution problem*.

In this context, authentication means verifying the identity of the communicating principals to one another. In general, if *source authentication* is not performed, it may be possible for an intruder to masquerade as the bona fide source and send false information to any station on the network, and if *destination authentication* is not realized, it may be possible for the intruder to pretend to be the intended receiver, and receive unauthorized information.

Network Management and Control, Vol.2, Edited by
I.T. Frisch *et al.*, Plenum Press, New York, 1994

There are several distributed and centralized key distribution protocols and authentication schemes proposed in the literature for these problems. However, there is a fundamental problem whenever the authentication of the session and the key exchange for the session encryption are two separate protocols [1]. A possible intruder who is capable of suppressing and forging messages will let the authentication data pass unchanged. But he will try to intercept a session key instead, so that his requests encrypted under this key appear authentic. If for example, the Diffie and Hellman key exchange protocol [2] were used, the intruder may perform the well known switching-in attack resulting in two keys, one between the user and intruder and the other between intruder and host. An effective way of dealing with this is to combine the key exchange and authentication protocols.

Our contribution in this paper is, providing a distributed[a] authentication scheme which, in addition to accomplishing authentication, also produces a session key for cryptographic applications at the end of the authentication procedure. This is achieved by combining a zero-knowledge authentication scheme and a public key exchange protocol. It is noteworthy that due to the combination, both protocols gain additional security against attacks that would otherwise be successful. The idea of integrating authentication with key exchanges was first proposed by Bauspieß's scheme is based on Beth's version of Diffie and Hellman scheme and rely on computing logarithms over finite fields. Our scheme is a combination of Fiat-Shamir [10] and Beth [14] and has two major advantages over Bauspieß.

Firstly, while Bauspieß has a component or an element of its key exchange protocol integrated with the authentication mechanism, it does not provide a session key at the end of the authentication procedure. It then uses the common authenticated element in the key exchange protocol to construct the session key. Our scheme provides a common authenticated session key at the end of the authentication procedure. Our scheme may be viewed as having one combined phase, instead of the two inter-related sequential phases in Bauspieß. One of the advantages of having a single phase is the reduction in time required to complete both authentication and generation of session keys. In communication networks, where the call set-up time is much smaller in relation to the conversation time, the reduction in message exchanges is highly beneficial, especially in radio networks or satellite links where the high propagation time is a key factor in the total call setup time[b]. The reduction in the number of messages exchanges in the whole exercise can significantly improve call setup times in such networks.

Secondly, in situations where only the authentication service is required (or when encryption key is not requested), our scheme can be easily modified to provide this service with the same computational complexity as Fiat-Shamir. This makes it possible for our scheme to provide the encryption service on demand, without expending computational time when encryption is not requested. In the scheme suggested by Bauspieß, the computational complexity for the similar *authentication only* service is much higher, as it involves modular exponentiation which takes time.

Our scheme is very general and is applicable to any digital network, and we believe that this scheme could be successfully implemented within the framework of the proposed

---

[a]Here, by *distributed* we mean a decentralized scheme, which does not require the maintenance of a data file of the other nodes in the network. i.e., two nodes are able to authenticate each other without prior arrangement or without further communication between the nodes and a central clearing house.

[b]The call setup time is the sum of the message propagation times and the computation times between both parties for authentication and key generation.

standard for cellular networks EIZ/TIA/IS/41 [3]. A preliminary scheme is outlined in our previous work [4][c].

This paper is organized as follows. In Section 2, we describe the authentication procedure used in cellular networks with an introduction to zero-knowledge schemes. This is followed by a brief look into the emerging technologies of smart cards. The public key exchange protocols are examined in Section 3. In Section 4, we describe our proposed scheme and give a proof. Section 5 describes an application. Performance is discussed in Section 6 and our conclusions are given in Section 7.

2. PROVING YOURSELF

Typically, in public telephone networks, the source authentication is confirmed by checking the telephone number to its assigned physical link. While such simple protocols may be sufficient for land networks (where land lines such as twisted pair, co-axial cable or fiber optic cable physically link the telephone or data terminal equipment to the distribution point of the local exchange), this is grossly inadequate in the case of *radio networks*. In terrestrial telephone networks, a physical link needs to be established (tap into a telephone line) in order for an impersonator to claim that he is an authorized user. Such access points are limited to the physical presence of these lines between the source and the exchange. But, in cellular radio networks, the link between the mobile unit and any base station is a *radio* link which is highly prone to information compromise. In addition, whereas each telephone unit is attached to a local exchange, the mobile unit may be allowed to roam even outside its home switching center giving the mobile unit a geographic limit bounded only by the coverage of the carrier. In essence the major differences are the high exposure due to the broadcast nature of the radio link and the vast extent of its coverage area due to its roaming ability.

The designer of an effective authentication scheme in such a *broadcast environment* is faced with additional challenges due to this *exposure of information* on the network. This problem becomes particularly challenging when the two parties (the prover A and verifier B) are adversaries and we want to make it impossible for B to misrepresent himself as A even after he witnesses and verifies arbitrarily many proofs of identity generated by A. For example, an intruder in a mobile radio network could impersonate the verifying base station and then, in turn pretend to be the mobile unit to the genuine base station. This could be avoided if A can prove to B that he is A, without giving away any more information than that was known before. Such identification processes are called *zero knowledge interactive proof systems*.

An interactive proof system for a language L is *zero-knowledge* if for each $x \in L$, (where x is a string) the prover tells the verifier essentially nothing, other than that $x \in L$; this should be the case even if the verifier chooses not to follow the proof system but instead tries to trick the prover into revealing something.

For example, if the prover is trying to prove to the verifier that y is a quadratic residue $mod\ x$, (Let x, y be integers, $0 < y < x$, such that $gcd(x,y)=1$; we say y is a quadratic residue $mod\ x$ if $y=z^2 \ mod\ x$ for some z; if not we say y is a quadratic non-residue) then certainly the verifier should not be able to trick the prover into revealing a square root of $y\ mod\ x$, or the factorization of x, or any information which would help the verifier to compute these quantities faster than before.

---

[c]Currently in mobile networks, source authentication is achieved by verifying a serial number programmed by the manufacturer, and the telephone number, both of which are transmitted by the mobile unit when powered. The serial number once known (by listening in) can be programmed into another mobile unit without much difficulty. Such non-stringent methods may provide an environment which is conducive for large scale fraud. New authentication schemes for these proposed standards are still under study.

Since these zero knowledge interactive authentication schemes verify the identities of the source and the destination without revealing any information, they seem well suited for applications in the broadcast environment.

In the next Section, we look into authentication schemes and the emerging technologies of smart cards. We then propose an authentication scheme which is an application of the identification scheme by Fiat and Shamir, and the key exchange protocol by Diffie and Hellman.

3. AUTHENTICATION SCHEMES

Authentication schemes can be broadly divided into two categories, central and distributed. Central is where an authentication server acts as the mediator and authenticates the source and the target to each other. The authentication between the server and the source or the server and the target could be via either conventional or public key cryptography. For large networks, a hierarchy of servers has been proposed[5]. Improvements on this scheme uses time stamps [6] and event markers [7] to avoid replay.

The concept of distributed authentication was first proposed by Diffie and Hellman in 1976 [2]. They revolutionized encryption by their idea of public keys. In such a scheme $(M^E)^D = (M^D)^E = M$, where E is a public key which is made public by the node, D a secret key known only by the node, M the message and M^E the output of the encryption of message M by the encryption key E. In this scheme, source authentication is achieved by a node A using its secret key D_A for the encryption of its message header. Then, not only would all other nodes be able to decrypt this message (using the public key E_A), but they would also know that *A and only A* could have encrypted this message (as only A has the secret key D_A needed for encryption). In a non-rigorous sense this is zero knowledge, as A was able to prove its identity without giving away any useful information.

Two of the popular public key algorithms are the RSA algorithm [8], which is based on the difficulty of factorization, and the Diffie and Hellman algorithm [2], which is based on difficulty of finding discrete logarithms in finite space. Then in 1984, Shamir [9] proposed public key crypto-system with an extra twist called the *ID based systems*. Here, instead of generating a random pair of public/secret key, the user chooses his id (name & address) as his public key. The corresponding secret key is computed by a key generation center and issued to the user in the form of smart card. Later in 1986, Fiat and Shamir proposed the first workable ID based protocol [10].

Fiat and Shamir protocol [10]: This identification scheme is a combination of zero knowledge interactive proofs [11] and identity based schemes [9]. It is based on the difficulty of extracting modular square roots when the factorization on n is unknown. The scheme assumes the existence of a trusted center which issues the smart cards to users after properly checking their physical identity. No further interaction with the center is required either to generate or to verify proofs of identity. An unlimited number of issuers can join the system without degrading its performance, and it is not even necessary to keep a list of all valid users. Interaction with the smart cards will not enable verifiers to reproduce them, and even complete knowledge of the secret contents of all the cards issued by the center will not enable adversaries to create new identity or to modify existing identities. Since no information whatsoever is leaked during the interaction, the cards can last a lifetime regardless of how often they are used.

Before the center starts issuing cards, it chooses and makes public a modulus n and a pseudo random function f which maps arbitrary strings to the range $[0,n]$. A pseudo random function is a one way trap-door function that is computationally indistinguishable from a truly random function by any polynomially bounded computation. For formal definition and family of functions which is provably strong in this sense see [11]. But in practice, it is believed that one can use simpler and faster functions without endangering the security of the scheme. The modulus n is the product of two secret primes p and q. The center keeps the factors p and q of modulus n a secret, and publishes the modulus n and the pseudo random function f.

This is different from the RSA scheme [8] where each user i keeps its modulus n_i and one prime factor p_i a secret and publishes the other factor q_i. The center can be eliminated if each user chooses his own n and publishes it in a public key directory. However, this RSA-like variant makes the scheme considerably less convenient [10].

Having picked and made public the modulus n and the function f, the center is now ready to start issuing smart cards. When an eligible user applies for the smart card, the center first prepares a string I which contains all the relevant information about the user (his name, address, ID number, etc.) and about the card (expiration date, limitations on validity, etc.). Since this information is the information verified by the scheme, it is important to make it detailed and to double check its correctness. The center then performs the following steps:

1. Compute the values $v_j = f(I,j)$ for small integer values of j for which v_j is a quadratic residue $mod\ n$ (typically $j \in [0,2^{16}]$) [12].
2. Pick k distinct values of j and compute the smallest square root, s_j of $v_j^{-1}\ (mod\ n)$.
3. Issue a smart card which contains I, the k secret (s_j values), and their corresponding indices.

Remarks:

1. To simplify the notation, the corresponding indices of the k s_j values are assumed to be the first k indices $j = 1,2, ...k$.
2. In typical implementations, k is between 1 and 18, but larger values of k can further reduce the time and communication complexities of the scheme.
3. n should be at least 512 bits long. Factoring such moduli seems to be beyond today's computers and algorithms, with adequate margins of safety against foreseeable developments.

The verification devices are identical stand-alone devices which contain a microprocessor, a small memory, and I/O interface. The only information stored in them are the universal modulus n and function f. When a smart card is inserted into a verifier, it proves that it knows $s_1, s_2 ... s_k$ without giving away any information about their values. The proof is based on the following protocol between the prover A and the verifier B:

1. A sends I to B.
2. B generates $v_j = f(I, j)$ for $j = 1, ... k$.
3. A picks a random $r_i \in [0,n]$ and sends $x_i = r_i^2\ (mod\ n)$ to B.
4. B sends a random binary vector $(e_{i1}, .. e_{ik})$ to A.
5. A sends to B:

$$y_i = r_i \prod_{e_{ij}=1} s_j \pmod{n} \tag{1}$$

6. B checks that

$$x_i = y_i^2 \prod_{e_{ij}=1} v_j \pmod{n} \tag{2}$$

The verifier B accepts A's proof of identity only if all the t (iterations $i = 1,..t$) checks are successful. In steps 5 and 6, the product is taken over all j such that $e_{i,j} = 1$. For a fixed k arbitrary t, this is zero knowledge proof. For a formal proof see [9].

The Diffie and Hellman key distribution scheme [2]: This key distribution scheme relies on the difficulty of computing logarithms over finite fields. The security of an implementation of this scheme to encrypt and decrypt messages is equivalent to that of the key distribution scheme. First, we review the distribution scheme.

Suppose that A and B want to share a secret K_{AB}, where A has a secret x_A and B has a secret x_B. Let p be a large prime and α be a *primitive root mod* p, both known. If $(\alpha, n) = 1$, α is the smallest positive power such that a^α is congruent to 1, and if α is equal to $\phi(n)$, then a is said to be a *primitive root* of n. Here $\phi(n)$ is the Euler's totient function giving the number of positive integers less than n, which are relatively prime to n. A computes and sends y_A, and B computes and send y_B where

$$y_j = \alpha^{x_j} \ (\text{mod } p), \quad j = A, B.$$

Then the secret K^{AB} is computed as

$$\begin{aligned} K^{AB} &\equiv \alpha^{x_A x_B} \ (\text{mod} \ \ p) \\ &\equiv y_A^{\ x_B} \ (\text{mod} \ \ p) \\ &\equiv y_B^{\ x_A} \ (\text{mod} \ \ p). \end{aligned} \tag{3}$$

Hence both A and B are able to compute K^{AB}. But, for an intruder, computing K^{AB} appears to be difficult. It is not yet proved that breaking the system is equivalent to computing discrete logarithms. For more details refer [2].

We review the public key crypto system based on this distribution scheme. Suppose that A wants to send B a message m, where $0 \leq m \leq (p - 1)$. First A chooses a number k uniformly between 0 and $(p - 1)$. Note that k will serve as the secret x_A in the key distribution scheme. Then A computes the "key"

$$K \equiv y_B^{\ k} \ (\text{mod} \ \ p) \tag{4}$$

where $y_B = \alpha^{x_B} \ (\text{mod} \ \ p)$ is either in a public file or is sent by B. The encrypted message (or ciphertext) is then the pair (c_1, c_2), where

$$c_1 \equiv \alpha^k \ (\text{mod} \ \ p) \ \ ; \quad c_2 \equiv K\, m \ \ (\text{mod} \ \ p) \tag{5}$$

and K is computed in equation (4).

Note that the size of the ciphertext is double the size of the message. Also note that the multiplication in equation (4) can be replaced by any other invertible operation such as addition mod p. The decryption operation splits into two parts. The first step is recovering K, which is easy for B since $K \equiv (\alpha^k)^{x_B} \equiv c_1^{\ x_B} \ (\text{mod} \ \ p)$, and x_B is known to B only. The second step is to divide c_2 by K and recover the message m.

The public file consists of one entry for each user, namely y_i for user i (since α and p are known for all users). It is possible that each user chooses his own α and p, which is preferable from the security point of view although that will triple the size of the public file

[13]. It is not advisable to use the same value k for enciphering more than one block of the message, since if k is used more than once, knowledge of one block m_i of the message enables an intruder to computer other blocks as follows. Let

$$c_{1,1} \equiv \alpha^k \pmod p \; ; \quad c_{2,1} \equiv m_1 K \pmod p$$
$$c_{1,2} \equiv \alpha^k \pmod p \; ; \quad c_{2,2} \equiv m_2 K \pmod p \quad (6)$$

Then

$$\frac{m_1}{m_2} \equiv \frac{c_{2,1}}{c_{2,2}} \pmod p$$

and m_2 is easily computed if m_1 is known.

Breaking the system is equivalent to breaking the Diffie-Hellman distribution scheme. First, if m can be computed from c_1, c_2 and y, then K can also be computed from y, c_1, and c_2 (which appears like a random number since k and m are unknown). That is equivalent to breaking the key distribution scheme. Second, (even if m is known) computing k and x from c_1, c_2, and y is equivalent to computing discrete logarithms as both x and k appear in the exponent in y and c_1.

An alternative possibility is to combine both authentication and key exchange protocols, and use data exchanged during the authentication protocol, and therefore authentic data, to construct the session key. Although Beth's [14] ID-based version of Diffie-Hellman's zero-knowledge authentication is well suited for this purpose [1], it has a major drawback when compared to Fiat-Shamir's scheme. Beth's scheme requires that each node keep a file of the published information y_l (which is the exponentiated value of the secret information x_j of node j) of all the other nodes in the network. This will be *a large and often changing data file)*. While Fiat-Shamir scheme does not require such a database, it is not suited for the key *distribution problem*. In the next Section, we propose a scheme, which is a combination of both Fiat-Shamir scheme as well as Beth.

4. KEY EXCHANGE BY AUTHENTICATION

We propose a scheme which is based on the Fiat-Shamir scheme, but derives the benefit of both the authentication scheme of Fiat-Shamir, as well as the key exchange protocol of Diffie-Hellman. This zero knowledge authentication scheme authenticates and distributes session keys *without requiring a data file of published keys and provides a session key for cryptographic applications* at the end of the authentication procedure. The proposed scheme works as follows.

Our scheme also assumes the existence of a trusted center which issues the smart cards. Before the center starts issuing cards, it chooses and makes public a modulus n, the pseudo random function f, and a primitive element $\alpha \pmod n$. The modulus n is the product of two secret primes p and q, which is kept secret by the center.

When an eligible user applies for a smart card, the center prepares a string I which contains all the relevant information about the user and about the card. The center then performs the following steps:

1. Compute the values $v_j = f(I,j)$ for small values of j.
2. Pick k distinct values of j for which v_j is a quadratic residue $(mod\ n)$ and compute the smallest square root s_j of $v_j^{-1} \pmod n$.
3. Issue a smart card which contains I, the k secret (s_j values), and their indices.

The verification devices are similar to the Fiat-Shamir devices and are identical stand-alone devices which contain a microprocessor, a small memory, and I/O interface. The only information stored in them are the universal modulus n, function f and the primitive element α. When a smart card is inserted into a verifier, it proves that it knows $s_1, s_2, .. s_k$ without giving away any information about their values and generates a term α^r, where r is a random number generated by A, which can be used as a session key for encryption between the source and the verifier. The proof is based on the following protocol:

1. A sends I to B.
2. B generates $v_j = f(I,j)$ for $j = k$ corresponding indices.

Repeat steps 3 to 6 for $i = 1..t$:

3. A picks a random $r_i \in [0,n)$ and sends $x_{1i} = r_i^2 \ (mod\ n)$ and $x_{21} = \alpha^{r_i}$ (mod n) to B.
4. B sends a random binary vector $(e_{i1}, e_{i2}, ... e_{ik})$ to A.
5. A sends to B:

$$y_{1i} = \left\{ r_i + \prod_{e_{ij}=1} s_j \right\}^3 \pmod{n} \quad y_{2i} = \alpha \prod_{e_{ij}=1} s_j \pmod{n} \tag{7}$$

6. B computes

$$\begin{aligned} \gamma_{1i} &= x_{1i} + 3\prod_{e_{ij}=1} v_j \pmod{n} \\ \gamma_{2i} &= 3x_{1i} + \prod_{e_{ij}=1} v_j \pmod{n} \end{aligned} \tag{8}$$

and checks that

$$\alpha^{y_{1i}} \equiv x_{2i}^{\ \gamma_{1i}} \ y_{2i}^{\ \gamma_{2i}} \pmod{n} \tag{9}$$

4.1 Security

Proposition: If A and B follow the above protocol and if all the checks are successful, the verifier B always accepts A's proof of identity and the session key as valid. And for a fixed k and arbitrary t, this is zero knowledge proof.

The proof is evident from the following lemmas:

Lemma 1: If A and B follow the protocol, B always accepts the proof as valid.
Proof: By definition

$$y_{1i} \equiv (r_i + \prod_{e_{ij}=1} s_j)^3 = r_i^3 + 3r_i^2 \prod_{e_{ij}=1} s_j + 3r_i^2 \prod_{e_{ij}=1} s_j^2 + \prod_{e_{ij}=1} s_j^3 \pmod{n} \tag{10}$$

$$\alpha^{y_{1i}} \equiv \alpha^{\left(r_i^3 + 3r_i^2 \prod_{e_{ij}=1} s_j + 3r_i \prod_{e_{ij}=1} s_j^2 + \prod_{e_{ij}=1} s_j^3 \right)} \tag{11}$$

$$\equiv \alpha^{r_i^3} \cdot \alpha^{3r_i^2 \prod_{e_{ij}=1} s_j} \cdot \alpha^{3r_i \prod_{e_{ij}} s_j^2} \cdot \alpha^{\prod_{e_{ij}=1} s_j^3} \tag{12}$$

$$\equiv (\alpha^{r_i})^{r_i^2} \cdot (\alpha^{\prod_{e_{ij}=1} s_j})^{3r_i^2} \cdot (\alpha^{r_i})^{3\prod_{e_{ij}=1} s_j^2} \cdot (\alpha^{\prod_{e_{ij}=1} s_j})^{\prod_{e_{ij}=1} s_j^2} \tag{13}$$

$$\equiv (\alpha^{r_i})^{r_i^2 + 3\prod_{e_{ij}=1} s_j^2} \cdot \left(\alpha^{\prod_{e_{ij}=1} s_j}\right)^{(3r_i^j + \prod_{e_{ij}=1} s_j^2)} = (x_{2i})^{r_{1i}} (y_{2i})^{r_{2i}} \tag{14}$$

Lemma 2: For a fixed k and arbitrary t, this is zero knowledge.

Proof: Since the rigorous analytical proof is intractable in such problems, we employ the proofs generally accepted in the literature. There are many difficult mathematical problems that have not been proved for its solvability, but are accepted as computationally infeasible. For example, factorization of large numbers, computing discrete logarithms in finite fields and finding modular roots of polynomials in finite fields are considered mathematically hard problems [14]. In cryptography, if a problem can be shown to be equivalent to solving one of the many mathematically difficult problems, then that problem is accepted as computationally infeasible.

The reason our zero knowledge scheme reveals no information whatsoever about the s_j is that y_1 contains an independent random variable r which *masks* the value of s. Computing r from x_1 is equivalent to finding the square root of $r \bmod n$, without knowing the factors of n. For large numbers this is infeasible. Computing r from x_2 (a random exponentiated value) is equivalent to computing discrete logarithms in a finite field. Once again, computing s_j from y_2 is equivalent to computing discrete logarithms as s_j appears in the exponent in y_1.

Lemma 3: Assume that A does not know the s_j and cannot compute in polynomial time the square root of $v_j \pmod n$ nor find the discrete logarithms in finite fields mod n. If B follows the protocol (and A performs arbitrary polynomial time computations), B will accept the proof as valid.

Proof: If A does not actually possess the s_j values and wants to cheat B by guessing, A would first send x_1 and x_2 and await the vector e. He would then try to compute y_1 and y_2 such that $\alpha^{y_1} = x_2^{\gamma_1} y_2^{\gamma_2} \pmod n$ is satisfied. This is equivalent to solving the congruence $\alpha^{y_1} = a\, y_2^{\,b} \pmod n$, for known a and b. For a given y_2, this congruence reduces to $\alpha^{y_1} = a \pmod n$, for known a. This is equivalent to finding discrete logs in finite fields. For a given y_1 this reduces to $y_2^{\,a} = b \pmod n$, for known a and b, which is equivalent to finding modular roots for a polynomial in a finite field. Both of these problems are considered computationally infeasible in polynomial time.

4.2 Notes: 1. A symmetric version of the above scheme is adopted for A and B to authenticate each other. The values $x_{A,1}$ and $x_{B,1}$ (the x_1 of A and B) are an integral part of the authentication protocol and therefore cannot be altered by an intruder without causing

the authentication to fail. Using it as a part of the session key is an effective way to combine authentication and key exchange
2. The protocol provides two keys $x_{A,1}$ and $x_{B,1}$, and this can be used to construct one key $\alpha^{r_A r_B}$, as they are powers of the same base α.
3. For mobile radio networks, the parallel version suggested by Fiat-Shamir [10] maybe adopted. In the parallel version, A sends all the x_i, then B sends all e_{ij}, and finally A sends all the y_i. This reduces the number of transmissions between the mobile unit and the base station.
4. The number of transmissions can be reduced further without affecting the security by sending all x_i along with I, the its corresponding indices.
5. Improvements in speed can be obtained by parallelizing the operations. A preparing x_{i+1} and y_{i+1} while B is still checking x_i and y_i values.
6. The time, space, communication and security of the scheme can be traded off in many possible ways, and the optimal choices of k, t and the e_{ij} matrix depends on the relative costs of these resources.

5. APPLICATION TO DIGITAL CELLULAR NETWORKS

In this section, we discus an application to the EIA/TIA-IS-41 [3], the proposed standard for networking between mobile carriers, and outline a preliminary scheme which can effectively integrate into the framework of this standard. For details see [4].

A *Cellular Subscriber Station (CSS)* is the interface equipment used to terminate the radio path at the user side. A *Base Station (BS)* is the common name for all the radio equipment located at one and the same place for serving one or several cells. A *Mobile Switching Center (MSC)* is an automatic system which constitutes the interface for user traffic between the cellular network and other public switched networks, or other MSCs in the same or other cellular networks. A *Home Location Register (HLR)* is the location register to which a user identity is assigned for record purposes such as subscriber information. A *Visitor Location Register (VLR)* is the location register other than the HLR used by an MSC to retrieve information for, for instance, handling of calls to or from a visiting subscriber. An *Authentication Center (AC)* is an entity which may manage the encryption keys associated with an individual subscriber, if such functions are provided for within the network.

5.1 Authentication

Here we propose an authentication scheme between a CSS and a MSC. First we look at the case where CSS is in its home region. Next we look at the case when a CSS is roaming in another service area.

5.1 Authentication within home MSC: For a call initiation between a CSS and its home MSC, the authentication procedure is sufficient. Since it does not have to be viewed as authentication between two mutually suspicious parties, one way identification verification would also be adequate. Given these conditions, our scheme can be down graded to a modified Fiat-Shamir authentication scheme while retaining its full potential. Here we assume that the trusted center which issues the smart cards is the home MSC. To simplify the notation, we assume t=1. The protocol works as follows:

1. *CSS* sends I to *MSC*.
2. *MSC* generates $v_j = f(I,j)$ for $j= k$ corresponding indices.

3. *CSS* picks a random $r \in [0,n]$ and sends $x_1 = r^2 \pmod n$ to *MSC*.
4. *MSC* sends a random binary vector $(e_1, e_2 \ldots e_k)$ to *CSS*.
5. *CSS* sends to *MSC*:

$$y_1' = r \cdot \prod_{e_{ij}=1} s_j \pmod n$$

6. *MSC* checks that

$$y_1'^2 \equiv x_1 \cdot \prod_{e_{ij}=1} v_j \pmod n$$

The verifier *MSC* accepts *CSS*'s proof of identity only if all the t checks are successful. For a fixed k and arbitrary t, this is zero knowledge proof. Note that

$$y_1'^2 \equiv (r_i \cdot \prod_{e_{ij}=1} s_j)^2 = r^2 \cdot \prod_{e_{ij}=1} s_j \equiv x_1 \cdot \prod_{e_{ij}=1} v_j \pmod n$$

5.1.2 Authentication outside home MSC: For authentication of a CSS which is outside of its home MSC, we outline both a centralized and decentralized authenticating scheme. Here, for brevity we denote the subscriber's home MSC as MSC-H, and the MSC in the roaming area as MSC-V.

Centralized authentication - First, the CSS sends the customer profile from its smart card to the MSC-V. Once MSC-V checks its HLR and finds that the CSS is not its customer, it will proceed to contact the MSC-H of the CSS. The MSC-V is unable to authenticate the roaming CSS as it would not possess the one way function f of the MSC-H. MSC-V could then proceed to send the authentication request to the MSC-H. At this point, the interactive transmission is between the roaming CSS, and its home MSC-H. Once the MSC-H has verified the identity of the CSS, the customer profile information is transferred from the HLR of the MSC-H to the VLR of the MSC-V. The simplified authentication procedure outlined above could be used, as no security or key exchange is required.

Decentralized authentication - This scheme is similar to the above operation except that each MSC would have to have a datafile of the one way function f_i and the modulus n_i of *MSC-V*$_i$, for all i on the network, stored in its *VLR*. This would enable any *MSC* to check the validity of the customer profile I_i as certified by *MSC-H*$_i$. The major drawback is the problem of keeping track of the changes in the network (lost or stolen cards). Unlike in the centralized scheme where the *MSC-H* is aware of the current status, the *VLR* of the *MSC-V* may not be current. To overcome this, the *VLR-R* needs to update its datafile periodically.

5.2 Security

Security within home MSC - When security between two *CSSs* in the same *MSC* is desired, for example the pit of a trading floor, our protocol offers a scheme which authenticates and provides a session key between the two *CSSs*. The *MSC-H* first authenticates each *CSS* with the simpler authentication suggested above and establishes a connection. Once communication between the two *CSSs* is achieved, they can execute our key exchange by authentication protocol. This is possible as the one way function f in the protocol is a common function within the *MSC*.

Security outside the home MSC - When security between two *CSSs* of different *MSCs* are desired, a modification in our protocol is necessary. Since the one way function f and the modulus n of the *CSSs* are different, they would have to obtain the function and the modulus from the *VLR* of their home *MSC-H*. Depending on the memory and computational power available at the *CSS*, variations of this scheme may be possible.

6. PERFORMANCE

Our scheme, when executed to provide authentication and key exchange, retains the computational complexity of Bauspieß, but *reduces* the number of messages exchanged between the prover A and the verifier B. Here, by message exchange we mean one way transmission between A and B. This is achieved at the cost of doubling the message length L (which is bounded by the number of bits in the modulus n). But as stated earlier, in radio networks and satellite links, where the propagation delay is high, the time required for message transmission is much smaller. Additionally, in very high speed digital networks, the ratio of the message transmission time to the propagation time is typically very small. For example take two nodes A and B separated by a distance of 1 mile, in a 10 Gbps LAN/MAN. For an L of 500 bits, the message transmission time is in the order of 50 nano-seconds, whereas the propagation time is in the order of 5 micro-seconds, making the transmission time insignificant.

In situations where only the authentication service is required, (or when the encryption key is not requested) our scheme can be easily modified to provide this service with the same computational complexity of Fiat-Shamir. In the scheme suggested by Bauspieß, the computational complexity is much higher as it involves modular exponentiation, which takes time. Our scheme can easily be down-graded to provide authentication service only (as opposed to authentication and key exchange for encryption), without any major modifications. This makes it possible for our scheme to provide the encryption service on demand, without expending computational time when encryption is not requested.

A comparison of message lengths, the number of message exchanges, and the computational complexity in terms of the number of modular exponentiation (exp), and multiplications (mul) between both schemes are shown in Table 1 and Table 2.

In Table 1, we see that when authentication service alone is requested, the computational complexity of our scheme equals that of the Fiat-Shamir scheme. This is possible because of the ability of our scheme to fall back to a modified version of the Fiat-Shamir scheme. Here, for the same number of message exchanges and message lengths, our scheme provides a much smaller computational time. In Table 2, we see that when authentication service and key exchange service are both needed, our scheme retains the same computational complexity as the Bauspieß scheme. The trade off here is the number of messages exchanged versus the length of the message L. In a two way protocol, our scheme requires a message of length $2L$ to be transmitted *six times*, whereas a message of length L is transmitted *eight times* in Bauspieß. The final trade off now becomes the propagation time for the message exchanged versus transmission time for message of length L. In networks where the propagation times are much larger than the transmission times, our scheme can reduce the total time taken to complete authentication and generate the session key for encryption.

7. CONCLUSION

We have proposed a cryptographic scheme that is a combination of a *zero-knowledge authentication scheme* and a *public key exchange protocol*. It is noteworthy that due to the combination, both protocols gain additional security against attacks that would otherwise be successful. This distributed authentication scheme does not require that each node maintain a datafile of public keys and provides a common authenticated session key at the end of the authentication procedure.

Table 1. Computational Complexity for Authentication Only

Authentication Only			
	Ours	Bauspieß	Fiat-Shamir
Computations			
To be verified	2 mul (mod n)	1 exp (mod n) 1 mul	2 mul (mod n)
To verify	2 mul	3 exp 3 mul	2 mul
2 way verification	4 mul	4 exp, 4 mul	4 mul
Message Length	L	L	L
No. of Exchanges	3 X 2	3 X 2	3 X 2

Table 2. Computational Complexity for Authentication and Key Exchange

Authentication and Key Exchange		
	Ours	Bauspieß
Computations	5 exp, 5 mul	5 exp, 4 mul
Message Length	2L	L
No. of Exchanges	3 x 2	4 x 2

Our scheme is very general and is applicable to any digital network, and preliminary scheme that can be implemented within the framework of the proposed standard for cellular networks EIA/TIA-IS-41 [3] was examined. The protocols makes the network center free from key management problems and enables two CSSs to authenticate each other and obtain a common key.

References

[1] F. Bauspieß and H. Knobloch, "How to Keep Authenticity Alive in a Computer Network," Proc. EUROCRYPT'89, Lecture Notes in Computer Science, Springer Verlag, 1989, pp. 38-46.

[2] W. Diffie and M. E. Hellman, "New Directions in Cryptography," IEEE Trans. Info. Theory, Vol. IT-22, No.6, Nov. 1976, pp.644-654

[3] Proposed EIA/TIA IS41 Standard, Document PN2078, July 1991.

[4] G. Coomaraswamy, "Security Issues in Broadcast Fiber Optic and Digital Radio Networks," Ph.D. Dissertation, Northwestern University, Evanston, Dec. 1992.

[5] R. M. Needham and M. D. Schroeder, "Using Encryption for Authentication in Large Networks of

Computers," Comm. ACM., Dec. 1978, Vol. 21, No.2, pp.993-999.

[6] D. E. Denning and G. M. Sacco, "Time stamps in Key Distribution Protocols," Comm. ACM, Vol. 24 No. 8, Aug 1981, pp. 533-536.

[7] R. K. Bauer, T. A. Berson, and R. J. Feiertag, "Event Markers in Distribution Protocols," ACM Trans. Comp. Syst, Vol.1, No. 3, August 1983, pp. 249 -255.

[8] R. L. Rivest, A. Shamir, and L. Adleman, "A Method for Obtaining Digital Signatures and Public Key Cryptosystems," Comm. ACM., Feb. 1978, Vol. 21, No. 2, pp. 120 - 125.

[9] A. Shamir, "Identity Based Cryptosystems and Signature Schemes" Proc. CRYPTO '84, Lecture Notes in Computer Science, no.196, Springer Verlag, 1985, pp.47 -53.

[10] A. Fiat and A. Shamir, "How to prove yourself: Practical solutions to identification and signature problems," Proc. CRYPTO '86, Lecture Notes in Computer Science, Springer Verlag, 1985, pp.186 - 194.

[11] S. Goldwasser, S. Micali and C. Rackoff, "The knowledge complexity of interactive proof systems," SIAM J. Comput. Vol. 18, no.1, pp. 186-208, Feb 1989.

[12] H. Knobloch, "A Smart card Implementation of the Fiat-Shamir Identification Scheme," Proc. CRYPTO '87, Lecture Notes in Computer Science, Springer Verlag, 1986.

[13] T. Elgamal, "A Public Key Cryptosystem and a Signature Scheme Based on Discrete Logarithms," IEEE trans. Info. Theory, Vol. IT-31, July 1985, pp.469-472.

[14] T. Beth, "Efficient Zero Knowledge Identification Scheme for Smart Cards," Proc. CRYPTO '87, Lecture Notes in Computer Science, Springer Verlag, 1986.

About the Authors

Srikanta P. R Kumar received his Ph.D degree from Yale University, New Haven, Connecticut in 1981 in engineering and applied science; and B.E and M.E degrees in electrical engineering from Indian Institute of Science in 1974 and 1976 respectively. He joined the electrical engineering and computer science department of Northwestern University, Evanston, Illinois, in 1985, and is currenlty an associate professor there. During 1982-85, he was an assistant profrssor in electrical engineering and computer science at Reneselaer Polytechnic Institute, Troy, New York; during 1981-82, he was on the faculty of State University of New York, Buffalo, New York; and during summer of 1983, he was a research fellow at University of California at Berkeley. His research interests and publications span various aspects of communication networks including wireless networks and PCS, security, failsafe protocols, and performance modelling.

Gnanesh Coomaraswamy received his B.S degree in electronics and telecommunication engineering from the University of Moratuwa, Sri Lanka in 1983.; and his M.S and Ph.D in electrical engineering from Northwestern University in 1988 and 1992, respectively. During the summer of 1988, he was a research intern at INTELSAT, Washington DC, where he studied error burst characteristics of satellite links. His areas of research are network security, architecture and performance. His recent publications include security in fiber-optic LAN/MAN systems, and cellular/wireless authentication and privacy.

SECURITY ARCHITECTURE IN THE UMTS NETWORK

A. Barba and J.L. Melús

ETSETB, Polytechnic University of Catalonia
c/Gran Capitan s/n, Module C-3 Campus Nord 08071 Barcelona, Spain
email: telabm@mat.upc.es.

ABSTRACT

The Universal Mobile Telecommunication System (UMTS) network is being developped within the European Commission's Research on Advanced Communications in Europe (RACE) for giving telephonic mobile support in Europe and the world in the 2000s. A description of the UMTS and the Global System for Mobile Communications (GSM) network with respect to the security aspects is provided. Several aspects related with call handling, location management and databases are specially studied.*

INTRODUCTION

The European mobile communications networks of the second generation, such as GSM or the Digital European Cordless Telecommunications (DECT), extensively and which will be used for the nineties have been studied. A effort to join these networks, and to provide universal access, leads to the third generation of mobile stations. In the last few years, at least four groups have been developing network architectures of third generation mobile networks. Task Group 8/1 of the International Consultative Committee on Radio (CCIR) (with the FPLMTS network), U.S. Bell Communications Research (Bellcore), Rutgers University Wireless Information Network Laboratory (WINLAB) [1] and, within the European Commission's Research on Advanced Communications in Europe (RACE) program, the Universal Mobile Telecommunication System (UMTS) [2].

The CCIR Task Group 8/1, with the development of the Future Public Land Mobile Telecommunications Systems (FPLMTS) has elaborated a functional model similar to the GSM model. Within the UMTS network requirements, compatibility with the FPLMTS is contained and an interface with Universal Personal Telecommunications (UPT) and BISDN provided.

The purpose of this paper is to trace the security architecture of the UMTS network in relation to the GSM network.

UMTS AND GSM NETWORKS

UMTS network description

The UMTS architecture is based on the idea of universality of the connections between all type of users in any place and at any time [3]. For providing these objectives, we have to study both the viability and the security requirements together.

In addition, UMTS offers personal mobility through the UPT interface. Its use is possible with the help of subscriber's identity devices (SIDs), which enable UMTS users to use different terminals within the UMTS network.

The compatibility of the UMTS with the fixed network will require us to provide some intelligent functions. In this way UMTS will contain an Intelligent Network (IN) with the signalling protocol supporting IN services between Mobile Control Nodes (MCNs), Information Storage Node (ISNs),

---

* This work has been supported in part by the Spanish Research Council (CICYT) projects: PRONTIC - INO399, TIC 1323-92/PB, TIC 90-0718 and TIC 92-1180.

management centers and the switching nodes as Local Exchange (LE) or Transit Exchange (TX) which will receive the service requests of the users. The IN will provide a flexible structure, distributed intelligence, sophisticated signalling, and the ability to deploy new services quickly.

Other main characteristics of the UMTS networks is related with the network environments. The Customer Premises Networks (CPNs) associated with each one of them will provide these differences. The Mobile Customer Premises Network (MCPN) representing a mobile PABX which can be either public or private equipment, there are also networks like Business Customer Premises Network (BCPN) and Domestic Customer Premises Network (DCPN) (see Fig. 1).

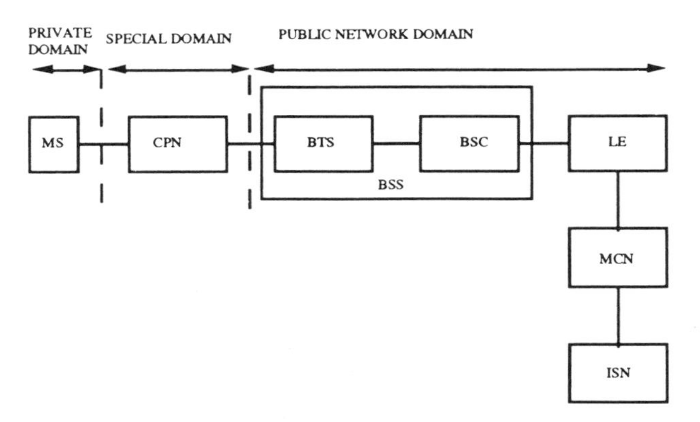

Figure 1. UMTS Architecture.

GSM network description

The GSM structure consists of three parts [4]: the Base Station Sub-system (BSS), the Network Subsystem (NSS) and the Network Management Subsystem (see Fig. 2).

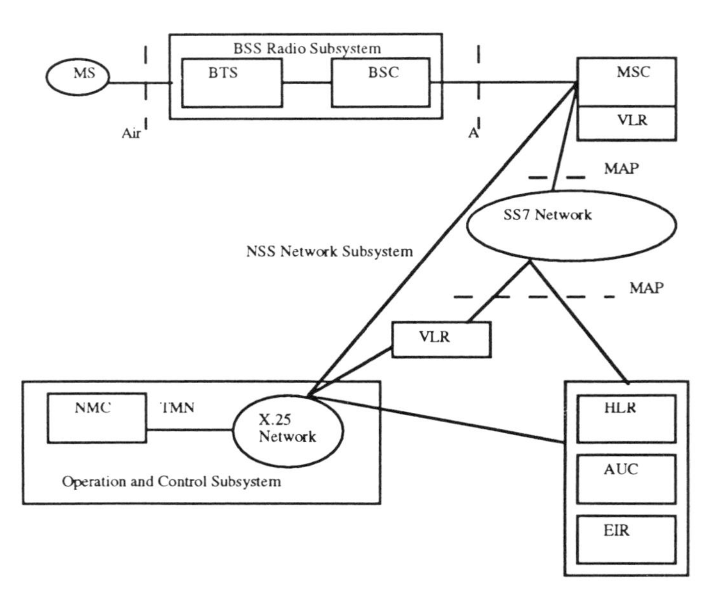

Figure 2. GSM Architecture.

The communication area is covered by cells, each with a base station (BTS). The Location areas (LA) are built from several of these cells. Every mobile station is registered in a LA depending on its position in a way that, when the mobile is attached, it must consult its LA and compare its parameters with the information saved in it. If this information is the same, start transmission and, if not, the system must be updated with the information concerning to the new location. Because of this, the mobile station calls and provides the IMSI (International Mobile Subscriber Identification) and the identity of the first LA to the new base station. The Base Station Controller (BSC) transmits this information to the Visitor Location Register (VLR). This procedure is longer if every Location area (LA) depends on a different VLR. In this case it will be necessary to change these data in the conformable Home Location Register (HLR).

ARCHITECTURES IN THE UMTS AND GSM NETWORK

Network functional components of the UMTS network

In this section the most relevant components of the UMTS network are explained.

• **Terminal Equipment (TE):** Terminal Equipment handles communications at the user - side of the user / network interface. Examples of terminal equipment include data terminals, telephones, personal computers, and digital telephones. TE devices provide protocol handling, maintenance functions, interface functions, and connection functions to other equipment. A TE is also appropriate for a UMTS network.

• **Mobile Terminal Equipment (TEm):** The TEm is the mobile terminal for UMTS network. It provides the functions for handling the air radiolink. From the point of view of security, it is defined a UMTS Subscription Identity Device (SID). In this case, the security services could be a Subscription Device or a smart card, or both, depending on what is more practical. For the services mentioned above, security measures have to be studied (such as the use of a PIN together with a secret/public Personal Subscription Identity in a smart card (SID) and so on. A SID is not only possible for a TEm, but might also be possible for a TE.

• **Network Termination 1 (NT1):** NT1 devices provide functions equivalent to layer 1 of the OSI model. These functions include signal conversion, timing, maintenance of the physical transmission line, and the physical and electrical termination of the network at the user end.

• **Network Termination 2 (NT2):** NT2 devices are more intelligent than NT1's and provide additional functions which may include multiplexing and protocol handling at the Data Link and Network levels (layers 2 and 3) of the OSI model.

• **Base Station. Base Station Subsystem (BSS):** It comprises both the transceiver and control functions in the base station configuration and provides an interactive communication between mobile terminals and the network using the cellular concept to increase the network capacity.

- **Base Transceiver Station (BTS).** This represents the physical transmissionequipment at the network base station site, and it comprises the functionality, the user signalling and the network to radio link interface. It will contain the transmitter and receiver equipment to control the radio interface in a cell. It could perform measurements on transmission quality.

- **Base Station Controller (BSC).** This represents the control functions between the network and the base stations subsystem, and it adapts the radio infrastructure to the U interface communication protocols.

• **Line Termination (LT):** These devices performs the line termination functions at the exchange end of the transmission line in the LE. (see Fig. 6).

• **Local Exchange (LE):** The local exchange contains the functionality to interact with the MCN (Mobile Control Node) in order to support the provision of mobile services.

• **Transit Exchange (TX):** Representing switching and compression functions related to the transit connections in the network and all control functionality already present in the IBCN TX.

• **Mobile Control Node (MCN):** It comprises specific mobile control functions, that are not present in the fixed network blocks. It contains functions in order to guide the LE and the radio network in the support of mobility features and functions to interact with the ISN to derive the information from the distributed databases and to send updated operational information coming from the radio infrastructure to the nodes ISN.

• **Information Storage Node (ISN):** These represent the possibility that there might be separate nodes in the network containing only data. It must be noted that data will also be stored in other network nodes, like for instance the MCN. The information can be in relation with UMTS network management and/or about directory (ISNd) or user data (ISNs) of the distributed databases. The information contained in these databases could be used for providing functions about reliability, availability, performance measurements, security, costing and billing.

• **Interworking Component:** These provide the connection with other networks such as N-ISDN, GSM, DECT and so on or interface with UPT or B-ISDN.

• **Management Center:** This center provides support to the tasks associated with network management. In order to be automatic, effective and efficient, this system needs to be intelligent.

Network functional components of the GSM network

In this part the most relevant components of the GSM network [5] are explained.

• **Mobile Switching Centre (MSC):** It is the interface between the mobile suscribers and the fixed network. It provides the exchange and control functions in the calls.

• **Base Station Controller (BSC):** Radiocontrol of the base station such as handover or power supply control.

• **Base Transceiver Station (BTS):** Handling of the links in the physical level to the mobile stations.

• **Home Location Register (HLR):** Database which contain the static or semistatic subscriber information, such as the present position of the mobiles.

• **Visitor Location Register (VLR):** Databases which temporarily contains the information about the location of the mobile stations in the network.

• **Network Management Center (NMC):** It provides network management functions.

• **Operation and Maintenance Center (OMC):** It provides operation and maintenance functions to the network.

Analysis of the architecture in both networks

The GSM network provides a centralized database architecture. It consists of a very simple structure, HLR-VLR (master-slave) with a lot of constraints. The number of HLRs is unique (one for each country), or very few, which has a negative impact on the international calls or the roaming of the mobile stations in the zones of different HLR, due to the great flows of signalling information over long distances [3], [6].

The distributed databases in the UMTS network are structured as a collection of distributed ISNs in the network. To determine the location of the information, there exist directory nodes (ISN_{Dn}, ISN_{DN}) which contain references to the data nodes (ISN_{S}) where the information is saved. In the case of the UMTS network, the use of distributed databases allows an improvement in the performance, reducing the signalling, improving fault tolerance, and giving more flexibility and availability in the network.

The use of exchange and control systems separated from the data is similar in both networks. In this case, the GSM network have the MSCs and the Location Registers while in the UMTS network exists the LE and MCNs in one side and several types of ISN for the data storage. There also exists a clear separation between the specific mobility functions and the functions performed by the standard fixed network.

The present concept of Intelligent Networks (IN) does not support distributed databases in spite of the current study for providing distributed SCPs. Furthermore, the IN is concentrated in the fixed network and it doesn't provide mechanisms for the activation of mobile entities or environments like CPNs.

FUNCTIONALITY OF THE UMTS AND GSM NETWORK: APPLICATION TO SECURITY

In this chapter the possible impact in the UMTS functions on the specification and requirement for security services will be identified. The most important requirement is that the provision of security must be maintained under all circumstances, especially during and after state transitions in the UMTS (e.g. during and after handover) [3].

UMTS Functions. Aspects of security services

Security services can be divided into two classes. One class has a 'discrete time' aspect, which means that the service is active on a certain instant (e.g. authentication and access control). The other class has a 'continuous time' aspect, which means that the service is active during a period of time (e.g. confidentiality and integrity). For discrete time services an aspect that has to be taken into account are the circumstances under which the result of the service remains valid. Aspects related with 'continuous time' services are due to the fact that in UMTS the point of attachment to the network (BS) may change during the call, and may even belong to another network operator. Other aspects that might influence security service specification would include the existence of several operator domains that have to interact to provide one instance of service provision.

Handover. The handover assures continuity of calls and it uses authentication to provide trustworthy entities in the communication. These entities will provide the confidentiality and integrity to the messages later.

Handover of a call to a new connection implies that the security functions need to be transferred to the new connection as well. Although security and handover are closely related, handover does not necessarily means that security functions are transferred immediately to the new connection. In this document handover of the connection (connection handover) and transfer of the security functions (crypto handover) are distinguished. Connection handover and crypto handover can occur simultaneously but it is also possible that they occur one after another or even that only connection handover takes place. From a security point of view there are three types of handover. The difference between these types becomes clear through analysis of the following properties for the various handover cases: The level at which handover takes place relative to the level at which the security functions are located and the timing relation between connection handover and crypto handover.

The types would be: Only connection handover, both, connection and crypto handover (crypto handover before connection handover and crypto handover possible after connection handover).

Different inter - cell handover situations can be identified. Furthermore, the parties involved can be different.

a) Between base stations controlled by the same BSC
b) Between BSC's controlled by the same MCN (or private MCN) and connected to the same LE or CEX.
c) Between BSC's controlled by the same MCN (or private MCN) but connected to different LEs or CEXs.
d) Between BSC's controlled by different MCN´s in one administrative environment.
e) Between administrative environments.

Location. It is used to locate terminals in the network and route calls to them. Location registration, which is due to the detection or selection of a new location area is called location updating since it implies a database update concerning the change of location area. In contrast we discriminate the "simple" location registration when a mobile terminal reports its current location, which does not necessarily indicate a change of location area. This discrimination is based on whether previous location information is available to the mobile terminal or not. Authentication is required before location registration. The UMTS location registration/updating is needed when:

- The mobile terminal is switched on for the first time
- Location updates between two location areas covered by one single ISNs
- Location updates between two location areas covered by two distinct ISNs, both belonging to one single network.
- Location updates between two location areas covered by two ISNs belonging to two distinct networks.
- Some abnormal cases occurr (loss of power, loss of radio coverage...)

Paging. It is used to page terminals in the location cell. Authentication would be applied between a network entity such as BSC and the terminal in the initial stage of the location management in the call handling.

Attach/Detach. It is used to avoid call setup attempts to inactive terminals. According to the different security policies of the network operator, the authentication services can be implemented /used or not.

GSM Functions. Aspects of security services

The calls in the GSM network are routed by the dialled subscriber number (MS-ISDN) towards the HLR. The HLR contains the address information of the VLR closest to the receiver. Finally, the VLR contains the receiver's LA.

In general, the GSM network has functionality similar to the UMTS network. However, it has lower performances and requirements related with the architecture and services which it supports. For instance, in the handover used in GSM network, the mobile doesn't leave the old connection to the base station until the network has not routed the call to the new base station. This is a secure but slow procedure. However, the subscriber identifier is not protected in all the cases.

Comparative analysis of the functionality in both networks: Aspects of security

As compared to the GSM network, UMTS provides the following characteristics: There exists a clear separation between the call control and the connection control, which permits it to organise better the network resources for providing the required teleservices for the user and establish the connections of the bearers by means of the exchange nodes and links control. In addition, it provides other advantages such as efficient and flexible support in multimedia and multiparty services or a better establishment of the call parameters. It is also more efficient in the setup because in UMTS a search for the called party is performed in the databases before the call establishment. The charging analysis is better in the UMTS network (it can be more complicated) because in the GSM network is based on the dialled number MS-ISDN analysis while in the UMTS network it is performed during the call or at the end of the call handling. The strategy of the call setup in the UMTS network is very different in relation to the GSM network. In GSM, for providing an outgoing call (see Fig. 3), we would have the following sequence:

1) MS (PSTN/ISDN) -> MSC1: The gateway receives an incoming call from the ISDN.
2) MSC1 -> HLR: The MSC1 interrogates the HLR.
3) HLR -> VLR: The HLR sends a query to the VLR in which the mobile station is roaming.
4)VLR -> HLR -> MSC1: The VLR provides information about routing.
5) MSC1 -> MSC2: Interrogation.
6)MSC2 -> VLR: The VLR is interrogated about the subscriber data.
7) VLR -> MSC2: The VLR answers to the MSC2.
8)MSC2 -> BSs -> MS: Paging broadcast to the mobile stations.
9) MS -> BS -> MSC2: Authentication and encryption.
10) MSC2-> BS -> MS: The call is setup to the mobile station.

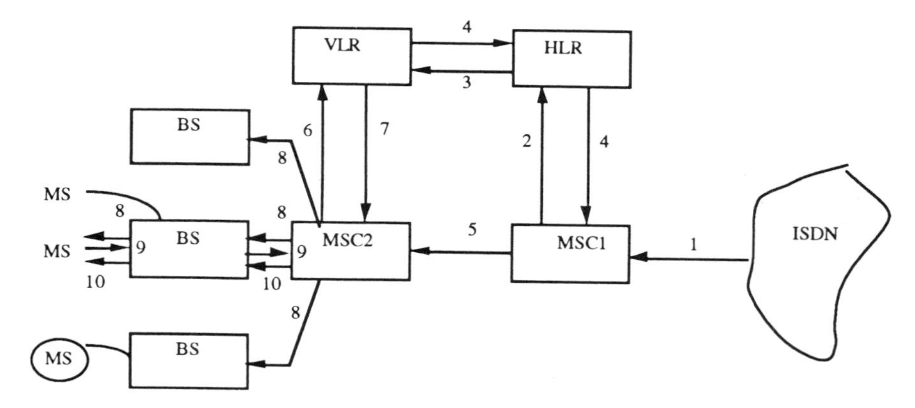

Figure 3. Outgoing call setup in the GSM network.

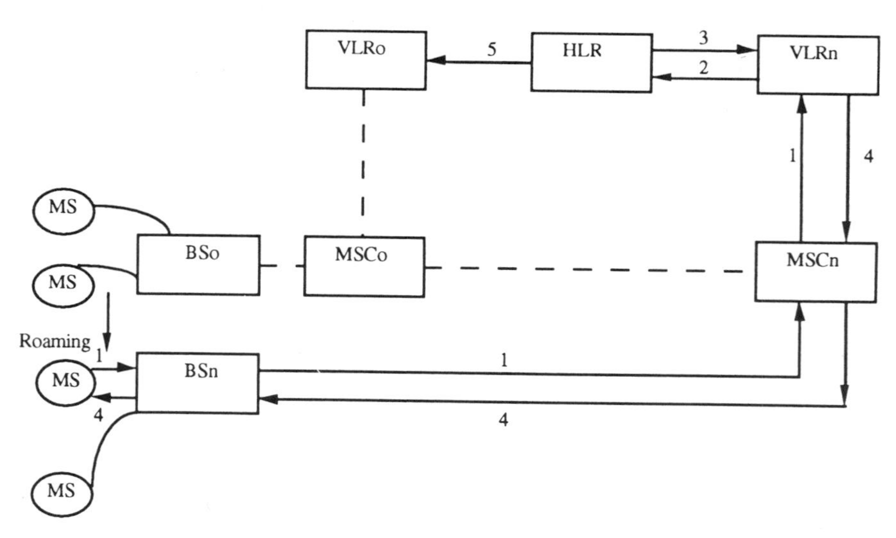

Figure 4. Location Updating in the GSM network.

The call setup in UMTS has the following parts: Interrogation, Paging, Signalling connection and service set-up, bearer provision and call completion. Still in study phase is the sequence of procedures to establish a call. Three cases are considered:

a) Alternating locating and call control.
b) Sequential locating and call control and
c) Integrated locating and call control.

The mechanism of Interrogation still requires important studies for an optimum functionning. Interrogation is needed to find the current location area of the called mobile terminal. Base stations are continuously transmitting environment information to the mobile terminals such as the identifiers of the Location area, base station, and so on. Based on this information, a mobile terminal can decide to do a Location Updating for providing the current location in the network. The sequence of the Location Updating in GSM would be (see Fig. 4):

1) MS -> BS -> MSCn -> VLRn: Location Updating request.
2) VLRn -> HLR: The message is sent to the HLR.
3) HLR -> VLRn: The HLR answers.
4) VLRn -> MSCn -> BS -> MS: Acknowledgement of the Location Updating.
5) In addition, the HLR would delete from the VLRo the old data.

The sequence of Location Updating in UMTS would be based in the distributed databases structure. In the case of a Location Updating between two different networks we would have (see Fig. 5):
1-3) TEm -> BS -> MCN1-> ISNs1: The new ISNs is able to find the ISN_{DN} of the old network using the old LAI (Location Area Identifier).
4) ISNs1 -> $ISN_{DN}0$: The ISN_{DN} is able to find the old ISNs using the old LAI.
5) $ISN_{DN}0$-> ISNs0: The old identity identifier and new ISNs are sent.
6) ISNs0 -> ISNs1: All information related to terminal and users registered on the terminal.
7-8) ISNs1 establish references about idetifiers in the top node in the new network and deletes all the information from the old nodes.
9-11)ISNs1 -> MCN -> BS -> TEm: The database finishes the Location Updating.

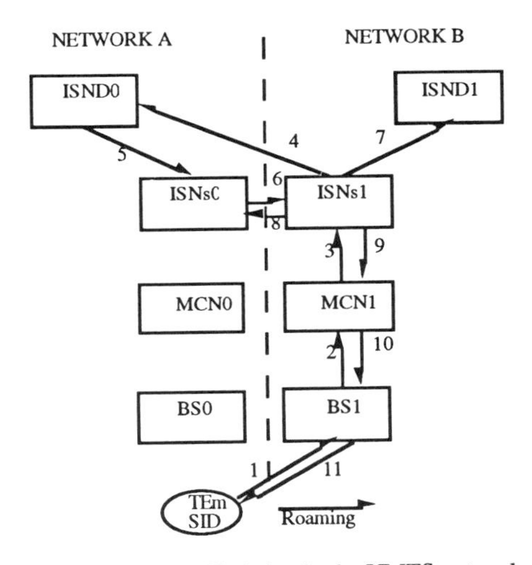

Figure 5. Location Updating in the UMTS network.

The handover used in UMTS permits the call control, supplementary services control and the charging control in the new link because the mobile station changes the connection point before the call is routed to the new connection point (forward handover). In this case, the new base station starts the change.

The handover used in the GSM network is slower because the mobile doesn't leave the old connection to the base station until the network has routed the call to the new base station. Nevertheless, the forward handover has two main problems. In case of a fault, the information is recovered with more difficulty and it is more difficult to have encryption continuity because the keys have to be transferred to the new BSC.

The UMTS network performs Location Updating many times and performs less pagings because the location information about users and mobile stations is found in the network.

Aspects about the IN in the GSM network as compared to the UMTS network: In the current standard of IN (INCS-1) the separation between call and connection is not performed and, as consequence, the Location Updating (LU) and Handover (H) have implementation problems. Due to the fact of the LU and H are not activated for the calls doesn't permit the interaction with the Service Switching Point (SSP) and in consequence are not processed by the IN. IN only supports one transaction at a time in the service logic while the UMTS functionality can be activate on parallel processes. In the IN, the possibility of activating a function repeatedly is not possible, because this affects the required simultaneity between H and LU. In general, the IN requires further development to provide correct functionality for the UMTS network.

Security services description in the UMTS network

The Security Services are described into five service classes: authentication, integrity, confidentiality, access control and incontestable charging; this is the subdivision generally accepted, and it is also useful for our purposes.

• **Confidentiality**. The purpose of confidentiality is to protect information from (deliberate or accidental) disclosure to an entity not authorized to have knowledge of that information. These services provide confidentiality of transferred user data, and stored/transferred signalling and management data in the air -interface and over the wired lines. Examples of signalling related data to be protected are: Identity (user/terminal), Location (user/terminal), Status (user/terminal), other information (user/terminal/network operator) and charging, security, other sensitive signalling data.

• **Integrity.** The maintenance of data's value is called integrity. It includes the implementation of the Integrity of transferred user data, and stored/transferred signalling and management data in the air -interface and over the wired lines.

• **Authentication.** Service used to establish the validity of a claimed identity. It includes the implementation of the following services. It can be provided by authentication of the person by the Security Identity Function (SIF) of the mobile equipment, network or service provider. The SIF can be located in the SID, TE, mobile base station and so on. Also, it could provide mutual authentication between the mobile equipment and the network/service provider.

• **Access control.** The prevention of unauthorized use of a resource, including the prevention of use of a resource in an unauthorized manner. It would be desirable to have access control in services, system databases, subnetworks and equipment.

• **Incontestable charging.** This service prevents disputes about charges by one of the entities involved in a communication which has participated in all or part of the communication. It would be desirable to introduce the Incontestable Charging between Suscriber, Network Operator and/or Service Provider

Security services description in the GSM network

In the GSM network the following Security Services are considered [7]:

• **Subscriber identity (IMSI) confidentiality.** The IMSI should not be made available or disclosed to unauthorized individuals, entities or processes.

• **Subscriber identity (IMSI) authentication.** Corroboration by the land-based part of the system that the subscriber identity, transferred by the mobile subscriber within the identification procedure at the radiopath, is the one claimed.

• **User data confidentiality on physical connections.** The user information exchanged on traffic channels should not be made available or disclosed to unauthorized individuals, entities or processes.

• **Connectionless user data confidentiality.** The user information which is transferred in a connectionless packet mode over a signalling channel should not be made available or disclosed to unauthorized individuals, entities or processes.

• **Signalling information element confidentiality.** A given piece of signalling information which is exchanged between mobile stations and base stations should not be made available or disclosed to unauthorized individuals, entities or processes.

Comparison between the Security Services provided

Within the service classes, the GSM network doesn't provide the mechanisms of digital signature and doesn't support Incontestable charging. Furthermore it doesn't develop services such as access control or information integrity.

The use of secret key algorithms in GSM and the probable use of public key algorithms (in study phase) in UMTS, constrain the distribution and key management in both networks. The public key gives a better solution for providing security services like authentication. For instance, mutual authentication is easier to use in the UMTS network by means of public keys.

Special components for Security purposes in the UMTS network

In the previous sections, the architecture, elements and other characteristics about UMTS have been analyzed. In this section some special and necessary components for assuring the overall system security in UMTS network will be explained. These elements are the Subscription Identity Device (SID), Authentication Center, Security Center and the Notarization Center.

• **Subscription Identity Device (SID):** The Subscriber Identity Device (SID) is the device which contains data that is very sensitive (e. g. user id, some keys) or which is always necessary (e.g. public key of home network or authentication center, if appropriate). The SID must be able to handle a PIN as requirement of user authentication by the terminal, the public keys or the encipherment secret keys (can be given by a smart card) and provides security functions to UMTS services. The SID would have the logic functions required by the cards: read, save, delete information, etc.

• **Security Modules (SM):** Several components in the UMTS network can allocate security features related with the security services implemented by security modules.

• **Authentication Center (AC):** Authentication center management may involve distribution of descriptive information, passwords or keys (using key management) to entities required to perform authentication. It would have functionality in relation to key management, for example: The generation of keys, back up, distribution, installation, update and destruction of keys. In the authentication center will be the authentication algorithm and the cipher key generating algorithm. It could contain the secret individual authentication keys of all subscribers of the UMTS.

• **Security Center (SC):** The most important tasks developed in the security environment would be [8]: The system security management, the security service management, and the security mechanism management. Security services include the management functions for the AC, NC and MC in the UMTS. It helps to provide more complex security functions to the network such as certificates, security labels, event handling management, security audit management and security recovery management.

• **Notarization Center (NC):** This component is used on the concept of a trustworthy third entity to assure certain properties about information exchanged between the UMTS network / Network Operator and the Mobile Terminal, such as its origin, its integrity, or the time it was sent or received or the time elapsed in a call. Also, it is used in relation to Non - Repudiation / Incontestable Charging security services.

Special components for Security purposes in the GSM network

According to [9], the mobile station must contain the security functions for providing Identity Subscriber Authentication, that means that a secret key for the authentication and a cryptographic algorithm have been allocated in the Subscriber Identity Module (SIM).

• **Subscriber Identity Module (SIM):** The SIM is a connectable module of the mobile station that can be implementable by means of a smart card [9]. It supports the following characteristics:
- Storage of subscriber related security information (e.g. IMSI, keys) of Rec. GSM 02.09 [7], and implementation of authentication and cipher key generation mechanisms of Rec. GSM 03.20.
- User PIN operation (if a PIN is required) and management.
- Management of mobile subscriber related information.

Other centers of the GSM network related with the security are the AUC, HLR and the EIR.

• **Authentication Center (AUC):** This Center provides the Authentication parameters in coordination with the HLR for the network security.

• **Equipment Identification Register (EIR):** It is the database of the mobile station for performing the equipment identification. It also contains the non authorized users or stolen equipment lists and it can be allocated in other network databases like HLR or security components.

Comparison between the Security Architectures in both networks

The AUC of the GSM network contains the secret keys of the subscribers and the generation algorithms of the secret keys while the AC of the UMTS network (assuming the use of public key algorithms) would incorpore the public keys that doesn't need additional protection, the secret key that would be in the SIM or smart card, and the secret key of the Location area (LA) used by the MCN. In fact, the structure of databases in the GSM network is more centralized to protect better the security and management entities: EIR, AUC, HLR. The distributed character of the information in the ISNs (UMTS) permits further redundancy and faults tolerance than in the case of the GSM structure, which is more centralized in a few HLRs.

SIGNALLING AND MANAGEMENT PROTOCOLS

Signalling and management protocols in the UMTS network

The UMTS network uses the OSI Application layer concept (OSI ALS), specified by ISO in the draft standard DIS 9545 for signalling and management protocols. Below, we give a short overview of them (see Fig. 6).

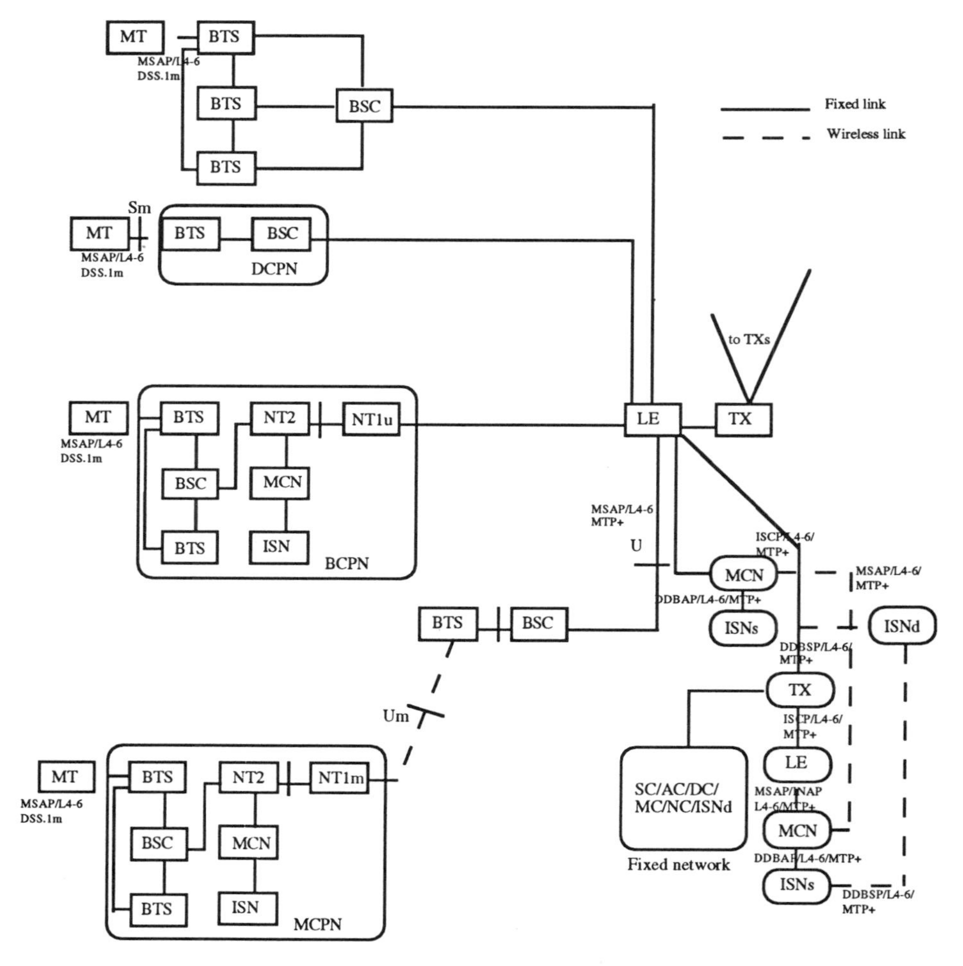

Figure 6. Signalling and Management protocols in the UMTS network.

• **Protocols at lower layers** The Subscriber access signalling uses DSS.1m, a modified version of the DSS.1 for providing mobile functionality. At the fixed network a new version of the Message Transfer Part (MTP+) of the SS7 is used, taking into account aspects of the BISDN and mobility.

• **Protocols at intermediate layers** The implementation of these layers is in study. If the layers 4-6 are used, it means a connection-oriented service, otherwise it means a connectionless service.

• **Application Layer protocols** The following protocols can be identified:
The Mobility Services Application Part (MSAP) is the future signalling protocol supporting all types of mobility; terminals, users, satellites and so on. It covers the subscriber access and the network signalling.

The IN Application Part (INAP) is the signalling protocol supporting IN services. The ISDN Signalling Control Part (ISCP) is the new specification for the ISUP. It provides connection establishment in the fixed network.

The System protocol to databases (DDBSP) connects the databases between themselves and the Access protocol to databases (DDBAP) could be the TCAP, RDA or DAP. The DDBAP and DDBSP together form the DDB Application Layer Protocol (DDBALP).

Signalling and management protocols in the GSM network

In the GSM network (see Fig. 7), the protocol which provide the signalling between the entities in the network side is the Mobile Application Part (MAP). This provides Application Service Elements (ASE) between two entities. It connects the MTP, SCCP and TCAP. It supports procedures for the location registration/cancellation and updating, handover, authentication, charging and set up. Other protocols used are the following: Call Control (CC), Mobility Management (MM), Radio resource management (RR) and BSS Management Application Part (BSSMAP). In the UMTS network there exists a clear separation between the call control and the connection control, in consequence, it requires a change in the specifications of the ISUP (GSM) towards a new protocol ISCP.

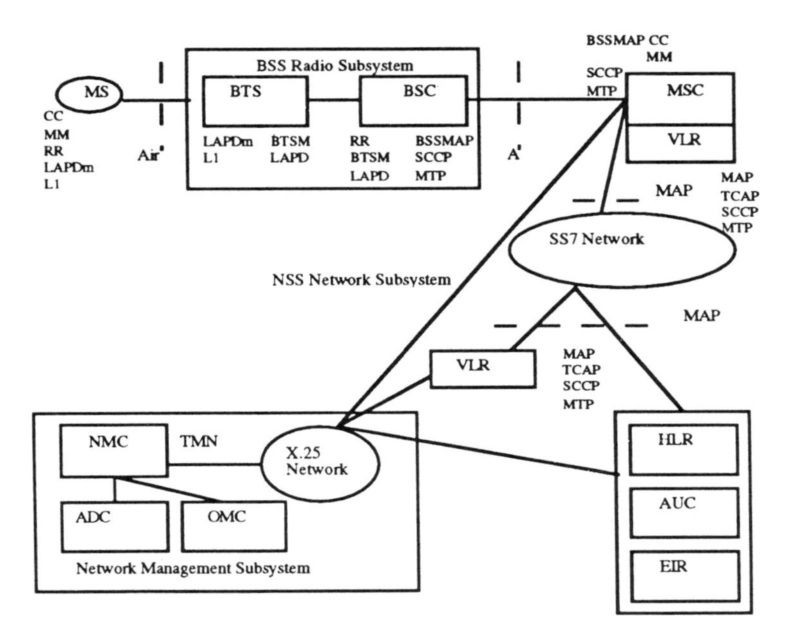

Figure 7. GSM Architecture and signalling protocols.

CONCLUSIONS

We briefly explained the architecture and functionalities related with mobility in the GSM and UMTS networks, giving comments about differences and similarities. The necessary security services, specially the necessary requirements for the Location Updating and call setup in relation to the databases, have been described in detail.

It has been shown that the UMTS network has better performance in general than the GSM network, is more complex, has more security requirements and the distributed databases improve the signalling. However, the UMTS network is not compatible with the current standard of IN and requires further study to improve the required performance.

ACKNOWLEDGEMENTS

The authors are currently working in the RACE 2066 MONET project in the workpackage MF3 related to Security Network together with ALCATEL SESA. The expressed opinions about the UMTS network are free. They would like to thank investigators of MF3 and a MONET for their collaboration.

ABOUT THE AUTHORS

Antonio Barba Martí was born in Tarragona (Spain) in 1963. He graduated from Polytechnic University of Catalonia (UPC), Barcelona, with a degree in telecommunication engineering in 1989.

In the same year, he joined the Matematica Aplicada y Telemática department in the Polytechnic University of Catalonia, where he is associate teacher in the area of Telephonic Systems and he is working towards his Ph. D. degree. He is currently engaged in research on the European Commission's Research on Advanced Communications in Europe (RACE) in the area of mobile systems (UMTS) and security. His research interests have spanned the areas of mobile systems, network management, security and protocols.

José Luis Melús Moreno received the M.S. and Ph. D. degrees in Telecommunication Engineering from the Polytechnic University of Catalonia (UPC), Barcelona, Spain, in 1980 and 1987 respectively.

In 1985 he joined the Applied Mathematics and Telematics Department of the UPC as an Assistant Professor, where he began working in modelling and simulation of multiprocessor systems. Since 1988 his research interests have spanned the areas of digital communications, packet communications, coding and cryptography. Currently he is a Professor and leads the research group on Cryptography and Data Security at the UPC.

REFERENCES

[1] David J. Goodman."Trends in Cellular and Cordless Communications", IEEE Communications Magazine, vol.29, No. 6, pp. 31-40, June 1991.

[2] H. de Boer et al. " Network aspects for the third generation mobiles", GLOBECOM '91, pp. 1517-1522.

[3] RACE 1043/RNL/FN12/DS/A/067/b1. Fixed Network Activities 1991.

[4] W. Weiss, M. Wizgall. "Sistema 900: el enfoque de RDSI para la radio móvil celular". Comunicaciones Eléctricas. Vol 63, nº4. pp. 400 - 408. 1989.

[5] M. B. Pautet and M. Mouly. " GSM protocol architecture: Radio sub-system signalling", 41st IEEE Vehicular Technology, pp. 326 - 332. 1991.

[6] M. Ballard et al. "La radio móvil celular como aplicación de redes inteligentes". Comunicaciones Eléctricas. Vol 63, nº4. pp. 389 - 399. 1989.

[7] Recommendation GSM 02.09. Security aspects. ETSI PT12.

[8] International standard ISO 7498 - 2. Security Architecture.

[9] Recommendation GSM 02.17. Subscriber identity modules, functional characterstics. ETSI PT12.

ACRONYMS

AC: Authentication Center (UMTS)
ADI: Access Domain Identifier
AUC: Authentication Center (GSM)
ASE: Application Service Elements
BCPN: Business Customer Premises Network
BSAI: Base Station Area Identifier
BSC: Base Station Controller
BSI: Base Station Identifier
BSS: Base Station Subsystem
BSSMAP: BSS Management Application Part
BTS: Base Transceiver Station
CC: Call Control
CPN: Customer Premises Network
DCPN: Domestic Customer Premises Network
DDBAP: Access Protocol to Databases
DDBALP: DDB Application Layer Protocol
DDBSP: System Protocol to Databases
DSS1: Subscriber Signalling System No.1
EIR: Equipment Identification Register
HLR o/n : Home Location Register old/new
ICSI: International Charging Subscriber Identifier
IMSI: International Mobile Subscriber Identifier
IMTI: International Mobile Terminal Identifier
IMTN: International Mobile Terminal Number
IMUN: International Mobile User Number
IMUI: International Mobile User Identifier
INAP: Intelligent Network Application Part

ISCP: ISDN Signalling Control Part
ISN: Information Storage Node
LAI: Location Area Identifier
LAIo: Previous (old) Location Area Information
LAIn: Current (new) Location Area Information
LE: Local Exchange
LT: Line Termination
MCN: Mobile Control Node
MCPN: Mobile Customer Premises Network
MM: Mobility Management
MS: Mobile Station
MSAP: Mobility Services Application Part
MSC: Mobile Services Centre
MTP: Message Transfer Part
NC: Notarization Center
NMC: Network Management Center
NT: Network Termination
PAI: Paging Area Identifier
RR: Radioresource Management
SC: Security Center
SCCP: Signalling Connection Part
SID: Subscription Identity Device
SIM: Suscriber Identification Module
TCAP: Transaction Capabilities Application Part
TE: Terminal Equipment
TEm: Mobile Terminal Equipment
TIDo: Temporary Identity of the MT in the Old Network
TIDn: Temporary Identity of the MT in the Current Network
TMSI o/n : Temporary Mobile Subscriber Identity old/new
TX: Transit Exchange
VLR o/n : Visitor Location Register old/new

SECURITY MANAGEMENT - AN OVERVIEW TO RELIABLE AUTHENTICATION PROCEDURES FOR AUTOMATIC DEBITING SYSTEMS IN RTI/IVHS ENVIRONMENTS

Rolf Hager, Peter Hermesmann*, and Michael Portz*

Institute of Computer Science IV
Aachen University of Technology
Ahornstr. 55, W-5100 Aachen, Germany
email: rolf@informatik.rwth-aachen.de

*Institute for Applied Mathematics especially Computer Science

ABSTRACT

The research and development in the area of short-range mobile radio networks have produced a new kind of applications and profiles of communication systems. With the concepts "Road Transport Informatics" (RTI) and "Intelligent Vehicle Highway Systems" (IVHS) much effort has been undertaken to develop and standardize mobile communication systems. In particular, the European projects PROMETHEUS and DRIVE and the American equivalent IVHS are promoting new technologies.

The particular application "Automatic Debiting" (AD) intends to increase driver's comfort. The driver will not have to stop and pay by hand at dedicated stations (to pay fees for highways, bridges, dedicated areas, parking sites etc.), but simply pass AD stations equipped with sophisticated communication systems and radio transceivers. Concerning the booking procedure, up to now only common security schemes have been implemented. The authors investigate specialized authentication schemes and evaluate them as dedicated part of the security management. The main result is the recommendation of the authentication scheme of Schnorr, together with initial guidelines to system designers.

INTRODUCTION

In the last five years much effort has been spent on research and development of sophisticated traffic information networks, where vehicles exchange specific data with other vehicles or with electronic beacons connected with infrastructure devices such as traffic management centers. Leading projects in this area are "Intelligent Vehicle Highway Systems" (IVHS) in the United States and the European equivalents "Programme for a European Traffic with Highest Efficiency and Unprecedented Safety" (PROMETHEUS) and "Dedicated Road Infrastructure for Vehicle Safety in Europe" (DRIVE). The prominent goals of these projects are traffic safety, efficiency, reduction of pollution and driver's comfort. To achieve these goals, dedicated applications like Route Guidance, Automatic Debiting, Intelligent Cruise Control, Emergency Warning, and Tourist Information have been designed. The basis of these applications is an integrated communication system with the capability to fulfil the typical requirements of short-range mobile radio networks (SR-MRNs), i.e. hard real time constraints, channels with high bit error rates, unpredictable load, mobility etc.

To increase the diver's comfort, the application "Automatic Debiting" (AD) has been developed. AD includes Road Pricing (payment of highways, roads, bridges), Toll Payment, Parking Fees and similar registrations. Obviously, specific security mechanisms within the communication systems are necessary. The basic functionalities of such communication systems like radio packet transmission, etc. are realised up to now.

Network Management and Control, Vol.2, Edited by
I.T. Frisch *et al.*, Plenum Press, New York, 1994

This paper focuses on specific security mechanisms covered by an integrated security management system of mobile communication systems.

AUTOMATIC DEBITING IN RTI/IVHS

Automatic payment with mobile stations assumes a variety of different components: the mobile communication systems, the fixed communication systems (electronic beacons connected with some infrastructure devices), and a facility capable of booking or crediting different amounts without losing the anonymity of the driver or vehicle. This facility can be realised by a smart card. Actually, the smart card will be identified, but this keeps the anonymity of the user and his vehicle. This principle is equal to telephone cards.

Vehicles pass dedicated beacons and communicate while driving through the communication area. The mobile communication system first announces its existence and intention to the beacon. The beacon then calculates the price and orders the corresponding booking procedure to the smart card facility. Other unequipped vehicles have to pass normal pay-stations to pay by hand. This causes time loss, traffic congestion and increased pollution.

From the communications point of view, only the layers 7 (application layer), 2 (data link layer) and 1 (physical layer) are necessary. These are now being discussed in the standardization process. On the other hand, from the security point of view, some effort has to be spend on the necessary security mechanisms within the communication system to protect the AD procedure against attacks, manipulation, denial etc. Mechanisms to protect a communication system belong to the security management functional area.

Security management is defined as the facility concerned with the protection of managed objects [8, 9, 10, 11]. It therefore establishes the requirements for security audit trails, including alarm delivery, selection analysis, event detection and journaling of these operations. Security mechanisms are Encipherment, Traffic Padding, Digital Signature, Data Integrity, Authentication Exchange, Access Control, Routing Control and Notarization.

For the study of an AD system (ADS), mainly authentication has to be considered. Authentication Exchange is a mechanism, wherein the identity of a party must be verified before access is granted to a resource. We focus in this paper on the consideration of possible candidates for this mechanism, keeping in mind the hard real time constraints of the underlying communication system caused by performance characteristics and mobility constraints (vehicles speed, communication range). Note that a frequency of 63 GHz is allocated by European standardization bodies for vehicle-to-vehicle communication and 5.8 GHz for vehicle-to-beacon communication. These bearer frequencies limit the transmission range of transceivers due to O_2 absorbtion etc. Additionally, one has to minimize the necessary communication range and therefore the transmission power of the beacons and vehicles in order to minimize interference [7].

The next paragraph defines Authentication Exchange and introduces different potential candidates.

AUTHENTICATION PROCEDURES

Security Management is subdivided into three areas: System Security Management, Security Services Management and Security Mechanisms Management. System Security Management covers all security aspects of the open communication system itself, whereas Security Services Management covers the administration of the security services needed e.g. to select suitable security mechanisms. Security Mechanisms Management covers specialised mechanisms in order to realize administration of keys, digital subscribtion, access control, authentication etc. For all these areas a well defined Security Management Information Base SMIB (logically a part of the Management Information Base, MIB) is necessary.

To protect the ADS from unauthorized use, authentication is necessary. Authentication is defined as definite proof of the identity. If a mobile station (A) wants to authenticate itself against an AD beacon (B), then A proves B its identity via the unique identification record I_A and its unique password s_A only known to A. Normally, calculation with I_A alone is not possible (record!), therefore each communication system needs a uniquely defined public key v_A to calculate the relationship between s_A and I_A. That is, A will be identified via the triple (I_A, v_A, s_A). The authentication procedure is subdivided into three phases:

1. Connection establishment: A transmits to B its identification record I_A;

2. B calculates v_A with a known public algorithm;

3. B uses v_A to calculate whether A knows s_A (this would be an accepted proof of identity):

   3.1 A transmits to B a value depending on a random number, to cover s_A;

   3.2 B stores this value and answers with a testnumber;

   3.3 A now transmits the proof value calculated from the first value and the testnumber. This proof value depends on s_A and the testnumber;

   3.4 B tests the relationship between the first value and testnumber by using v_A. The verification of this relationship is performable without the knowledge of s_A. To this procedure, only the proof value, the first value and v_A are necessary.

The phases 1 and 2 are independent of the chosen authentication method. Generally, two possibilities exist:

1. Phase 1 and 2 as explained above;

2. In phase 1 A transmits I_A and v_A, B verifies within phase 2 the relationship between I_A and v_A. The advantage of this alternative is the possible parallel calculation by B of v_A, therefore this phase is a faster method than alternative 1. The disadvantage is the larger packet size caused by the transmission of both I_A and v_A.

Phase 3 can be repeated in t loops in order to verify the identity of A. In addition, the following facts have to be mentioned:

- The protocol is interactive, both partners choose random numbers (A chooses a value, B chooses the testnumber). The unpredictability of these numbers is important to the security grade.

- Note that phase 3 involves a certain amount of parallelism, but it is essential to the security of this procedure is that the shown sequence should be maintained.

- B only uses public information, while A uses the private key s_A. Normally, Trapdoor-Functions are used [13]. Calculation of the function is easily done, the inversion of this function is impracticable without the trapdoor s_A.

- B does not need any private key, i.e. per mobile station only one secret key is necessary and this key has to be stored at the mobile station. In addition, no secret information needs to be protected at a key management center hierarchically allocated above the mobile stations. No keys need to be exchanged. New mobile stations can directly be added to the network without any further effort.

- The probability P for successful deceit by A depends on the number of loops of phase 3.
  A can successfully deceive B, if A guesses the testnumber or A knows the number of B. In this case A is able to choose value and proof value correctly without the knowledge about s_A.
  If p is the probability to guess the right testnumber, then the successful deceit probability P after t loops is $P = p^t$.

- The formula $P = p^t$ shows, that t determines every desirable probability border, i.e. every desired security level can be reached by adjusting t.

- B is never capable of calculating s_A.

For the last phase three types are possible: zero-knowledge procedures, signature schemes, and symmetric algorithms. The latter type is the fastest method. On the other hand the zero-knowledge procedures offer a variety of desirable security properties. We investigated the zero-knowledge methods of Fiat-Shamir [5],

Micali-Shamir [12], Brickell-McCurley [3], Guillou-Quisquater [6], Schnorr [13], Beth [2] and Chaum-Evertse-van de Graf [4].

The evaluation criteria were

- basic function (discrete logarithm or factorization),
- requirements for the system (reliable key management center),
- necessary parameters,
- initialization of the system,
- data exchange,
- necessary calculations of transmitter and receiver (time, complexity),
- security.

We think that the method of Schnorr is the most appropriate one for AD systems. The following section gives a short outline to the method of Schnorr and explains the advantages against other strategies listed above.

SELECTED METHOD

The authentication method of Schnorr [13] is an improved method of the authentication procedure introduced in [4]. The method is a zero-knowledge procedure and fulfils the criterias listed above in phase 3. We collected the following characteristics:

- **Basic function**
  The basic function in Schnorr's method bases on the difficulty of finding the discrete logarithm with base a (mod p).

- **Requirements**
  Reliable key management center Z providing all beacons with the necessary parameters (see below), and additional signature scheme is necessary. The signature scheme is used because of the second alternative in phase 1 and 2 listed above. The key management center does not need to know the secret keys s_A of the mobile stations. The public keys need to be confirmed with the signature scheme to guarantee the authenticity.

- **Parameters**
  Primes $p \geq 2^{512}$, $q \geq 2^{140}$, q is factor of p-1, $\alpha \in \mathbf{Z}_p$ with order q, (q is smallest x with $\alpha^x \equiv 1$ mod p);

  security parameter t with $t \approx 72$ [13];

  for the signature scheme, a private key x_Z and a public key $y_Z = \alpha^{x_Z}$ are needed.

- **Initialization**
  k-times of modular exponentiation needed.

- **Data exchange**
  For connection establishment mobile station A necessarily transmits one element of $\mathbf{Z}_p$ and $\mathbf{Z}_q$ to B. Additionally, B transmits one element of $\{1, ..., 2^t\}$ to A.
  To outline the time and effort of this method, consider the following table with our estimated calculation efforts and data sizes, in order to determine the upper bound on the time needed for the whole authentication:

Table 1. Time and action plan to Schnorr's Scheme

t	Action of Mobile Station	Packet	Action of Automatic Debiting Beacon
t_0	identity I_A, public key v_A, signature S	(I_A, v_A, S) ——>	Verification of subscription to (I_A, v_A) with S; valid up to t_5
t_1	Selection of $r \in \{1, .., q\}$, $x \equiv \alpha^r \bmod p$ (preprocessing possible)	$x \in \mathbf{Z}_p$ ——> $\approx$ 512 bit	Selection $e \in \{1, ..., 2^t\}$ (preprocessing possible)
t_2		<—— e 72 bit	preprocessed verification:
t_3	Calculation $y = r + s*e \pmod q$ Effort: 1 multiplication + 1 addition		$(v_A)^e$ Effort: 1 exponentiation
t_4		y ——> $\approx$ 140 bit	$k = \alpha^y*(v_A)^e \bmod p$ $x = k$? Effort: 1 exponentiation + 1 multiplication + 1 comparison
t_5		Feedback <——	

This table contains the necessary information needed by system developers to integrate authentication into a vehicle-to-beacon communication system. Together with the mentioned parameters and constraints (see section 2) a calculation of the establishment phase duration is possible. In [1, p.78] there has been shown the computational power of a dedicated multiplier architecture to compute exponentiation of maximum module length n = 593 bits. This architecture is created for the field $GF(2^n)$. Obviously, the Schnorr scheme has to be adaptated. The measured throughput of the device was approximately 300 kbps using a 15 MHz clock rate and exponents with an average Hamming weight of 150. This architecture would give an exponentiation duration of two milliseconds. This duration is fairly acceptable for our system [7].

Obviously, the total duration strongly depends on other parameters of the mobile communication system, i.e. collision durations, connection delays and throughput of the channel have to be taken into account. To get more insight consider [7].

- **Necessary calculations**
  See table 1 for caculations of stations A and B.

- **Security**
  Probability of deceit $\leq 2^{-t}$. See additionally section 3 and [13] for further insights.

CONCLUSION

The advantage of Schnorr's scheme is the that the procedure listed in table 1 needs only to be performed once to achieve the same security level as other schemes do. This advantage is well suitable for RTI/IVHS environments because of the hard real time constraints. In the communication area (AD beacon contact area) the mobile station A just needs to calculate one multiplication and one addition and therefore needs less computation power, i.e. the system costs will be smaller for A in comparison to other schemes.

In addition, the RTI/IVHS environment contains a lot of mobile stations and a few AD beacons. Schnorr's scheme puts the computational burden on the AD beacons, i.e. the whole system costs are smaller than using the other schemes where both stations, A and B, have to spend the same computation power.

The calculations of A are restricted to a subgroup of order q and not to the multiplicative group of $\mathbf{Z}_p$ as in the other schemes. That means, smaller numbers need to be exchanged and stored. This leads to smaller packet sizes and shorter calculations.

Section 4 has shown that much of the arithmetic to be done by mobile station A can be done in a preprocessing phase, using idle time of the processor. This saves valuable computation and therefore communication time. It shortens the whole authentication procedure.

The authors conclude that Schnorr's scheme is well suited to short-range mobile radio networks, especially to the usage for communication between mobile stations and automatic debiting beacons. The required real time constraints are explicitely fulfilled through this scheme. In comparison to the other mentioned schemes, Schnorr's scheme is to be preferred.

ABOUT THE AUTHORS

Rolf Hager (rolf@informatik.rwth-aachen.de), born 1965 in Dormagen, Germany, received his MS degree 1991 in Computer Science from the Aachen University of Technology, Germany. He is with Prof. Spaniol at the Institute of Computer Science IV, Aachen University of Technology where he works on his Ph.D. thesis on mobile communication. His main interest areas are packet switching in mobile radio networks and network management. He is project manager within the European projects PROMETHEUS and DRIVE. Rolf Hager is member of IEEE, the German Society of Computer Science (GI) and the German Institute for Standardization (DIN), subgroups NI 21.4 (OSI Network Management) and GK717/Ak9 (Dedicated Short Range Communication).

Peter Hermesmann, born 1967 in Iserlohn, Germany, studies mathematics at the Aachen University of Technology. He currently works on his master thesis with the emphasis on security in mobile radio networks. His main interests are cryptography, especially authentication mechanisms.

Michael Portz (michaelp@terpi.informatik.rwth-aachen.de), born 1962 in Rheydt, Germany, received his MS degree in 1988 in Computer Science from the Aachen University of Technology. There he works at the Institute of Applied Mathematics especially Computer Science, Prof. Oberschelp. In 1993 he finished his Ph.D. thesis on the use of interconnection networks in cryptography. His main interest areas are cryptographic protocols and blockciphers. Michael Portz is member of the I.A.C.R.

BIBLIOGRAPHY

[1] Agnew, G.B., Mullin, R.C., Onyszchuk, I.M., Vanstone, S.A., "An Implementation for a Fast Public-Key Cryptosystem", Journal of Cryptology, Springer, vol. 3, no. 2, 1991

[2] Beth, T., "Efficient zero-knowledge identification scheme for smart cards", Eurocrypt'88, Lecture Notes in Computer Science, vol. 330, Springer-Verlag Berlin 1988, pp.77-86

[3] Brickell, E.F., McCurley, K.S., "An interactive identification scheme based on discrete algorithms and factoring", Journal of Cryptology, vol. 5, 1992, pp. 29-39

[4] Chaum, D., Evertse, J.H., van de Graaf, J., "An improved protocol for demonstrating possession of discrete logarithms and some generalizations", Eurocrypt'87, Lecture Notes in Computer Science, vol. 304, Springer-Verlag Berlin 1988, pp. 127-41

[5] Fiat, A., Shamir, A., "How to prove yourself: Practical Solution to Identification and Signature Problems", Advances in Cryptology - Proc. of Crypto'86, Lecture Notes in Computer Science, vol. 263, Springer-Verlag Berlin 1987, pp. 186-99

[6] Guillou, L.C., Quisquater, J.J., "A practical zero-knowledge protocol fitted to security microprocessor minimizing both transmission and memory", Eurocrypt'88, Lecture Notes in Computer Science, vol. 330, Springer-Verlag Berlin 1988, pp.123-28

[7] Hager, R. et al., "Network Management in Short-Range Mobile Radio Networks: Performance Tuning optimizes Collision Behaviour", IEEE Vehicular Technology Society Conference VTC'92, Denver, USA, May 1992

[8] ISO/IEC DIS 16410-7, Information Technology - Open Systems Interconnection - Systems Management - Part 7: Security Alarm Reporting Function, May 1990

[9] ISO/IEC DIS 16410-8, Information Technology - Open Systems Interconnection - Systems Management - Part 8: Security Audit Trail Function, May 1990

[10] ISO/IEC DIS 16410-9, Information Technology - Open Systems Interconnection - Systems Management - Part 9: Objects and Attributes for Access Control, May 1990

[11] ISO/IEC 7498-4, International Standard, Information Processing Systems - OSI Basic Reference Model, Part 4: Management Framework, 1989

[12] Micali, S., Shamir, A., "An improvement of the Fiat-Shamir identification and signature scheme", Crypto'88

[13] Schnorr, C.P., "Efficient Identification and Signatures for Smart Cards", Advances in Cryptology - Proc. of Crypto'89, Lecture Notes in Computer Science, vol. 435, Springer-Verlag Berlin 1990, pp. 239-52

MANAGEMENT OF HIGH-SPEED DIGITAL NETWORKS

INTRODUCTION

High speed networks and the services and applications they support will become common place as we near the close of this decade. Gigabit testbeds exist; Fast packet services such as Switched Multi-megabit Data Service (SMDS), Frame Relay Service, and Cell Relay Service are being deployed; and Broadband Asynchronous Transfer Mode (ATM) based networks are emerging in public as well as private networks. It is essential that high speed networks be managed in such a way as to make effective use of resources while maintaining performance objectives. The papers presented in this section touch upon some of the many aspects in the management of high speed networks and services.

High speed data services in the public network rely on coordination among the Local Exchange Carriers (LECs) and Interexchange Carriers (ICs). The paper "Adding Network Management Tools to an Interexchange Carrier's Operations Toolkit for Switched Multi-megabit Data Service," by Hansen, describes a set of tools that extends an IC's operations capabilities by providing operations information originating from the LEC network. A LEC's Exchange Access Operations Management (XA-OM) Agent and associated applications allow an IC to receive event notifications, initiate tests, and retrieve information. These capabilities provide part of an end-to-end management solution for SMDS.

A layered architecture for adaptive traffic control in ATM networks is presented in a paper by Gersht, Shulman, Vucetic, and Keilson entitled "Dynamic Bandwidth Allocation, Routing, and Access Control in ATM Networks." The presented architecture provides for traffic control at the Virtual Path (VP), Virtual Channel (VC), and Cell Levels. In the architecture, at least one VP exists for each source-destination pair. Admission control, routing decisions, VC bandwidth reservations, and cell-level congestion control are performed at the boundary of the backbone network. An optimal VP bandwidth allocation algorithm that minimizes total rejected bandwidth and cell loss while satisfying maximum cell loss and delay requirements is presented.

Network engineers and planners need to evaluate the performance of high speed data networks. The model presented in "A Modeling Approach for the Performance Management of High Speed Networks," by Sarachik, Panwar, Po, Papavasiliou, Tsaih, and Tassiulas, allows the performance of different networks and network configurations to be compared, and network performance during specific intervals to be evaluated. Another paper in this section, "Allocation of End-to-end Delay Objectives for Networks Supporting SMDS," by Lin, considers the end-to-end delay objective allocation problem for networks supporting SMDS. End-to-end delay objectives are allocated to network elements in such a way that if each network element meets its allocated delay objective, the end-to-end delay objective is also met.

A key factor for the success of high speed networks is providing customer access in a cost effective manner. The use of ATM based Passive Optical Networks (ATM PONs) to provide low cost access to Broadband Integrated Services Digital Network (B-ISDN) is described in the paper by Venieris, Mourelatou, Theologou, and Protonotarios, entitled

"VPI/VCI Management in B-ISDN Access Optical Networks." In this scheme, a VP connection is maintained to each ATM PON network termination, and a service class is associated with a pool of VC Identifier (VCI) values within the VP. A primary focus of this paper is configuration management of ATM PONs.

Andrew J. Mayer
Bellcore

ADDING NETWORK MANAGEMENT CAPABILITIES TO AN INTEREXCHANGE CARRIER'S OPERATIONS ENVIRONMENT FOR SWITCHED MULTI-MEGABIT DATA SERVICE

Shannon Hansen

Bellcore
331 Newman Springs Road
Red Bank, NJ 07712

INTRODUCTION

Local Exchange Carriers (LECs) and Interexchange Carriers (ICs) are in the process of interconnecting their networks to provide a national SMDS service. To help ensure the quality of SMDS, ICs and LECs will need network management capabilities and procedures that support traffic engineering and administration, facilitate prompt problem resolution, and allow for efficient utilization of network resources. These capabilities can be shared among carriers and added to their existing operations environment.

The sharing of the network management capabilities among carriers is referred to as Intercarrier Operations Management.[1] Intercarrier Operations Management encompasses the following relationships:

- Relationship between a LEC and an IC when the LEC provides the IC with Exchange Access SMDS (XA-SMDS),
- Relationship between an IC and an International Carrier (INC) when the IC interconnects with the INC in providing international SMDS,
- Relationship between two LECs when the LECs interoperate to provide Exchange and Exchange Access SMDS, and
- Relationship between LECs providing SMDS in separate Local Access and Transport Areas (LATAs).

This paper focuses on the first relationship listed above (the relationship between a LEC and an IC) and the network management capabilities that might be added to the IC's operations environment as part of this relationship. These network management

---

1 Intercarrier Operations Management is described in further detail in the SMDS Interest Group (SIG) document SIG-IS-003/1993.[1]

Network Management and Control, Vol.2, Edited by
I.T. Frisch *et al.*, Plenum Press, New York, 1994

capabilities are referred to as SMDS Exchange Access Operations Management (XA-OM) service.[2] In particular, this paper focuses on the initial XA-OM capabilities that have been targeted for the 1994 time frame.

PURPOSE AND MOTIVATION

An IC uses XA-SMDS when it provides interexchange SMDS to an end-user directly connected to a LEC network as illustrated in Figure 1.

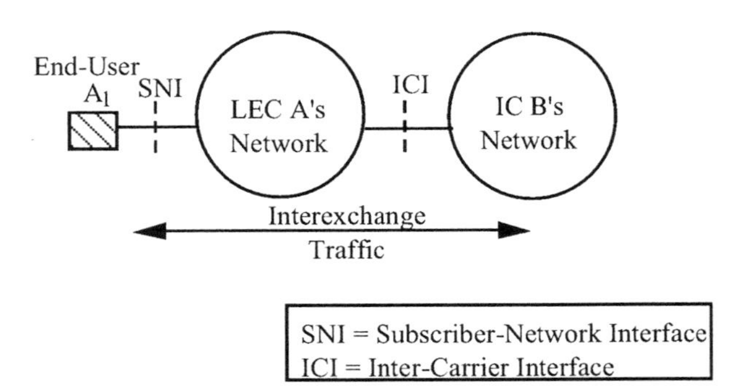

Figure 1. Example of XA-SMDS

In this example, LEC A provides XA-SMDS to IC B when its network transports SMDS traffic from End-User A_1's Subscriber-Network Interface (SNI) to IC B's Inter-Carrier Interface (ICI) or vice versa. For more information on XA-SMDS, refer to Bellcore Technical Reference TR-TSV-001060.[3]

Since LEC network facilities can affect an IC's interexchange service, a LEC may choose to offer an SMDS XA-OM service to an IC. This service consists of network management capabilities that will allow an IC to directly manage aspects of its XA-SMDS service. As illustrated in Figure 2, these capabilities will initially allow an IC to obtain information related to its XA-SMDS service. It is envisioned that an IC may use this information as an aid in identifying or sectionalizing a trouble or to supplement its traffic and engineering applications. Taking these uses into account, Bellcore has defined the majority of XA-OM information in terms of counts. These counts have already been defined for LEC operations and follow Bellcore Technical Reference specifications.[3] For example, a LEC may want to obtain for a given ICI the number of Inter-Carrier Interface Protocol (ICIP) packets discarded because of protocol errors within a 15-minute interval.

---

2 SMDS XA-OM service is specified in Bellcore Technical Advisory TA-TSV-001237.[2]

3 Note that some processing may be performed by SMDS XA-OM functionality before a count is provided to an IC. For example, an XA-OM count may result from the summation of several SMDS Switching System (SS) counts.

Since an IC may use XA-OM capabilities for identification or sectionalization of troubles, Bellcore has incorporated the event-based approach instead of the polling-based approach for network management into the SMDS XA-OM service description. Using an event-based approach, an IC will be notified through XA-OM that a given event has occurred. For example, an IC may want to be notified when the count of ICIP packets discarded because of protocol errors for a given ICI exceeds a threshold within a 15 minute interval. If XA-OM were to use the polling-based approach for network management, an IC would have to periodically inquire about the state of its XA-SMDS service. For example, an IC would retrieve the above discard count every 15 minutes to determine whether the count had crossed a given threshold. Not only does the polling-based network management approach increase the amount of communications between LEC and IC Operations Systems (OSs) but it also conflicts with LEC maintenance procedures that tend to use the event-based approach.

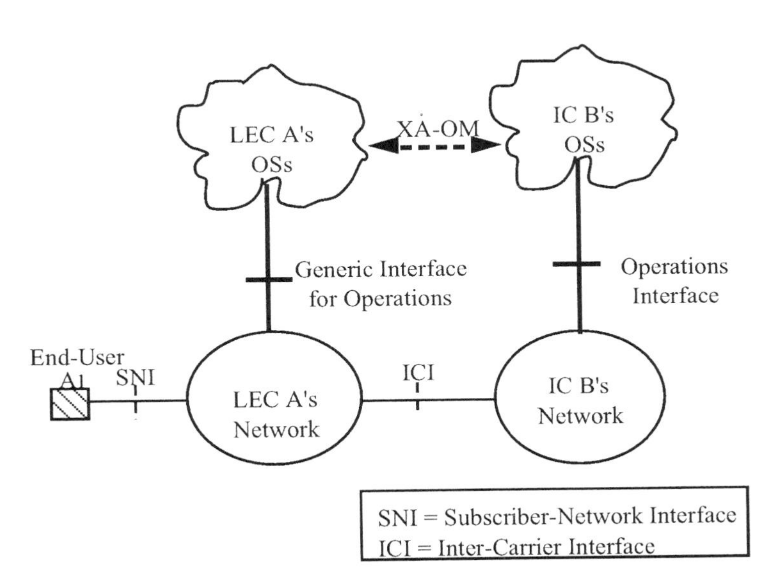

Figure 2. SMDS XA-OM Capabilities

SERVICE FRAMEWORK

Following the current data management environment and taking into account the wide variety of XA-OM information, Bellcore has proposed the framework illustrated in Figure 3 for the initial offering of SMDS XA-OM service.

As illustrated, SMDS XA-OM service has three dimensions: (1) IC access to SMDS XA-OM service capabilities, (2) the service capabilities, and (3) XA-OM information used by the service capabilities.

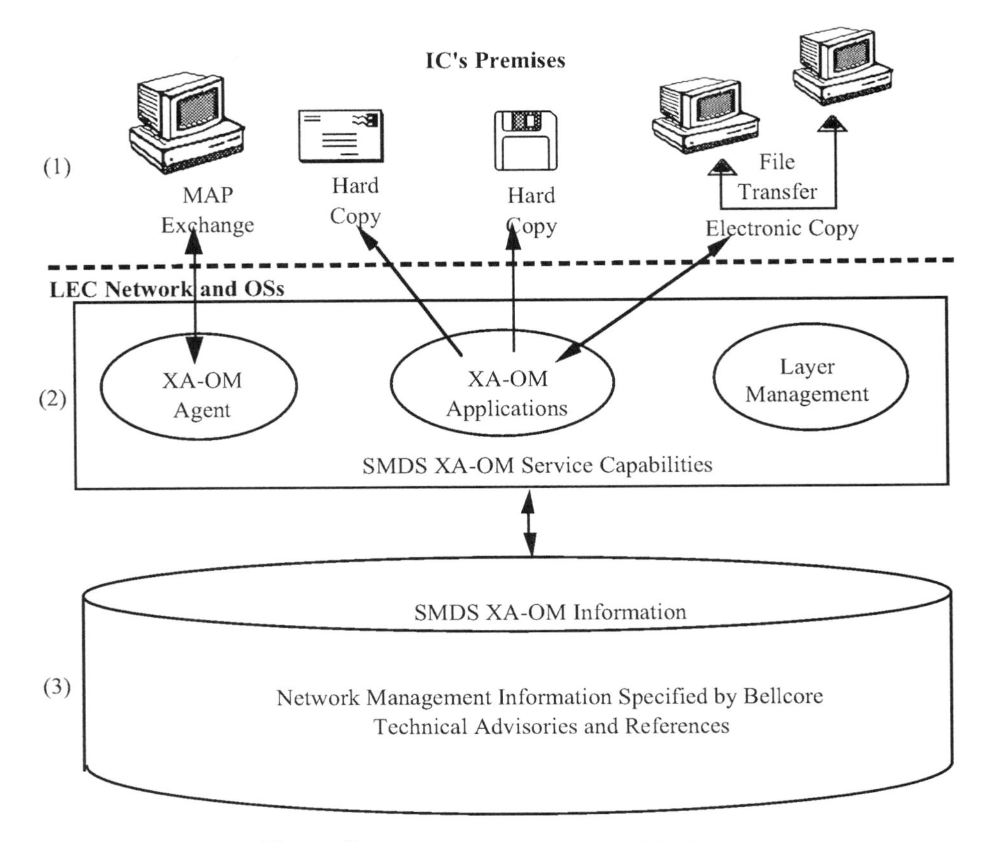

Figure 3. Framework for SMDS XA-OM Service

As illustrated in Figure 3, a LEC may support IC access to SMDS XA-OM capabilities through an XA-OM agent application.[4] An agent provides near-real time access to XA-OM information that has been summarized in some fashion and generates event notifications that are sent to an IC's NM application for further processing. An XA-OM agent will communicate with an IC's NM application using a Management Application Protocol (MAP) exchange. The MAP specified for initial XA-OM capabilities is Simple Network Management Protocol version 1 (SNMPv1).

A LEC may support IC access to SMDS XA-OM capabilities through one or more applications. Three applications have been identified for the initial offering of SMDS XA-OM:

- Hard-copy access to performance information,
- Electronic-copy access to performance information, and
- Access to XA-SMDS usage information.[5]

---

4 Throughout the remainder of this paper, "agent application" is referred to as "agent."

5 This application supports both electronic- and hard-copy access.

The purpose of these applications is to periodically provide an IC with a large amount of XA-OM information in either an electronic-copy or hard-copy form.
As illustrated in Figure 3, layer management functions are included in the XA-OM service framework. These functions exist whether or not XA-OM service is provided. Layer management functions monitor and maintain any given layer of a data network architecture. As illustrated in Figure 4, LEC and IC networks will use layer management protocols to exchange information related to the status and performance of an ICIP physical layer. These protocols are referred to as monitoring and surveillance layer management protocols.

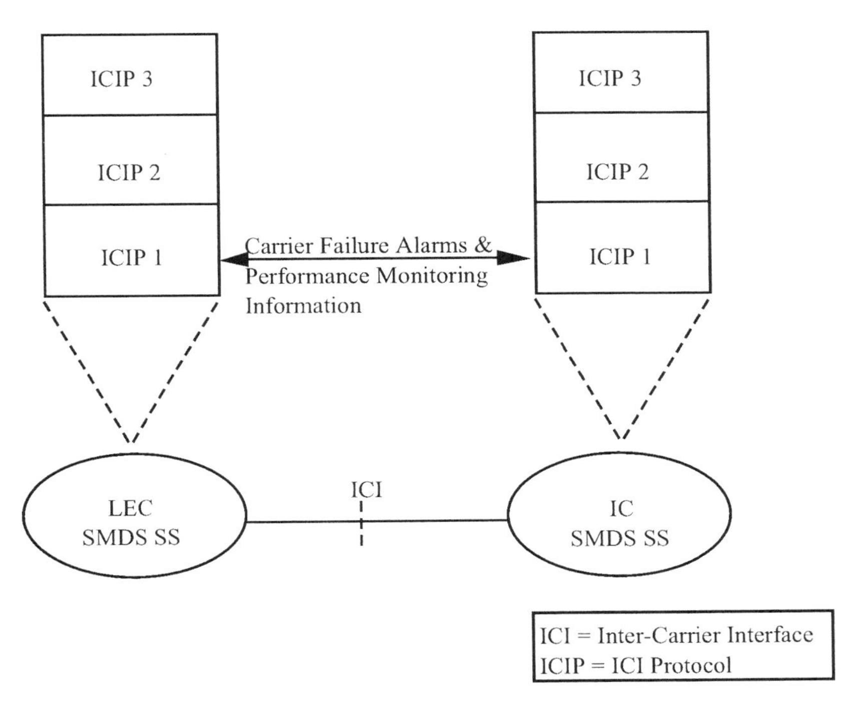

Figure 4. ICIP Physical Layer Management

SERVICE CAPABILITIES

Seven capabilities have been specified for the initial offering of SMDS XA-OM service and are briefly described below. As this service matures, additional capabilities may be provided.

- *Receive Event Notifications* provides an IC with the ability to receive unsolicited event notifications. An IC may receive an event notification when:
  - the status of a parameter in an IC's subscription profile has changed,
  - an ICI-related count has exceeded a threshold, or
  - an ICI is brought out-of-service or returned-to-service.

  These event notifications are unsolicited, can aid an IC in isolating a failure, and can be sent directly to an IC's network management (NM) application.
- *Initiate Tests* supports the Internetwork Access Arrangement Verification test specified by the Intercarrier Working Group of the SIG in SIG-IS-003/1993.[1]

The objective of this test is to remotely verify the condition of an end-user's access arrangement by checking for the existence of alarms or protocol errors on the access line associated with the involved SNI and by checking the operational status of SMDS line termination equipment (i.e., the SMDS line card). A check is also made to determine whether any service disagreements were recorded for the SNI over the current 15-minute interval and several recent intervals. An IC initiates this test through use of the of the *retrieve intercarrier-SMDS related information* and the *retrieve general SMDS Intercarrier Operations Management information* service capabilities.

- *Retrieve Subscription Profile Information* provides an IC with the ability to retrieve information that describes how its XA-SMDS service is currently provisioned.
- *Retrieve Intercarrier SMDS-Related Information* provides an IC with the ability to retrieve information related to its interoperating with an LEC to provide interexchange SMDS.
- *Retrieve General SMDS XA-OM Information* provides an IC with the ability to retrieve information referring to the SMDS XA-OM agent and the general administrative status of an IC's ICI.
- *Retrieve Performance Information* provides an IC with the ability to retrieve information related to the performance of a given ICI.
- *Retrieve Usage Information* provides an IC with the ability to retrieve the XA-SMDS usage information that is forwarded to a LEC billing system.

These capabilities are designed to support trouble identification and sectionalization, and to supplement an IC's traffic and engineering applications. Examples of how an IC could use these capabilities are as follows.

Trouble Identification and Sectionalization Example: After receiving an end-customer reported trouble, an IC could conduct the Internetwork Access Arrangement Verification test to verify the end-user's access arrangement. The IC would conduct this test using the *retrieve intercarrier-SMDS related information* and the *retrieve general SMDS Intercarrier Operations Management information* service capabilities. The IC would access these capabilities through the XA-OM agent using SNMPv1.

Supplementing Traffic and Engineering Applications Example: An IC could use the *retrieve performance information* service capability to obtain ICI-related information collected over the past month for: a) the number of segments (i.e., Level 2 Protocol Data Units (L2_PDUs)) sent across a given ICI and b) the number of segments that were destined for a given ICI but were discarded due to congestion. The IC would most likely access this capability through an XA-OM application that provides either hard-copy access or electronic-copy access to performance information.

SUMMARY

A LEC may offer an IC SMDS XA-OM capabilities that it can add to its operations environment. These capabilities are shaped by two key applications: 1) trouble identification and sectionalization and 2) traffic and engineering applications. To fit into the IC's operations environment, these capabilities need to follow the event-based approach to network management. By adding XA-OM capabilities, an IC can directly manage aspects of its XA-SMDS service thereby helping to ensure the quality of the end-to-end SMDS service offered to end-customers.

ABOUT THE AUTHOR

Shannon Hansen (shannon@cc.bellcore.com) is a Member of Technical Staff at Bellcore. Her current responsibilities include defining network management capabilities that complement the exchange access, broadband data services offered by a LEC to an IC. Shannon received her BSEE from the University of Southwestern Louisiana in 1986 and her MSEE from Columbia University in 1988.

REFERENCES

1. *Guiding Principles for SMDS Intercarrier Operations Management*, SMDS Interest Group Document, Revision 1, May 1993.
2. TA-TSV-001237, *SMDS Generic Requirements for Initial Operations Management Capabilities in Support of Exchange Access and Intercompany Serving Arrangements*, Issue 1 (Bellcore, June 1993).
3. TR-TSV-001060, *Switched Multi-megabit Data Service Generic Requirements for Exchange Access and Intercompany Serving Arrangements*, Issue 1 (Bellcore, December 1991) plus Revision 1, August 1992 and Revision 2, March 1993.

ACRONYMS AND ABBREVIATIONS

ICI	Inter-Carrier Interface
ICIP	ICI Protocol
IC	Interexchange Carrier
L2_PDU	Level 2 Protocol Data Unit
LATA	Local Access and Transport Area
LEC	Local Exchange Carrier
MAP	Management Application Protocol
NM	Network Management
OS	Operations System
SIG	SMDS Interest Group
SNMPv1	version 1 of Simple Network Management Protocol
SNI	Subscriber-Network Interface
SMDS	Switched Multimegabit Data Service
SS	Switching System
XA-OM	Exchange Access Operations Management
XA-SMDS	Exchange Access SMDS

DYNAMIC BANDWIDTH ALLOCATION, ROUTING, AND ACCESS CONTROL IN ATM NETWORKS

Alexander Gersht,[1] Alex Shulman,[1] Jelena Vucetic,[1] and Julian Keilson[2]

[1]GTE Laboratories Incorporated, 40 Sylvan Road, Waltham, MA 02254
[2]GTE Laboratories Incorporated, 40 Sylvan Road, Waltham, MA 02254
and University of Rochester, Rochester, NY 14627

ABSTRACT

We propose a layered architecture for adaptive traffic control in ATM networks that utilizes the virtual path (VP) concept. The proposed control guarantees quality of service and simplifies call and cell processing. The control is organized in three levels: a VP level, a virtual circuit (VC) level, and a cell level.

The VP level represents the top level of the traffic control. It efficiently provisions network resources to meet bandwidth demand and satisfy performance requirements on VC and cell levels. The VP-level control logically assigns a virtual bandwidth to each VP. The assignments are periodically adjusted to relatively slow dynamics of bandwidth demand and network VC load. We present an efficient algorithm for the optimal VP bandwidth assignment that equalizes cell losses over the VP set and maximizes call and cell throughput. The VC level handles call processing in a decentralized fashion according to the provisioning specified by the VP-level control. The VC-level control makes call admission and routing decisions and logically reserves the required bandwidth for an admitted call. This is done at the call's source node independently from all other nodes. That significantly simplifies the call setup. The total reserved rate on the VP cannot exceed the VP's virtual bandwidth. The VC-level control reserves the peak rate for admitted calls carrying real-time traffic and less than a peak rate for admitted calls carrying non-real-time bursty traffic. The cell-level control handles cell processing and is also decentralized. We focus on the cell-level VP congestion control (VPCC). The cell-level VPCC is done only for non-real-time traffic. The control is executed at the VP's source node. The control maintains the cell access rate to each VP below the VP's virtual bandwidth. This precludes cell congestion inside the VP logical pipe and sustains quality of service in the backbone network.

INTRODUCTION

ATM multiservice public networks [1,2] aim to provide a variety of services using a single switching fabric for both broadband and narrowband applications. These services are different in bandwidth and quality of service requirements. They also have different traffic statistics. Some broadband services, such as live video, generate a constant bit rate (CBR) real-time traffic. The others, such as LAN-to-LAN communications, have very bursty patterns and generate a variable bit rate (VBR) traffic. The VBR applications can be both real-time and non-real-time. In our scheme, to perform traffic control, the services are divided into classes based on their bandwidth and quality of service requirements, revenues, and other factors.

Network Management and Control, Vol.2, Edited by
I.T. Frisch *et al.*, Plenum Press, New York, 1994

The variety of services and their traffic characteristics — non-exponential and unknown traffic statistics, high speed of cell switching and transmission, and volatility of traffic demand — make multiservice traffic management a very challenging problem in ATM networks (see [3,4]).

The main advantages of ATM networks are flexibility in accommodating, in a single switching fabric, a wide variety of different types of applications; simplification of protocol and control; and VC network interface. Standard recommendation I.371 [5] states that "the primary role of Traffic Control and Congestion Control parameters and procedures is to protect the network and the user in order to achieve Network Performance objectives. An additional role is to optimize the use of network resources. The design of an optimum set of ATM layer traffic controls and congestion controls should minimize network and end-system complexity while maximizing network utilization." Although bandwidth saving is desirable in ATM networks, it is not the main priority, especially as fiber optics becomes less expensive.

We propose a layered architecture for adaptive traffic control which utilizes the VP concept [6] in interoffice ATM networks. The suggested control guarantees quality of service and simplifies call and cell processing through advance allocation of network resources and distribution of the traffic control functions to network edges. This simplification is done at the expense of some bandwidth utilization which is optimized within the constraints of the suggested architecture. In this way, the proposed control complies with standards recommendation I.371. The control is organized in three levels: a VP level, a VC level, and a cell level. We use the terms VCs and calls interchangeably.

The VP level represents the top level of traffic control. It efficiently provisions network resources to meet the multiclass bandwidth demands and satisfy the performance requirements on the VC and cell levels. The VP-level control logically assigns a virtual bandwidth to each VP. The VP virtual bandwidth is defined as the maximal permitted rate (MPR) of cell transmission for this VP connection. Contrary to the circuit-switching-based network architecture, in which time slots are reserved to individual circuits, the link equipment transmits all cells on a first-in-first-out (FIFO) basis, regardless of their VP identifiers (VPIs). The bandwidth assigned to a VP is sufficient to guarantee the quality of service (cell loss and delay) for all admitted calls with this VPI. The VP-level control is invoked periodically. It adapts to relatively slow dynamics of multiclass bandwidth demand and network VC load. The control can be implemented in either a centralized or decentralized fashion. We present the optimal VP bandwidth allocation algorithm for the VP-level control. The algorithm minimizes the total rejected bandwidth and cell loss while satisfying the maximal cell loss and delay requirements. In addition, the algorithm equalizes, on the minimal possible level, the total cell losses on the VP set of each source-destination (SD) pair.

The VC-level control handles all call admission and routing decisions and reserves bandwidth for admitted calls. All of these functions are performed at the call's access node. A call is admitted if a sufficient virtual bandwidth for its class is available on one of the VPs of the VP set of the call's SD pair. Otherwise, the call is rejected at the source node. The routing decision for each admitted call is done in a form of VPI assignment. The VC-level control logically reserves, within the VP corresponding to the call's VPI, the bandwidth required for the VC connection. It also keeps track of bandwidth reservations and releases for all VCs corresponding to each VP. The reservation and VPI assignment are done at the source node independently from all other nodes. This is possible because the VP virtual bandwidth is preassigned, and bandwidth reservations and releases are recorded for each VP at its source node. The rejection of calls is done by the VC-level control according to a multiclass virtual bandwidth sharing (MBS) strategy. The parameters of the MBS strategy are periodically adjusted by the VP-level control at the beginning of the control period. In this paper, we describe two MBS strategies.

The cell-level control handles cell processing in a decentralized fashion. We focus on the cell-level congestion control which is exercised on the VP basis. It is applied to non-real-time traffic locally accessing each VP at the VP's source node. The control dynamically regulates the total cell rate (from all VCs allocated on the VP) that locally accesses the VP and precludes the congestion that may occur inside the VP logical pipe due to the burstiness of the traffic. We propose the cell rate regulation strategy, which represents a modified version of the congestion control strategy suggested in [7,8]. Parameters of the strategy are adjusted, for each control period, by the VP level according to the MPR assigned to the VPs. The described congestion control scheme does not require any interaction between a call's source node and any other nodes in the network.

The rest of the paper is organized as follows. First we outline the suggested control architecture for ATM networks. The next section describes the optimal VP bandwidth allocation algorithm. We then derive the formula for calculating the cell loss probability as a function of the total VP rate logically assigned on a physical link. Finally, we summarize the results of this work.

TRAFFIC CONTROL ARCHITECTURE FOR ATM NETWORKS

We propose a layered architecture (outlined in Figure 1) for adaptive traffic control in ATM networks that utilizes the VP concept. In this architecture, at least one VP is established for each SD pair. The traffic control is organized in three levels: a VP level, a VC level, and a cell level. All levels are designed to guarantee quality of service and simplify call and cell processing. In the following three subsections, we describe control mechanisms for VP, VC, and cell levels, respectively.

VP-Level Control

The VP-level control provisions and performs the optimal logical assignment of VP bandwidth for an upcoming control period. The logical bandwidth assignment is defined as an

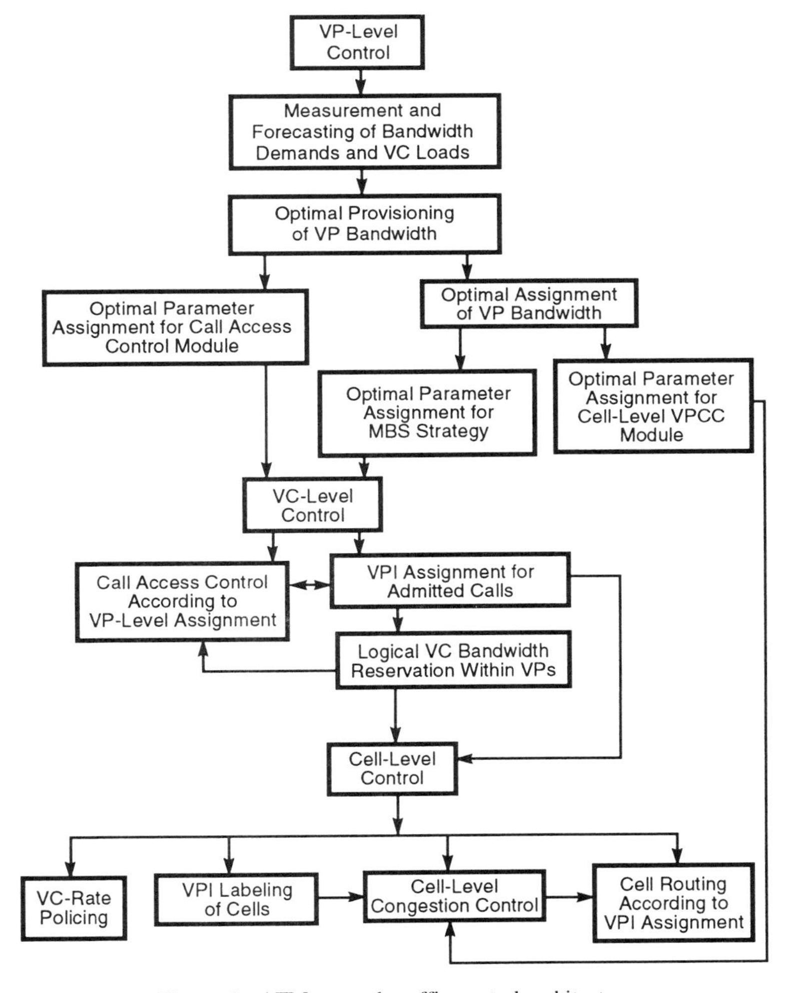

Figure 1. ATM network traffic control architecture.

assignment of the MPR of cell transmission for a VP. Contrary to the circuit-switching-based network architecture, in which time slots are reserved to individual circuits, the link equipment transmits all cells on a FIFO basis regardless of their VPIs. Since the physical link is shared by all VPs passing through the link, the sum of the MPRs for all VPs passing through the link cannot exceed its capacity. The MPR for a VP is set (1) to satisfy the end-to-end delay and cell loss requirement for the VP and (2) to maximize the allocated bandwidth demand and cell throughput. To simplify the protocol, we impose the same cell loss requirement for all classes of traffic and make it equal to the strictest loss requirement among them. In ATM networks, the cell loss is required to be small. This sets a limit for the sum of the MPRs of all VPs passing through the link that is tighter than the limit imposed by the link capacity.

The VP-level control collects real-time network measurements and forecasts bandwidth demand and network VC load for an upcoming control period. The length of the control period is empirically chosen to permit the adaptation of VP MPR to the dynamics of bandwidth demand and network VC load. It also represents a tradeoff between the processing and measurement costs incurred in obtaining the required information and the resulting savings in bandwidth or improvements in throughput.

An effective forecasting technique applicable to a broad range of traffic patterns was presented in [9]. This technique allows us to dynamically assign correct MPRs to the VPs, thus efficiently utilizing network bandwidth. Numerical results in [9] show that the control period equal to half of the average holding time results in 5% projection error, which is sufficient for achieving efficient bandwidth utilization. If the bandwidth is inexpensive and demand is relatively stable, the control period can be made longer, for instance, it can be set hourly.

Based on the forecast, the VP-level control determines the optimal logical VP bandwidth assignment satisfying objectives (1) and (2). It computes the MPR for each VP and determines, for each SD pair, the amount of bandwidth demand that cannot be allocated in the upcoming period. The VP-level control then passes the rate to the VPI assignment module of the corresponding source node, thus producing a logical reservation of VP bandwidth for a control period. The rejection of a call that should not (or cannot) be allocated is done by the VC-level control according to a MBS strategy. For each control period, the strategy parameters are adjusted by the VP-level control and passed to the call admission module of the VC level (see Figure 2).

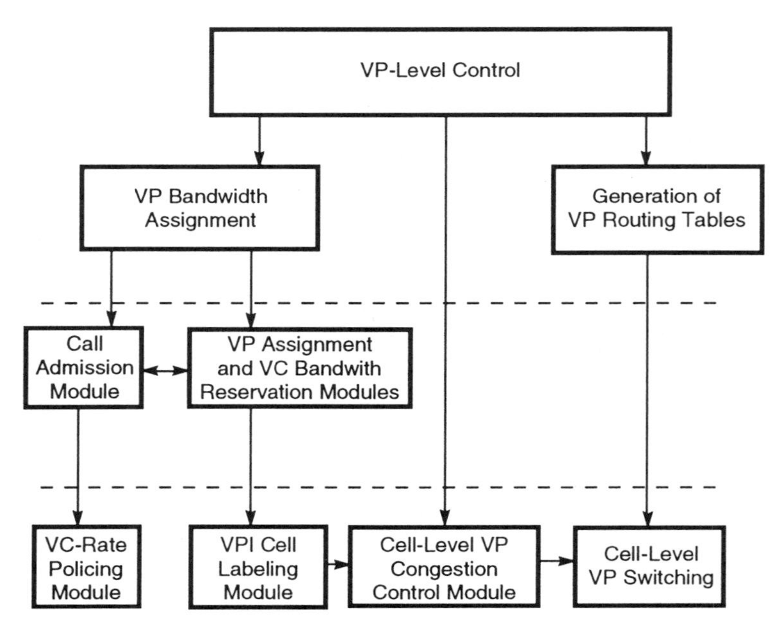

Figure 2. Simplified traffic control architecture: the VC-level and the cell-level modules are located at a source backbone node.

The VP-level control provisions how multiclass traffic shares the VP set of each SD pair. It decides how to allocate real-time and non-real-time demands on the VP set and sizes the parameters of the cell-level VPCC strategy. These parameters are then passed to the VC-level VPI assignment and to the cell-level VPCC modules, respectively (see Figure 2). In addition, VP strategy may also include rules for load balancing of very long calls or for allocation of the calls consuming very large amounts of bandwidth. In both cases, the calls can be allocated, for instance, on the minimum-hop VPs of the VP set of the call's SD pair.

VC-Level Control

The VC-level control performs call admission and processing according to the provisioning specified by the VP-level control for an upcoming period. The control makes all call admission and routing decisions and bandwidth reservations at the access node independently of all other nodes. The routing decision for each admitted call is done in the form of VPI assignment by a VPI assignment module located at the call's access node (see Figures 2 and 3). The VPI assignment module logically reserves the bandwidth required for the VC connection on the VP corresponding to the call's VPI. The amount of logically reserved bandwidth remains constant for a call's duration (another possibility is described in the Comments and Generalizations section). For each VP originated at a node, the VPI assignment module keeps track of the currently reserved VC bandwidth. This is done separately for the total VP real-time and non-real-time traffic. A call cannot be accommodated on a VP if the call allocation results in exceeding the MPR assigned to the VP. To implement the VP set sharing policies, the VPI assignment module has look-up tables for VP routing. These tables are adjusted at the beginning of a control period. Since the VPI assignment module is located at each source node, the routing tables are based only on destinations.

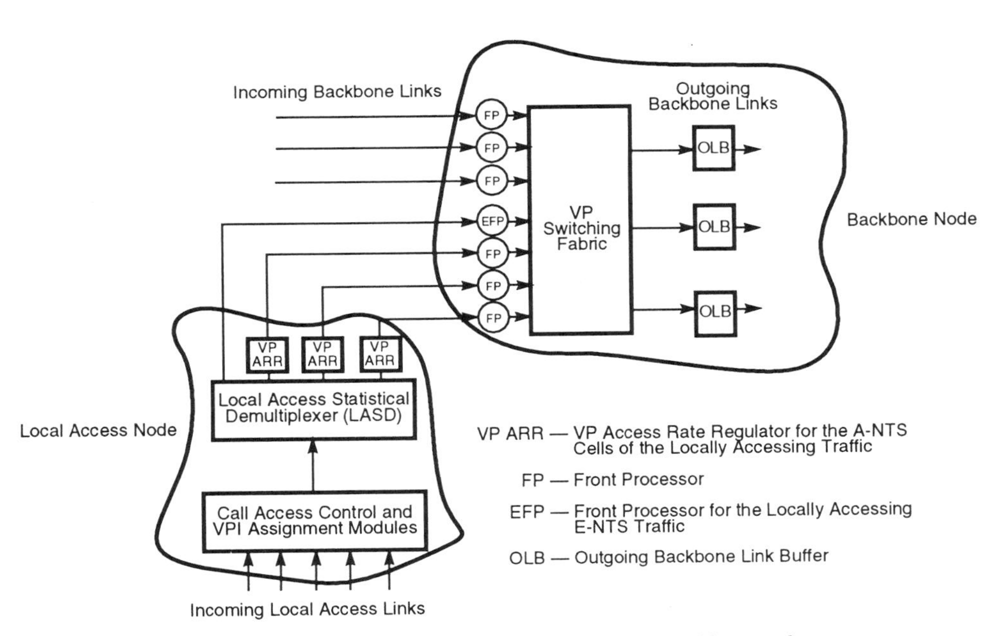

Figure 3. Simplified architecture of local access and backbone nodes.

The call admission and routing decisions are done taking into account the following: (1) traffic descriptors incorporated in the connection request ID; (2) the amount of the bandwidth currently reserved on the VP set for the call's SD pair; and (3) MPRs assigned for real-time and non-real-time traffic for the VP set. The call admission is performed according to either threshold or profitability strategies.

The first strategy, suggested in [10] for ATM networks, is a threshold-based strategy that allows a simple implementation in hardware. The threshold for call admission is set for each class of traffic and is used by VC-level control to permit or reject the call. The optimal thresholds are engineered to maximize the revenue or call throughput that would be obtained by allocating demand on the VP set of each SD pair. An efficient optimization algorithm for threshold engineering was developed in [10]. In [11], it is also shown that if the goal is to maximize the revenue, the strategy favors classes with higher profitability. The class profitability is defined as a ratio of revenue to resource consumption per call of the class. Under the second strategy, if a rejection of some demand is provisioned by the VP-level control, the arriving calls are rejected with probability inversely proportional to the call's profitability [9]. In this case, the arriving call may be rejected even when a sufficient bandwidth for its allocation is available. If no rejection of demand is provisioned for an SD pair, all calls with this SD are admitted when a sufficient bandwidth is available.

Following [7,8], we suggest two types of Network Transport Service (NTS): express service (E-NTS), appropriate for real-time traffic; and first-class service (A-NTS), appropriate for non-real-time traffic (another type of NTS is described in the Comments and Generalizations section). For E-NTS calls, the maximum cell end-to-end delay and peak-rate transmission are guaranteed. For A-NTS calls, only the average VP delay (including delay at VP access) and some rate below the peak rate are guaranteed. The peak rate of connection and NTS type are specified by the user in the connection request ID. In addition, if the user selects A-NTS, he also specifies/negotiates the guaranteed rate. The guaranteed rate can be 0 (the least expensive alternative). That means that the user is ready to stop the transmission at the network's request. The user choice depends on the capability of customer premises equipment (CPE) to transmit at different rates and tolerance to delay at CPE introduced by the choking procedure. In the absence of the cell-level congestion, the A-NTS user is allowed to transmit at the peak rate (see the Cell-Level Control section).

According to the user's choice, the VC-level control assigns an NTS for the admitted call. To provide the required quality of service for NTSs, the VC-level control reserves the peak rate for the E-NTS calls and some rate between the peak and guaranteed rates for A-NTS. The rate reserved for A-NTS calls depends on the selected guaranteed rate and is determined by the network provider to satisfy the average VP delay requirement (see the details in [7,8]). Due to the bursty nature of A-NTS traffic, the scheme can use statistical multiplexing to utilize network resources more efficiently. This allows to reserve for A-NTS calls a rate below the peak rate and still meet the average delay requirement. Remember that cell loss requirements are satisfied since they are uniformly limited for all classes by the VP logical bandwidth assignment.

Cell-Level Control

The cell-level control handles cell processing in the interoffice network and is the fastest. We focus on the cell-level VPCC (see Figures 2 and 3) and do not consider other cell-level control functions, such as policing (see [6]), cell retransmission, etc. The cell-level VPCC is executed for each VP at its access node. It limits, at the VP local access, the total cell rate of non-real-time traffic. The main goals of this regulation are as follows:

- To maintain, at the VP local access, the total peak rate of A-NTS cells below the maximum VP permitted rate assigned to this NTS.
- To save bandwidth by reserving for A-NTS calls the rate between its peak and guaranteed rates.

Since for E-NTS cells the peak rate is reserved, the total MPR on the VP is not exceeded. This precludes congestion inside each VP logical pipe and sustains the quality of service. To avoid the VP local access overflow, the cell-level control chokes A-NTS users if the VP access for A-NTS cells is getting congested. In the absence of congestion, A-NTS traffic will be transmitted at a peak rate. On the other hand, when the congestion at the access node is detected, the transmission rate of A-NTS traffic has to be reduced to the guaranteed rate. The detection is performed by observing the queue of the A-NTS cells locally accessing the VP source node. No additional cell-level congestion control is applied at the intermediate nodes, and no information exchange between the source and other nodes is required.

Figure 3 displays a simplified node architecture for cell processing in the interoffice network. The VPI assignment module assigns a VPI label to the cells generated by an admitted call. The local access statistical demultiplexer (LASD) routes the arriving cells according to the cells' VPIs. The A-NTS cells are directed to the VP access rate regulators (VP ARR). The E-NTS cells are routed directly to the VP switching fabric. For E-NTS and A-NTS cells to be distinguished at the LASD, they must have different VPIs (see the Comments and Generalizations section). The VP ARR maintains its output peak rate below the MPR assigned to the VP for A-NTS. When the VP ARR is getting congested, the ARR chokes all A-NTS users of the VP. Choking of the traffic is stopped when the congestion is abated. From the VP ARR, the A-NTS cells are directed to the VP switching fabric. This self-routing fabric switches all cells arriving from the local access and incoming backbone links according to the cells' VPI and the VP routing tables. The cells are then directed to the outgoing link buffers (OLBs). In the OLB, the cells belonging to the VPs passing through the link are placed in the same queue regardless of their traffic classes. The outgoing links transmit cells on a FIFO basis. The OLBs are small: they are engineered to satisfy the maximum delay requirement of E-NTS traffic. Since the maximal VP rate determined by VP-level control is based on the peak rate for E-NTS traffic, and since the VP ARR limits the peak VP access rate for A-NTS cells, no congestion will occur within the backbone network.

Figure 4 displays a VP ARR implementation. The VP ARR consists of a rate regulator (RR) and a buffer (B) for queuing the incoming A-NTS cells. The cells coming from the LASD are buffered in B and then transmitted on a FIFO basis. If the queue length exceeds the threshold T, the RR chokes all A-NTS users of this VP connection. When the congestion in B is abated below the threshold A, the RR allows the users to restart the transmission. The VP-level control sizes the thresholds T and A to limit the average VP delay and choking/releasing frequency for A-NTS cells. The sizing of T and A depends on the MPR that was assigned for A-NTS traffic on each VP and can be done, for example, by using the algorithm developed in [7].

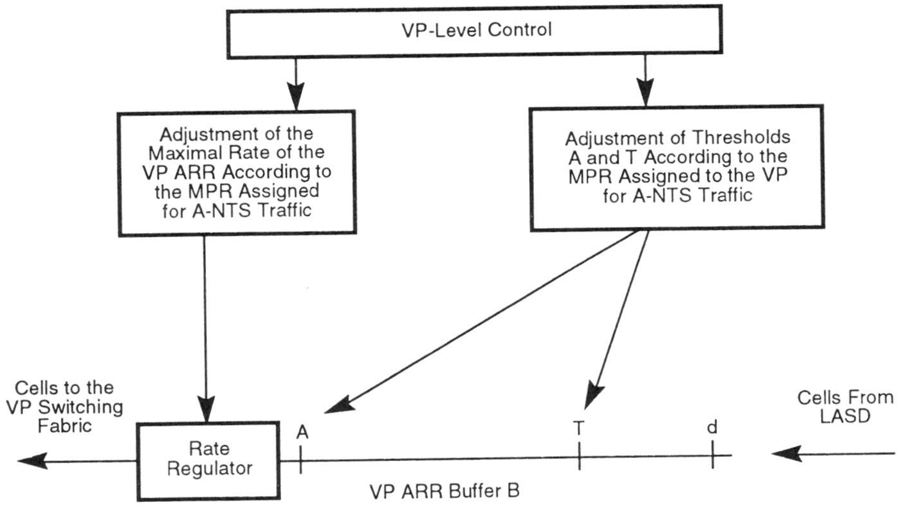

VP-Level Control

Adjustment of the Maximal Rate of the VP ARR According to the MPR Assigned for A-NTS Traffic

Adjustment of Thresholds A and T According to the MPR Assigned to the VP for A-NTS Traffic

Cells to the VP Switching Fabric

Rate Regulator

A

T

d

Cells From LASD

VP ARR Buffer B

Figure 4. Parameter assignment for the VP ARR by the VP-level control.

Comments and Generalizations

The VP set (number of hops, OLB sizes) is engineered for meeting the maximum and average VP delay requirements. For E-NTS traffic, the VP control selects the VP set that satisfies the maximum end-to-end delay requirement D^E:

$$OLBD * NH + D_p \leq D^E, \qquad (*)$$

where D_p is the propagation delay on each VP, OLBD is the maximum OLB delay, and NH is the number of VP hops.

Similarly, for A-NTS traffic, the VP control selects the VP set that satisfies the average VP delay requirement D^A:

$$D_a + DVP + D_p \leq D^A, \qquad (**)$$

where D_a is the average VP cell access delay and DVP is the average cell delay (due to queueing in OLB) on the VP.

All OLBs have the same size, which is engineered to satisfy Eq. (*) for VPs with the maximum permitted number of hops. The selected buffer size is used to determine VP MPR assignment satisfying the loss requirements. The reservation rate for A-NTS traffic is selected by decreasing the peak rates in the same proportion for all classes of A-NTS traffic. The decrease ratio is chosen to satisfy Eq. (**) .

Since $D^A > D^E$, the control may select for non-real-time traffic the VPs with the larger number of hops. This will result in a more efficient use of the VP set. Alternatively, to simplify the control, the VP set may be used uniformly by all types of traffic. In this case, since $D^A > D^E$, a larger access delay D_a is permitted, leading to a smaller amount of bandwidth that needs to be reserved for A-NTS calls (see [7,8]).

E-NTS and A-NTS calls must be assigned different VPIs. It can be achieved, for example, by assigning E-NTS and A-NTS calls to different VPs. Alternatively, the VP-level control may assign them to the same VP with different VPIs. Regardless of the assignment method, it is still desirable to minimize the number of VPIs required for E-NTS and A-NTS traffic separation.

The current congestion control scheme obviates a need for priority processing of E-NTS traffic at the source node and thus requires less cell-level processing than the strategy presented in [7,8]. On the other hand, the NTS separations require some additional processing on the VP level at the beginning of the control period and on the VC level for each call. Contrary to the scheme described in [7,8], the present strategy cannot use the statistical fluctuations of E-NTS traffic to save some bandwidth.

For simplicity, to explain the concept, we described only two types of NTS. For both NTSs, the amount of bandwidth to be reserved for a call remains constant for the duration of the call. The advanced assignment of VP bandwidth and distribution of VC-level control to the source nodes makes it possible to efficiently introduce another type of NTS. We will call it B-NTS. B-NTS is suitable for users that generate very long bursts of non-real-time traffic separated by silent periods (see [6]). The B-NTS call uses bandwidth reservation only during the burst. At the beginning of a session, a user requests a connection specifying the burst peak rate. If the connection is granted, the user requests permission to send a burst. The VC-level control at the source node checks to see if there is sufficient VP bandwidth to handle the burst with the required quality of service. When the bandwidth is available, the control reserves, at the source node, the burst peak rate on the VP and sends a positive acknowledgment. The user then transmits the burst. The last cell of the burst is a token that acknowledges the source node with the end of the burst transmission on the outgoing link. Then the reservation is released. Thus, the effective reservation rate for the duration of the B-NTS call is less than the peak rate for each of its bursts. The average/maximal delay within the backbone network for this NTS is guaranteed. The strategy can be modified to guarantee the average burst delay at the network access.

The bandwidth reservation for each burst is done at the user-network interface. No other backbone nodes are involved in the process. This leads to a smaller reservation overhead. It also makes the reservation process independent of propagation delay, which may be quite substantial in wide area networks. Traffic control schemes that are not based on the advanced assignment of VP bandwidth and in which the control is not done at the source node result in a higher overhead and slower burst admission to the network.

The node architecture shown in Figure 3 only illustrates the concept; other implementations are possible. For example, the VP ARR can be placed after the switching fabric, thus eliminating the need for the LASDs.

Note that B-NTS and E-NTS do not require the installation of LASD and VP ARRs. On the other hand, B-NTS introduces more user-network interactions.

The routing tables may also take into account the presence of multipoint connections. In [12], the performance of multipoint routing for interoffice networks was evaluated for a variety of criteria, including computational complexity, dynamic reconfiguration, robustness to node failures, and bandwidth efficiency. The conclusion of the study was that the use of multiple point-to-point connections for multipoint calls in interoffice networks is a viable solution in many cases. Even bandwidth efficiency of multiple point-to-point connections is good if the number of network backbone nodes involved in multipoint connection is less than five or if the interoffice network is highly connected. That conclusion was reached assuming that the load consists of multipoint connections only. If the multiple point-to-point connection demand represents only a fraction of the total traffic, the bandwidth efficiency of the multipoint connections become less important. In this paper, we focus on point-to-point traffic only, assuming that the multipoint demand within the interoffice network is allocated using the multiple point-to-point connection strategy.

OPTIMIZATION OF THE VP-LEVEL CONTROL

In this section, we consider the optimization of the VP-level control for backbone ATM networks. We assume that cell transfer time through a backbone node is much less than transmission time through a link, and thus the nodes are not the bottlenecks in network performance.

For presentation simplicity, we assume that E-NTS and A-NTS traffic are allocated on the same VC set and that the set satisfies requirements (*) and (**). We also assume that the control period is relatively short. More specifically, the control period is assumed to be less than the average holding time for all classes of traffic but much longer than the maximum end-to-end cell delay. To provide efficient utilization of network resources, it is then sufficient to adjust the VP maximal permitted rate over the same VP set at the beginning of each control period. The VP set can be adjusted a lot less frequently. Since the control period is relatively short, it is sufficient to optimize the VP rate adjustment by allocating the predicted demand (for an upcoming period) over the existing VC network load. The bandwidth demand and VC loads prediction techniques applicable for this case were presented in [9]. The optimization scheme can be easily modified to handle longer control periods.

Problem Formulation and the Main Properties of the Optimal Solution

Let us introduce some notations.

R_{ij} = the VP set for the backbone SD pair (i,j)

Z_{ij}^{s} = the total bandwidth that has been logically reserved for all VCs allocated on the sth VP ($s \in R_{ij}$) at the beginning of the upcoming control period

$Z_{ij} = \sum_{s \in R_{ij}} Z_{ij}^{s}$ = the total bandwidth that has been logically reserved at the beginning of the upcoming control period for SD pair (i,j)

$\tilde{Z}_{ij}$ = a portion of the total bandwidth Z_{ij} projected to be released during the control period and thus can be reused for bandwidth reservation

B_{ij} = the total projected bandwidth demand (for the upcoming period) for the logical bandwidth reservation for SD pair (i,j)

$\tilde{B}_{ij}$ = a portion of the total projected demand B_{ij} that exceeds the total released (for the period) bandwidth $\tilde{Z}_{ij}$

$$\tilde{B}_{ij} = \max(0, B_{ij} - \tilde{Z}_{ij})$$

h_{ij}^{s} = the portion of demand $\tilde{B}_{ij}$ to be reserved on s^{th} VP, $s \in R_{ij}$

h_{ij}^{0} = the portion of demand $\tilde{B}_{ij}$ that cannot be allocated on VP set R_{ij} due to insufficient capacity of VP set R_{ij}

$$H = (\{h_{ij}^s\}, \{h_{ij}^0\}) = (h, h^0)$$

The following flow conservation rule holds:

$$\sum_{s \geq 0} h_{ij}^s = \tilde{B}_{ij} \tag{1}$$

$X_l^0 =$ the total VP load on link l at the beginning of the upcoming control period
$X_l =$ the total VP rate to be assigned to link l for the upcoming control period

$$X_l = \sum_{ij} \sum_{s \in R_{ij}} h_{ij}^s \delta_{ij}^s (l) \tag{2}$$

where

$$\delta_{ij}^s(l) = \begin{cases} 1 \text{ if route s passes through link l} \\ 0 \text{ otherwise} \end{cases}$$

$q_l(X_l + X_l^0)$ = steady state probability of the cell loss on link
C_l = capacity of link l
p = cell loss requirement for a VP
$F_A = \sum_{i,j} h_{ij}^0$ – a fraction of total projected demand $\sum_{ij} \tilde{B}_{ij}$ which cannot be allocated for the period.
Q_{ij}^s = the sum of cell loss probabilities on route $s \in R_{ij}$

$$Q_{ij}^s = \sum_l q_l(X_l + X_l^0)\delta_{ij}^s(l) \tag{3}$$

With these notations, the capacity constraints and cell loss requirements can be written as follows:

$$X_l + X_l^0 \leq C_l \tag{4}$$

$$Q_{ij}^s \leq p \tag{5}$$

Let

$$F_L(h) = \sum_l \int_0^{X_l} q_l(y_l + X_l^0) dy_l \tag{6}$$

We introduce the following objective function for the optimal VP bandwidth assignment problem (VP BAP):

$$F_0(H) = F_L(h) + pF_A \tag{7}$$

The function $F_0(H)$ is a weighted sum of the total rejected demand F_A and a monotonically increasing convex function of the total load assigned to each link. Note that pF_A represents provisioning of the total cell loss corresponding to the maximal allowed cell loss probability equal to p.

In our scheme, the VP bandwidth assignment allocates the bandwidth demand $\tilde{B}_{ij}$ as a logical overlay to the VC load that existed at the beginning of the control period. Thus, the VP BAP can be formulated as follows.

Given network topology, demand $\tilde{B}_{ij}$, VP set R_{ij}, link capacities C_l, and initial link loads X_l^0, *minimize* $F_0(H)$ *subject to* flow conservation constraints (1), capacity constraints (4), and loss requirement (5).

Since $q_l(X_l)$ is a monotonically increasing function of X_l and $p << 1$, the capacity constraint (4) follows from the stronger requirement (5). Thus, the constraint (4) may be omitted in the formulation. The advantage of function F_0 is that the cell loss probability requirement (5) can also be omitted (see Theorem further in the text). Thus the VP BAP can be rewritten as follows:

$$\min_{H} F_0(H) \tag{8}$$

s.t. conservation flow (1).

Let us introduce a notion of a slack VP for each pair (i,j). This VP has an unlimited capacity to be used for allocation of the rejected traffic h_{ij}^0 and is assigned an index $s = 0$. VP length L_{ij}^s is calculated by the following equation:

$$L_{ij}^s = \partial F_0/\partial h_{ij}^s = \begin{cases} Q_{ij}^s & \text{if } s > 0 \\ p & \text{otherwise} \end{cases} \tag{9}$$

Let

$$\bar{Q}_{ij} = \min_{s>0} Q_{ij}^s \tag{10}$$

Theorem: The optimal solution of the VP BAP has the following properties.

1) h_{ij}^k can be allocated on VP $k > 0$ if and only if the VP length L_{ij}^k is minimal and satisfies the cell loss requirement:

$$L_{ij}^k = \bar{Q}_{ij} \le p, \qquad k \in R_{ij} \tag{11}$$

2) No bandwidth demand is rejected ($h_{ij}^0 = 0$) if

$$\bar{Q}_{ij} < p \tag{12}$$

3) Bandwidth demand is rejected ($h_{ij}^0 \neq 0$) only if

$$\bar{Q}_{ij} = p \tag{13}$$

Proof

The objective function (7) is a convex, separable function of flow X_1. The conservation flow constraint (1) is linear. The result then follows directly from the theorems 9.6.2 and 3.C.2 in [13].

The theorem shows (see equation (11)) that the optimal VP rate assignment allocates the demand satisfying the cell loss requirement (5). Thus, constraint (5) may be omitted.

Optimization Algorithm for VP Logical Bandwidth Assignment Problem

We will use an Accelerated Flow Deviation algorithm similar to the one described in [14] for solving the VP BAP. The algorithm assigns to all VPs the virtual bandwidth sufficient to accommodate as much of bandwidth demand $\tilde{B}_{ij}$ as feasible. It also calculates the portion of the demand to be rejected at the source node. We do not require constraints (5) to be satisfied during the execution of the algorithm. The algorithm convergence has been proven using the technique described in [14].

Algorithm

Step 1. Set all $X_l(0) = 0$ and accommodate the bandwidth $\tilde{B}_{ij}$ on the slack routes.

Step 2. Compute Q_{ij}^s as a function of X_l using Eq. (3).

Step 3. Find the minimum length VP for each SD pair using Eqs. (3) and (9) by executing a shortest path algorithm in metric (3), (9).

Step 4. Update $X_l(m)$ by the following formula:

$$X_l(m+1) = (1-\gamma)\, X_l(m) + \gamma \sum_{s \in R_{ij},\, s>0} \tilde{B}_{ij} \Theta_{ij}^s(m) \delta_{ij}^s(l) \tag{14}$$

where $0 < \gamma < 1$ and

$$\Theta_{ij}^s(m) = \begin{cases} 1 \text{ if } L_{ij}^s(m) = \bar{Q}_{ij}(m) \leq p,\ s > 0 \\ 0 \text{ otherwise} \end{cases}$$

$$\sum_{s \geq 0} \Theta_{ij}^s(m) = 1$$

Step 5. Perform the feasibility and stability test. If passed, GOTO Step 6; otherwise GOTO Step 2.

Step 6. Compute h_{ij}^s for $s > 0$ as follows:

$$h_{ij}^s = \frac{\sum_{k=1}^{n} \theta_{ij}^s(m+k)\, \tilde{B}_{ij}}{n}$$

where n is a preset number (see [14] for details).

$$h_{ij}^0 = \tilde{B}_{ij} - \sum_{s>0} h_{ij}^s$$

The step size γ should slowly decrease with iterations to provide a sufficient accuracy (see (15,16]). The algorithm can also be implemented in a decentralized fashion using the scheme similar to the one described in [16,17].

MBS Policy

The VP bandwidth assignment algorithm calculates the excess demand h^0 that cannot be admitted into the network. This excess demand has to be split among traffic classes to be rejected at their corresponding source nodes. We suggest to split this demand according to the profitability b_k of its class. Let $h_{ij}^0(k)$ be the amount of the rejected demand h_{ij}^0 to be assigned to class k. Then

$$h_{ij}^0(k) = \frac{b_k^{-1}}{\sum_r b_r^{-1}} h_{ij}^0 , \tag{15}$$

$$b_k = \frac{a(k)}{m(k)} u(k) , \tag{16}$$

where

a(k)	—	Revenue generated per class k call.
m(k)	—	Bandwidth reserved for class k calls.
u(k)	—	An estimate of class k service rate (see [9]).

The total maximal VP rate V_{ij}^s to be in effect for an upcoming control period is composed from the existing VC load Z_{ij}^s and the additional logical overlay rate h_{ij}^s:

$$V_{ij}^a = Z_{ij}^s + h_{ij}^s, \tag{17}$$

where h_{ij}^s is the solution produced by the optimization algorithm.

The VP-level control assigns E-NTS and A-NTS to the VPs, thus splitting the rate V_{ij}^s between these two types of traffic. In this process, we have to take into account the breakdown of Z_{ij}^s between these two types of NTSs. The VP-level control downloads V_{ij}^s and h_{ij}^s to VC and cell-level controls.

ANALYSIS OF CELL LOSS PROBABILITY

The VP assignment algorithm uses the cell loss probability q_l in computing (14). Statistics of cell arrivals for new services depend on information transfer content and are often unknown. This feature and the high speed of ATM networks make it very important to obtain a reliable estimate of cell loss probability q_l in order to guarantee quality of service. For that reason, we derive conservative and robust estimates of cell loss probability.

We make the following assumptions:

1. Local discrete time of a time slot is equal to cell transmission time on a physical link.
2. Cells arrive in bursts of size L, and there is no more than one burst per slot.
3. There is no correlation between bursts' arrivals.

Selecting L sufficiently large, one can obtain a conservative and robust estimate of packet loss probability in an OLB. Note that any network design and sizing imposes certain restrictions on the maximum burst size L.

We consider a discrete Markov model for queuing behavior in OLB of size d, which is engineered to satisfy maximum delay requirements (*). The states v, $0 \le v \le d$ of the Markov chain correspond to the number of cells in the queue. Since the cell loss probability of interest is sufficiently small, we use the infinite buffer model for estimation of buffer overflow probability q_l for OLB.

The offered traffic ρ_l on physical link l is defined as the average number of cell arrivals in a time slot. Under the assumptions 1–3, we have

$$\rho_l = L u_l \le 1, \tag{18}$$

where u_l is a probability of arrival of burst on link l in a time slot. Thus,

$$u_l = \rho_l/L. \tag{19}$$

ρ_l satisfies the steady state conservation law and is computed by the following formula:

$$\rho_l = \frac{X_l^0 + X_l}{C_l(1 - q_l)} \approx \frac{X_l^0 + X_l}{C_l}. \tag{20}$$

Average number N_l of lost cells is

$$N_l = \sum_{k=0}^{(L-2)} u_l \cdot (L - k - 1)\, q_l^{d-k}\ .$$

Average number of arrived cells is ρ. Thus the cell loss probability q_l is

$$q_l = \frac{N_l}{\rho_l} \approx \frac{1}{L} \sum_{k=0}^{(L-2)} (L - k - 1) q_l^{d-k}.$$

where $u_l = \rho_l/L$ [see (19)].

Generating function $G_l(z)$ is defined as

$$G_l(z) = \sum_{\nu=0}^{\infty} q_l^{\nu} z^{\nu} \tag{21}$$

and is equal, after some algebra, to

$$G_l(z) = \frac{1 - \rho_l}{1 - (\rho_l/L) \sum_{i=0}^{L-1} z^i}\ . \tag{22}$$

From Eq. (22), it follows that if $L = 1$, then $q_l^{\nu} = 1$, i.e., the OLB is always empty. Let

$$\beta_l(L) = \frac{\rho_l/L}{1 - \rho_l/L} < 1. \tag{23}$$

Then for $L = 2$, we have, after some algebra,

$$q_l^{\nu} = \frac{\beta_l^{\nu}(1 - \rho_l)}{1 - \rho_l/2}\ , \tag{24}$$

where $\beta_l = \beta_l(2)$. Thus, the cell loss probability is

$$q_l = q_l^d = \frac{\beta_l^d(1 - \rho_l)}{2(1 - \rho_l/2)}\ . \tag{25}$$

Note that for a finite buffer case and $L = 2$, one can obtain a very similar expression:

$$q_l = q_l^d = \frac{\beta_l^d(1-\rho_l)}{2(1-\rho_l/2)(1-\beta_l^{d+1})} = \frac{\beta_l^d(1-\beta_l)}{2(1-\beta_l^{d+1})} \tag{26}$$

From Eq. (25), one can see that if $q_l^d << 1$, then $\beta_l^{d+1} << 1$, and expression (26) is a good approximation of formula (24). For $L \geq 3$, probabilities q_l can be obtained numerically from the following recursive procedure:

$$q_l^v = \beta_l(L) \sum_{i=1}^{L-1} q_l^{v-i} \Delta_{v-i}, \qquad v \geq L-1 \tag{27}$$

$$\Delta_{v-i} = \begin{cases} 1 & \text{if } v \geq i \\ 0 & \text{otherwise} \end{cases} \tag{28}$$

$$q_l^0 = \frac{1-\rho_l}{1-\rho_l/L} \tag{29}$$

In addition, if d is very large, an asymptotic expression can be used for q_l^v when $d \rightarrow \infty$. This asymptote is obtained for $v \rightarrow \infty$:

$$q_l^v = \frac{(1-\rho_l)(\gamma_l - 1)}{\gamma_l^v(\rho_l \gamma_l^{L-1} - 1)}, \tag{30}$$

where $\gamma_l > 1$ and is defined as a root of the following equation:

$$\frac{1}{L} \sum_{i=0}^{L-1} \gamma_l^i = 1/\rho_l \tag{31}$$

Note again that the equation (20) connects the total amount of VP rates assigned on link l and the offered traffic ρ_l for this link. This ties together the optimization algorithm and the calculation of cell loss probability. Cell loss probability is calculated for each iteration of the algorithm (see step 2 of the optimization algorithm) and can be done for each link independently in a parallel fashion.

Finally, from (22) one can obtain the following upper bound on probability q_l for $L > 2$:

$$q_l^v \leq [\beta_l(L)(L-2)]^v \frac{1-\rho_l}{1-\rho_l/L}, \tag{32}$$

where $\beta_l(L)(L-2) < 1$ and average queue length $E(v_l)$

$$E(v_l) = \frac{\rho_l(L-1)}{2(1-\rho_l)}. \tag{33}$$

CONCLUSION

We proposed the layered architecture for adaptive traffic control in ATM networks. The traffic control guarantees quality of service and simplifies call and cell processing. In the proposed architecture, many important control functions, such as call admission and routing decisions, VC bandwidth reservations, and cell-level congestion control, are efficiently performed at the boundary of the backbone network. These functions do not require the source nodes to communicate with other nodes. To transport information efficiently, all classes of traffic are divided into three NTSs. The suggested traffic control permits less than a peak rate reservation for two of these NTSs that are suitable for non-real-time bursty traffic. We also presented an efficient algorithm for VP dynamic bandwidth assignment that equalizes cell losses over the VP set and maximizes call and cell throughput. Finally, we developed a simple model for calculation of cell loss probability.

ABOUT THE AUTHORS

Alexander Gersht received the BS and MS degrees in applied mathematics from Voronezh and Gorky University in 1956 and 1958, respectively, and the PhD degree in operations research and statistics from the Academy of Science of the USSR, Moscow, in 1970.

He conducted research until 1979 at the Computer Center of the Academy of Science, Leningrad, in the area of performance analysis, routing and design of circuit switched networks, automata theory, distributed adaptative control, and operations research. Since 1981, he has been with GTE Laboratories Incorporated, Waltham, Massachusetts. As Principal Investigator, he led projects on large network analysis, design, and planning. His current research interests focus on broadband networks and include traffic and fault management, congestion control, and distributed control for B-ISDN.

Alexander Shulman is a Principal Member of Technical Staff at GTE Laboratories Incorporated, Waltham, Massachusetts. He received the BS and MS degrees from Bielorussian University in 1977 and 1978, respectively, and the PhD degree from Boston University in 1986. He is a member of ORSA. His research interests include capacity expansion of telecommunication networks and network design, routing, and congestion control in broadband networks.

Jelena Vucetic has been with GTE Laboratories Incorporated, Waltham, Massachusetts, since 1990. Her research interests focus on routing in communication networks, including routing in multidomain networks; modeling and performance evaluation of communcation and computer systems; development and analysis of hardware and software (all layers of protocols) in a multidomain integrated services digital network; and resource allocation in the second- and third-generation wireless networks.

Julian Keilson received the PhD in physics from Harvard (1950), was at Harvard (1950–1952), Lincoln Laboratory (1952–1956), and Sylvania's Applied Research Laboratory (1956–1966). From 1966 to 1986, he was Professor of Statistics and Operations Research at the University of Rochester. From 1986 to 1993, he divided his time between GTE Laboratories Incorporated as Senior Scientist in the Network Architecture and Services Laboratory and the Sloan School at the Massachusetts Institute of Technology as Adjunct Professor. In 1993 he returned to teach at the University of Rochester. He is the author of *Green's Function Methods in Probability* and *Markov Chain Models, Rarity and Exponentiality,* is past Editor of *Stochastic Processes and their Applications,* and past president of the ORSA/TIMS Applied Probability College.

REFERENCES

[1] J.P. Coudreuse, P. Adam, and P. Gonet, "Asynchronous Time-Division Switching: The Way to Flexible Broadband Communication Networks," *Proceedings of the International Zurich Seminar on Digital Communications '86,* March 1986.

[2] J.S. Turner, "Design of an Integrated Services Packet Network," *IEEE JSAC,* SAC-4, No. 8, pp. 1373–1380, November 1986.

[3] Special Issue on Congestion Control in High Speed Packet Switched Networks, *IEEE JSAC,* Vol. 9, No. 7, September 1991.

[4] Special Issue: Congestion Control in ATM Networks, *IEEE Network, the Magazine of Computer Communications,* Vol. 6, No. 5, September 1992.

[5] Recommendation 1.371. "Traffic Control and Congestion Control in B-ISDN," Geneva, 1992.

[6] Jean-Yves Le Boudec, "The Asynchronous Transfer Mode: A Tutorial," *Computer Network and ISDN Systems 24,* pp. 279–309, 1992.

[7] A. Gersht and K.J. Lee, "A Congestion Control Framework for ATM Networks," *IEEE JSAC,* Vol. 9, No. 7, pp. 1119–1130, September 1991.

[8] A. Gersht and K.J. Lee, "A Congestion Control Framework for ATM Networks," *Proceedings of Infocom '89,* pp. 701–710, April 1989.

[9] A. Gersht and S. Kheradpir, "Integrated Traffic Management in SONET-Based Multiservice Networks," *Proceedings of the 13th ITC Congress,* Copenhagen, 1991.

[10] A. Gersht and K.J. Lee, "Virtual Circuit Load Control in Fast Packet-Switched Broadband Networks," *Proceedings of Globecom '88,* pp. 7.31–7.37, December 1988.

[11] A. Gersht and K.J. Lee, "A Call Admission Control Strategy in ATM Networks," submitted for publication in *IEEE Transactions on Communications.*

[12] K.J. Lee, A. Gersht, and A. Friedman, "Multipoint Connection Routing," *International Journal of Analog and Digital Communication Systems,* Vol. 3, pp. 177–186, 1990.

[13] C.B. Garcia and W.I. Zangwill, *Pathways to Solutions, Fixed Points and Equlibria,* Prentice-Hall Inc., Englewood Cliffs, 1981.

[14] A. Gersht and A. Shulman, "Optimal Routing in Circuit Switched Networks," *IEEE Transactions on Communications,* Vol. 37, November 1989.

[15] A. Gersht and S. Kheradpir, "Real-Time Decentralized Traffic Management Using a Parallel Algorithm," *Proceedings of Globecom '90,* pp. 408–414, December 1990.

[16] A. Gersht and S. Kheradpir, "Real-Time Decentralized Traffic Management Using a Parallel Algorithm," *IEEE Transactions on Communications,* Vol. 41, No. 2, February 1993.

A MODELING APPROACH FOR THE PERFORMANCE MANAGEMENT OF HIGH SPEED NETWORK [1]

P. Sarachik, S. Panwar, P. Liang, S. Papavassiliou,
D. Tsaih and L. Tassiulas

Polytechnic University
Center for Advanced Technology in Telecommunications
333 Jay Street
Brooklyn, N.Y. 11201

INTRODUCTION

As data networks have become larger, higher speed and more complex, there has arisen a growing need for advanced modeling, management and planning tools which will assist network operators maintain and improve network performance. This paper describes the analytical techniques used in a system tool that will simulate and analyze packet switched networks carrying bursty traffic and which can be used for a variety of networks and services such as frame relay, SMDS or B-ISDN. This network tool is based upon models of network elements. The information needed to drive the tool and describe its elements will vary from detailed to sketchy. Some information will be derived from historical measurements and some from expected characteristics of users based on responses to a questionnaire answered at the time of subscription to the data service.

By using the tool based on the analytical methods described in this paper, operators will be able to alter network elements and configurations in the model, analyze performance data, and then identify the consequences of network changes, including the evaluation of proposed configurations. The tool will also be used to compare the performance of different networks, to compare different configurations of the same network, and to evaluate network performance during specified intervals. In this paper, we describe the analytical models that were developed for this tool.

A PACKET SWITCHED NETWORK

In general a packet switched network can be represented as a set of nodes interconnected by a set of links. This gives a topological characterization of the network. Customers can send traffic over the network by connecting to one of the nodes via an input interface and sending messages to an output interface at the same or some other node. This is illustrated in Figure 1. The nodes are the network switches, the links are the trunks connecting them or to customer access lines.

[1]This work was supported in part by NYNEX Science and Technology and by the New York State Foundation for Science and Technology as part of its Centers for Advanced Technology Program and by the National Science Foundation under Grants NCR-9003006 and NCR-9115864.

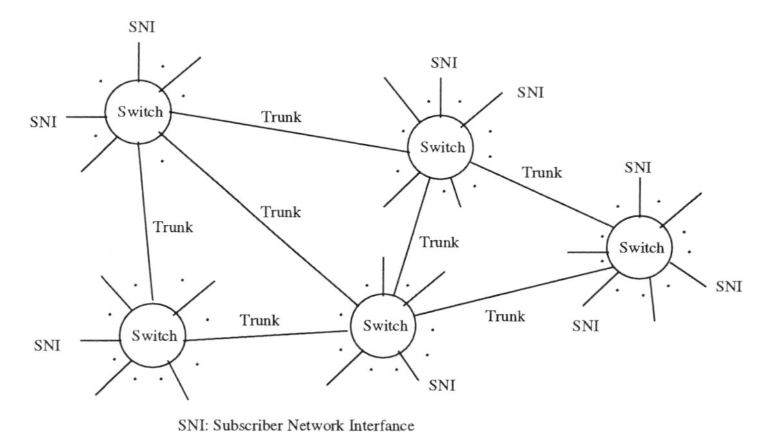

Figure 1. A Packet Switched Network

The input parameters that affect the performance of the entire network are described in items 1–3 below.

1. The traffic requirements that the entire network must accommodate. This can be described by a traffic matrix

$$\overline{R} = \begin{bmatrix} \overline{R}_{11} & \overline{R}_{12} & \dots & \overline{R}_{1N} \\ \overline{R}_{21} & \overline{R}_{22} & \dots & \overline{R}_{2N} \\ \vdots & \vdots & \dots & \vdots \\ \overline{R}_{N1} & \overline{R}_{N2} & \dots & \overline{R}_{NN} \end{bmatrix}$$

   where the sub matrix $\overline{R}_{\ell n} = [\overline{R}_{\ell n}(i,j)]$ is itself a traffic matrix whose elements $\overline{R}_{\ell n}(i,j)$ give the average rate of traffic entering the network at the ith interface module of switch ℓ which must be delivered to the jth interface module of switch n.

2. The burstiness of the arriving traffic. This can be characterized in a number of ways. A characterization of burstiness includes some statistical description of idle and bursty periods and the intensity of bursts. One way it can be characterized is in terms of the ratio of the variance to the squared mean of the inter-arrival times. This parameter v_a^2, called the squared coefficient of variation for inter-arrivals, is used by the QNA method[1][2] discussed later. An alternate description is in terms of the three parameters peak rate R_{peak}, utilization ρ and the average burst length b. This three parameter approach is used by the equivalent capacity method[3].

3. The associated packet length distribution matrices

$$\overline{L} = \begin{bmatrix} \overline{L}_{11} & \overline{L}_{12} & \dots & \overline{L}_{1N} \\ \overline{L}_{21} & \overline{L}_{22} & \dots & \overline{L}_{2N} \\ \vdots & \vdots & \dots & \vdots \\ \overline{L}_{N1} & \overline{L}_{N2} & \dots & \overline{L}_{NN} \end{bmatrix}$$

   which gives the packet lengths for the traffic above and a corresponding matrix of squared coefficients of variation for packet lengths.

By routing the traffic load, described above, through the network we can obtain the traffic load on each of the N switches. In order to perform this routing it is necessary to first model the delays and packet losses associated with the switches.

CLASSIFICATION AND CHARACTERIZATION OF TRAFFIC

In high-speed packet switched network architectures, several classes of traffic streams with widely varying traffic characteristics are statistically multiplexed and share common switching, buffering and transmission resources. Because of the potentially dramatic differences in the statistical behavior of connections and the associated problems posed for bandwidth management and traffic control, we will classify the different users and applications that require packet switching service.

We will identify a number of different classes of traffic, which should include most network users (either directly or as a mixture of such traffic types, so that their overall traffic will be determined by the percentage of each of the different classes they use), according to their traffic patterns, behavior and characteristics. For each of these classes we will determine traffic parameters which will help us to develop a realistic model for the source traffic in the analytical model. We present a number of services in association with different users that require packet switching service. We identify expected traffic patterns, behavior and characteristics for these users.

1. LAN Interconnection Traffic: LANs support a wide variety of users and applications, with different performance requirements and traffic characteristics. Possible applications include distributed file systems and databases, host-peripheral connections, demand paging, parallel processor interconnection, computer servers, disk less work stations, electronic mail, file transfer, etc. These applications are supported by the Network Disk (ND) protocol, the TCP and the UDP protocols among others. According to recent results of LAN traffic analysis studies conducted at Bellcore[4][5][6], the protocols above have the following traffic characteristics.

Protocol	Traffic
TCP	80% of 50 bytes, 10% of 500 bytes , 10% of 1Kbyte
UDP	80% of 150 bytes, 10% of 1.5 Kbytes
ND	80% of 1550 bytes, 20% of 50 bytes

The utilization due to ND traffic is comparable to that of UDP traffic, while TCP traffic represents a small fraction of the total network traffic. The combination of the above traffic types results in a "trimodal " distribution for the overall traffic. The largest percentage of packets are less than 200 bytes in size while there are considerable spikes at lengths of 1 Kbyte and 1550 bytes.

2. Video applications

Video Telephony, Multimedia Teleconferencing: The data rate ranges from 1.5 Mbits/sec to 140 Mbits/sec depending on the encoding, the compression, and the service quality requirements.

Compressed Real Time Packet Video: Transmission of good quality compressed video requires at least 1.5 Mb/sec with small video packet loss and delay [7]. Given the study in reference [8] and for 1.5 Mbits/sec transmission rates, we conclude that a video frame can be segmented into 4 to 5 packets of 1500 bytes with inter-arrival times of approximately 8 msec.

3. Teletex(Correspondence exchange): With a transmission rate of 9.6 Kbps, page data volume of 20 Kbits and packet sizes of 1500 bytes, if we consider one page as a burst, then the mean burst period is about 2 secs while the packet inter-arrival time is 1 sec.

4. Facsimile: A bit rate of 2 Mbits/s (for compressed data) and more (for uncompressed data) is desirable in order to achieve acceptable transmission times for compatibility with the operation speeds of document preparation equipment of the near future.

5. High-resolution graphics – Electronic Imaging (Medical Imaging, CAD–CAM applications): Based on the typical transmission speed of work stations which is about 0.5 Mbits/sec and using as a packet size the maximum allowable Ethernet packet of about 1530 bytes, while assuming that a single image or graph represents a burst, we conclude that the burst length is on the order of minutes and the mean packet inter-arrival time is approximately 25 msec.

THE QNA METHOD OF NETWORK ANALYSIS

The Queuing Network Analyzer (QNA) is a software tool developed at AT&T Bell Telephone Laboratories to evaluate the performance of a queuing network. It is based on work described by W. Whitt [1][2]. Each node is viewed as a single GI/G/m queue. The analysis is approximated by considering the first and second moments of the inter-arrival time, packet length and service time distribution. It gives an approximate performance measure at the network level, instead of investigating the performance of particular queues at each node. We decided to use this approach in our modeling effort because it allows the characterization of traffic merged from many component streams as well as traffic split from a common stream.

The parameters that characterize the flow of packets to a switch are the means and squared coefficients of variation of the inter-arrival time and packet length; namely $E\{T_a\}$ (or the average arrival rate $\lambda = 1/E\{T_a\}$), $E\{L\}$, v_a^2 and v_L^2. Here v^2 is the variability parameter or squared coefficient of variation, which is defined to be $\mathrm{Var}(T)/E^2(T)$. The service is characterized by an average service time $E\{X\}$ (or the average service rate $\mu = 1/E\{X\}$) and v_X^2. The service discipline is assumed to be first-in-first-out.

The basic methodology is to merge all traffic flows (characterized by the above four parameters) entering an input interface of a node or queue into one combined flow. After service, the parameters of the departure flow, different from those of the input, can also be approximately characterized. This four parameter characterization can be used for all flows within the network. The output of QNA gives the mean and variance of the delay at each queue by using approximate GI/G/m formulas.

Let
W be the packet waiting time of a queue;
λ be the packet arrival rate to the queue;
X be the packet service time of the queue, which is equal to the packet length L (in bits or bytes) divided by the service rate R_s (in bits or bytes per second) of the queue, L/R_s;
$\rho = \lambda E\{X\}$ be the load on the queue;
v_T^2 be the squared coefficient of variation of the packet inter-arrival time;
v_X^2 be the squared coefficient of variation of the packet service time, which is equal to that of the packet length v_L^2.

The mean waiting time of the queue is approximated by

$$E\{W\} = g\frac{\rho E\{X\}(v_T^2 + v_X^2)}{2(1-\rho)} \tag{1}$$

where $g = \exp\left(-\frac{2(1-\rho)(1-v_T^2)^2}{3\rho(v_T^2+v_X^2)}\right)$ if $v_T^2 < 1$ and $g = 1$ if $v_T^2 > 1$, is a function of the traffic and service parameters.

Once the traffic parameters for a queue have been estimated, we can use formula (1) to determine an approximation to the mean waiting time. After the queue service, we want to know the departure traffic parameters, which are input parameters to the next queue. If there is no loss in the queuing system, the packet rate, the mean packet length and the variability parameter of the packet length are the same as those of the input traffic parameters, but the variability parameter of the packet inter-departure time, v_D^2, is changed. This can be estimated by Marshall's formula [9],

$$v_D^2 \approx v_T^2 + 2\rho^2 v_X^2 - \frac{2\rho(1-\rho)E\{W\}}{E\{X\}} \tag{2}$$

which can be further approximated as

$$v_D^2 \approx (1-\rho^2)v_T^2 + \rho^2 v_X^2 \tag{3}$$

simply by letting $g = 1$ in the estimation of $E\{W\}$ and substituting it into (2). Equation (3) is the approximation used by [1].

If the departure traffic is split into many traffic streams after the queue service, the QNA method also gives a way to calculate the parameters for the split traffic stream. The packet rate, packet length and the variability parameter of the packet length of each stream are easily calculated, because those parameters are dependent on the traffic to each stream and independent of the queue server. Again, we have to estimate the variability parameter of the packet inter-arrival time to each stream. Suppose that a packet goes to the ith output stream with probability P_i, then, the variability parameter of the packet inter-arrival time, $v_{T_i}^2$, is

$$v_{T_i}^2 = 1 - P_i + P_i v_D^2 \tag{4}$$

These parameter estimation methods and the method for combining traffic are the basic techniques used in the QNA method. We will refer to these formulas quite often in the later discussion.

MODELING DELAY

Many switches are based on shared bus or polling architectures and can be modeled by using a round robin service discipline (i.e., the server visits each station in order and serves at most one packet during each visit). A typical switch of this type, used for Frame Relay Service, is shown in Fig.2. It consists of three parts: an input module, a switch module and an output module. An input packet (or a frame of data) from a Customer Premises Equipment (CPE) access line or a trunk connected to the switch is detected and verified by an Access Processor (AP) in the input module. Valid packets are forwarded to the transmit buffer of the cyclic server and wait for the switch server to switch them to their destination output module. Packets arriving at the output module are stored in the transmit cache for the AP to perform the HDLC processing. They are then multiplexed and transferred to the output line. The switch discipline is round robin and at most one packet is switched when a queue is polled. In the output module, the packet service rate depends mainly on the transmission rate of the output line. A queuing model for the switch module is shown in Fig. 3.

There are three queues which must be modeled; the input queue, the switch queue and the output queue. There are also some packet transfer delays in or between the modules which are proportional to the packet size and can be placed anywhere in the queuing model. For simplicity, we place all transfer delays which occur before the switch module into the input module and the transfer delays which occur after the switch module into the output module.

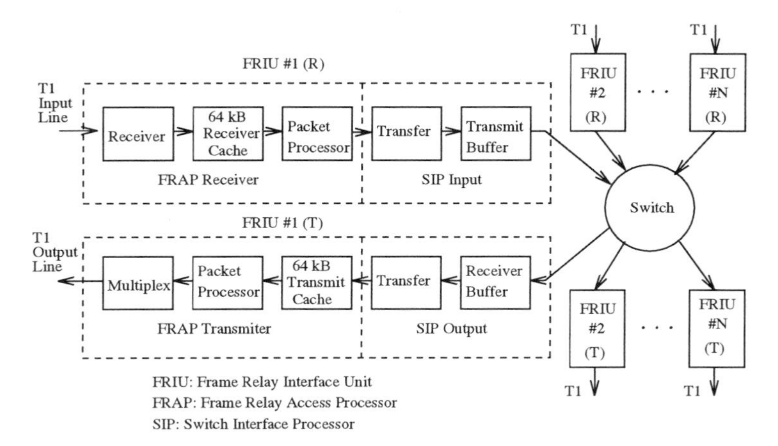

Figure 2. Switch Transport Process

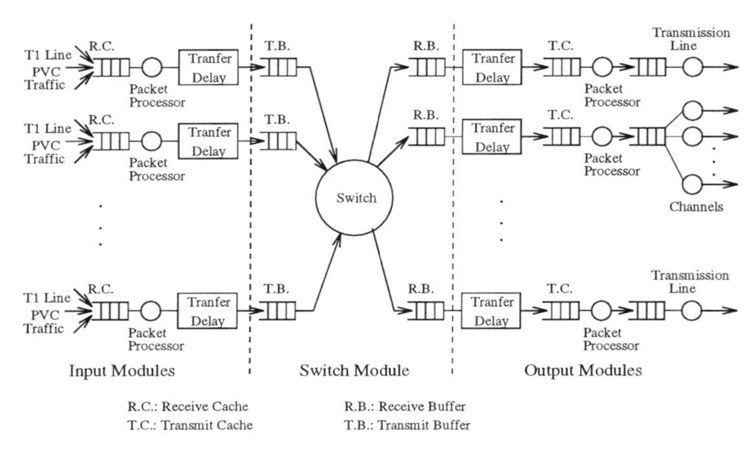

Figure 3. Equivalent Model for the Switch

The transfer delays at the input and output modules are assumed to be equal, because similar data transfers are necessary before and after the switch module.

Input Traffic Description

Traffic to a switch node either comes from a previous node through a trunk or enters at the node through an access line from CPEs. A source CPE sends its traffic through the network via a Permanent Virtual Circuit (PVC) to the destination CPE. PVC k (denoted by PVC_k), which originates at input line (or channel) i of a node and goes to output line (or channel) j, has four parameters: the packet arrival rate α_k, the squared coefficient of variation of packet inter-arrival times $v_{A_k}^2$, the mean packet length $E\{L_k\}$ and the squared coefficient of variation of the packet length $v_{L_k}^2$. Those four parameters are observed before the traffic is transmitted to the input module, hence, when many PVCs share an input line, the input traffic is the PVCs' combined traffic with four parameters: the packet arrival rate λ_{1i}, the squared coefficient of variation of the packet inter-arrival time $v_{T_{1i}}^2$, the mean packet length $E\{L_{1i}\}$ and the squared coefficient of variation of the packet length $v_{L_{1i}}^2$. A simple way to estimate the parameters of the combined traffic is by the QNA method [1].

If an input line is a trunk or a full access line, all PVCs coming through the line are merged in the line with a line service rate of 1.344 Mbps (for a T1 one line).

When PVCs come through channel m of input module i, similarly, the combined four parameters, λ_{1im}, $E\{L_{1im}\}$, $v^2_{T_{1im}}$ and $v^2_{L_{1im}}$ for that channel can be found by considering each channel as an input line with the service rate of the channel. The channel traffic streams are further combined and form the traffic of input module i. Thus, there are two merges for the input traffic through channelized lines.

Delay Model of the Input Module

An input module i has a packet processor, which processes each incoming packet with a constant processing time $E\{X_{1i}\}$. The load on the queue is $\rho_{1i} = \lambda_{1i}E\{X_{1i}\}$. The waiting time of the packet processor is[1,2]

$$E\{W_{1i}\} = g_{1i}\frac{\rho_{1i}E\{X_{1i}\}(v^2_{T_{1i}} + v^2_{X_{1i}})}{2(1-\rho_{1i})} \tag{5}$$

where $g_{1i} = \exp\left(-\dfrac{2(1-\rho_{1i})(1-v^2_{T_{1i}})^2}{3\rho_{1i}(v^2_{T_{1i}} + v^2_{X_{1i}})}\right)$ if $v^2_{T_{1i}} < 1$ and $g_{1i} = 1$ if $v^2_{T_{1i}} > 1$ and the variability parameter $v^2_{X_{1i}} = 0$ in this case.

The transfer delay is proportional to the packet size. Let the transfer rate be R_t, which depends on the switch fabric and can be measured or calculated on a real system. Because the transfer rate is typically very high, there is no queuing delay but there is a delay in addition to the packet delay. If there are M transfers in the input module, the total input transfer delay is ML/R_t, where L is the packet length.

The packet departure rate and the packet length parameters after the processor are the same as for the input to the processor, but the squared coefficient of variation of the inter-arrival time has been changed. By the QNA method, the variability parameter, $v^2_{T_{2i}}$, is estimated by

$$v^2_{T_{2i}} \approx (1-\rho^2_{1i})v^2_{T_{1i}}. \tag{6}$$

The four parameters for the input to the switch are then determined.

The Switch Module

The switch module is modeled as a non-exhaustive cyclic queue with at-most-one packet served per polling of the queue. In this part, we combine some basic methods and try to develop a better delay model. We first discuss the switch model for Poisson input traffic and then modify it for general input traffic.

The Conditional Cycle Times

Define C''_i to be the conditional cycle time random variable for queue i, given that a packet from queue i is served in the cycle; C'_i to be the conditional cycle time random variable for queue i given that no packet from queue i is served in the cycle.

Kuehn [10] has approximate formulas for mean cyclic times, $c''_i = E\{C''_i\}$ and $c'_i = E\{C'_i\}$. Let $c''_{\sim i}$ and $c'_{\sim i}$ be Kuehn's approximate values, then

$$c''_{\sim i} = \frac{s_0 + h_i}{1 - \rho_0 + \rho_i} \tag{7}$$

$$c'_{\sim i} = \frac{s_0}{1 - \rho_0 + \rho_i} \tag{8}$$

where $\rho_i = \lambda_{2i} h_i$ is the load contributed to the switch by the queue i traffic and $\rho_0 = \sum_{i=1}^{N} \rho_i$ is the total switch load for N inputs. λ_{2i} is the packet arrival rate to the switch input queue i, which is equal to λ_{1i}, if there is no packet loss at the input module; h_i is the average packet service time, which is equal to the packet length $E\{L_{1i}\}$ over the switch rate R_w; and s_0 is the sum of the mean polling times s_i for each queue i.

Equations (7) and (8) give extreme approximations for the conditional cycle times. If $c_0 = s_0/(1 - \rho_0)$ is the average cycle time, the relationship among the cycle times is

$$c''_{\sim i} \geq c''_i \geq c_0 \geq c'_i \geq c'_{\sim i}. \tag{9}$$

We have found another approximation method for c''_i and c'_i. Let $c''_{\simeq i}$ and $c'_{\simeq i}$ be these approximate values where

$$c''_{\simeq i} = s_0 + h_i + \sum_{j \neq i} \delta''_j h_j \tag{10}$$

$$c'_{\simeq i} = s_0 + \sum_{j \neq i} \delta'_j h_j. \tag{11}$$

Here $\delta''_j = \lambda_{2j} c''_{\simeq j}$ is the probability that a packet from queue j is served in the cycle, given that a packet from queue j is served in the previous cycle; $\delta'_j = \lambda_{2j} c'_{\simeq j}$ is the probability that a packet from queue j is served in the cycle, given that no packet from queue j is served in the previous cycle.

The solutions to the linear equations (10) and (11) are

$$c''_{\simeq i} = \frac{s_0 + h_i(1 - d_1) + d_3}{1 - \rho_0 + \rho_i(1 - d_1) + d_2} \tag{12}$$

$$c'_{\simeq i} = \frac{s_0}{1 - \rho_0 + \rho_i(1 - d_1) + d_2} \tag{13}$$

where $d_1 = \sum_{j=1}^{N} \frac{\rho_j}{1 + \rho_j}$; $d_2 = \sum_{j=1}^{N} \frac{\rho_j^2}{1 + \rho_j}$ and $d_3 = \sum_{j=1}^{N} \frac{h_j \rho_j}{1 + \rho_j}$. Equations (12) and (13) also are upper and lower bound approximations for the conditional cycle times.

By combining Kuehn's approximations and our approximations, a better approximation for the conditional cycle times can be acquired. Let $\epsilon''_{ij} = \min(\lambda_{2i} c''_{\sim j}, \lambda_{2i} c''_{\simeq j}, \delta''_i, 1)$ and $\epsilon'_{ij} = \min(\max(\lambda_{2i} c'_{\sim j}, \lambda_{2i} c'_{\simeq j}, \delta'_i), 1)$, which gives a better approximation for the conditional probabilities that a packet from queue j is served in a cycle, given that a packet from queue i is served in the cycle or given that no packet from queue i is served in the cycle. The approximations for the cycle times are

$$c''_{\cong i} = s_0 + h_i + \sum_{j \neq i} \epsilon''_{ij} h_j \tag{14}$$

$$c'_{\cong i} = s_0 + \sum_{j \neq i} \epsilon'_{ij} h_j \tag{15}$$

which give better approximations for the conditional cycle times, and can improve the queuing delay approximation.

Let $c''^{(2)}_{\cong i}$ and $c'^{(2)}_{\cong i}$ be the second moments of these conditional cycle times, then from [10], we get

$$c''^{(2)}_{\cong i} = \sum_{j=1}^{N} (s_j^{(2)} - s_j^2) + h_i^{(2)} - h_i^2 + \sum_{j \neq i} (\epsilon''_{ij} h_j^{(2)} - \epsilon''^2_{ij} h_j^2) + c''^2_{\cong i} \tag{16}$$

$$c'^{(2)}_{\cong i} = \sum_{j=1}^{N}(s_j^{(2)} - s_j^2) + \sum_{j\neq i}(\epsilon'_{ij}h_j^{(2)} - \epsilon'^2_{ij}h_j^2) + c'^2_{\cong i} \tag{17}$$

where $s_j^{(2)}$ is the second moment of the polling time and $h_i^{(2)}$ is the second moment of the packet service time for the ith queue of the switch, which is the second moment of L_{1i}/R_w.

In the above discussion, we assume that the switch is stable. For unstable systems, an additional step must be taken to identify the unstable queues and treat them differently.

The Mean Waiting Time of the Cyclic Queues

The mean waiting time of Poisson input traffic queuing systems depends on the mean residual time and the traffic load. A packet upon arrival to an input queue will find either that there is a head-of-line (HOL) packet or there is none. It sees different residual times in those two cases. Based on the renewal theory and M/G/1 queue theory, we can derive Kuehn's formula for the mean waiting time for queue i,

$$E\{W_{2i}\} = \frac{c_i'^{(2)}}{2c_i'} + \frac{\lambda_{2i}c_i''^{(2)}}{2(1-\lambda_{2i}c_i'')}. \tag{18}$$

Kuehn used his approximations for the first and second moment of the cycle times and had a delay approximation formula.

Another method developed by Boxma and Meister [11] is based on two assumptions:
1. $p_i = \lambda_{2i}c''_{\sim i}$ is the utilization observed at queue i;
2. All arrival packets see approximately the same mean residual time r.
The second assumption is good if the traffic load is light or the queues are not very unbalanced, otherwise a large error exists. Based on those assumptions, the mean waiting time is approximated by

$$E_{BM}\{W_{2i}\} = \frac{r}{1-\lambda_{2i}c''_{\sim i}} \tag{19}$$

where r is a parameter to be determined. Boxma and Meister used the conservation law, which was first developed by Watson [12], to evaluate r. This gives r and the waiting time approximation for queue i as

$$r \approx \frac{1-\rho_0}{(1-\rho_0)\rho_0 + \sum_{i=1}^{N}\rho_i^2}C_{NE} \tag{20}$$

$$E_{BM}\{W_{2i}\} \approx \frac{1-\rho_0+\rho_i}{1-\rho_0-\lambda_{2i}s_0}r \tag{21}$$

where

$$C_{NE} = \rho_0\frac{\sum_{i=1}^{N}\lambda_{2i}h_i^{(2)}}{2(1-\rho_0)} + \rho_0\frac{s_0^{(2)}}{2s_0} + \frac{s_0}{2(1-\rho_0)}(\rho_0 + \sum_{i=1}^{N}\rho_i^2).$$

Equation (21) gives a closed form formula for the queue waiting time approximation which, for moderate switch loads and slightly unbalanced queues, gives more accurate results than the approximation for (18) found by Kuehn's method. Later, some numerical results will show that when the switch load is high and queues are very unbalanced (the load of the maximally loaded queue is more than twice the load of the minimally load queue), large errors can exist, which are even worse than Kuehn's approximation. The errors produced are due to the two assumptions. If either one could be improved, the errors would be reduced.

The first method we introduced is to use $c''_{\cong i}$ as the approximation for the conditional cycle time c_i'' instead of $c''_{\sim i}$ in (19), then the residual time and the mean waiting time are approximated by

$$r_0 \approx \frac{(1-\rho_0)C_{NE}}{\sum_{i=1}^{N} \frac{\rho_i(1-\rho_0-\lambda_{2i}s_0)}{1-\lambda_{2i}c''_{\cong i}}} \tag{22}$$

$$E_1\{W_{2i}\} \approx \frac{r_0}{1-\lambda_{2i}c''_{\cong i}}. \tag{23}$$

Equation (23) gives a better approximation than (21), especially when the switch load is high and queues are highly unbalanced, due to a more accurate approximation for the mean cycle time c''_i.

Table 1. Waiting Times for the First Queue

Load	Simulation	Method 1	Method 2	BM	Kuehn
0.2	.4047(.012)	.4077(.74)	.4135(2.1)	.4073(.64)	.3860(-4.6)
0.4	.9959(.029)	.9975(.16)	1.026(3.0)	.9932(-.27)	.8683(-13.)
0.6	2.548(.068)	2.479(-3.)	2.562(.54)	2.436(-4.4)	1.941(-3.1)
0.8	11.13 (.25)	10.40 (-3.)	10.62(-2.)	9.602(-10.)	6.917(-35.)
0.85	22.18 (.65)	21.62(-2.5)	21.94(-.1)	18.82(-15.)	13.43(-39)
0.9	139.9(23)	163.2 (17.)	164.4(18.)	95.7(-32)	91.7(-34)

Table 2. Waiting Times for the Second Queue

Load	Simulation	Method 1	Method 2	BM	Kuehn
0.2	.3814(.013)	3.800(-.5)	.3766(-1.3)	.3801(-.34)	.3576(-6.2)
0.4	.8225(.031)	.8383(1.9)	.8219(-.07)	.8411 (2.3)	.7282(-11)
0.6	1.651(.044)	1.735(5.1)	1.692(2.48)	1.768(6.7)	1.421(-16.)
0.8	3.469(.086)	4.325(25.)	4.272(23.5)	4.935(42.)	2.800 (8.7)
0.85	4.255(.038)	5.959(40.)	5.931(39.0)	7.790(83.1)	5.856(37.6)
0.9	5.335(.056)	9.006(69.)	9.069(70.)	22.83(328.)	11.88(123.)

Table 3. Waiting Times for the Third and Fourth Queues

Load	Simulation	Method 1	Method 2	BM	Kuehn
0.2	.3780(.013)	.3665(-3.0)	.3584(-5.2)	.3671(-2.9)	.3413(-9.7)
0.4	.7487(.024)	.7655(-2.2)	.7277(-2.8)	.7709(-2.9)	.6478(-13.)
0.6	1.307(.023)	1.441(10.3)	1.336(-2.2)	1.484(13.5)	1.140(-13.)
0.8	2.259(.022)	2.906(28.6)	2.655(17.5)	3.422(54.8)	2.567(13.6)
0.85	2.569(.024)	3.621(40.9)	3.314(29.)	4.911(91.2)	3.653(42.2)
0.9	2.910(.027)	4.723(62.)	4.351(50.)	12.31(323.)	6.270(115.)

A second method we derived is to use equation (18) as a preliminary approximation for the mean waiting time by substituting c''_i, c'_i, $c''^{(2)}_i$ and $c'^{(2)}_i$ with $c''_{\cong i}$, $c'_{\cong i}$, $c''^{(2)}_{\cong i}$ and $c'^{(2)}_{\cong i}$. Let $E_K\{W_{2i}\}$ be this preliminary approximation for the waiting time of the ith input line. We then scale $E_K\{W_{2i}\}$ by a factor b so that Watson's conservation law is satisfied, to obtain $E_2\{W_{2i}\}$, the corresponding approximation of the mean waiting time.

$$E_2\{W_{2i}\} = b \cdot E_K\{W_{2i}\} \tag{24}$$

where b is given below,

$$b = \frac{C_{NE}}{\sum_{i=1}^{N} \rho_i(1-\lambda_{2i}c_0)E_K\{W_{2i}\}}. \tag{25}$$

When b is found, the mean waiting time $E_2\{W_{2i}\}$ is given by equation (24). In most situations, the second method gives better approximations than the other methods.

Tables 1, 2, 3 and 4 show some numerical results for delay in a 4 queue system with constant polling times $s_i = 0.05$ and exponential packet service times with means $h_i = 1$. The arrivals to the queues are Poisson with average rates $\lambda_1 = 2\lambda_2 = 4\lambda_3 = 4\lambda_4$. These values were chosen to yield an unbalanced system for which the load at queue 1 is twice that at queue 2 and four times the load at queues 3 and 4. Tables 1, 2 and 3 give the waiting time of the first queue, the second queue, and the third and fourth queues respectively. The first column is the total switch load for N inputs. Simulation results from a special purpose simulation program are given in the second column, with the 95th percentile confidence interval range shown in parentheses. The mean waiting time delay for each method and their percentage errors (compared with the simulation results) are shown in the remaining columns. The results show that improved methods one and two give better waiting time approximations than both Boxma-Meister's method and Kuehn's method. The difference between method 1 and 2 is not large.

In the above discussion, the mean waiting times of the cyclic server queues are for Poisson arrival traffic. If the input traffic is general with two parameters, λ_{2i} and $v^2_{T_{2i}}$, the formula for the waiting time must be modified. According to the GI/G/1 theorem[1], the mean waiting time, $E_G\{W_{2i}\}$, for general input traffic can be approximated by

$$E_G\{W_{2i}\} = g_{2i}\left(\frac{c''_{\cong i}\rho_{2i}(v^2_{T_{2i}}-1)}{2(1-\rho_{2i})} + E_2\{W_{2i}\}\right) \tag{26}$$

where $g_{2i} = \exp\left(-\frac{2(1-\rho_{2i})(1-v^2_{T_{2i}})^2}{3\rho_{2i}(v^2_{T_{2i}}+v^2_{L_{2i}})}\right)$ if $v^2_{T_{2i}} < 1$ and $g_{2i} = 1$ if $v^2_{T_{2i}} > 1$; $\rho_{2i} = \lambda_{2i}c''_{\cong i}$ is the utilization observed at queue i. Equation (26) gives us the waiting time for general input traffic.

Table 4 shows the results for the same system as Tables 1 to 3 except that the arrivals are not Poisson. The inter-arrival times have a squared coefficient of variation equal to 2. For simulation, a hyper-exponential distribution was used to generate the inter-arrival times. No comparison can be made to the Kuehn and Boxma-Meister results since these are not applicable to non-Poisson traffic. The results are not as accurate as for the Poisson arrival case, but are adequate for engineering purposes.

Table 4. Waiting Times for Non-Poisson Arrivals

Load	Q1 Sim	Method 1	Method 2	Q2 Sim	Method 1	Method 2	Q3,4 Sim	Method 1	Method 2
0.2	.560(.015)	.510(-8.9)	.516(-7.9)	.446(.016)	.433(-2.9)	.430(-3.7)	.407(.024)	.393(-3.4)	.385(-5.4)
0.4	1.66(.067)	1.32(-21)	1.35(-19)	1.11(.023)	1.01(-9.0)	.990(-11)	.887(.016)	.850(-4.1)	.812(-8.4)
0.6	5.74(.226)	3.39(-41)	3.47(-40)	2.74(.084)	2.19(-20)	2.14(-22)	1.71(.032)	1.66(-2.9)	1.55(-9.1)
0.8	29.7(6.17)	14.4(-52)	14.6(-51)	6.98(.628)	5.73(-18)	5.67(-19)	3.24(.094)	3.47(7.1)	3.22(-.62)
0.85	61.3(8.31)	29.8(-51)	30.1(-51)	9.48(.451)	7.96(-16)	7.93(-16)	3.80(.089)	4.36(15)	4.05(6.6)

Model of the Output Module

Packets coming to the output module go though two tandem queues for service. One is for packet processing and the other is at the output line. Packet processors perform the same function as in the input queue module. The output line is a T1 line, which may be partitioned into channels. Therefore, there are three possible transmission rates, 56 kbps, 384 kbps or 1.344 Mbps for the output transmission. The queuing model is simple. The only problem is to calculate the four incoming traffic parameters to each output module. We will estimate the traffic parameters to the output queues in the following subsections.

Traffic Parameters of the Output Modules

Let:
λ_{3j} be the packet arrival rate to the packet processor queue of the output module j;
$E\{L_{3j}\}$ be the average packet length of the traffic to the queue;
$v_{L_{3j}}^2$ be the squared coefficient of variation of the packet length;
$v_{T_{3j}}^2$ be the square coefficient of variation of the packet inter-arrival time to the queue.

Still assuming that all packets destined to output queue j arrive there without loss, then from [1], we get

$$\lambda_{3j} = \sum_{PVC_k \in \text{ output } j} \alpha_k \tag{27}$$

$$E\{L_{3j}\} = \sum_{PVC_k \in \text{ output } j} E\{L_k\}\alpha_k/\lambda_{3j} \tag{28}$$

$$v_{X_{3j}}^2 = v_{L_{3j}}^2 = \sum_{PVC_k \in \text{ output } j} \frac{\alpha_k E^2\{L_k\}(1+v_{L_k}^2)}{\lambda_{3j} E\{L_{3j}\}^2} - 1 \tag{29}$$

These three parameters are calculated independently of the previous queue information, but $v_{T_{3j}}^2$ depends on that information and requires more calculation. It is derived in the following steps.
a) Let $v_{SW_i}^2$ be the squared coefficient of variation of the input queue i packet inter-departure time at the output of the switch module. Then,

$$v_{SW_i}^2 = \rho_{2i}^2 v_{X_{2j}}^2 + (1-\rho_{2i}^2) v_{T_{2i}}^2. \tag{30}$$

b) Traffic from input queue i will split among the output queues. The probability, γ_{ij}, that a packet from queue i goes to output queue j is, approximately,

$$\gamma_{ij} = \frac{\sum_{PVC_k \in (i,j)} \alpha_k}{\lambda_{2i}} \tag{31}$$

where (i,j) is defined to be a path from input queue i to output queue j. Thus, the variability parameter, $v_{SW_{ij}}^2$, of the packet inter-arrival time from input queue i to output queue j is,

$$v_{SW_{ij}}^2 = 1 - \gamma_{ij} + \gamma_{ij} v_{SW_i}^2. \tag{32}$$

c) The traffic from input queues to the output queue j is combined and forms the traffic to the output module. If we assume that the traffic from different queues are independent, we can use the QNA's traffic-merge method to combine the traffic.

Let θ_{ij} be the fraction of traffic from input queue i to output queue j, that is

$$\theta_{ij} = \frac{\sum_{PVC_k \in (i,j)} \alpha_k}{\lambda_{3i}}. \tag{33}$$

Hence, the variability parameter of inter-arrival to the output module j is

$$v_{T_{3j}}^2 = (1 - w_{3j}) + w_{3j} \sum_{i=1}^{N} (\theta_{ij} v_{SW_{ij}}^2) \tag{34}$$

where $w_{3j} = 1/[1 + 4(1-\rho_{3j})^2((\sum_{i=1}^N \theta_{ij}^2)^{-1} - 1)]$ and $\rho_{3j} = \lambda_{3j} E\{X_{3j}\} = \lambda_{3j} E\{X_{1j}\}$ is the load of the output processor j. The input traffic parameters to the output packet processor are then completely specified by formulas (27) through (34).

Delay of the Output Module

There are five sources of delay for an output module: packet processing delay, waiting time delay in the receive cache, packet transfer delay, waiting time delay for multiplexing and transmission and the transmission delay.

The first queue in the output module is the packet processor queue, which performs the same function as in the input queue. Changing the subscript $1i$ of (5) into $3j$, gives the formula for the mean waiting time, $E\{W_{3j}\}$, of the output packet processor. The transfer delay is the same as that of the input module. The packet inter-departure variability parameter, $v^2_{T_{4j}}$ after the packet processor can be calculated, which is an input parameter to the transmission queues. Due to the lossless assumption, the other parameters for transmission queue j (i.e., λ_{4j}, $E\{L_{4j}\}$ and $v^2_{L_{4j}}$) are the same as those for the input to the output packet processor (i.e., the values given by (27) to (29)). $E\{W_{4j}\}$ is then calculated using (5) with these parameter values.

The last queue of the frame relay switch is at the output transmission line. The transmission line can be a full T1 line or channelized. In the former case, the arrival traffic is known, it is a simple GI/G/1 queue; in the later case, the traffic to output module j is split onto the channels. We consider that it is randomly split. The QNA method can be applied to calculate the parameters and delays of the traffic streams.

Switch Delay of PVC Packets

The PVC packet delay through the switch is defined to be the interval from the time a PVC packet arrives to the input module of the switch till it leaves the output module of the switch. If PVC_k goes through the switch via input module i and the output module j, the average switch delay, $E\{D_k\}$, of the PVC_k packets is the sum of the waiting time delay and the service time delay of the queues it passes through in the switch. The average waiting time delay is the same for any PVC going through the same path; but the service time delay is different for PVCs whose traffic has different average packet lengths. Let $E\{W_{ij}\}$ be the waiting time delay of the path from input module i to output module j and $E\{S_k\}$ be the average service time delay for PVC_k packets going through the switch. Then, if PVC_k is via path (i, j), the PVC_k packet delay is,

$$\begin{aligned} E\{D_k\} &= E\{W_{ij}\} + E\{S_k\} \\ &= E\{W_{1i}\} + E_G\{W_{2i}\} + E\{W_{3j}\} + E\{W_{4j}\} \\ +E\{X_{1i}\} &+ E\{X_{3j}\} + E\{L_k\}\left(\frac{2M}{R_t} + \frac{1}{R_w} + \frac{1}{R_{4j}}\right) \end{aligned} \quad (35)$$

The first four terms are the waiting time delay; the remaining terms are the service time delays. $E\{X_{1i}\}$ and $E\{X_{3j}\}$ are due to the input and the output packet processors; R_t is the packet transfer rate of the input or output modules; R_w is the service rate of the cyclic queue switch and R_{4j} is the service rate of the transmission line.

CONCLUSIONS

This paper has described an approach being used to develop a tool to enable network operators to better manage large communication networks. This tool is based on the ability to model the network switches for delay as well as customer traffic for both its average demand and burstiness. These models are then used in a general routing procedure, described in [13], to obtain optimal paths along which to route traffic through the network. This routing algorithm takes into account delay to route the traffic. A graphical interface, not reported on here, is also being developed in conjunction with the models which enables the network operator to see the network and its performance and to accept or override the actions proposed by the tool. We are currently working on extending the applicability of the model to new services and switches.

ABOUT THE AUTHORS

Philip E. Sarachik received his AB, BS, MS and PhD degrees from Columbia University in 1953, 1954, 1955 and 1958 respectively. From 1958 to 1960 he was a Staff Engineer at the IBM Research Laboratories and from 1960 until 1964 he was on the Electrical Engineering faculty at Columbia University. He was a Professor of Electrical Engineering at New York University from 1964 until 1973 when he joined the Polytechnic. He has held visiting positions at MIT, IBM T.J. Watson Research Center and Tel Aviv University.

In 1991 he was elected a Fellow of the IEEE for his contributions to modern control.

His research interests and publications have included optimal and adaptive control, stability and optimization theory, computational methods, systems identification, state estimation, routing problems in networks, self-organizing networks and performance modelling in data networks.

Shivendra S. Panwar received the B.Tech. degree in electrical engineering from the Indian Institute of Technology, Kanpur, in 1981, and the M.S. and Ph.D. degrees in electrical and computer engineering from the University of Massachusetts, Amherst, in 1983 and 1986, respectively. He joined the Department of Electrical Engineering at the Polytechnic Institute of New York, Brooklyn (now Polytechnic University), where he is now an Associate Professor. He is currently the Associate Director of the Center for Advanced Technology in Telecommunications. He has been a Special Consultant to AT&T Bell Laboratories and a Visiting Scientist at the IBM T.J. Watson Research Center. His research interests include the performance analysis, design and control of high speed networks.

Po Liang was born in Guang Dong, China on January 28, 1958. He received his B.Sc. and M.Sc. degrees in Electrical Engineering from Shanghai Jiao Tong University, Shanghai, China, in 1982 and 1986, respectively and the Ph.D. degree in Electrical Engineering from Polytechnic University, Brooklyn, New York, in 1994.

From 1982 to 1983, he worked at Guang Zhou Shipyard as an engineer for communication systems. From 1986 to 1989, he was a lecturer in the Mobile Communications and Microcomputer Applications Laboratory, Shanghai Jiao Tong University, involved in mobile communication network design, modulation theory and frequency-hop systems. From 1990 to 1993, he was a research fellow and research assistant of Polytechnic University, working on high speed network modeling, simulation and network management.

Derchian Tsaih received his BS from Chinese Culture University at Taipei in 1983 and MS from George Washington University at Washington DC in 1989.Since 1989, he has been a Ph.D student in the Department of Electrical Engineering, Polytechnic University.

Symeon Papavassiliou was born in Athens Greece, on December 1967. He received the Diploma in Electrical Engineering from the National Technical University of Athens, Athens, Greece, in October 1990, and the M.S. degree in Electrical Engineering from Polytechnic University, Brooklyn, N.Y., in January 1992. He is currently pursuing the Ph.D degree at Polytechnic University. During his stay at Polytechnic University he has been a Research Fellow in the Center for Advanced Technology in Telecommunications(CATT).

His main research interests lie in the mobile radio communication system design, network management and optimization of stochastic systems.

Mr. Papavassiliou is a member of the Technical Chamber of Greece.

Leandros Tassiulas was born in 1965, in Katerini, Greece. He obtained the Diploma in Electrical Engineering from the Aristotelian University of Thessaloniki, Thessaloniki, Greece in 1987, and the M.S. and Ph.D. degrees in Electrical Engineering from the University of Maryland, College Park in 1989 and 1991 respectively.

Since September 1991 he has been in the Department of Electrical Engineering of Polytechnic University, Brooklyn, NY as an Assistant Professor.

His research interests are in the field of computer and communication networks with emphasis on wireless communications and high-speed networks, in control and optimization of stochastic systems and in parallel and distributed processing.

REFERENCES

[1] W. Whitt "The Queueing Network Analyzer" The Bell System Technical Journal, Vol.62, No.9, November 1983.

[2] W. Whitt "Performance of the Queueing Network Analyzer" The Bell System Technical Journal, Vol.62, No.9, November 1983.

[3] R. Guerin, H. Ahmadi and M. Naghshineh, "Equivalent Capacity and Its Application to Bandwidth Allocation in High-Speed Networks,", IEEE J. Select. Areas Commun., Vol. 9., No. 7., pp. 968-981, Sept. 1991.

[4] K.M. Khalil, K.Q. Luc and D.V. Wilson, "LAN Traffic Analysis and Workload Characterization," Proc. 15th Conf. on Local Comp. Netw., pp. 112-122, Oct 1990.

[5] H.J. Fowler and W.E. Leland, "Local Area Traffic Characteristics, with Implications for Broadband Network Congestion Management," IEEE J. Select. Areas Commun. Vol. 9, No. 7, pp. 1139–1149, September 1991.

[6] W.E. Leland and D.V. Wilson, "High Time-Resolution Measurement and Analysis of LAN Traffic: Implications for LAN Interconnection," Proc. INFOCOM '91, pp. 1360-1366, April 1991.

[7]D. S. Lee, B. Melamed, A. Reibman and B. Sengupta, "Analysis of a Video Multiplexer using TES as a Modeling Methodology," IEEE GLOBECOM'91, vol. 1, pp. 16-20.

[8] J. Beran, R. Sherman, M. S.Taqqu, and W. Willinger, "Variable Bit-Rate Video Traffic and Long-Range Dependence," Pre-publication Copy.

[9] K.T. Marshall, "Some Inequalities in Queueing," Oper. Res., Vol. 16, No. 3, pp. 651-665, May-June 1968.

[10] P.J. Kuehn, "Multiqueue Systems with Nonexhaustive Cyclic Service," The Bell System Technical Journal, Vol. 58, No. 3, pp. 671-698, March 1979.

[11] O.J. Boxma and B. Meister, "Waiting-Time Approximations for Cyclic-Service Systems with Switch-over Times," Performance Eval. Rev. vol.14 pp. 254-262, 1986.

[12] K.S. Watson, "Performance Evaluation of Cyclic Service Strategies – a Survey," Performance'84, ed. E. Gelenbe, pp. 521-533, North-Holland, Amsterdam, 1984.

[13] Z. Flikop, " Routing Optimization in Packet Switching Communication Networks," European J. of Oper. Res., Vol. 19, pp. 262-267, 1985.

SERVICE MANAGEMENT AND CONNECTION IDENTIFICATION IN SHARED MEDIUM ACCESS NETWORKS

I.S. Venieris, K.E. Mourelatou, N.D. Kalogeropoulos, M.E. Theologou, and E.N. Protonotarios

National Technical University of Athens
Department of Electrical and Computer Engineering
Division of Computer Science
157 73 Zografou, Athens, Greece

ABSTRACT

Passive Optical Networks employing the ATM principle (ATM PONs) are considered the best candidate for providing B-ISDN services with low cost to small business and residential customers. Simplicity and economical reasons impose several burdens in the ATM PON design. The lack of signalling capability in a Virtual Path (VP)-based operation chosen for the ATM PON necessitates a flexible Virtual Path Identifier (VPI) allocation scheme for supporting basic functions like routing and addressing. This VPI allocation scheme should be kept transparent for the standardized ATM interfaces which leads to the introduction of a VPI translation function at the entrance and exit of the ATM PON. This in turn calls for the design of a new (ATM PON) - (B-ISDN - Local EXchange (LEX)) fast interaction protocol which should serve the ATM PON specific requirements and also be general enough to be used with other PONs.

INTRODUCTION

The forthcoming B-ISDN is intended to offer a wide range of services with varying bandwidth demands and diverse Quality of Service (QoS) requirements for all big business, small business and residential customers.

While big customers with high bandwidth demands can afford dedicated fibers for access to the core network, small business and residential customers foresee to low prices of the offered broadcast services comparable to the one these customers currently pay in traditional single service networks.

To guarantee the provision of services in low cost, flexible means for providing cheap access to the core B-ISDN is currently investigated since the cost of the access takes

Network Management and Control, Vol.2, Edited by
I.T. Frisch *et al.*, Plenum Press, New York, 1994

usually the largest share of the total cost of the network[9,10]. In this respect, a shared medium is expected to drastically reduce the initial investment in the subscriber local loop taking advantage of the low bandwidth demands experienced by small business and residential customers. Thus, Passive Optical Networks employing the Asynchronous Transfer Mode principles (ATM PONs) based on components implementing traffic multiplexing and concentration are to be introduced in the local loop (see figure 1).

The main objectives in the design of access systems based on shared components are the following[8,10]:

- fair treatment of all subscribers
- high utilization of the network resources
- accommodation of diverse Quality of Service (QoS) requirements.

These requirements are actually met in any ATM network. The difference however is that the ATM PON has to be cost effective which leads to solutions that exclude signalling capability and connection handling operations on a Virtual Channel level. This together with the extra addressing and routing requirements arising due to the topology of the ATM PON (e.g. tree topology) impose serious design burdens.

Shared bandwidth pushes to the development of bandwidth allocation and access control mechanisms that find a compromise between the aforementioned design objectives. Note that all internal ATM PON operations should be kept transparent to the standard network; i.e. the specific topology and the associated mechanisms of autonomously developed and administrated local loops should ideally be hidden from the standardized T and V interfaces[2] of figure 1.

A typical broadband access network implementation incorporating all the characteristics outlined above is an ATM PON operating on a VP basis.

In this paper we put our focus on the routing, addressing and service identification problems accompanying this solution. VP bandwidth management and ATM PON Medium

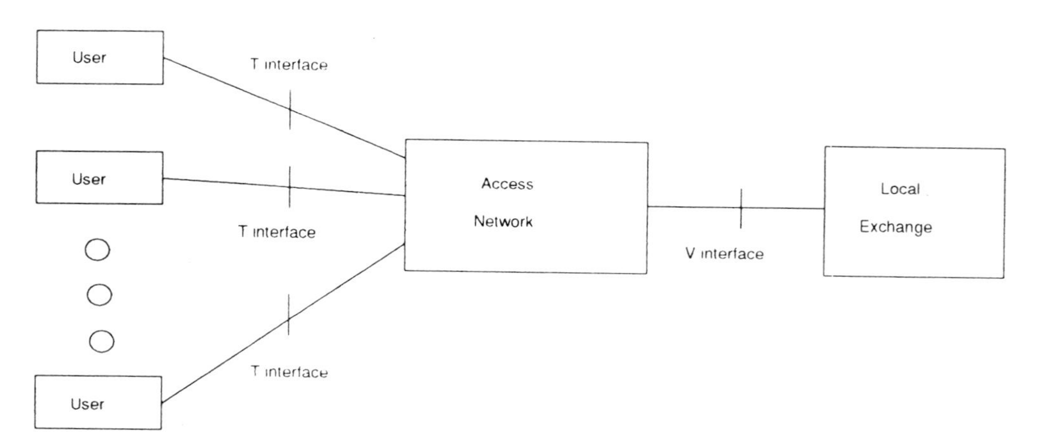

Figure 1. Access Network Reference Configuration

Access Control (MAC) protocols are outside the scope of this paper. The interested user is referred to [4,6] and [7] for general methods of VP bandwidth allocation, to[3] for the ATM PON case and to [1] for MAC protocol alternatives and their performance.

We propose an ATM PON internal VPI allocation scheme. This scheme provides the information required for the proper delivery of cells downstream; i.e. to the appropriate Network Termination (NT) indicated as user in figure 1, and for identifying the type of service supported. The latter allows the realization of access priorities at the NTs for upstream ATM PON traffic. We present the T, V interface characteristics based on the I. series standards[2] and give the motivation as well as a technique for VPI translation at the entrance and exit of the ATM PON. An ATM PON - LEX Interaction Protocol is also developed in terms of primitives and associated parameters for facilitating the implementation of the VPI translation mechanism.

A VPI ALLOCATION SCHEME SERVICING ROUTING, ADDRESSING AND SERVICE IDENTIFICATION IN AN ATM PON.

A VP based ATM PON does not provide any signalling capability. Furthermore NTs interpret only the VPI field of the cell header. In this respect the basic functions of ATM related to the handling of Virtual Channel Connections (VCCs) i.e. service related operations, are not supported[10]. These operations would allow routing and addressing inside the ATM PON, access priority realization and policing in the points of access to the ATM network. Considering the VP based the ATM PON appropriate mechanisms must be employed that will facilitate the implementation of the above basic functions. Moreover, the design of these mechanisms should aim primarily on satisfying the demand for flexible access systems in the sense that the ATM PON internal topology, features and procedures can be kept transparent to the extent possible to the fixed ATM network.

In this section, a VPI management policy is presented[8] to fulfil the objective of accommodating the functions required by the ATM PON.

VPI values will be allocated within the ATM PON in the following manner:

- Specific VPI values are allocated to each NT (figure 2).

This policy enables routing and addressing of information within the ATM PON. The allocation of specific VPI values to each NT enables the recognition and gathering of ATM cells destined to the specific NT. Each NT is informed about its corresponding VPI value at call set-up as the ATM PON internal part of the ATM PON - LEX Interaction Protocol running each time a new VPI is allocated by the LEX or the bandwidth it carries changes. In a static approach; i.e. VPIs are semi-permanently allocated to NTs this control protocol can be replaced by slow management procedures activated at large time scale.

- Some bits out of the twelve VPI bits will be used as an indicator of the service class (figure 3).

Service classification has been widely used within telecommunication services and system engineering as a tool for defining the common functions of classes of services [2,4,5]. The criteria used for service classification are usually specific to the particular study. In our approach the selected criteria for service classification impact on the performance requirements of each class in terms of access delay and cell loss at the access point.

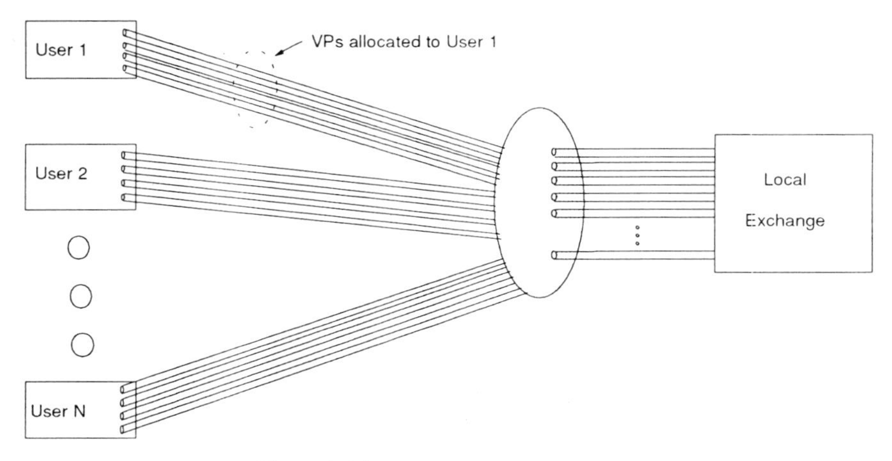

Figure 2. Virtual Paths in Access Networks

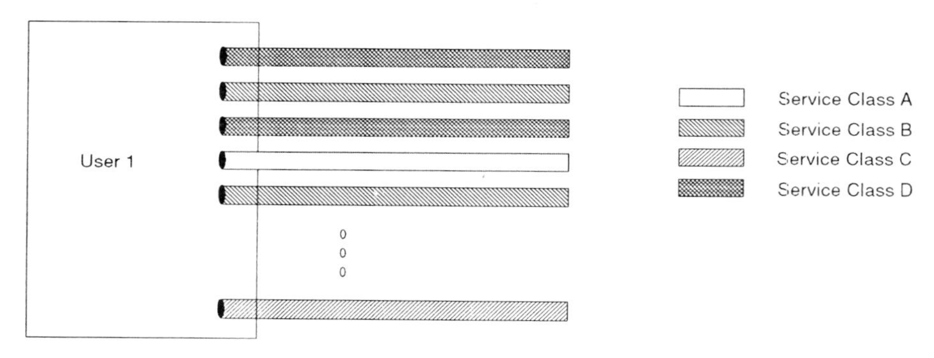

Figure 3. Connection identification in Access Networks

If four service classes are used, then two bits of the VPI field are used. Hence, all connections conveying information of services within a service class are accumulated in VPIs with at least two bits in common. In effect the VPIs of the ATM PON are organized in four pools, each pool consisting of each NT VPI of a specific service class. VPIs within pool are treated the same by both the ATM PON centralized functions; i.e. bandwidth allocation methods, and the NT local functions; i.e. priority schemes. The minimum number of VPI values that must be allocated to each NT equals to the number of the adopted service classes.

THE NEED FOR VPI TRANSLATION FUNCTIONALITY

Flexible policies should be employed in order to support ATM PONs internal needs such as routing and addressing taking into account security requirements imposed by the shared nature of the medium. The VPI allocation scheme presented above is a promising policy for accomplishing these tasks keeping in mind that the whole operation is an internal local loop matter. This means that the ATM network does not have to adopt the same VPI allocation scheme or even know the ATM PON internal VPI values. Instead it should be free to employ any VPI allocation policy. As a consequence, the necessity to translate the VPI values while entering and exiting the local loop arises (see figure 5).

The independence of both the T and the V standard interfaces from the specific topology and associated mechanisms employed by the autonomously developed and administered local loops should be preserved[2,3]. Moreover, the inconsistencies found in the T and V interfaces structure and capacity impose the translation of VPI values. The interface structure and the capacity for each interface type are given in Table 1.

Table 1. Interface characteristics.

I/F	VPI field (bits)	Capacity (Mbits/sec)	VPI values available
V	12	149.76	4096
U	12	600	4096
T	8	149.76	256

The U interface in Table 1 is associated with the ATM PON and can be regarded as an internal interface between the NT and the ATM PON Line Termination (LT) not shown in figure 1. The V interface is placed between the LT and the LEX (see figure 1).

At both the T and V interfaces the VPI values will be translated to ATM PON VPI values which are unique within the ATM PON. The arising requirement is that the number of different VPI values of the ATM PON should be greater than the number of supported VPI values in all T or V interfaces.

The different size of the VPI field in the T and V interfaces illustrated in figure 4 necessitate the following translations: the 8-bit VPI values of each T interface must be translated to a unique 12-bit ATM PON VPI value and vice versa; i.e. the combination of the particular T interface and the VPI value is unique. The 12-bit ATM PON VPI value must be mapped to the 12-bit VPI values of each V interface and vice versa. The combination of the particular V interface and the VPI value is unique.

The VPI translation tables placed at the entrance and exit of the ATM PON (see figure 5) should be dynamically updated. This is due to the dynamic way the LEX allocates the VPI values to call set-up requests. Thus, a protocol between the ATM PON and the LEX is required in order to guarantee that the output of the translation is always consistent with the VPI values the LEX has allocated to the specific T interface.

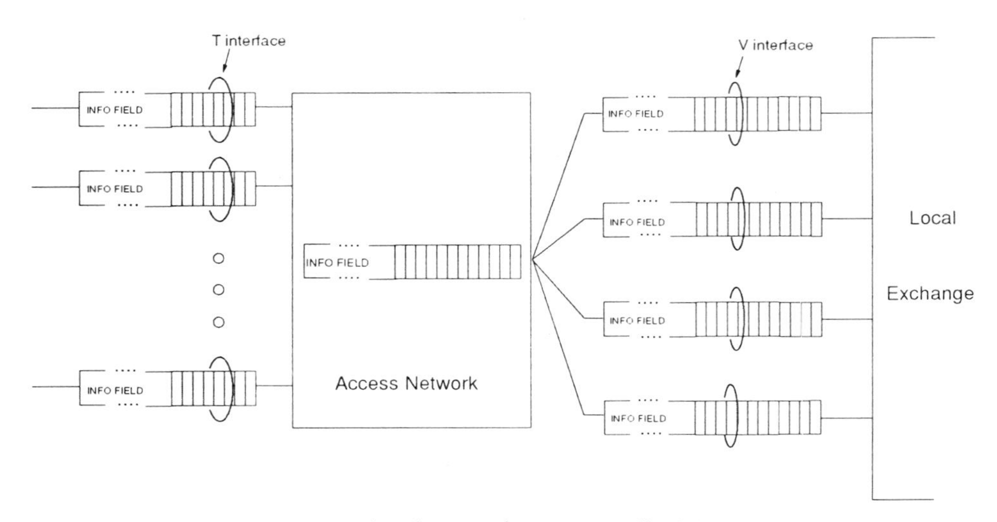

Figure 4. T, V interfaces requirements regarding VPI values

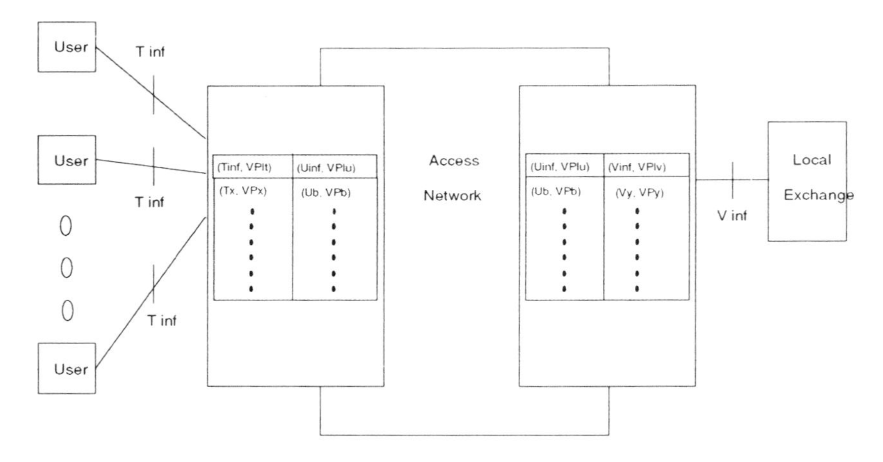

Figure 5. VPI Translation Functionality

ATM PON - LEX INTERACTION PROTOCOL

In the following, we specify the kind of information required to be exchanged between the LEX and the ATM PON, so that the translation mechanism of the ATM PON operates dynamically. The parameters to be exchanged between ATM PON and LEX for the establishment of a new connection and proper routing of cells to the correct destination are given in Table 2.

Table 2. Parameters of the ATM PON - LEX Protocol.

Parameter	Information
Tinf	The specific T interface
Vinf	The specific V interface
VPIt	The VPI value at the T interface
VPIu	The VPI value at the U interface
VPIv	The VPI value at the V interface
Class	The service class of the call
BWr	The bandwidth required for the specific call

It would be useful to distinguish the information maintained by the LEX from that maintained by the ATM PON. This, in addition to the information needs of both the LEX and the ATM PON, permit the identification of the information exchange which occurs between the ATM PON and the LEX.

Considering that the internal functions of the ATM PON must be kept as transparent as possible to the LEX, the LEX should recognize only the subscribers i.e. Terminal Equipment (TE), addresses, indicated as users in all figures. Connections are established using a signalling procedure between the LEX and the subscriber. Upon acceptance of a new connection by the Connection Admission Control (CAC)[2] of the LEX, the pairs (Vinf, VPIv) and (Tinf, VPIt) are specified. Only the VPIt value is declared to the TE.

The ATM PON is aware of the supported NTs which are in one-to-one correspondence with a Tinf value. The ATM PON also knows the supported V interfaces which are identified by their Vinf values. It is then able to translate a non-unique VPI value at a T interface to a unique VPI value of the common U interface; i.e. a (Tinf, VPIt) is translated to a VPIu. At the exit of the ATM PON this VPIu is translated to a VPIv which is unique within the specific V interface (Vinf).

In figure 6, we can see the message flow for the establishment of a conection. These messages are analyzed in the following:

The SETUP message is transferred from the subscriber to the LEX through the ATM PON. In fact it is a signalling message containing the request for the establishment of a new connection and it is transparent to the ATM PON. This message contains the identity of the user (Tinf), the service type of the connection (Serv), the class it belongs (Class), the bandwidth required (BWr), the destination and other information however not essential for the ATM PON-LEX protocol as it will appear below.

If the connection set-up request is accepted by the CAC located in the LEX, the LEX sends the ALLOC message to the ATM PON Management. In this message, the LEX provides to the ATM PON all the necessary information considering this call request. More specific, it contains the identity of the user (Tinf), the VPI values allocated by the LEX to both T and V interfaces (VPIt, VPIv), the V interface (Vinf), the service class (Class) and the bandwidth requested (BWr). With this information the ATM PON will be able to update both the VPI translation tables as well as VP bandwidth use status.

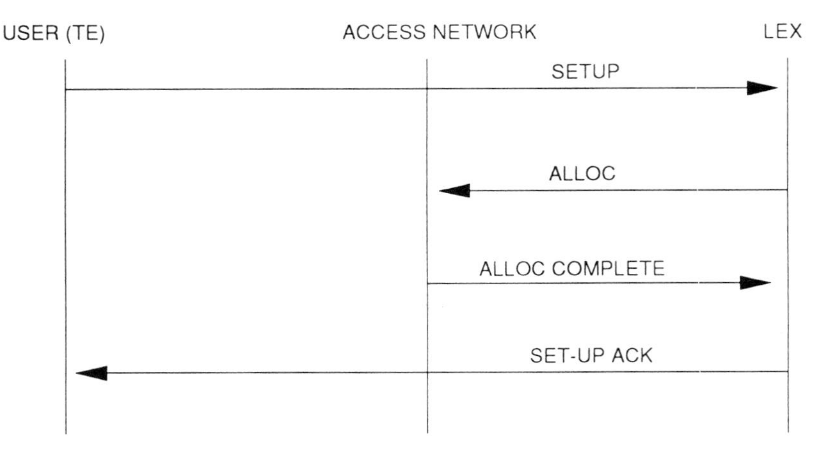

Figure 6. Message flow for call establishment.

The ALLOC COMPLETE message is sent from the ATM PON Management to the LEX in response to the ALLOC message. The ALLOC COMPLETE message may contain one of the following parameters:

- *null*, in case no bandwidth problem is faced within the ATM PON.
- an *alarm* indicating the extreme cases where the requesting NT has exceeded the bandwidth allocated to it, resulting in the violation of the upper limit set by the ATM PON according to the resource allocation methods employed within the ATM PON[3] or the ATM PON is in a fault condition. Note that the LEX will underestimate the actual bandwidth use in VPs (either peak or effective) due to the presence of the access control mechanism[1, 3]. It is then possible that the LEX sees free bandwidth while the ATM PON is overloaded. Such situations can be accommodated by allowing VPs to seize a part of a "spare bandwidth" never allocated to a specific VP. When almost the full spare bandwidth becomes utilized the ATM PON sends an alarm to signal that no new call can be accepted.

Finally, a signalling SET-UP ACK message is sent from the LEX to the user's TE informing it that the connection has been accepted and indicating him the specific VPI, VCI value assigned to it, i.e. the (Tinf, VPIt). Messages are summarized in Table 3.

Table 3. Messages of the ATM PON - LEX Interaction Protocol.

Message	Information
SETUP	Tinf, Class, BWr
ALLOC	Tinf, VPIt, Vinf, VPIv, Class, BWr
ALLOC COMPLETE	*null* or *alarm*
SET-UP ACK	Tinf, VPIt

CONCLUSIONS

Both the ATM PON inability to interpret signalling information and the necessity to preserve standardized interfaces and procedures have been considered in the study of the ATM PON resource management. Flexible methods for allocating both logical resources; i.e. VPI values, and physical resources; i.e. bandwidth, have been proposed in order to fulfil routing, addressing and QoS requirements within the ATM PON. We have specified a translation mechanism ensuring the proper routing of the information and a protocol between the ATM PON and the standardized LEX which enables the operation of the above mechanisms.

ACKNOWLEDGEMENTS

This work was partially funded by the European Community RACE Project R2024 Broadband Access Facilities (BAF). The opinions appearing in this manuscript are those of the authors and not necessarily of the other members of the consortium.

ABOUT THE AUTHORS

Iakovos S. Venieris is a research associate in the Telecommunications Laboratory of the National Technical University of Athens (NTUA). He received the Dipl.-Ing. degtree from the University of Patras in 1988 and the Ph.D degree from NTUA in 1990 all in electrical engineering. He is involved in several research projects in the area of broadband, ATM-based communications. He is the author of over thirty scientific papers and serves as a reviewer in major IEEE Journals. Dr. Venieris is a member of IEEE and the Technical Chamber of Greece.

Konstantina E. Mourelatou works towards a Ph.D degree in the Computer Science Division of the National Technical University of Athens (NTUA). She received a diploma degree in Physics from the University of Athens in 1990. Her research interests involve all aspects of network management and object oriented techniques for integrated servises high speed networks. She is currently involved in several european research projects in telecommunications.

Nikos D. Kalogeropoulos is a postgraduate student in the National Technical University of Athens (NTUA). He received the Dipl.-Ing. degree from the Electrical and Computer Enginneering Department of NTUA in 1991. His work involves service engineering, signalling, network managment and access control for integrated services networks. He is a member of the Technical Chamber of Greece.

Michael E. Theologou is an Assistant Professor at the Department of Electrical and Computer Engineering of the National Technical University of Athens (NTUA). His research interests are in the fields of Integrated Broadband Communications Networks, routing, flow control, quality of service and mobile communications. He has several publications in the above areas and has participated in several research projects of the European Union. Dr. Theologou is a member of IEEE and the Technical Chamber of Greece.

Emmanuel N. Protonotarios is the Director of the Telecommunications Laboratory of the National Technical University of Athens (NTUA) where he has been a Professor since 1973. He has contributed significantly to the modernization of electrical engineering and computer science while at the Bell Laboratories, Columbia University, New York and NTUA. Dr. Protonotarios has received numerous prizes for academic achievement and has served as General and Technical Committee Chairman in several conferences. He has published over 100 scientific papers and supervises many doctoral students working in several areas of Telecommunications and Computer Networks engineering.

REFERENCES

1. L. Verri, T. Toniatti, C. Blondia, O. Casals, J. Garcia, J.D. Angelopoulos, I.S. Venieris, "Performance of Shared Medium Access Protocols for ATM Users Concentration", accepted for publication in *European Transactions on Telecommunication.*
2. CCITT SG XVIII Recommendations I.371, I.121, I.150, I.413, I.311, I.361, I.362, Geneva, June 1992.
3. I.S.Venieris, K.E.Mourelatou, N.D.Kalogeropoulos, "Bandwidth Allocation and Service Class Priority handling in a VP-based Optical Access Network", *EFOC & N'93*, The Hague, June 1993.
4. Y.Sato, K.I.Sato, "Virtual Path and Link Capacity Design for ATM Networks", *IEEE Journal on Selected Areas in Communications*, Vol.9, No 1, January 1991.
5. M.Wernik, O.Aboul-Magd, H.Gilbert, "Traffic Management for B-ISDN Services", *IEEE Network Magazine,* September 1992.
6. E.D.Sykas, K.M.Vlakos, I.S.Venieris, E.N.Protonotarios, "Simulative Analysis of Optimal Resource Allocation and Routing in IBCN's", *IEEE Journal on Selected Areas in Communications*, Vol.9, No 3, April 1991.
7. S.Ohta, K.I.Sato, "Dynamic Bandwidth Control of the Virtual Path in an Asynchronous Transfer Mode Network", *IEEE Transactions on Communications*, Vol.40, No. 7, July 1992.
8. I.S.Venieris, K.E.Mourelatou, M.E.Theologou, "VPI/VCI Management in B-ISDN Access Optical Networks", *IEEE First International Workshop on Systems Management*, LA, April 1993.
9. D.B.Keck, A.J.Morrow, D.A.Nolan and D.A.T.Thompson, "Passive Components in the Subscriber Loop", *IEEE Journal of Lightwave Technology*, Vol. 7, No. 11, Nov. 1989.
10. *RACE Project R2024* (Broadband Access Facilities), Deliverable 1, "High Level Specification", Sept. 1992.

ALLOCATION OF END-TO-END DELAY OBJECTIVES FOR NETWORKS SUPPORTING SMDS

Frank Y.S. Lin

Bell Communications Research
Piscataway, New Jersey 08854

ABSTRACT

In this paper, the end-to-end delay objective allocation problem for networks supporting Switched Multi-megabit Data Service (SMDS) is considered. Traditionally, for engineering tractability, end-to-end service objectives are allocated to network elements in such a way that, if the allocated service objective for each network element is satisfied then the end-to-end service objectives are satisfied. Such an objective allocation strategy is referred to as a *feasible* objective allocation strategy.

For networks supporting SMDS, the delay objectives state that 95% of the packets delivered from the origin Subscriber Network Interface (SNI) to the destination SNI should be within a given time threshold. From network monitoring point of view, this percentile type of delay objectives makes it complicated to compute feasible allocation strategies so that network elements instead of each origin-destination pair should be monitored. From network planning point of view, these end-to-end percentile-type delay objectives usually impose an excessively large number of nonconvex and complicated constraints (delay distributions are convolved, assuming delays are mutually independent).

The emphasis of this paper is three fold: (i) to propose an efficient and generic approach to replacing the set of end-to-end percentile-type delay constraints by a simpler set of network element utilization constraints (small size and convex), (ii) to investigate how this approach could be adopted in conjunction with a number of possible allocation schemes and (iii) to compare the relative effectiveness (in terms of the utilization thresholds determined) using $M/M/1$ and $M/D/1$ queueing models. The significance of this work is to provide a general and effective way to calculate engineering thresholds on the network element utilization factors when percentile types of end-to-end delay objectives are considered. This work lays a foundation to help the system planners/administrators monitor, service, expand and plan networks supporting SMDS.

INTRODUCTION

Switched Multi-megabit Data Service (SMDS) is a high-speed, connectionless, public, packet switching service that will extend Local Area Network (LAN)-like performance beyond the subscriber's premises, across a metropolitan or wide area. SMDS has been recognized as the first step towards the Broadband Integrated Services Digital Network (BISDN), and is commercially available. To support the planning and engineering functions of networks supporting SMDS (referred to as SMDS networks), it is important for the system planners/administrators to be able to efficiently and effectively consider the service performance objectives.

Network Management and Control, Vol.2, Edited by
I.T. Frisch *et al.*, Plenum Press, New York, 1994

Since end-to-end performance is users' direct perception about the service quality, like many new services, e.g. Frame Relay Service (FRS), Asynchronous Transfer Mode (ATM) and Advanced Intelligent Network (AIN) services, performance objectives are specified on an end-to-end basis for the SMDS service. When network planning and engineering functions are performed, these end-to-end service objectives are typically difficult to handle principally due to the excessively large number of constraints they impose. (When the link set sizing problem for SMDS networks is considered[1], the number of original end-to-end delay constraints equals the total number of simple paths in the network.)

For networks supporting SMDS, the delay objectives state that 95% of the packets delivered from the origin Subscriber Network Interface (SNI) to the destination SNI should be within a given time threshold[2]. The percentile nature of the end-to-end delay performance objectives for SMDS networks makes it more difficult to handle the constraints, since each of the constraints is highly nonlinear (involving convolution when delays are assumed to be mutually independent) and nonconvex. The nonconvex property can be illustrated by the following example. Consider a network with only one network element. For a given time threshold, the probability that an arbitrary packet will not be delivered cross the network element within the time threshold, referred to as the *overdue* probability, shall increase with the utilization and approaches 1 asymptotically. The overdue probability is thus clearly not a convex function of the utilization.

To make the problem tractable, there is a need to circumvent the above two difficulties -- volume and nonconvexity of end-to-end delay constraints. One possible, and traditional, approach is to allocate the end-to-end performance objectives to each network element in such a way that, if the performance objective allocated to each network element is met then the end-to-end performance objectives are satisfied[3]. In addition, for monitoring purposes, the derived (surrogate) service objective for each network element is further converted into a threshold on the network element utilization.

The allocation approach has the following significance. First, it reduces the number of constraints greatly (from the number of simple paths to the number of links in the network). Second, this approach decouples the decision variables (utilization) in the constraint set (each constraint involves only one decision variable). Third, this approach makes the delay constraints linear. Fourth, if properly handled as will be discussed in the next section, the new set of constraints will define a feasible region which is a subset of the feasible region of the original constraints. In other words, any feasible solution to the surrogate problem is a feasible solution to the original problem.

This allocation approach is also commonly adopted when the network monitoring problem is considered. For a feasible allocation strategy, when the utilization of each network element is no greater than the allocated threshold, it can be sure that the end-to-end delay objectives be satisfied. Otherwise, the network planner/administrator may need to calculate/measure the end-to-end delay for every origin-destination (O-D) pair at the same time to evaluate the delay objectives.

To satisfy the feasibility property (a feasible solution to the surrogate problem is a feasible solution to the original problem) mentioned earlier, in this paper we propose a general approach to replacing the end-to-end percentile-type delay objectives by a set of link constraints in such a way that the end-to-end percentile-type delay objectives are satisfied under any routing assignment as long as the link constraints are satisfied. This approach is suitable for heterogeneous networks. How to implement this approach in a network planning/engineering problem to achieve the best result is also discussed.

Following the general approach, a number of possible allocation schemes are discussed. They include

1. longest delay control
2. complete decomposition

3. GI/G/1 bounding scheme
4. Markov inequality
5. Chebyshev inequality and
6. normal approximation.

Two allocation schemes for homogeneous networks are also discussed. They are

1. Convolution scheme and
2. Chernoff bounding scheme.

Each of the above schemes requires different input, e.g. interarrival time distribution, service time distribution, delay distribution, packet blocking probability and so on, and different degree of computational complexity. Those differences are discussed. In addition, the relative effectiveness in terms of the utilization determined in a homogeneous network is compared using $M/M/1$ and $M/D/1$ queueing models.

The significance of this work is to provide a general and effective way to calculate engineering thresholds on the network element utilization factors when percentile types of end-to-end delay objectives are considered. This work lays a foundation to help the system planners/administrators monitor, service, expand and plan networks supporting SMDS.

The remainder of this paper is organized as follows. In the second, a general approach to allocating the end-to-end percentile-type delay objectives to network elements in a heterogeneous environment is proposed. In the third section, 6 schemes following the general approach proposed in the second section are discussed. In the fourth section, 2 allocation schemes suitable in a homogeneous environment are discussed. In the fifth section, two case studies are given to compare the schemes using $M/M/1$ and $M/D/1$ queueing models. The last section summarizes this paper.

GENERAL APPROACH

In this section, we propose a general approach to replacing the end-to-end percentile-type delay objective constraints by a simpler set of link constraints in such a way that, under any routing assignment, if the link constraints are satisfied then the end-to-end percentile-type delay objectives are satisfied. This approach is suitable for heterogeneous networks, and thus for homogeneous networks as special cases. In addition, this approach is applicable when end-to-end mean delay or loss rate constraints are considered. How to implement this approach in a network planning/engineering problem to achieve the best result is also discussed.

An SMDS backbone network is modeled as a directed graph $G(V,L)$ where delay elements (e.g., a trunk or a switch fabric) are represented by directed links and the junctions between delay elements are represented by nodes. The nodes (junctions) do not incur any delay. Let L be the set of links and V be the set of nodes in the graph (network).

Consider a planning/engineering problem for a heterogeneous SMDS network denoted by problem (P). The goal is to replace the original intractable and nonconvex end-to-end delay constraints into $|L|$ convex link constraints. More importantly, any solution satisfying the link constraints should satisfy the original end-to-end delay constraints. A procedure is proposed below.

The first step is to identify a set of performance indicators $m^1, m^2, \ldots, m^n$ (each of which is a k-vector), e.g. the overdue probability and the time threshold, such that for a path consisting of links 1 to k the following condition

$$O^j(m_1^j, m_2^j, \ldots, m_{k-1}^j, m_k^j) \leq M^j \qquad \forall j = 1,2,\ldots,n \tag{1}$$

guarantees the overdue probability for this path be no greater than 5%, where O^j is an operator on performance indicators $\{m_l^j\}_l$ (e.g. O^j is $\sum$, then $O^j(m_1^j, m_2^j, ..., m_{k-1}^j, m_k^j) = \sum_{l=1}^{k} m_l^j$), m_l^j is the allocated type-j performance indicator to link l and M^j is a prespecified upper bound on the aggregate effect (through operator O^j) of type-j performance indicators $\{m_l^j\}_l$.

Assume the longest-hop path involves K hops (or alternatively a hop constraint can be imposed). Then, let m_l^j, $\forall l \in L$, $j = 1,2,...,n$, be the solution to the following equation

$$O^j(m_l^j, m_l^j, \ldots, m_l^j, m_l^j) = M^j \tag{2}$$

where the left hand side has K arguments. Note that the left hand side of (2) is the K-fold O^j operation on m_l^j and that for all $l \in L$, m_l^j's are the same. It is clear that with this assignment strategy, if all m_l^j's are satisfied then the end-to-end delay constraints are satisfied. Otherwise, given an $\{m_l^j\}_{l,j}$ assignment, we may need to identify those paths where the end-to-end delay constraints are violated and exclude them from the feasible set, which is usually intractable. The next step is for each link l to calculate the highest utilization where the allocated $\{m_l^j\}_j$ are satisfied.

This general approach can be generalized to allow non-even allocation of performance indicators, which can best be illustrated by the following example. If a particular network element l is much more expensive than the others and/or is obviously the bottleneck of the network, then we may want to increase its allowable utilization than the allocated value by the original approach. Then, we can increase m_l^j to be greater than the solution to Equation (2) and set m_i^j, $\forall i \in L$, $i \neq l$, to be the solution to

$$O^j(m_i^j, m_i^j, \ldots, m_i^j, m_l^j, m_i^j, \ldots, m_i^j, m_i^j) = M^j$$

where the left hand side has K arguments.

Next, we describe an implementation procedure to achieve the best result when solving problem (P). Let problem (P_K) be problem (P) where the above general allocation approach is used to simplify the constraints and a hop constraint which requires each active path not involve more than K hops is added. Note that with this hop constraint, a feasible solution to problem (P_K) is still a feasible solution to problem (P). In a solution procedure to (P_K), a shortest-path algorithm is usually incorporated. The shortest-path problem seems to be the only part impacted by this additional hop constraint. However, the hop-constrained shortest-path problem is still solvable in polynomial time by the Bellman-Ford algorithm. The impact of K can be grossly described below. When K is increased, the allocated utilization for each link becomes smaller, which tends to increase the objective function value, but the routing flexibility is increased, which tends to decrease the objective function value. A series of problem (P_K)'s, where K may typically ranges from the network diameter to $|V| - 1$, are then solved. The objective function values for the problems are recorded and the best is chosen.

ALLOCATION SCHEMES FOR HETEROGENEOUS NETWORKS

In this section, based upon the general allocation approach proposed in the previous section, we discuss a number of possible allocation schemes for heterogeneous networks. In the next section, 2 allocation schemes suitable for homogeneous networks are discussed.

Approach 1: Longest Delay Control

One possible approach to controlling the network element loads so that the delay objectives will be met is to limit the longest packet delay on each network element. More precisely, a packet will either be transmitted strictly within the given delay threshold (by properly choosing the buffer size) or be dropped (due to buffer overflow). Note that for those packets discarded due to buffer overflow, the corresponding delays are infinite from the standpoint of an SMDS network. As such, the end-to-end packet loss probability should be no greater than 5%. However, a more stringent end-to-end packet loss performance objective is specified in [2], which requires that the end-to-end packet loss probability be no greater than 10^{-4}. Assume that the buffer overflow probability of queue i can be expressed as a function of the utilization factor ρ_i and the buffer size J_i (in packets), denoted by $B_i(\rho_i, J_i)$. Let t_i be the service time of the longest packet on server i. Let $\bar{t}$ be the end-to-end delay threshold. Consider a path comprising queues 1 to k. Then the problem is to find a feasible solution to the following system:

$$\begin{cases} \prod_{i=1}^{k} (1 - B_i(\rho_i, J_i)) \geq 1 - 10^{-4} \\ \sum_{i=1}^{k} J_i \, t_i \leq \bar{t}. \end{cases}$$

To apply the general approach described in the previous section, for each link i we set

$$B_i(\rho_i, J_i) = 1 - (1 - 10^{-4})^{\frac{1}{K}}$$

$$J_i \, t_i = \frac{\bar{t}}{K}.$$

If $B_i(\rho_i, J_i)$ is a monotonically increasing function of ρ_i, ρ_i can be determined from the above system using standard line search techniques.

This approach is simple but may involve the following difficulties. First, a mechanism to control the buffer size may not be available for an SMDS Switch System (SS) through an Operations Support System (OSS). Second, as the number of hops increases, ρ_i may decrease fast due to the joint effect of decreased J_i and decreased B_i.

Approach 2: Complete Decomposition

If the probability density function (pdf) of the delay on each network element is known and can be expressed as a function of the utilization factor, we may allocate the end-to-end delay objectives by properly allocating the overdue probability and the time threshold. Let T_i be the delay on network element i (a random variable). Let $f_{T_i}(t_i, \rho_i)$ be the pdf of T_i, where ρ_i is the utilization factor of network element i. Let $F_{T_i}(t_i, \rho_i)$ be the probability distribution function (PDF) of T_i. Let $F_{T_p}(t_p, R_p)$ be the PDF of the end-to-end delay T_p $(= \sum_{i \in h_p} T_i)$ when path p is considered, where h_p is the set of links on path p and R_p is the vector of ρ_i $\forall i \in h_p$. The following lemma provides one possible way of allocating the end-to-end delay objectives.

Lemma 1: *If $F_{T_i}(b_i \, t, \rho_i) \geq 1 - a_i$, $b_i \geq 0$, $\sum_{i=1}^{k} b_i \leq 1$, $\forall i = 1,2,\ldots,k$, where T_i's are mutually independent, then $F_{T_p}(t, R_p) \geq \prod_{i=1}^{k} (1 - a_i)$.*

Proof: $F_{T_p}(t, R_p)$ is the integral of $f_{T_1}(t_1, \rho_1) \circledast f_{T_2}(t_2, \rho_2) \circledast \cdots \circledast f_{T_k}(t_k, \rho_k)$ over the subspace $S = \left\{ (t_1, t_2, \ldots, t_k) \mid \sum_{i=1}^{k} t_i \leq t \right\}$ where the symbol $\circledast$ is used to denote the convolution operator. This subspace contains $S' = \left\{ (t_1, t_2, \ldots, t_k) \mid t_i \leq b_i \, t, \forall i = 1,2,\ldots,k \right\}$. Take the integral over subspace S' and then we have $F_{T_1}(b_1 \, t, \rho_1) \times F_{T_2}(b_2 \, t, \rho_2) \times \cdots \times F_{T_k}(b_k \, t, \rho_k)$. Use the preconditions and then the result follows. $\square$

Following the general approach described in the previous section, for each link i we set

$$a_i = 1 - 0.95^{\frac{1}{K}}$$

$$b_i = \frac{1}{K}.$$

This approach has an advantage of simplicity. However, one potential drawback of this approach is that over-conservative decisions may be made, especially when K is large. A quality indicator of the utilization threshold determined is the space defined by S' divided by that defined by S. When the ratio is low, the determined utilization threshold tends to be low. As an example, when all b_i's are chosen to be $1/K$ (the ratio is maximized), the ratio becomes $K!/K^K$, which approaches zero as K goes to infinity.

Approach 3: GI/G/1 Bounding Scheme

In this scheme, a $GI/G/1$ queueing model is considered and the required input is the interarrival time distribution and the service time distribution. This scheme is based upon a result due to Kingman[4]. Using another result also due to Kingman[5], a similar allocation scheme can be developed where the input is the first two moments of the interarrival time distribution and the service time distribution.

We briefly summarize the result in [4] below. Let $A_i^*(s)$ be the Laplace transform of the interarrival time distribution for link i. Let $B_i^*(s)$ be the Laplace transform of the service time distribution for link i. Let $C_i^*(s)$ be $A_i^*(-s) \times B_i^*(s)$. Let W_i be the waiting time on network element i. Let $F_{W_i}(t,\rho_i)$ be the probability distribution function of W_i. Then the result states

$$1 - F_{W_i}(t,\rho_i) \leq e^{-s_0 t} \tag{3}$$

where s_0 is found from

$$s_0 = \sup\left\{ s > 0 : C_i^*(-s) \leq 1 \right\}. \tag{4}$$

It is also worth mentioning that the average waiting time $\overline{W}_i$ is upper bounded by $1/s_0$. This result may be useful for other networks where the performance objective is the mean end-to-end delay.

From Equation (3), $F_{W_i}(t,\rho_i)$ is lower bounded by the PDF of an exponentially distributed random variable with mean s_0. In this allocation scheme, the performance indicator is s_0. For a path with K hops, let the allocated value of s_0 be the same, denoted by v. For simplicity and illustration purposes, we assume deterministic service times. Therefor, to consider the overdue probability, we can consider the total waiting time and adjust the time threshold accordingly (the original time threshold minus K times the service time).

Consider the following result.

Lemma 2: *If* $F_{T_1}(t,\rho_1) \leq F_{T_2}(t,\rho_2)$ *and* $F_{T_3}(t,\rho_3) \leq F_{T_4}(t,\rho_4)$ $\forall t$, *then* $F_{T_1+T_3}(t,\rho_1,\rho_3) \leq F_{T_2+T_4}(t,\rho_2,\rho_4)$ $\forall t$.

Proof:

$$\begin{aligned}
F_{T_1+T_3}(t,\rho_1,\rho_3) &= \int_0^t f_{T_1}(t-y,\rho_1) F_{T_3}(y,\rho_3)\,dy \\
&\leq \int_0^t f_{T_1}(t-y,\rho_1) F_{T_4}(y,\rho_4)\,dy \\
&= \int_0^t f_{T_4}(t-y,\rho_4) F_{T_1}(y,\rho_1)\,dy \\
&\leq \int_0^t f_{T_4}(t-y,\rho_4) F_{T_2}(y,\rho_2)\,dy \\
&= F_{T_2+T_4}(t,\rho_2,\rho_4). \quad \square
\end{aligned}$$

This lemma immediately leads to the following proposition.

Proposition 1: *If* $F_{T_i}(t,\rho_i) \le F'_{T_i}(t,\rho_i)\ \forall t \ge 0, i = 1,2,\ldots,k,$ *then* $F_{T_1}(t,\rho_1) \circledast f_{T_2}(t,\rho_2) \circledast f_{T_3}(t,\rho_3) \circledast \cdots \circledast f_{T_k}(t,\rho_k) \le$ $F'_{T_1}(t,\rho_1) \circledast f'_{T_2}(t,\rho_2) \circledast f'_{T_3}(t,\rho_3) \circledast \cdots \circledast f'_{T_k}(t,\rho_k).$

Then the PDF of the end-to-end delay is lower bounded by the PDF corresponding to the following Laplace transform

$$\frac{1}{s}\left[\frac{v}{s+v}\right]^K.$$

Calculating 1 minus the inverse Laplace transform of the above expression yields

$$e^{-vt}\left[1 + vt + \frac{v^2 t^2}{2!} + \cdots + \frac{v^{(K-1)} t^{(K-1)}}{(K-1)!}\right]. \tag{5}$$

The next step is to calculate v such that the above expression equals 5%. It can be verified that the above expression is a monotonically decreasing function of v. Therefore, standard line search techniques can be applied.

Once the value of s_0 $(= v)$ for each link is calculated, we may apply Equation (4) to calculate the utilization threshold. Two examples are given below to demonstrate this step.

Example 1: $M/M/1$

Consider an $M/M/1$ queue with mean arrival rate λ and mean service rate μ. We first determine the maximum s that satisfies $\frac{\lambda}{s+\lambda}\,\frac{\mu}{-s+\mu} \le 1$. It is easy to find that $s_0 = \mu - \lambda$. Therefore, the utilization threshold is set to $1 - s_0/\mu$.

Example 2: $M/D/1$

The arrival process is Poisson with parameter λ. Let the service time for each packet be t_{max}. We first determine the maximum s that satisfies $\frac{\lambda}{s+\lambda} e^{s\, t_{max}} \le 1$ or equivalently $\frac{\rho}{s\, t_{max} + \rho} \le e^{-s\, t_{max}}$. It can be shown that at s_0 the equality holds and there are two roots (one is at 0) if the service rate $(1/t_{max})$ is greater than the arrival rate. We can use standard line search schemes to find the second root.

Approach 4: Markov Inequality

If the mean delay on each link is available and can be expressed as a function of the link utilization factor, the following allocation scheme based upon the Markov inequality is proposed. Consider path p. Let $\overline{T}_l(\rho_l)$ be the mean delay on link $l \in h_p$. Let t be the end-to-end time threshold. Then by the Markov inequality

$$1 - F_{T_p}(t, R_p) \le \frac{\sum_{l \in h_p} \overline{T}_l}{t}.$$

Setting $\sum_{l \in h_p} \overline{T}_l / t \le 5\%$, the problem is then reduced to allocating $0.05t$ to the links on path p. Then the maximum link utilization factors can be calculated through $\overline{T}_l(\rho_l)$.

Following the general approach proposed in the previous section, we set

$$\overline{T}_l(\rho_l) = \frac{0.05\, t}{K}.$$

Since $\overline{T}_l(\rho_l)$ is typically a monotonically increasing function of ρ_l, ρ_l can be calculated using standard line search techniques, if not analytically.

One comment on this scheme is given below. If $\bar{T}_l(\rho_l)$ is a convex function and the number of candidate paths is manageable, we may apply the formulation and solution approach in [6] to improve the effectiveness. More precisely, 0.05 t is not evenly preassigned to each link. Instead, for each path p we consider the following constraint

$$\sum_{l \in h_p} \bar{T}_l(\rho_l) \leq 0.05\ t. \tag{6}$$

This alternative treatment is attributed to the convex property associated with Equation (1) in this scheme (shown in Equation (6)), which is unique among the schemes discussed in this section.

Approach 5: Chebyshev Inequality

If the mean and variance of delay on each link are available and they can both be expressed as functions of the link utilization factor, an allocation scheme based upon the Chebyshev inequality is possible. As will be shown shortly, this scheme has similarity to the normal approximation scheme to be discussed in the next subsection. However, one major difference between these two schemes is that the Chebyshev inequality scheme guarantees a feasible allocation strategy while the normal approximation scheme does not.

Let T be a random variable. The Chebyshev inequality makes use of the mean $\bar{T}$ and variance σ_T^2; it states that for any $t > 0$

$$P[\,|T - \bar{T}\,| \geq t] \leq \frac{\sigma_T^2}{t^2}.$$

If (i) the mean delay $\bar{T}_l(\rho_l)$ and the variance of delay $\sigma_{T_l}^2(\rho_l)$ on link $l \in h_p$ are known, (ii) the delays are mutually independent, and (iii) the end-to-end time threshold is t, then by the Chebyshev inequality

$$\begin{aligned} 1 - F_{T_p}(t,R_p) &\leq 1 - F_{T_p}(t,R_p) + F_{T_p}(t - 2 \sum_{l \in h_p} \bar{T}_l(\rho_l), R_p) \\ &= P[\,|T_p - \sum_{l \in h_p} \bar{T}_l(\rho_l)| \geq t - \sum_{l \in h_p} \bar{T}_l(\rho_l)] \\ &\leq \frac{\sum_{l \in h_p} \sigma_{T_l}^2(\rho_l)}{(t - \sum_{l \in h_p} \bar{T}_l(\rho_l))^2}. \end{aligned} \tag{7}$$

In this scheme, we use the mean and variance of delay as the performance indicators. Let the mean and the variance of delay allocated to each link be D and V, respectively. If the network is homogeneous, then for each link the mean and the variance have the same relationship. Therefore, the utilization threshold can be determined by solving

$$\frac{K\ V}{(t - K\ D)^2} = 0.05. \tag{8}$$

However, for heterogeneous networks, the relationship between the mean and the variance may not be the same for all links. Under this condition, a uniform/representative relationship must be described, e.g. $V = a\ D^2 + b\ D + c$ where a, b and c are constants. This relationship together with Equation (8) can be used to determine V and D. For each link, a utilization threshold can be determined by V and D, respectively. The smaller value is chosen.

Approach 6: Normal Approximation

If the first two moments of each link delay is known, then the standard normal approximation technique can be applied. Let $T_1, T_2, \ldots$ be independent identically distributed random variables having mean $\bar{T}$ and finite nonzero variance σ^2. Set $S_k = \sum_{i=1}^{k} T_i$. Then from the Central Limit Theorem

$$\lim_{k \to \infty} P\left[\frac{S_k - k\bar{T}}{\sigma\sqrt{k}} \le t\right] = \Phi(t), \qquad -\infty < t < \infty$$

where

$$\Phi(t) = \int_{-\infty}^{t} \frac{1}{(2\pi)^{1/2}} e^{-y^2/2}\, dy.$$

The Central Limit Theorem strongly suggests that for large k we can make the approximation

$$P\left[\frac{S_k - k\bar{T}}{\sigma\sqrt{k}} \le t\right] \approx \Phi(t), \qquad -\infty < t < \infty$$

or equivalently

$$P[S_k \le t] \approx \Phi\left(\frac{t - k\bar{T}}{\sigma\sqrt{k}}\right), \qquad -\infty < t < \infty. \tag{9}$$

Since we require that the overdue probability be no greater than 0.05, by a table look-up we have

$$\frac{t - k\bar{T}}{\sigma\sqrt{k}} \ge 1.645.$$

It is clear that this equation has the same structure as that of Equation (8). Therefore, as mentioned earlier, the procedure developed for the Chebyshev inequality scheme can be applied. We therefore omit the description of the procedure for the normal approximation scheme.

ALLOCATION SCHEMES FOR HOMOGENEOUS NETWORKS

In this section, 2 allocation schemes for homogeneous networks are discussed. Both schemes require the knowledge of the delay distribution for each link.

Approach 7: Convolution Scheme

With full information about the link delay distribution and the assumption of mutual independence among link delays, this approach exactly calculates the end-to-end delay distribution (and thus the overdue probability) by convolution. By letting the utilization threshold be the same for each link, we consider a path with K hops and can express the overdue probability as a univariate monotonically increasing function of the utilization threshold. Numerical procedures for convolution (with high computational complexity) and line search can be applied to calculate the utilization threshold such that the overdue probability equals 0.05.

To establish the validity of this approach, we provide the following result.

Lemma 3: *If $F_{T_l}(t, \rho_l)$ is a monotonically decreasing function of ρ_l $\forall l \in \{1,2\}$ and $\rho_1 \ge \rho_1'$, where T_1 and T_2 are independent, then $F_{T_2+T_1}(t, \rho_2, \rho_1) \le F_{T_2+T_1}(t, \rho_2, \rho_1')$ $\forall t$.*

Proof:

$$\begin{aligned}
F_{T_2+T_1}(t, \rho_2, \rho_1) &= \int_0^t f_{T_2}(t-y, \rho_2) F_{T_1}(y, \rho_1)\, dy \\
&\le \int_0^t f_{T_2}(t-y, \rho_2) F_{T_1}(y, \rho_1')\, dy \\
&= F_{T_2+T_1}(t, \rho_2, \rho_1'). \qquad \square
\end{aligned}$$

The monotonicity assumption made in Lemma 3 states that for a given time threshold the overdue probability increases as the utilization factor increases. For a typical queueing system, this assumption should be valid.

Proposition 2: *The exact overdue probability for path p is upper bounded by* $1 - F_{T_p}(t_p, \rho, \rho, \ldots, \rho)$ *where* $\rho = \max_{l \in h_p} \rho_l$.

Proof: Apply Lemma 3 $|h_p| - 1$ times. Then the result follows. □

This proposition basically states that if any $\rho_l < \rho$, $l \in h_p$, then the end-to-end delay objective for path p is still satisfied.

Approach 8: Chernoff Bounding Scheme

A rather sophisticated means for bounding the tail of the sum of a *large* number of independent, identically distributed random variables is available in the form of the Chernoff bound. It involves an equality similar to the Markov and Chebyshev inequalities, but makes use of the entire distribution of the random variable itself. Again, line search techniques are required to calculate the utilization threshold. Due to the rather complicated form of the bounds, we do not show the inequality here. The interested reader is referred to Kleinrock[7] for details.

One major advantage of this approach is that legitimate upper bounds on the link utilization factors for a given time threshold are provided. Compared with the convolution scheme, the same amount of information is required, but lower computational complexity is involved. On the other hand, a disadvantage of this approach is that the Chernoff bound tends to be loose when the number of random variables is small.

COMPARISONS AMONG APPROACHES -- TWO CASE STUDIES

In this section, two comparisons among different schemes using $M/M/1$ and $M/D/1$ models, respectively, are made. Several points need to be emphasized regarding the comparisons. First, these comparisons should be deemed as an illustration of a number of theoretic results (e.g., the relative bound quality and the behavior of the bounds as the number of hops increases). Second, the criterion used in the comparisons is solely the bound quality. Third, to compare all schemes discussed, a homogeneous network is considered. Fourth, these comparisons certainly favor the convolution scheme because the convolution scheme calculates exact overdue probabilities and thus serves as a benchmark.

Comparison Using $M/M/1$ Models

In this set of performance test, $M/M/1$ queueing models are adopted. It is assumed that non-pipelining SSs are used. In other words, there is segmentation and reassembly operation performed in intermediate SSs. As such, for each network element the system time should be considered.

For the complete decomposition scheme, we want to find the minimum r such that $e^{-r\,t/k} \le a$ ($b_i = 1/k$), where $(1-a)^k = 95\%$. Then, $e^{-r\,t/k} \le 1 - 0.95^{1/k}$. Consequently,

$$r\,t \ge -k \ln(1 - 0.95^{1/k}).$$

The minimum $r\,t$ (residual capacity and time threshold product) satisfying the above inequality is plotted in Figure 1.

For the Markov inequality scheme, $|h_p|/[(\mu - \lambda)\,t] \le 5\%$. Consequently, $(\mu - \lambda)\,t \ge 20\,|h_p|$.

For the Chebyshev inequality scheme,

$$\frac{\frac{|h_p|}{r^2}}{\left(t - \frac{|h_p|}{r}\right)^2} \le 5\%.$$

where $r = \mu - \lambda$ is the residual capacity. After simple algebra, $r\,t \ge \sqrt{20\,|h_p|} + |h_p|$.

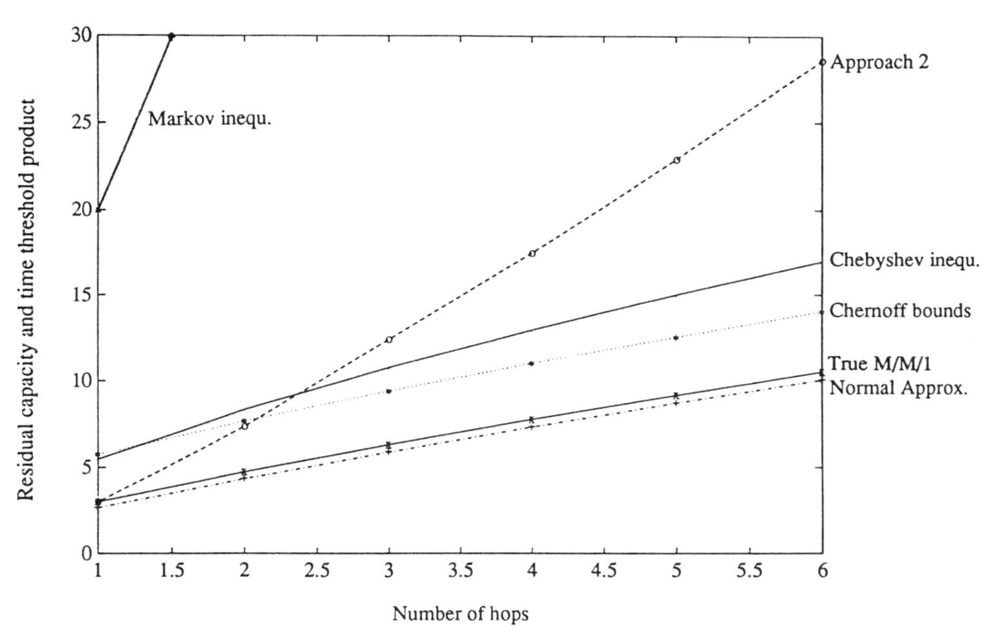

Figure 1. Comparison among different allocation schemes using M/M/1 models

For the normal approximation scheme,

$$1 - F_{T_p}(t,R_p) \approx 1 - \Phi\left[\frac{t - |h_p|/r}{r\sqrt{|h_p|}}\right] \tag{10}$$

where r is again the residual capacity for each link on path p. Setting the RHS of (10) to 5% yields $r\,t = 1.645\,\sqrt{|h_p|} + |h_p|$

Applying the Chernoff bounding scheme, we solve the following equation

$$\exp\left[|h_p| - r\,t + |h_p|\,\ln\left(\frac{r\,t}{|h_p|}\right)\right] = 0.05.$$

When applying the convolution scheme, Expression (5) can be used to characterize the overdue probability where v in (5) is replaced by the residual capacity r. We then applied the bisecting search method (without using derivatives) to find the minimum $r\,t$.

For $M/M/1$ models, results by different schemes are summarized in Figure 1. The first observation from Figure 1 is that the curve corresponding to the complete decomposition scheme is convex in the observed range and tends to diverge from the exact curve (associated with the convolution scheme). Therefore, the complete decomposition scheme is not suggested when the number of hops is large. Second, bounds obtained by using the Markov inequality are loose and therefore only the first two points (for k equals 1 and 2) are plotted in Figure 1. One property which is not shown in the figure is that the curve associated with the Markov inequality scheme is linear. Third, the approaches using the Chebyshev inequality and the Chernoff bounds have comparable performance when the number of hops is small. Moreover, the Chebyshev inequality scheme degrades faster with the number of hops. Fourth, the curve corresponding to the normal approximation is close to the exact curve. In addition, the percentage error (from the true values) improves as the number of hops increases. However, the normal approximation does not guarantee that the residual capacities (or utilization factors) obtained will satisfy the end-to-end delay objective, as shown in this case. Last, the curve corresponding to the convolution scheme (true $M/M/1$) is concave.

Comparison Using $M/D/1$ Models

For $M/D/1$ queues, we assume that the link capacities are 45 Mbps and that the packet size is 83 Kbits. It is assumed that pipelining SSs are used. In other words, there is no segmentation and reassembly operation performed in intermediate SSs. Also because the length of a Level 2 Protocol Data Unit, L2_PDU, (53 octets) is small compared with a Level 3 Protocol Data Unit, L3_PDU, the packet (L3_PDU) transmission time will not be counted in the intermediate SSs. The delay threshold (after considering the deterministic service time) is assumed to be 15.625 msec.

For analyzing the packet loss probability of the longest delay control scheme, we consider a slotted $M/D/1/J$ system, where each time slot equals the packet (deterministic) service time and packets arriving during a slot cannot be transmitted until the beginning of the next slot. It is clear that this slotted system overestimates the packet loss probability, especially when the system load is low. A procedure described in [8] can be applied to calculate the packet loss probability of the slotted $M/D/1/J$ system.

For a number of the discussed schemes, the pdf of the delay on each $M/D/1$ queue must be known. This can be obtained by the following analysis. Let the number of packets in the system (in service and in the queue) be n. The probability mass function (pmf) of n is given below[9]

$$\begin{cases} p_0 = 1 - \rho \\ p_1 = (1-\rho)(e^{\rho} - 1) \\ p_n = (1-\rho) \sum\limits_{k=1}^{n} (-1)^{n-k} e^{k\rho} \left[\dfrac{(k\rho)^{n-k}}{(n-k)!} + \dfrac{(k\rho)^{n-k-1}}{(n-k-1)!} \right] & (n \geq 2) \end{cases}$$

where p_n is the probability that there are n packets in the system and the second factor in p_n is ignored for $k = n$. When a new packet arrives, it will potentially experience two delays. One is the residual service time for the packet in service. The other is the waiting time for the packets that are ahead in the queue. It is easy to verify that the residual service time is uniformly distributed in $[0, t_{max}]$. Given the condition that n packets are in the system ($n-1$ packets are in the queue) upon the arrival of the new packet, then the waiting time w is uniformly distributed in $[(n-1)\, t_{max}, n\, t_{max}]$. Removing the condition on n (using p_n), the pdf of the waiting time can be obtained. To calculate the exact overdue probability, a numerical procedure was developed to conduct the convolution operation based upon the pdf obtained above.

For a number of the proposed approaches, the mean and variance of the waiting time on an $M/D/1$ queue are required. The mean and variance can both be obtained by the P-K formula. For the convenience of the reader the variance is given below

$$\frac{3x^2\rho^2 + 4x^2\rho(1-\rho)}{12(1-\rho)^2}$$

where x is the (mean) service time.

The comparison of different approaches using $M/D/1$ queues is shown in Figure 2. In general, Figure 2 shows the same relative performance among the schemes compared in Figure 1. However, a number of new observations are obtained from Figure 2. First, the $GI/G/1$ bounds are the closest among the legitimate ones except for the convolution scheme which serves as a benchmark. In addition, the quality of the $GI/G/1$ bounds tends to degrade (slightly) as the number of hops increases. Second, the longest delay control scheme is very conservative and the bound quality degrades fast with the number of hops.

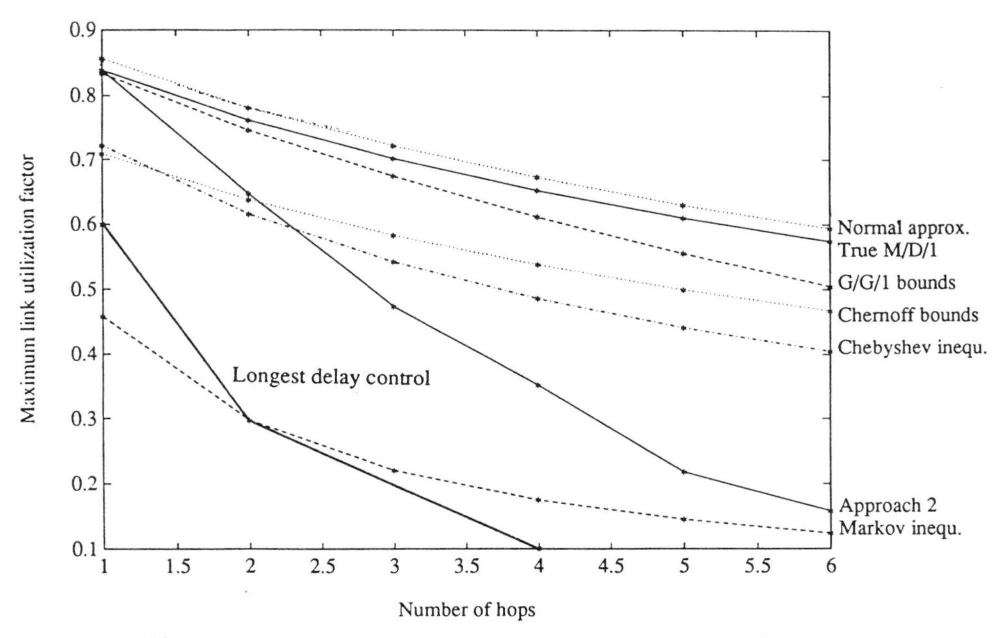

Figure 2. Comparison among different allocation schemes using M/D/1 models

From the above two comparisons we may draw the following conclusions. First, we may consider the longest delay control scheme, the complete decomposition scheme and the Markov inequality scheme as secondary candidates when the bound quality is the major concern. Second, although the normal approximation scheme provides the smallest absolute error bound, the allocation strategies determined by this scheme are not feasible for these two test cases. Third, the Chernoff bounding scheme performs consistently with the number of hops in terms of the bound difference. Fourth, the Chernoff bounding scheme and the Chebyshev inequality scheme are comparable, although the Chebyshev inequality scheme degrades faster.

SUMMARY

This paper deals with a very important problem in network planning and engineering: how to circumvent the difficulty of (i) the excessively large number of end-to-end performance objective constraints and (ii) the nonconvexity of each of these constraints particularly resulting from the percentile nature of the end-to-end delay constraints for the SMDS service. A general approach is to replace the set of end-to-end delay constraints by a simpler set of link utilization constraints in such a way that, if the link utilization constraints are satisfied then the end-to-end delay constraints are satisfied.

For end-to-end percentile-type delay constraints, to compute an allocation strategy satisfying the aforementioned feasibility criterion is nontrivial, especially for heterogeneous networks. We then propose a general allocation approach using the concept of allocating a set of performance indicators as an intermediate step. For each link, the highest utilization satisfying the allocated performance indicators can be calculated. We also address the issue of how to apply this general allocation approach in a network planning/engineering problem to achieve the best result.

Based upon the general allocation approach, a number of allocation schemes for heterogeneous and homogeneous networks, respectively, are discussed. Their relative effectiveness in terms of the utilization thresholds determined is compared for a homogeneous network using $M/M/1$ and $M/D/1$ queueing models.

From the case studies, it is indicated that a number of the discussed schemes are over-conservative and may be impractical. However, more studies should be performed in the future to confirm/reject this result when more information about traffic characteristics and available traffic measurements is available.

The general allocation approach proposed in this paper is applicable for many new services, such as the Frame Relay service, Asynchronous Transfer Mode and the Advanced Intelligent Network services, where end-to-end percentile-type performance objectives are considered. This work lays a foundation to help network planners/administrators perform various network planning and engineering functions for a wide class of services.

ACKNOWLEDGEMENTS

The author would like to thank Vincent Calcagno, Jack Kichura, Arthur Lin, Jonathan Lu, Marco Mostrel, Jonathan Wang, Ellen White, and Mehmet Ulema for valuable information and fruitful discussions.

ABOUT THE AUTHOR

Frank Y.S. Lin received his B.S. degree in Electrical Engineering from National Taiwan University in 1983 and his Ph.D. degree in Electrical Engineering from University of Southern California in 1991. Since 1991, he has been with Bell Communications Research where he is involved in developing network capacity management algorithms for advanced technologies and services. His current research interests include computer network optimization, high-speed networks, distributed algorithms and performance modeling.

REFERENCES

1. F.Y.S. Lin, "Link Set Sizing for Networks Supporting SMDS," *Second IEEE Network Management and Control Workshop,* September 1993.

2. Bell Communications Research, "Generic System Requirements in Support of SMDS," Bellcore technical reference, TR-TSV-000772, Issue 1, May 1991.

3. J.L. Wang, "An Integrated Methodology for Supporting Network Planning and Traffic Engineering with Considerations to SMDS Service," *Proc. IEEE Globecom,* December 1991.

4. J.F.C. Kingman, "Inequalities in the Theory of Queues," *Journal of the Royal Statistical Society,* Series B, 32, pp. 102-110, 1970.

5. J.F.C. Kingman, "On Queues in Heavy Traffic," *Journal of the Royal Statistical Society,* Series B, 24, pp. 383-392, 1962.

6. F.Y.S. Lin and J.R. Yee, "A Real-time Distributed Routing and Admission Control Algorithm for ATM Networks," *Proc. IEEE Infocom'93,* April 1993.

7. L. Kleinrock, *Queueing Systems,* Volume 1, New York: Wiley Interscience, 1975.

8. J.F. Hayes, *Modeling and Analysis of Computer Communications Networks,* 1984.

9. D. Gross and C.M. Harris, *Fundamentals of Queueing Theory,* John Wiley & Sons, 1974.

IV

NETWORK MANAGEMENT PROTOCOLS

INTRODUCTION

As the information technology age progresses, the need for greater bandwidth, multimedia and the connectivity between dispersed networks grows. As the network becomes more sophisticated, the demand for open interface protocols for the management of operational, performance configuration, connection, and billing related parameters of the multi-vendor network elements become high. It is a key consideration in the introduction and deployment of new technology in the network. This session on network management protocols examines current technologies and architecture, as well as those in development and soon to be deployed. It focuses on:

1. Object oriented paradigm and its use in building relationships into OSI management information. It stresses the importance of relationship objects and how they are used in managing the network. The paper also addresses the service primitives used for adding, changing, deleting, and querying relationship objects.

2. Signaling and protocols for establishing broadband connection services.

3. Evaluation of broadband user network interface signaling protocol techniques.

4. Managed objects semantic and modeling tools.

5. Distributed MIB repository for supporting integrated network management.

6. Connection control protocols in broadband network services.

Rajan Rathnasabapathy
Senior Product Planner
NEC America, Inc.

FACTORS IN PERFORMANCE OPTIMIZATION OF OSI MANAGEMENT SYSTEMS

Subodh Bapat

16441 Blatt Blvd, Mail Stop 203
Ft. Lauderdale, FL 33326

ABSTRACT

This paper discusses some techniques for optimizing the performance of an OSI network management system. The primary objective of performance optimization is to improve the response time of any productized network management system, so that it responds to external management requests (modeled using CMIP operations) within constraints imposed by specifications. This paper shows how the response characteristics of a network management system vary with the utilization pattern which that system is subjected to. We present footprints indicative of typical utilization patterns, and suggests how a management system may be optimized in the presence of such utilization patterns. While there are many aspects to performance optimization, this paper will primarily focus on the aspect of optimal distribution of MIB information.

INTRODUCTION

Because networks are a critical infrastructural component of today's society, network management systems must also be considered equally critical. This implies that network management systems must themselves possess a degree of high availability, fault tolerance and performance which is commensurate with the characteristics of the networks they manage. A highly reliable, high-performance managed network is a poor operating risk unless its network management system also has a corresponding degree of reliability and performance.

The capability specification for a network management system must not only describe its function, but also specify its performance. This is especially the case if the managed network is large and is required to have a quick reconfiguration response and fast reconstitution times in the event of failure. Examples of such networks include public telephone networks, and private mission-critical networks such as those used for space telemetry, on-line financial transactions, and transportation operations such as air traffic

Network Management and Control, Vol.2, Edited by
I.T. Frisch *et al.*, Plenum Press, New York, 1994

control. The performance required of the management systems controlling these networks is typically specified in terms of several real-time parameters, such as the number of configuration operations required per second, the average alarm processing rate, the maximum burst alarm processing rate, and so on.

There are many well-known techniques in computer systems design which can be used to ensure that a network management system product delivers reliable and highly available operation. In this paper, we will focus primarily on the *performance* aspects of network management systems, and demonstrate some issues involved in tuning a network management system for optimal performance.

Before we proceed with the discussion of technical aspects, we would like to define the contours of this paper which delimit the terms of discussion as follows:

- We classify the choices involved in building a real-time network management system product in two major categories: *technology choices* and *architecture choices*. The technology choices in building a real-time network management system primarily involve the selection of the proper components for the product platform. These include hardware with appropriate performance (speed and MIPS available from the CPU(s), average disk I/O time, etc.) as well as software with the desired characteristics (operating system context switching latency, interprocess communication delay, average database transaction delay, etc.). Clearly, the proper choice of technologies is important while selecting the product platform for a real-time network management system. However, even after incorporating the best *technologies*, it is the internal *architecture* of the network management system which is critical in determining overall system performance. In this paper, *we will assume that appropriate technology choices have already been made*, and will concentrate on some of the architecture choices involved in constructing a high-performance network management system.

- The target audience of this paper comprises the *architects, designers, engineers, programmers, implementers* and *marketers* of network management products. The recommendations of this paper are not targeted at *users* of these products. This paper identifies and addresses some architectural issues in the design of network management systems which influence system performance. Because typically users have little influence over the internal architecture of a system once it has been productized, the recommendations of this paper will not be relevant to that audience. In particular, this paper does not address the *tuning* and *configuration* issues which may help *users* of a product to incrementally refine its performance. Rather, it addresses *architectural* issues which may help *vendors* of network management systems -- who have control over how management solutions are implemented -- to improve their design procedures.

- When undertaking performance optimization, attention must be paid to response time both on the side of the *managed object* and on the side of the *managing system*. This implies that the processing time both on the side of the *agent* and the side of the *manager* must be examined. An agent represents some real physical or logical network resource which must be controlled. The process of optimizing the response time for network management operations in generic managed resources is highly resource-specific, and is not discussed in this paper. Here, we concentrate on optimizing the performance on the side of the *managing system* (which is typically implemented on stand-alone or networked computer systems such as workstations, servers and mainframes), which normally issues CMIP requests to be serviced by an agent, and receives responses and event reports from the agent. However, it is also possible for a managing system to play an *agent role*, which it does when responding to a request from a peer management system. In this paper, we examine the nature of management operations experienced by such a managing system, regardless of whether it sees them in an agent role or manager role. We will examine its processing components of such a managing system (e.g. the information repository) and show how their performance may be optimized. Since the components affecting the performance of agent-only implementations representing generic managed entities could be very different, the optimization guidelines recommended in this paper do not apply in that regime.

- This paper does not attempt to demonstrate how *individual CMIP operations* may be made faster. CMIP requests issued by a manager or serviced by an agent can vary widely in their response characteristics, depending on the number of objects scoped, the nature of the operation, and the inherent processing speeds of the managed resource and/or the platform hosting its agent implementation. Merely ensuring that CMIP operation execution is fast will not necessarily improve the response times as perceived by the manager. Bottlenecks could exist at the level of the network management *applications* which issue and interpret these CMIP requests. Applications often play a large part in determining overall response time, and protocol performance tuning may, under certain circumstances, be almost irrelevant. Rather than addressing the speed of individual CMIP operations, this paper looks at *large-scale statistical trends* in the overall mix of CMIP transactions experienced by a management system, and suggests methods of optimizing its performance in the presence of such trends.

With these caveats in mind, we proceed with the technical discussion.

INTERNAL SYSTEM IMPLEMENTATION FOR PROCESSING MANAGEMENT OPERATIONS

An OSI management system receives stimuli from many external sources. Acting in an agent role, it may receive CMIP requests such as M-SET, M-GET, M-CREATE, etc. from peer management systems across an interoperable interface [CCITT X.711]. Acting in a manager role, it may receive CMIP requests such as M-EVENT-REPORTs from agents associated with real resources in the managed network.

These requests perform abstract operations against a virtual Management Information Base (MIB) which is conceptually resident within the network management system. This is shown in Figure 1: when CMIP operations are executed against the MIB, real operations such as store, update, retrieve, and remove are performed on these database and memory elements.

The Management Information Base is the conceptual repository of all information required for network management functions. The MIB specification generally consists of abstract templates [CCITT X.722] which specify the objects which are to be managed and their attributes which are of management interest. Although the MIB is a virtual collection

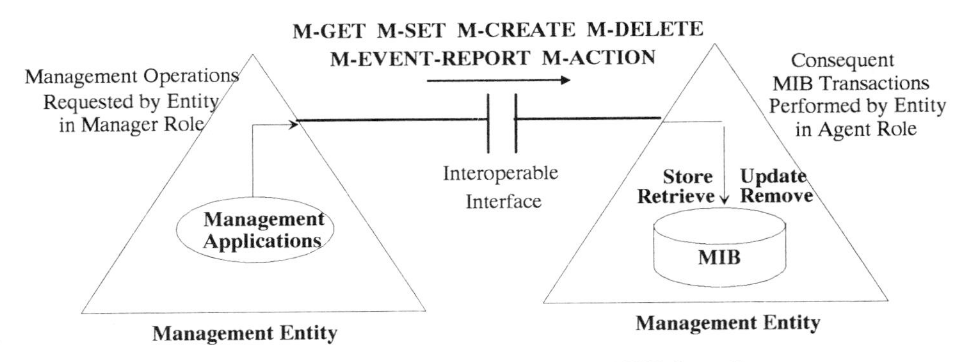

Figure 1. MIB Transactions Resulting from CMIP Operations

of information about managed objects, in a real implementation it is usually manifested as a distributed information repository, consisting of various types of databases, memory and firmware elements. In this paper, we will use the term *Management Information Base* largely to indicate the *implemented form* of such a *repository*, which is an information store representing many managed object instances in an actual managed network.

Because the MIB is a distributed information store, the pattern of distribution of this information determines the speed and ease of access of MIB data for network management operations. The overall performance of the network management system is very sensitive to the speed with which these operations can be executed; therefore, the architectural choices made in distributing MIB information strongly affect the overall system performance characteristics. Hence, the objective of performance optimization is to design an architecture which facilitates the optimal internal distribution of this MIB information, which allows the fast processing of CMIP operations.

In this paper we will show that there is no single pattern of distributing MIB information which leads to optimal performance. Rather, the internal distribution pattern of MIB information must take into account the manner in which the network management is used. This is because the utilization pattern dictates the nature and frequency of operations performed on the Management Information Base. Therefore, each different utilization pattern gives rise to different considerations and priorities for distributing MIB information for optimal performance.

INTERNAL DISTRIBUTION OF MIB INFORMATION

The internal implementation of a MIB requires a mapping from the attributes of the abstract object model to real records, fields, registers and memory segments in the database, firmware, and software elements resident within both the network management station and real network resources.

Many network management stations contain an internal repository which stores MIB information in a persistent fashion. This repository may be a relational, object-oriented, flat-file or proprietary database [Codd90, Kim90]. This database typically contains information such as [Bapa91a]:

- System-wide information, such a directory of all managed objects
- Topology and connectivity information for hardware objects
- Deployment, licensing and version information for software application objects
- Current configuration profile information (attribute values, strap values)
- Possible future configuration profiles (other permitted attribute and strap values which may not be currently instantiated)
- Previous configuration profile histories
- Object creation and replication templates (e.g. Initial Value Managed Objects)
- Alarm and event histories
- Aging policies for expiration of historical information
- Alarm filtering information
- Automated restoral and reconfiguration rules triggered by alarm processing
- Security permissions and policies for network administrators (e.g. resource access permissions in a multi-operator network management system)
- Security permissions and policies for network users (e.g. configured closed user groups or policy-routing administrative domains for a packet or datagram network)
- Other automation information (e.g. automated time-of-day-based reconfiguration, automated report generation, etc.)
- Database mechanisms for tracking trouble report creation, escalation, resolution and closure.

The database may also contain copies of information normally resident within the real resource, for purposes of redundancy.

MIB information may also be stored as downloaded data or firmware within a managed hardware resource. Dynamic attributes, such as protocol entity counters, are usually recovered from register values within real resources or memory locations within software entities. Other software-based representations of MIB information may include management application processes, layer entities, agent processes, or the local instrumentation of the resource. Occasionally such information may be stored in an intermediate management entity, such as a proxy agent, or a mediation device in a Telecommunications Management Network. Fragments of the MIB may be stored in different places in the overall network management solution; this is shown in Figure 2.

Clearly, there are many choices as to where any particular item of MIB information should be stored: in the actual managed resource, in an intermediate management entity, in a management station repository, or in multiple locations involving some combination of these. There is no single correct architectural choice for the pattern of distribution of MIB information. The process of optimizing the performance of a network management system must take into account the nature of usage. The internal architecture of the network management system must be designed such that it is flexibly allows the suitable tailoring of information distribution within the MIB in accordance with the expected usage pattern.

MIB TRANSACTIONS RESULTING FROM NETWORK MANAGEMENT OPERATIONS

To execute a CMIP operation, MIB information internally distributed within the management solution needs to be accessed. If an application requires information about a managed object, more than one location may potentially need to be accessed in order to reconstruct the required information from object fragments. In the general case, to reconstruct the entire object from its fragments, all the different locations which contain its attribute information must be accessed. The speed of access, therefore, depends on the nature of the actual information distribution within the management solution.

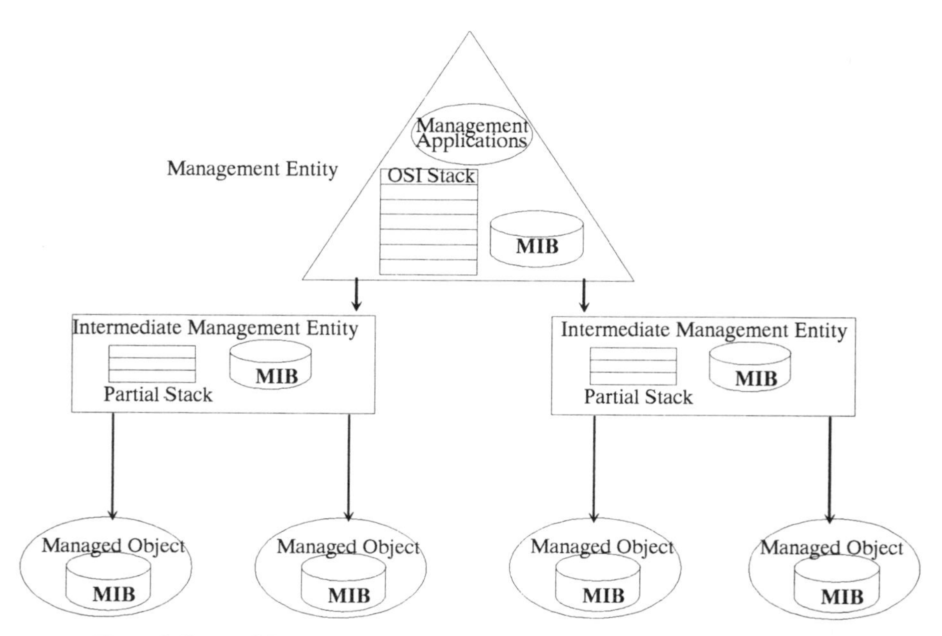

Figure 2. Internal Distribution of MIB Information Affects System Performance

The management station repository is a particularly important focus of performance optimization, as much MIB information is often stored within it. This repository must be amenable and customizable for performance tuning. In particular, it must exhibit fast response times when frequent CMIP operations are executed against it.

Consider a situation where the MIB is implemented in a relational database repository in a network management workstation [Teor90, Bapa91b]. Suppose that this system is an element manager controlling a subset of the real resources deployed in a large network, and often responds to network management requests received from other peer-level management systems by making appropriate changes to its resources and to its database representation of those resources. Thus, an externally received network management request could translate into an SQL transaction on a relational database. Such a typical mapping between a management operation and its corresponding database transaction is represented in the following table.

Table 1.

M-GET	MIB Information Retrieval (e.g. SQL SELECT)
M-SET	MIB Attribute Value Update (e.g. SQL UPDATE)
M-CREATE	MIB Object Record Creation (e.g. SQL INSERT)
M-DELETE	MIB Object Record Deletion (e.g. SQL DELETE)
M-EVENT-REPORT	MIB Event Instance Creation (e.g. SQL INSERT)
M-ACTION	MIB Attribute Value Update (e.g. SQL UPDATE)

A CMIP operation such as an M-GET, after being decoded by the CMIP protocol machine and mapped to the local information schema by a management application, may result in a SELECT primitive in SQL being issued against one or more tables within the relational repository. Similarly, other CMIP operations correspond to other SQL primitives being executed against the database schema. Therefore, optimizing the repository with respect to its primitive store (write), retrieve (lookup), update (modify) and remove (delete) operations, has a significant effect on the processing rate for CMIP operations, and thus on the overall performance of the network management system.

TRANSACTION MIX DISTRIBUTION ANALYSIS

To determine the distribution of MIB information which delivers optimal performance, one must study the nature and frequency of the operations perceived by the network management system. This is referred to as *transaction mix*. Because each type of transaction requires different architectural support for efficient processing, it is the overall transaction mix experienced by the network management system which determines its performance. A network management system tuned to deliver optimum performance under one transaction mix distribution may perform poorly if it experiences a different transaction mix distribution.

The data presented in the following sections has been acquired from the private operations network of a large transportation carrier The managed network which was studied for collecting this data was a moderately large network with over 50,000 managed objects. The managed objects were largely hardware resources consisting of customer premises equipment, carrier-based services, and mainframe-based data networking equipment such as large computer systems, mid-range systems, front-end processors, cluster controllers and terminals. About 13% of the managed objects also included campus LANs and internetworking equipment managed indirectly through other systems using proprietary interoperability interfaces, and about 6% of the managed objects represented software applications. The data was collected and analyzed in units of TMOPD (thousand management operations per day) and averaged over a period of sixty days in order to discern statistical trends. For reasons of confidentiality protection of this client's commercial data, we cannot present actual numbers of network operations; we can, however, present the relative frequencies. These percentages are presented below, and are sufficiently illustrative in demonstrating performance optimization concepts.

The transaction mix data is plotted to show the distribution of *object-operations*. Each data item represents the total of similar operations, with each operation being weighted by the number of managed objects affected by it. This weighting is necessary because a single CMIP operation may affect multiple managed object instances, due to scoping and filtering. It is possible for a network management system acting in an agent role to process just one CMIP request, and yet experience considerable load if the scope of the operation is broad enough to affect a large number of managed objects. Therefore, studying transaction mix distributions merely on the basis of the number of operations will be erroneous for performance tuning, unless each operation is multiplied by the number of its operand object instances to provide a more accurate quantification of the offered load. In collecting the following data, the scoping and filtering software routine was modified to log to a data file the number of actual operand objects which were scoped and remained after filtering. This was correlated with the actual operation type for conducting the performance analysis.

Under this analysis, we identified the footprints of three distinct types of transaction mix distributions which were experienced by the network management system:

- *Configuration-and-Testing-Intensive Utilization*
- *Alarm-Collection-Intensive Utilization*
- *Report-Generation-Intensive Utilization*

The sections below present the graphs for each, and discuss an appropriate performance optimization strategy for each type of transaction mix distribution. In these graphs, the vertical axis represents, for each transaction type, the percentage of the total object-operations experienced by the system. Each of these graphs serves as the footprint of a particular utilization pattern and, when present, suggests a particular mode of optimization for the management solution.

CONFIGURATION-AND-TESTING-INTENSIVE UTILIZATION OF A MANAGEMENT SYSTEM

Figure 3 demonstrates the transaction mix distribution for a network management system whose primary utilization is for *network configuration*, *topology validation*, and *confidence and diagnostic testing*.

Under this utilization pattern, a fairly even distribution of some CMIP operations is obtained. M-CREATE, M-SET, and M-GET operations constitute a significant percentage of the total number of operations during the set-up phase where the network configuration is being defined. Managed objects are created within the network management system to represent real resources using M-CREATE operations. Although these are often created using Initial Value Managed Object templates, in almost every case some of their attribute values need to be refined using M-SETs. M-GETs are primarily used to validate configuration settings. Finally, a large number of M-ACTIONs is used to invoke tests and other procedures to validate the resource's operation within the network.

Under this utilization of a network management system, performance tuning would require fast processing of M-ACTION confirmation and linked replies. Set-up, configuration and testing procedures are often conducted interactively through a human user interface; thus, a reasonable response time is required from the M-GET, M-SET and M-CREATE operations.

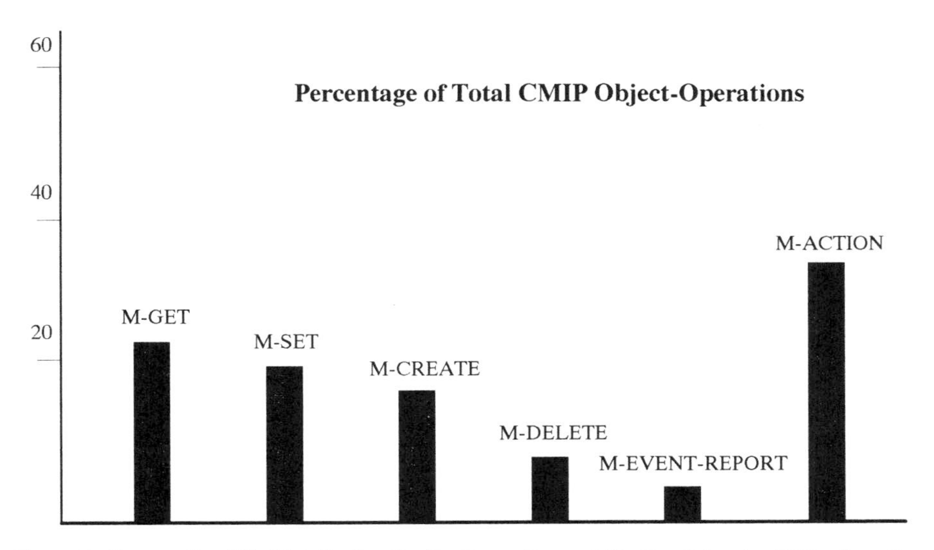

Figure 3. Transaction Mix Distribution for Configuration-and-Testing-Intensive Utilization of a Network Management System

ALARM-COLLECTION-INTENSIVE UTILIZATION OF A MANAGEMENT SYSTEM

If the transaction mix distribution follows a different pattern, a different performance tuning strategy is required. Figure 4 demonstrates the transaction mix distribution where the primary utilization of the network management system is for *alarm collection.*

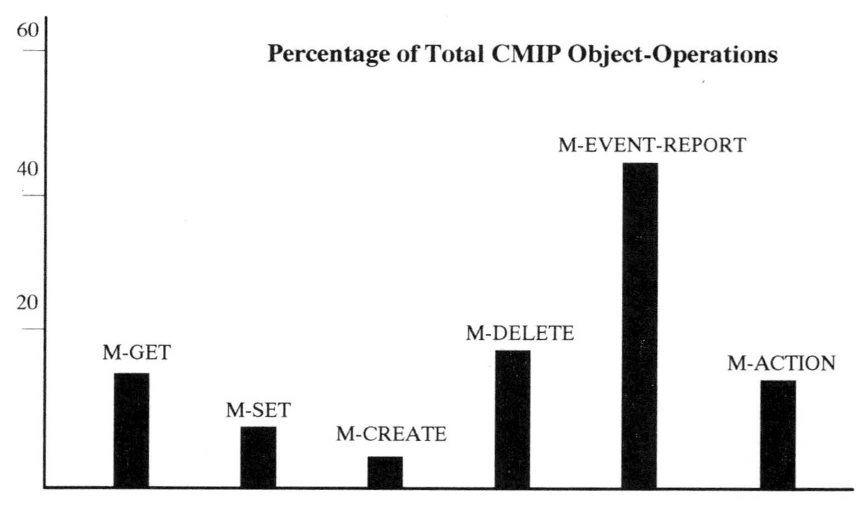

Figure 4. Transaction Mix Distribution for Alarm-Collection-Intensive Utilization of a Network Management System

Under this utilization pattern, an overwhelming number of CMIP operations is M-EVENT-REPORT operations arriving from agent processes. Since alarm collection is a round-the-clock activity and often proceeds unattended by a human observer, alarm information must be stored in the management station repository for later inspection by a management operator or for report generation.

Performance tuning under this utilization pattern requires tuning the repository for fast insertion of event information. Under circumstances of catastrophic network failure, a burst of M-EVENT-REPORT requests may be expected from many resources. In order not to lose any significant operational data, the repository must respond with performance similar to a high-demand OLTP (On-Line Transaction Processing) database.

Since alarm management through human interaction sometimes involves retrieving associated parameters, performing diagnostic testing, and deleting event instance records, some M-GET, M-DELETE and M-ACTION activity is also seen.

REPORT-GENERATION-INTENSIVE UTILIZATION OF A MANAGEMENT SYSTEM

Some OSI management systems are used to describe network activity through periodic *report generation*. Occasionally, these systems may not themselves be used for alarm collection, but may depend on a common networked repository, which is indirectly populated by another management system primarily used for alarm collection. Figure 5 depicts the transaction mix distribution one such report-generation-intensive management system.

Under this utilization pattern, the number of M-GETs is significantly high. Many M-GETs in this situation have a large number of operand objects because report generation activity sometimes tends to scope large portions of the Management Information Tree. However, this is not always the case; often, customized and specialized reports find it difficult to use a single M-GET scoping the entire containment hierarchy. This is because the parameters of interest in every class of managed object are different, and must be specified separately in multiple M-GET requests. Further, the order in which information is retrieved from the containment hierarchy is not necessarily the order suitable for human presentation in the report; thus, user-friendly report generation is best achieved using multiple M-GETs in a sequence approximating the format for report presentation.

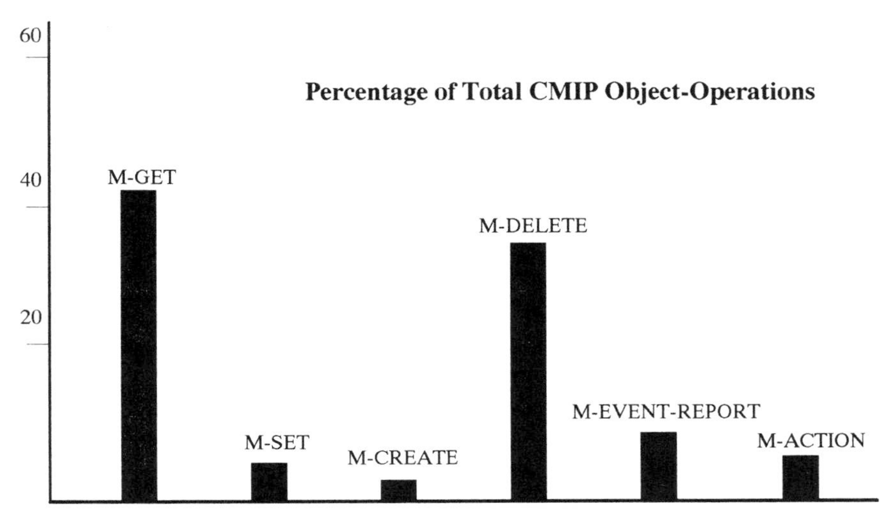

Figure 5. Transaction Mix Distribution for Report-Generation-Intensive Utilization of a Network Management System

Routine statistical information is often deleted from the repository, either immediately after report generation or after expiration following some aging policy. Therefore, a significant number of M-DELETEs is also seen. It should be noted that the information affected by the M-DELETE is not the same information which is requested in M-GETs at the same time. The information retrieved by the M-GET is typically used to generate a report on the current state of the network; the information expunged by the M-DELETE is usually that which has been routinely stored in the repository for its aging period and is not significant enough for long-term archival.

CONSIDERATIONS IN PERFORMANCE OPTIMIZATION

Because CMIP operations could affect many attributes of a managed object instance, and because object fragments may be distributed over a management system, the nature of the actual distribution has a profound effect on system performance. By repartitioning managed objects to achieve a more optimal distribution of object fragments, system performance can be greatly improved.

It should be emphasized that a particular distribution of MIB information is only optimal relative to a given mix of CMIP transaction queries. Should the pattern of CMIP queries change, the MIB information needs to adapt itself to a different distribution in order to continue delivering optimal performance.

There are three types of changes in the transaction load mix which a network management could experience:

- Short-term, cyclical, predictable changes in usage patterns
- Long-term, evolutionary changes in usage patterns
- Unpredictable and unanticipated changes in usage patterns.

Predictable changes are ones that can be anticipated, such as daily or weekly changes in the pattern of use in a management system. Some management systems, for example, are used for alarm collection during peak hours of operation, and for report generation during off-peak hours. Long-term changes in the system usage pattern could occur with time as operators become familiar with the network management system's capabilities and begin exercising the system in different ways.

Because system performance and response characteristics are highly sensitive to the distribution of MIB information, maintaining performance under changing utilization patterns often requires redistribution. The architectural strategy for distributing MIB information should be able to adapt and evolve with changes in the transaction mix distribution experienced by the system.

For changes in the utilization pattern which are cyclical and predictable, the network management system should employ a strategy of periodically redistributing MIB information. *Data migration* is an important tactic in this regard: MIB data should be migrated to wherever it is most immediately useful at the anticipated time. This can be accomplished by in-memory caching from secondary storage or data download to a remote resource. Depending on the periodicity of the cycle of the usage pattern, the secondary copy of data can subsequently be destroyed, invalidated or refreshed to maintain consistency.

For long-term changes in the usage pattern, it is necessary to monitor the system long-term to assess evolutionary changes in the transaction mix distribution. *Trend analysis* is an important tactic in this regard. On the basis of trend analysis results, a new distribution strategy for MIB information can be employed.

DISTRIBUTION-BASED PERFORMANCE OPTIMIZATION PRINCIPLES

To arrive at an optimal distribution of MIB information, the transaction mix distribution analysis must not only track the nature of CMIP operations, but must also identify which attributes are frequently accessed. It is also important to identify the nature of this access for each frequently accessed attribute -- whether it is a GET-access (lookup) or a REPLACE-access (update). The frequency of GET- and REPLACE-accesses for each attribute should be quantified. Further, the study must indicate which managed object classes are most frequently M-CREATEd and M-DELETEd.

Such analysis can yield important results. If some subset of MIB information is GET-accessed with high demand, it must be optimized for fast lookups. This could involve in-memory caching or even (for attributes which are never updated) reimplementation in firmware.

Heuristics for optimal distribution of MIB information may be determined based on the above study. If an object is frequently GET-accessed for reading attribute values, it should be ideally stored closest to where it is most often required, even if this means storing multiple redundant copies. For example, an eventReportingSieve object could be directly stored inside any real network resource which has the capability of local storage, so that sieving may be done locally. This will lead to optimal performance even if a second copy of the eventReportingSieve exists in the management station repository. If the sieve is stored only in the repository, all events must be reported up to the management station where they are sieved, and no sieving can be performed locally in the resource. This may result in an unnecessary increase in the traffic load on the link between the resource and the management station.

By the reverse token, if a network management application performs a GET-access to read an attribute value which is stored only with the managed resource, and if the delay in querying the managed resource exceeds response time requirements, then it may be better to redundantly store a copy of that attribute value locally within the management station repository itself. This is sometimes more efficient if the resource must be queried over a wide-area network.

On the other hand, if an object is frequently created or its attributes are frequently REPLACE-accessed, performance will be improved if only one copy of that object exists. This is because the existence of multiple copies of information which is frequently updated requires the overhead of a synchronization protocol to assure consistency between the multiple copies. Because distributed multiple-phase commit protocols are often time-expensive, frequently updated objects should not be redundantly stored.

UTILIZATION-BASED PERFORMANCE OPTIMIZATION PRINCIPLES

The process of performance tuning for an OSI Management System needs to take into account the manner in which the system is utilized, since any single performance parameter profile is only optimal relative to a given transaction mix distribution. After studying the utilization pattern for any given system, it can be matched to one of the above footprints, and an appropriate mode of optimization can be selected.

If the footprint indicates that the primary utilization is alarm collection, the management station repository must be tuned for fast insertion. For example, in a relational database, a uniqueness index on a table assures consistency of information, but slows down insertion because every new record must be checked for uniqueness against every other record in the table [Elms89]. Further, even if the index is not a unique index, the maintenance of the index itself (which acts as a sort tree or hash table for fast lookups) is an activity which requires updating the index with every new record insertion. For fast insertion, indexes must be removed, with consistency checks done by an offline process separately invoked under low-load conditions. Other repository maintenance activities, such as free space reclamation and garbage collection, must also be postponed for similar background, offline processes.

If the footprint indicates that the primary utilization is report generation, the repository must be tuned for fast retrievals, which may mean construction of unique indexes or hash tables based on attributes used as lookup keys. Unique indices also assure consistency in generated reports. Since report generation is often concomitant with large M-DELETE sequences, and since instance deletion also requires lookup, such hash tables would speed up M-DELETEs as well. Reclaiming repository space freed up by M-DELETEs is best done by processes which run immediately after report generation, since much routine information is often expunged as soon as its report has been generated.

If the footprint indicates that the primary utilization is configuration and testing, the repository must be tuned to respond to fast update operations due to the effects of reconfiguration and testing on attribute values. If the configuration activity involves the creation of many network objects (for example, during the network set-up phase), the repository will be frequently populated with new data. In this instance, it must be tuned for fast insertion, similar to the alarm collection pattern described above. Further, during intensive, network-wide testing, M-ACTIONs may produce many side-effects (state modifications, linked replies) which may need to be speedily recorded; a repository tuned for fast insertion would be beneficial under these circumstances as well.

CONCLUSION

Optimizing the performance of a network management system is a process which must account for the nature of usage experienced by the system. Depending on the transaction mix distribution of CMIP operations experienced by the system, different strategies must be employed to distribute MIB information within the internal architecture of the network management system. This paper has described the footprints for some typical transaction mix distributions experienced by a system managing a private data network, and has attempted to illustrate the performance optimization strategies appropriate to each such utilization pattern.

ABOUT THE AUTHOR

Subodh Bapat is a network consultant and has worked with several medium and large-sized companies in the architecture and design of network management systems. While a Principal Software Engineer for Racal Datacom, he was a lead architect in the implementation of standards-based network management platforms for data internetworking and telecommunication networks. In that role, he pioneered the use of applying transaction-processing concepts to network management systems, leading to the development of several useful performance evaluation and optimization tools and techniques. His interests lie in the areas of network management systems architecture, information modeling for management information bases, and the development of network management software. He has published several articles on these and related subjects in leading technical journals, and has presented papers at major industry conferences. His work has been frequently cited in the literature on network management. Subodh holds a Bachelor's degree from the Indian Institute of Technology, Bombay, India, and two Master's degrees in engineering and computer science from Syracuse University, Syracuse, New York. He has been awarded a number of patents in the area of implementing network management and control software.

REFERENCES

[Bapa91a] Bapat, Subodh, "OSI Management Information Base Implementation", *Integrated Network Management II: Proceedings of the IFIP TC6/WG6.6 Second International Symposium on Network Management*, North-Holland/Elsevier, New York, NY, 1991; pp. 817-832.

[Bapa91b] Bapat, Subodh, "Mapping C++ Object Models to SQL Relational Schema", *Proceedings of the C++ At Work Conference*, Santa Clara, California, November 1991.

[CCITT X.711] "Information Technology - Open Systems Interconnection - Common Management Information Protocol, Part 1: Specification", CCITT Recommendation X.711, 1992.

[CCITT X.722] "Information Technology - Open Systems Interconnection - Structure of Management Information, Part 4: Guidelines for the Definition of Managed Objects", CCITT Recommendation X.722, 1992.

[Codd90] Codd, E.F., "The Relational Model for Database Management, Version 2," Addison-Wesley, 1990.

[Elms89] Elmsari, Ramez and Navathe, Shamkant, "Fundamentals of Database Systems", Benjamin Cummings, 1989.

[Kim90] Kim, Won, "Introduction to Object-Oriented Database Systems", The MIT Press, 1990.

[Teor90] Teorey, Toby J, "Database Modeling and Design: The Entity-Relationship Approach", Morgan Kaufman, 1990.

INCORPORATING RELATIONSHIPS INTO OSI MANAGEMENT INFORMATION

Alexander Clemm[1]

Munich Network Management Team, Institut für Informatik
Universität München, Leopoldstr. 11b, 80802 München, Germany
E-mail: clemm@informatik.uni-muenchen.de

ABSTRACT

Relationships form the very essence of any network management information model because through them the interworking, interdependencies, and conceptional connections between the modeled resources are grasped. This information is essential for being able to perform any kind of network management function. The modeling of relationships and their incorporation into a management information base is therefore of great importance. While OSI network management standardization is well under way, the concepts offered to cover these aspects seem still immature and, especially from a specification perspective, leave a lot to be desired.

This paper proposes therefore a formal means for the specification of relationships independent of managed object classes. Services to access relationships are defined. A method is presented how thus specified relationships can be integrated into management information by mapping them onto managed object classes, attributes, and name bindings according to a specified scheme. The realization of relationship services as object management functions is described. This makes it possible to use the concepts presented with no modifications or extensions on top of the existing OSI management information model.

MOTIVATION

Managed objects (MOs) in a management information base (MIB) do not exist in isolation of each other; rather they are engaged in a lot of different relationships expressing interdependencies and conceptual connections between them. Information about relationships between MOs is essential for a wide variety of network management tasks because through them the interworking between the modeled resources is grasped.

---

[1] A. Clemm's work is also supported by IBM European Networking Center, Heidelberg, Germany

- In fault management, often symptoms are observed in one place when the true cause of a fault lies in another place, e.g. as symptom a severe degradation of a connection's quality of service is experienced when the true cause is that an underlying line was unplugged. To locate and identify a fault, possibly correlating multiple symptoms, causality chains have to be traced [7; 12]. These causality chains represent paths of relationships between the MOs where the symptoms are observed and the MOs where the fault that caused them occurred.

- For correct configuration of a network, knowledge of the dependencies, i.e. relationships, between network components is necessary e.g. to configure compatible versions [20].

- Performance characteristics of a network can be improved and bottlenecks eliminated if information about layering relationships within a system and communication relationships between peers are known because this makes a better adjustment and tuning of parameters possible, e.g. of window sizes with respect to link speeds and buffer sizes [2].

In the context of OSI network management, relationships between MOs are so far handled in the following way:

- Containment relationships are represented through naming, where MOs are arranged in a naming, i.e. containment tree [10].

- All other relationships are represented through attributes, where one MO has an attribute with values serving as pointers to the other MOs it is in relationship with [9].

This approach has several deficiencies due to the fact that relationships are not treated as a separate concept in their own right [4], including the following:

- Relationships span across multiple MOs and do not belong to any one of the participating MOs individually. This makes them different from "ordinary" attributes, such as counters. Relationship modeling through attributes of the relationship participants is therefore misleading.

- Relationships are subjected to relationship specific constraints, such as cardinalities. For a precise model, formal specification means for the modeling of such constraints should be provided. This would also make it possible or at the very least make it easier to enforce the consistency of models (i.e. MO class and relationship definitions) as well as of MIBs (i.e. instantiated management information).

- New kinds of relationships between MOs can become of interest as new management functions and applications are introduced. Specification of relationships inside the MOs makes it impossible to add new kinds of relationship later without having to change existing managed object class definitions.

- Other drawbacks include: unclean treatment of relationship attributes (compare with fallback relationship [9]); consistency problems due to the scattering of relationship information across relationship participants without an appropriate transaction concept; difficulties involved in MO migration and location transparency [15]; and more.

It might be argued that separate modeling of relationships through dedicated MO classes (MOCs) is not prohibited by the standards, although it would run against the current practice which they imply. Such practice can indeed alleviate some problems, and one noteworthy attempt is described in [13]. However, the basic underlying problem which lies in that relationships are really an independent, separate, **different** concept cannot be resolved in a satisfactory manner using only the current OSI management information model. As long as no formal means are provided for the specification of relationship specific constraints and everybody can use their own convention, precise relationship specification cannot be enforced and depends solely on the discipline of the modeler. Common and generic mechanisms to support network model consistency and also MIB consistency are hard, if not impossible to supply.

As a consequence, it is necessary to come up with alternatives and improvements over the existing scheme. Essentially needed is explicit relationship support, both for the modeler of networks and the users and applications operating on these models. For this purpose, in this paper a general and comprehensive framework for the specification and management of relationships in the context of OSI network management is presented. It is depicted in Figure 1 and was designed with the following ideas in mind: A supplement to the information model is needed that allows to model relationships separately (1). This supplement will allow modelers to think more naturally in terms of the domain to be modeled, which includes both the objects and the relationships between them. By the same token, also services have to be defined for the access and modification of relationships that will allow users and applications to operate on the resulting network models in the same natural and intuitive way (2). In addition to the relationship specification and definition of services, provision has to be made for instantiation of relationships as part of actual management information (3) and for the implementation of the defined services by the pass-through services of the object management functions [8] (4). It is desirable that all this takes place within the established and generally accepted OSI management concepts.

The abstract specification of relationships between MOs can be conceived as an additional modeling layer on top of the basic information model. With a mapping according to a fixed scheme between the two defined, uniform treatment of relationships as part of the MIB and realization of relationship services becomes possible, allowing applications to operate on a "value-added" MIB with explicit relationships which is based on a conventional MIB without.

The rest of this paper outlines these concepts in more detail. A formal framework for the specification of relationships is put forward in section 2. Section 3 presents defined relationship services for their access and modification. A method for instantiating relationship information in an MIB independent from the MOs that are put into relationship is presented in section 4. The core of this method is to maintain relationship information in dedicated MOs which at the same time provide mechanisms to ensure consistency of relationship information. Also a fixed scheme is indicated that allows to derive the respective MOCs and name bindings from the relationship specifications. Section 5 briefly describes how the relationship services are realized using standard object management functions. As also ISO has lately begun picking up on this subject, related new trends in standardization are explored in section 6. Section 7 arrives at some conclusions.

A MODELING FRAMEWORK FOR THE SPECIFICATION OF RELATIONSHIPS

As was argued above, relationships form an important concept in their own right separate from the MOs involved and should therefore also be modeled separately. For this purpose, appropriate means for specification are needed, appropriate meaning that at least the most important characteristics inherent to relationships must be able to be expressed, preferrably in a formal manner. In fact, their modeling should even be enforced, as they can reflect important constraints on the consistency both of the model and of the instantiated MIB and thus make

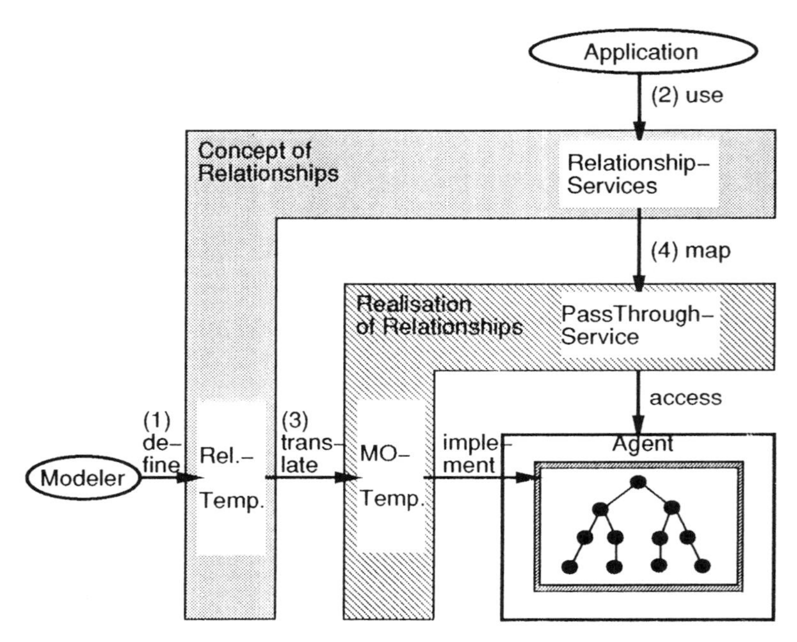

Figure 1. Adding relationships to OSI management information

model and MIB much more precise. At the same time, specification means should be simple enough so that they do not overstrain modelers and thus reduce acceptance and introduce too much complexity into models. In order to be able to come up with appropriate specification means and to judge different approaches, a clear understanding of the concept of relationships is necessary. For this purpose, a catalogue with the different aspects that are involved in the concept and the management of relationships has to be systematically derived. A list without claim for completeness of some of the more important aspects that have to be taken into consideration follows below; it is an excerpt from a more detailed analysis that would be beyond the scope of this paper.

- Different types of relationships, or *relationship classes*, exist that can be instantiated, just as there are different classes of MOs.

- Participants in a relationship are in general not equal but play different *roles*. For instance, in a service relationship we can distinguish between the roles of the service user and the service provider. An exception are *symmetric relationships* where only one kind of role can be distinguished. For instance, in a peerConnection relationship (indicating that the underlying resources of the respective MOs maintain a connection over which they exchange data) no difference has to be made between the roles of the peer connection endpoints. Roles per se do not imply a direction from one "source" participant to another "target" participant. Rather, in a relationship involving two participants, both directions can be of interest, depending on the application.

- There can be constraints on the number of MOs a MO is allowed to be in relationship with, or conditions that state that a MO is obliged to maintain relationships with a certain number of other MOs. For example, a certain type of service provider may only be able to serve a certain number of service users at a time, i.e. the number of MOs acting in a service user role that a service provider MO can be in service relationship with is limited. At the same time, a service provider must provide a service, or to be more precise, a service provider MO must always be in an according serviceProvision relationship with at least one service MO. These constraints are referred to respectively as *upper* and *lower bounds* on the *relationship cardinality*.

- Depending on the modeling paradigm, relationships can also have *attributes*, e.g. priority information in a fallback relationship or a timestamp.

- To play a given role, MOs may be required to possess certain characteristics, such as a certain behaviour or certain attributes, i.e. *static characteristics* that are related to their class, or even *dynamic characteristics* such as having a certain state. For instance, in a backup relationship [9], the MO in the back-up role must have an administrative state attribute whose value must in addition be "unlocked".

- There may be restrictions on the *combinations of relationship participants* allowed, e.g. only systems can serve as fallback for systems whereas for instance SAPs cannot.

- The participation in a relationship may have certain *impacts on MOs*. An example would be a case when in which a log MO is to be deleted which causes the log record MOs that it is in containment relationship with to be deleted as well.

- Just as other management information, relationships may be *subject to modification* by management operation or be read-only.

These and other aspects cannot be specified adequately using only the concepts provided in the basic OSI information model. A supplement is therefore needed that allows and enforces separate and explicit specification of relationships. The templates sketched below fulfill that purpose. They use GDMO-like (Guidelines for the Definition of Managed Objects, in [10]) notation and are to a certain extent self-explanatory. A brief explanation follows below.

```
<relationship-label> MANAGED RELATIONSHIP CLASS

  [DERIVED FROM <relationship-label>[, <relationship-label>]*
    [WITH ROLE ASSIGNMENTS
      <role-name> AS <derived-from-role-name>
      [, <role-name> AS <derived-from-role-name>]*];]

  [ROLES
    [<role-name> ROLE
      CARDINALITY (<min_cardinality>..<max_cardinality>)
      [SYMMETRIC]
      [REQUIREMENTS
        [<package-label>[, <package-label>]*];
        [<relationship-label> <role-name>
            [, <relationship-label> <role-name>]*]]]+;]

  [CHARACTERIZED BY <package-label> [, <package-label>]*;]

  [CONDITIONAL PACKAGES
        <package-label> PRESENT IF condition-definition
        [, <package-label> PRESENT IF condition-definition]*;]

REGISTERED AS <object-identifier>;
```

The first template which is depicted above is used to specify managed relationship classes (MRCs). A MRC models the relationship as such, without regard as to which kinds of managed objects will actually participate in the relationship. The parts that make up a MRC definition are listed below.

- Refinement of relationships is possible through inheritance, with the option of refining also roles of superior MRCs. A relationship-top has been defined (not shown here) which like the top for MOCs serves as the ultimate super relationship class from which all other MRCs are directly or indirectly derived.
- Like for MOCs, allomorphism is supported for MRCs by a corresponding conditional attribute inherited from relationship-top that is present if an implementation supports allomorphism.
- An arbitrary number of roles can be defined, with a keyword "symmetric" devoted to indicate symmetric roles or relationships, respectively. Roles are specific to the MRC and are therefore not defined independently.
- Upper and lower cardinality bounds have to be specified for each role.
- Static requirements that a MO in a certain role is required to possess can be specified. This includes both characteristics of its class as well as its participation in other relationships.
- Behaviour, attributes, notifications, and possibly actions can be defined in packages.
- Packages may be conditional in order to accomodate e.g. for flexibility with respect to the scope of implementations.

```
<relationship-binding-label> RELATIONSHIP BINDING

   MANAGED RELATIONSHIP <relationship-label>
   [BEHAVIOR <behavior-definition-label> [,<behavior-definition-label>]*;]

   RELATED CLASSES <class-label> [AND SUBCLASSES] WITH ROLE <role-name>
        [CARDINALITY RESTRICTION (<min_cardinality>..<max_cardinality>),]
        [DELETE [delete-modifier) [, <parameter-label>]*]
     [, <class-label> [AND SUBCLASSES] WITH ROLE <role-name>
        [CARDINALITY RESTRICTION (<min_cardinality>..<max_cardinality>),]
        [DELETE [delete-modifier] [,<parameter-label>]*]]*;

[DYNAMIC REQUIREMENTS <dependency-definition> [, <dependency-definition>]*;]
[MODIFY [ <parameter-label>[, <parameter-label>]*;]

 REGISTERED AS <object-identifier>;
```

The other template depicted above is used to specify relationship bindings. Relationship bindings serve as link between MRCs and MOCs and are therefore the glue of the model. Their function is that they specify what combinations of MOCs are allowed for the relationship. This allows for independent specification of MRCs and MOCs while at the same time maintaining model accuracy. In addition, relationship bindings allow for some degree of refinement of the relationship with respect to the particular object classes involved. This serves to avoid excessive numbers of relationship subclasses where each subclass has really the same semantic purpose and includes only minor variations to account for the classes of the MOs to be bound in the relationship. In particular, the template allows to specify the following:

- Further refinements of relationship behaviour specific to that binding can be made.
- It must be specified which MOCs may be bound in the relationship, and if this applies also to their subclasses.
- Cardinality bounds may be further constrained due to technical limitations of the MOCs involved.
- What to do in case one of the relationship participants is to be deleted must be specified, e.g. whether deletion is prohibited or not.
- Dynamic requirements that allow to express consistency conditions between relationship participants can be formally specified (format not shown here).
- It has to be specified if modification of relationship information by management operation shall be possible.

This modeling approach has been used to model generic as well as specific relationships, in particular to supplement the generic network model described in [18] as well as specific ones, such as MIB-II-OIM [16].

RELATIONSHIP SERVICES

Services for the access and the manipulation of relationships include the following. The names of the services should be self-explanatory in the sense that they indicate what their respective purposes are.

establish-relationship: Parameters include MRC, participants in their respective roles, and possibly attributes.

terminate-relationship: Parameters include MRC and participants in their respective roles.

terminate-all-relationships (of a given MO): Parameters include the identifier of the MO of which all relationships are to be deleted.

query-relationship: Parameters include MRC and one or more participants in the respective roles. Returned are all relationships with these participants and possibly the corresponding relationship attributes.

query-all-relationships (of a given MO): Parameters include the identifier of the MO of which all relationships are to be queried. Returned are all relationships it takes part in.

cancel-query: Parameters include the invoke identifier of the respective query.

modify-relationship-attribute: Parameters include MRC, participants in their respective roles, and the attribute list.

relationship-action: Parameters include MRC and participants in their respective roles as well as the respective action.

Applications operating on the "value-added" models mentioned above do so using these services. Identifiers for instances of relationship information are not needed as they are always identified in an associative manner in terms of the MRC and/or some or all of their respective participants.

INTEGRATION OF SPECIFIED RELATIONSHIPS INTO AN MIB

Managed relationship class and relationship binding templates provide powerful specification means that are very important for the development of rich and precise network models which provide a clear understanding and picture of the networks to be managed. However, it is necessary to allow for actual instantiation of these models to make them accessible for management applications. Within OSI management, this is not directly possible as managed relationship classes and relationship bindings are not part of the established information model. Supplementing the information model with managed relationship classes and relationship bindings must therefore be considered and treated as an additional modeling layer on top. Independently of whether or not such an additional modeling layer exists, relationship information will have to be maintained as part of the current information model.

To instantiate managed relationships as part of actual management information, i.e. to integrate them into a MIB, they have to be mapped onto the basic information model as managed objects, attributes, and name bindings. A suitable scheme to maintain relationship information as part of the current information model is therefore needed. Various schemes are conceivable. However, for an adequate scheme one very important requirement in connection with separate relationship specification is that it must be possible to formulate a systematic method that allows to map MRCs and relationship bindings onto it, losing as little semantical information in the process as possible. By requiring all relationships to be modeled using the concepts of MRCs and relationship bindings and subsequently applying this method of mapping, a uniform representation and treatment of all different kinds of relationships on the level of the basic information model can thus be achieved. This in turn helps to keep complexity in managing relationship information low and makes it also easier to offer generic relationship implementation support by management development environments.

Several considerations have to be made when attempting to integrate relationship information into an MIB, the most important of which are listed below:

- The integration of relationships happens in the context of the object oriented OSI information model. In object oriented design in general, relationships between objects should be modeled explicitly, as failure to do so leads to hidden dependencies and adds complexity to the model [17]. This is true also for the OSI information model. Accordingly, also in MIBs relationship information should be kept separate from the MOs involved and made explicit. This means it should be represented through own MOs.
- As was indicated above, relationships have characteristics and underly specific constraints that makes them uniquely different from MOs. Nevertheless it appears necessary to make them part of management information as MOs. While in the OSI information model there are no explicit formal means for relationship specification, a proposed integration should nevertheless allow that the according constraints can still be expressed at least through (informal) behaviour specifications. This means that in the process of a mapping of MRCs and relationship bindings onto the basic information model, no aspects should be lost. Concepts analogous between MOs and relationships should be exploited as much as possible in order to make use of existing generic mechanisms for the management of MOs to make the realization of correct relationship behaviour easier.
- Concepts for the integration should be kept as simple as possible to ensure that they are easy to use and that their implementation is feasible. This is true not only for relationship information itself but also for the realization of relationship services for their access and modification.

To be sure, in order to represent relationships as part of concrete management information, current practice could be kept, i.e. maintain information on relationships in attributes of the respective relationship participants. However, this leads to several problems, for instance the inability to treat relationship bindings adequately as additional bindings cannot be accomodated later without changing existing specifications, but also consistency problems due to the scattering of relationship information across MOs. The key is therefore again to capture relationship information explicitly and separately as dedicated managed objects of their own.

Several approaches for this are poss·ble. One candidate approach is e.g. to draw on the already mentioned work in [13]. There, relationships are interpreted as graphs that are kept in MOs. Access and manipulation of relationship information takes place through special actions. Problems lie again in the mapping of relationship bindings, but also in the complexity of additional actions that have to be implemented.

Therefore, a different solution is proposed which is well suited for the task and takes into account the considerations that were previously discussed. It is depicted in Figure 2. The basic idea is to represent relationship information through two kinds of MOCs: MOCs for relationship objects (ROs) and MOCs for relationship anchors (RAs). Both classes together realize the concept of managed relationship classes. The actual instantiated relationship information is kept in the ROs, which keep references to the related MOs in attributes. ROs of one kind are conceptionally contained in a RA which is unique within agent systems. When thinking of relationships as a table, ROs would be the table entries and the RA the table itself. So in a way, this scheme resembles also the representation of logs.

The following points outline some more technical aspects. At the same time, they also indicate how the mapping of MRCs and relationship bindings to this scheme takes place. Many more details are involved that cannot be discussed here without going beyond the scope of this paper; however, the general principles should become clear.

- Each RO holds one instance of relationship information. For each MRC, an according relationship object MOC is defined. For each role, an attribute is defined whose value is a reference (meaning it contains an identifier) to the MO playing that particular role. The attribute name is generated according to a fixed convention from the role name. These role attributes display for instance the behaviour that they always refer to MOs that actually exist. Further constraints on attribute values regarding what kinds of object they may refer to are expressed in the name bindings between the MOCs for RO and RA (see below). Relationship object classes also include the packages from the corresponding managed relationship classes.

- RAs serve as containers for ROs. For each managed MRC, an according relationship anchor MOC is defined. There are constraints on the number of ROs with given attribute values that an RA may contain, reflecting upper cardinality bounds for the respective roles. Constraints on the number of ROs with given attribute values they must contain can also exist, reflecting lower cardinality bounds.

- Name bindings between ROs and RAs reflect the concept of relationship bindings. For each combination of relationship participant MOCs, a corresponding name binding between ROs and RAs exists. Therefore, several name bindings between the same pair of MOCs (i.e. RO and RA) can exist which differ mainly in their behaviour part. Among other things, the behaviour part specifies the classes of MOs that the respective attributes in the RO bound according to this name binding are allowed to refer to. Creation and deletion of ROs by management operation with a given name binding is allowed iff the respective relationship binding allows for modification of relationship information.

The solution has been successfully implemented for several relationships described in [4], using the OSIMIS (OSI Management Information Service) development environment from University College London [14]. A detailed discussion of the implementation and related considerations is given in [19].

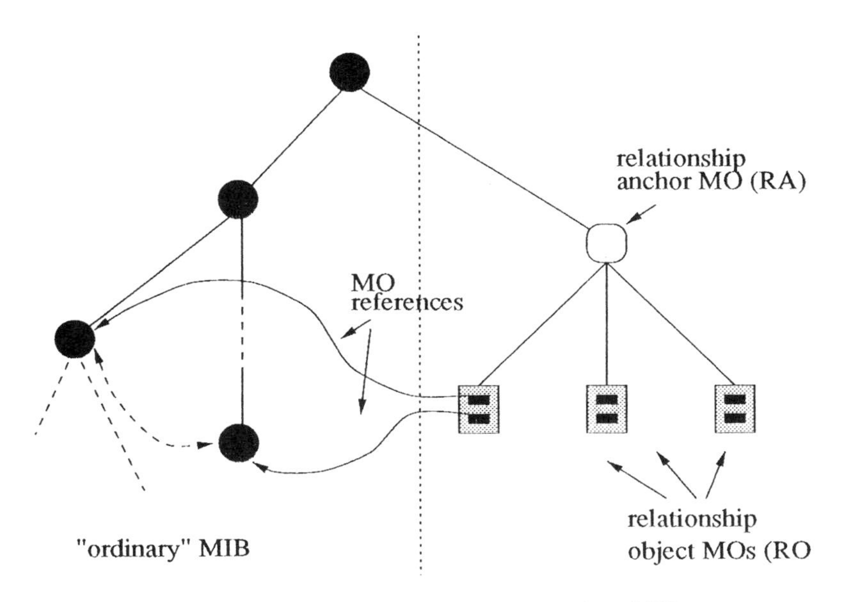

Figure 2. Relationship instantiation in a MIB

REALIZATION OF RELATIONSHIP SERVICES

Services for access and manipulation can be mapped to object management functions [8] that use the RA as base object, obliviating the need to define own actions. Direct addressing of relationship objects can be done without, using simple scoping. Essentially, establishment of relationships corresponds to the creation of ROs, termination to their deletion, while relationship querying can be performed using scoped and filtered "Get"-operations.

NEW TRENDS IN STANDARDIZATION

ISO has apparently recognized the deficiencies with respect to relationship in the information model. Therefore, lately work has begun to follow in the same direction that has meanwhile reached committee draft for comments stage. In it, templates for relationship classes and relationship bindings are proposed as well [11]. At the current stage there are still several unresolved issues [5]; because of that and because the work will continue to evolve it is too early to base any concrete work on them. However, in any case several major differences to the work presented in this paper exist. The following list will point out two of them; it is intended for the reader with a good knowledge of ISO's work.

- The ISO approach proposes that the way relationship information can be instantiated must be specified in the model. For this purpose, a second kind of cardinality is introduced, called role cardinality. It specifies the number of MO references present for each role in any one instance of relationship information, i.e. relationship information treated as an entity. One consequence of this is that also the behaviour of relationship instances must be specified, e.g. what to do if an MO wishes to dynamically enter or leave a relationship instance. Accordingly, also the issue of relationship instance naming must be addressed.

On the contrary, the approach presented in this paper keeps the instantiation of relationships transparent from the modeler and the user of relationship information. Relationship information is accessed in an associative manner, as was indicated in section 3. This helps to keep the complexity of the modeling process and resulting models low and avoids distraction by instantiation matters which are basically handled on a lower layer of abstraction. Modeling of relationship instantiations also involves the danger that relationships get modeled from a particular application perspective instead of independently from specific applications. An example would be a many-to-many (referring to upper cardinality bounds) relationship that gets modeled from a 1-to-many perspective, which is a modeling practice actually encouraged by the concept of role cardinalities. While convenient in some cases, the resulting model would be awkwardly to use for applications having to access relationship information in the other direction, or in the worst case lead to duplicate modeling of the same relationship from another perspective. To illustrate, consider the situation in Figure 3 where two different perspectives (one for each relationship role) of the same basic relationship are depicted. In the client perspective, for instance c3, s1, and s3 would form a relationship instance, whereas from the server perspective an example would be a relationship instance with s1, c1, c2, and c3. It is not obvious which perspective should be considered "right" by the modeler, or if maybe even the two should both be modeled.

For cases in which the modeling of such perspectives on relationships under certain viewpoints is indeed desirable, it should be briefly mentioned that so-called "extended relationship views" can also be additionally defined, which has not been shown here. They represent essentially virtual relationship instances whose existence depends directly on that of the underlying "basic" relationships of which they offer a view.

- According to the ISO proposal, the integration of relationships into management information is also the responsibility of the modeler. It is suggested although in no way mandatory that one of the objects participating in a relationship should act in a relationship coordinator role that ensures the relationship is managed in the proper way. Thus, the mechanism by which a relationship is to be managed is specified as part of abstract relationship definitions. In a way, relationship specifications can therefore also be said to serve as additional documentation for MO classes involved in the management of relationships.

The intention of the approach presented in this paper is different. It keeps the mechanism for relationship integration transparent from modelers and users in order to offload them. Contrary to the ISO proposal it is considered part of a lower layer of modeling abstraction, with a uniform scheme to maintain relationship information in a MIB already provided along with a mapping onto it. MO classes involved in the management of relationships are generated from relationship specifications instead of being documented by them. As a consequence, multiple mechanisms for the management of relationships are avoided and homogeneous treatment ensured. This makes it also easier for management agent architectures to provide generic relationship support.

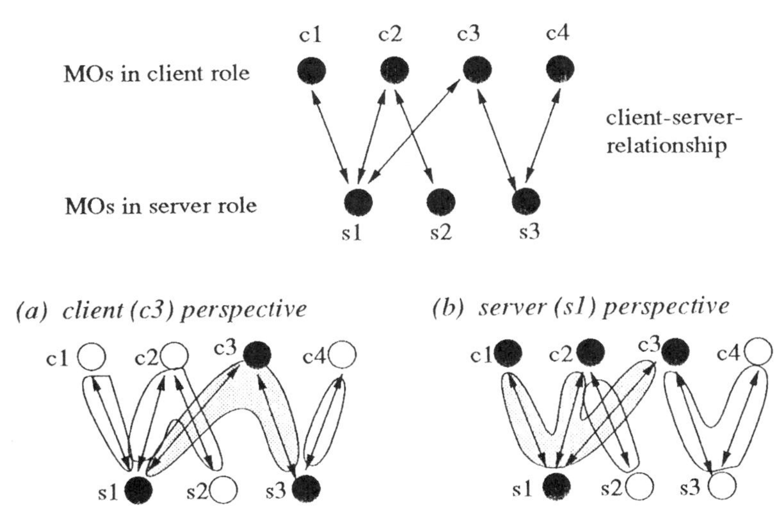

Figure 3. Different perspectives on relationships

It should be briefly mentioned that recently also in other places the need to address the issue of relationships in network management has been recognized and is considered an area requiring further research [6]. In [3], some examples for generic network relationships are presented on a very abstract level. In [1], even a simple GDMO-like notation for relationships is introduced; however, the focus is less on a comprehensive concept for relationship management but on another aspect of the OSI information model. Issues like relationship services and relationship instantiation are left open or to the modeler; to this respect, ISO's work is reflected and analogous comments apply.

CONCLUSIONS

Richer information models are needed for adequate treatment of relationships in OSI management which lie at the very core of network management tasks. At the same time it must be ensured that these can be accomodated for within the existing OSI information model framework. In this paper it was shown that this can be accomplished by supplementing the information model with templates for the separate specification of relationships that can also be regarded as an additional modeling layer. It was also shown that this modeling layer can be mapped onto the OSI management information model in a uniform manner to assure integration into an MIB. The principles presented have been successfully implemented in a prototype; it is planned to use the implementation in a concrete OSI management scenario. This will allow applications to be designed to operate on more appropriate network models which the author believes will in turn prove in addition to the conceptional and theoretical also the practical benefits of the presented approach.

ACKNOWLEDGEMENTS

For valuable discussions and insightful comments on this topic I wish to sincerely thank Prof. Dr. Heinz-Gerd Hegering and Dr. Andreas Mann. Andreas

Mann and also Brigitte Bär provided in addition comments on earlier versions of this paper that were helpful to improve it. For continuing interesting discussions I want to thank also the other members of the Munich Network Management Team and of IBM ENC's System and Network Management Department.

ABOUT THE AUTHOR

Alexander Clemm received his M.S. degree from Stanford University in 1990. He is a member of the Munich Network Management (MNM) Team, a group of network management researchers at the Munich Universities and the Bavarian Academy of Sciences, and as part of a cooperation also affiliated with IBM European Networking Center, Heidelberg. Mr. Clemm has authored or coauthored a number of technical papers in the area of network management and is currently in the process of finishing his Ph.D. studies at the University of Munich. His research interests include relationships between MOs, management gateways, information modeling, and AI applications.

REFERENCES

[1] S. Bapat, *"Richer Modeling Semantics for Management Information,"* 3rd IFIP/IEEE International Symposium on Integrated Network Management, San Francisco, CA, 1993.

[2] L. Bennett, W. Chou, *"An Expert System for Diagnosing Performance Problems in SNA Networks,"* Network Management and Control, edited by A. Kershenbaum, Plenum Press, New York, 1990.

[3] Ch. Benz, M. Leischner, *"A High Level Specification Technique for Modeling Networks and their Environments Including Semantic Aspects,"* 3rd IFIP/IEEE International Symposium on Integrated Network Management, San Francisco, CA, 1993.

[4] A. Clemm, B. Schott, *"Ein ISO/OSI-basierter Ansatz zur Modellierung von Netzen und Netzbeziehungen,"* Workshop Entwicklungstendenzen in Rechnernetzen, Selected Papers Volume 1, Technische Universität Dresden, 1991.

[5] A. Clemm, *"Comments on General Relationship Model, 3rd WD,"* Submitted comment to ISO/IEC JTC 1/SC 21/WG 4, University of Munich, 1992. (copy available from the author)

[6] H.-G. Hegering, S. Abeck, *"Integriertes Netz- und Systemmanagement,"* Addison-Wesley, Bonn, Paris, 1993.

[7] P. Hong, P. Sen, *"Incorporating Non-Deterministic Reasoning in Managing Heterogeneous Network Faults,"* 2nd IFIP/IEEE International Symposium on Integrated Network Management, Washington DC, 1991.

[8] International Standards Organisation, *"Information Technology - Open Systems Interconnection - Systems Management - Object Management Functions,"* ISO IS 10164 Part 1, 1991.

[9] International Standards Organisation, *"Information Technology - Open Systems Interconnection - Systems Management - Attributes for Representing Relationships,"* ISO IS 10164 Part 3, 1991.

[10] International Standards Organisation, *"Information Technology - Open Systems Interconnection - Systems Management - Structure of Management Information,"* ISO IS 10165 (parts 1-4), DIS (part 5), 1992.

[11] International Standards Organisation, *"General Relationship Model,"* ISO CD for Comments 10165 Part 7, 1993.

[12] D. Jordaan, M. Paterok, *"Event Correlation in Heterogeneous Networks Using the OSI Management Framework,"* 3rd IFIP/IEEE International Symposium on Integrated Network Management, San Francisco, CA, 1993.

[13] K. Klemba, M. Kosarchyn, *"A Model for Object Relationship Management,"* 2nd IFIP/IEEE International Symposium on Integrated Network Management, Washington DC, 1991.

[14] G. Pavlou, *"Implementing OSI Management,"* Tutorial at the 3rd IFIP/IEEE International Symposium on Integrated Network Management, San Francisco, CA, 1993.

[15] P. Putter, J. Roos, *"Modelling Management Policy Using Enriched Managed Objects,"* 3rd IFIP/IEEE International Symposium on Integrated Network Management, San Francisco, CA, 1993.

[16] Internet Activity Board (L. Labarre ed.), *"OSI Internet Management: Management Information Base,"* RFC 1214, 1991.

[17] J. Rumbaugh et al., *"Object-Oriented Modeling and Design,"* Prentice Hall, Englewood Cliffs, NJ, 1991.

[18] B. Schott, A. Clemm, U. Hollberg, *"An ISO/OSI Based Approach for the Modeling of Heterogeneous Networks,"* 4th IFIP International Conference on Information Networks and Data Communications, Espoo, Finland, 1992.

[19] S. Vaak, *"Ein Management-Agent zur Unterstützung expliziter Beziehungsinformation in einer OSI Management Information Base,"* Master's Thesis, Department of Computer Science, Technical University of Magdeburg, 1993.

[20] R. Valta, *"Design Concepts for a Global Network Management Database,"* 2nd IFIP/IEEE International Symposium on Integrated Network Management, Washington DC, 1991.

SEMANTIC MODELING OF MANAGED INFORMATION

Yechiam Yemini[1], Alex Dupuy[2], Shmuel Kliger[2], and Shaula Yemini[2]

[1]Distributed Computing and Communications Lab
450 Computer Science Building
Columbia University, New York, NY 10027
[2]System Management Arts (SMARTS)
199 Main Street, Suite 900
White Plains, NY 10601

ABSTRACT

This paper presents a preliminary report on a semantic model of managed information, ERC. This model enriches the traditional Entity-Relationship (E-R) model with object oriented abstractions, constraints and event-correlation information. The model has much more expressive power than the CMIP management information model, (which in turn is more expressive than SNMP's model), yet is simpler overall, and is more efficient to implement. It is thus particularly suitable for supporting complex event management systems. The paper presents the model, illustrates it through example applications, and briefly describes an efficient, protocol-independent implementation of the model used in a distributed management system.

INTRODUCTION

The main components of a typical network management system are depicted in figure 1. Instrumentation routines at devices collect information on operational behaviors and provide means to configure device parameters and control its behavior. Agents, embedded in devices, organize this instrumentation data in a management information base (MIB). Platform manager applications utilize a management protocol (e.g., SNMP, CMIP [5]) to query these databases and bring MIB data to the platform. This data is typically presented to operations staff who are responsible to interpret and act on it. Communications among the platform and agents is supported by a management network. In the case of LANs, the management network is typically identical with the managed network. In a WAN, the management network is sometimes separate from the managed network.

Network Management and Control, Vol.2, Edited by
I.T. Frisch *et al.*, Plenum Press, New York, 1994

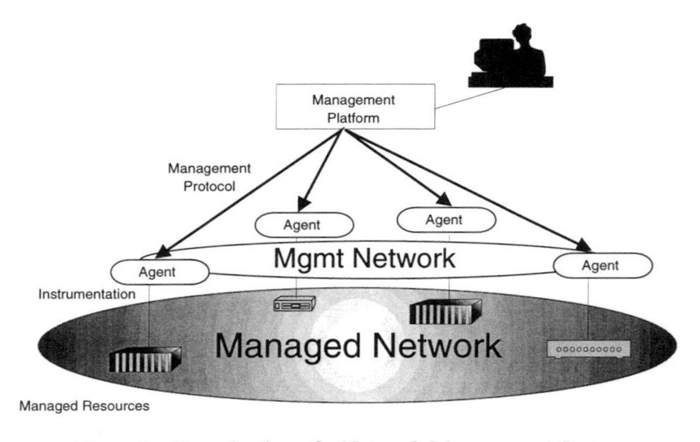

Figure 1. Organization of a Network Management System

From the point of view of traditional database theory[1] management protocols provide a unified logical database mechanism to access the physical (instrumentation) database layer. The organization of and access to management information is syntactically unified under the protocol framework. This syntactic framework consists of a data definition language (DDL) and a data manipulation language (DML). The DDL of SNMP is specified by the structure of managed information (SMI) [3] while the DDL of CMIP is specified by the guidelines for the definition of managed objects (GDMO) [2]. The DMLs are given by the respective protocol primitives of SNMP and CMIP.

The syntax of managed information provides means to organize and access managed data. Applications that need to process this data, however, require a semantic model of its meaning. In the absence of a semantic model, the interpretation of MIB data and the corresponding control of the network are left to expert operations staff. The main goal of this paper, is to present a semantic model of managed information that can be processed by programs to automate many management tasks that are currently done manually.

For example, consider the problem of building an application to manage the performance of the network layer. Such an application needs to collect traffic data from multiple devices at the network layer (e.g., traffic sources and destinations, routers, switches etc.). It needs to analyze the relationships among these devices (e.g., connectivity) and relationships between network layer entities and entities at higher and lower layers. The data required by such an application is currently spread through multiple entities and is difficult to identify in its entirety from current information models. Management protocols provide only means for the application to access and retrieve this data once it knows where to find it Sophisticated applications require a richer semantic model of the meaning of the data in order to correlate and analyze it.

Traditional database design techniques use an entity-relationship (ER) model to provide a uniform semantic model of data, which is independent of the underlying database organization. The same ER model can be stored in a hierarchical, network or relational database. This paper extends the ER model to support semantic modeling of management information. Section 2 describes an object-oriented Entity/Relationships/Constraints (ERC) model of managed information. Section 3 describes implementation issues. Section 4 examines ERC applications to configuration and fault management. Section 5 provides a summary and conclusions.

---

[1]For a database background the reader should consult any general reference book sucl as [6].

2. ERC: AN EXTENDED ENTITY RELATIONSHIP MODEL

We first define the goals of semantic modeling of management information as required by management applications.

- Management information must model entities (e.g., nodes, links, interfaces and protocol entities) and the relationships among them (e.g., contained-in, peer-of, connected-to).
- An abstraction/inheritance mechanism is valuable for handling the enormous complexity and diversity of managed entities. An abstraction mechanism permits a performance management application, for example, to view multiple different intermediate devices as "nodes", abstracting from the details of individual devices created by different vendors.
- Protocol MIBs were designed to optimize transmission over a network rather than supporting computation. A semantic model of management information, must instead, be optimized to support computation by management applications while still supporting remote access through standard protocols.
- The attributes of a managed entity are of two types: configuration parameters and real-time behavior data. Configuration parameters are explicitly changed by a person or a program. Real time behavior changes implicitly as a side effect of managed objects' operation. Distinguishing these types of attributes provides insight into event management.
- Events play a central role in management processing and must be captured by the semantic model. Most events can be described in terms of expressions (constraints) on properties of managed entities. An event model needs to represent such constraints and to capture the causality among events to facilitate their correlation.

The Entity-Relationships-Constraints (ERC) model seeks to address these goals through respective extensions of the ER model. The fundamental elements of the model are depicted in figure 2 below. Entities are depicted as rectangles. Configuration attributes are depicted as rounded rectangles while activity (real-time) attributes are depicted as triangles. Relationships are depicted as diamond shapes.

Constraints model the mechanism that generates events. We distinguish between synchronous and asynchronous constraints.

Synchronous constraints, depicted as circles, are attached to configuration or relationship attributes. They are evaluated synchronously whenever a respective attribute is changed. Synchronous constraints can be used to identify configuration inconsistencies. For example, suppose a token ring interface card is configured to run at 4Mb/s (a configuration attribute of the interface entity). If the interface card is connected_to a token-ring configured at 16Mb/s, a configuration inconsistency would arise. A configuration constraint attached to the respective attributes (bandwidth attributes and "connected_to" relationship), can be evaluated to identify the inconsistency event. Synchronous constraints and their respective events may thus be viewed as mechanisms to protect managed systems against configuration inconsistencies.

Asynchronous constraints model the generation of performance events. An asynchronous constraint is a logical expression in activity attributes (possibly involving configuration attributes too). Such constraints are evaluated periodically and asynchronously with the changes in these parameters. For example, a threshold condition "error_rate<tolerance" may be a constraint attached to an error counter. Periodically, the counter will be sampled and its value used to compute the error_rate and compare with the tolerance threshold. Should the constraint be violated, an event will be generated to signal it.

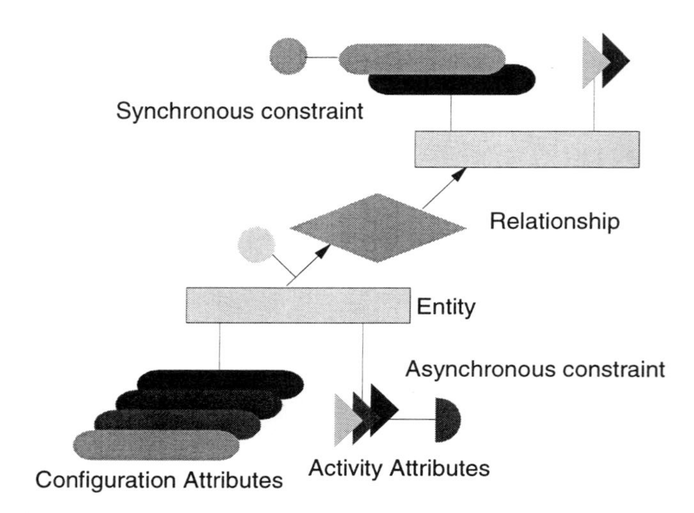

Figure 2. Basic components of the ERC model

The graphical rendering of the ERC model does not reflect two additional constructs: inheritance and event correlation information.

The ERC class hierarchy builds on the class model of NetMate [1]. [Readers are referred to [1] for details of the NetMate class model and its application.] The top of the ERC class inheritance tree is partially depicted in figure 3 above the dashed line. It provides generic classes such as Resource, Node, Link, etc., used to model managed entities. Generic classes are used to model objects related through generic relationships. For example, objects of Resource type may be Part_of other Resources. Link and Node objects can be Connected_to other Nodes and Links and so on.

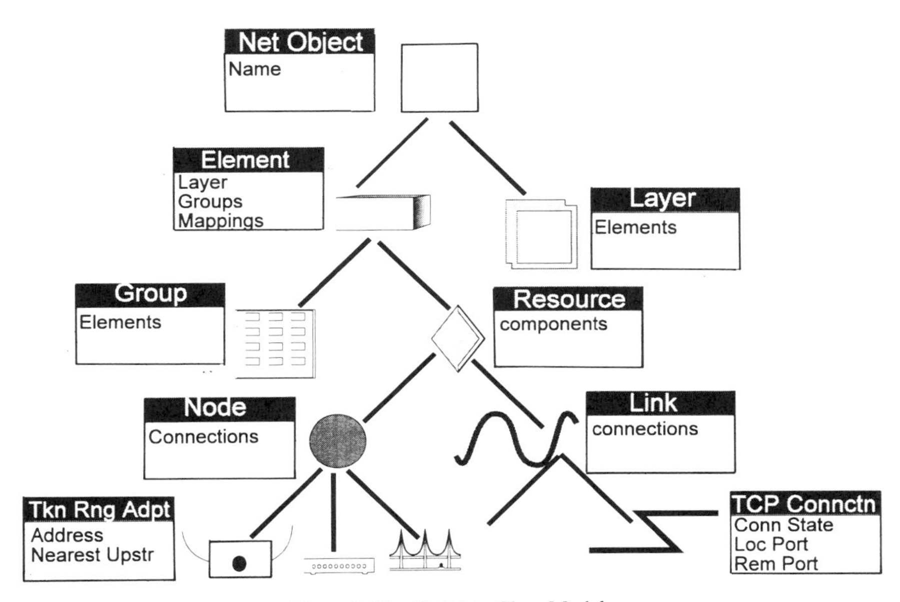

Figure 3. The NetMate Class Model

Both Node and Link objects inherit relationships from their parent class, Resource. Thus, in particular, Node and Link objects may be Part_of other Resource objects. Also, as Elements, they may be Member_of a Group. Inheritance permits both abstraction and specialization of the relative roles defined by the generic relationships. Specialized entities (depicted below the dashed line) such as Token Ring Adapter and TCP connection are defined as special cases of Node and Link objects respectively. This means that they inherit the relationships in which Node and Link may be involved. A total of only 10 generic classes involving 8 generic relationships is used in ERC to model a large number of networks and configurations.

A second type of information captured by ERC, and not reflected in the graphical model of figure 2 is event causality and correlation. One can view correlation as analysis of events observations to determine their common causes. Events signaled by synchronous constraints typically result from configuration changes so their correlation is simple. Events signaled by asynchronous constraints are related to perfomance. Correlation is necessary to identify their causes, and may be very complex.

The problem of correlation is complex because events may propagate among entities via relationships. For example, a failure of a link may cause events at attached nodes. The information required for correlation can be captured in terms of a causality graph (or a matrix) representing the propagation of events.

The ERC model is described by a data definition language, similar to the GDMO [GDMO91], as part of a broader Management Information Class Definition Language. It includes constructs for defining:

- Managed object classes and their inheritance
- Configuration and activity attributes
- Relationship attributes
- Synchronous and asynchronous constraints
- Event correlation information

For the purpose of this paper we use an abbreviated form of this language to illustrate a simplified ERC model in figure 4 below. The first line defines a Token_Ring_interface object as a kind of Node. It inherits the relationships of Node (e.g., connected_to). The second line describes two configuration parameters associated with an interface: bandwidth and addressing mode. The third line describes various activity attributes. The fourth line describes new relationships in which the interface is involved. In this case there are none.

```
Node CLASS Token_Ring_interface
CONFIGURATION {bandwidth Enumeration(high, low),    addressing Enumeration (short, long)}
ACTIVITY { input_error Counter, output_error Counter, token_rotation Time,...}
RELATIONSHIPS { }
SYNCHRONOUS {[bandwidth_mismatch:= bandwidth≠r(connected_to).bandwidth WHEN
        connect_to, bandwidth], ...}
ASYNCHRONOUS {[faulty_connector:= rate(input_error)>max_error_tolerance EVERY 60 ]
.PROPAGATION { [Token_Ring_Link VIA connect_to:> token_loss,....}
CAUSALITY { [faulty_connector:>token_loss, high_error,...], ...}
END token_ring_interface
```

Figure 4. Simplified partial definitions of ERC managed object

The fifth and sixth lines describe constraints. A constraint has a name (e.g., bandwidth_mismatch, faulty_connector), a logical expression that must be evaluated and a specification of when the evaluation is to occur. Synchronous constraints are evaluated WHEN certain configuration attributes change. These attributes are specified as part of the WHEN statement. For example, the bandwidth_mismatch constraint is evaluated whenever the bandwidth attribute or the connect_to relationships are changed. Asynchronous constraints are evaluated periodically using a time interval specified by the EVERY (in seconds). The PROPAGATION statement describes events imported from related objects via a relationship. The CAUSALITY statement describes the relations among problems and the symptom events that they cause.

It is useful to compare the ERC model to the information model of the GDMO. ERC extends the GDMO model in three areas:

1. Drawing distinction among configuration and activity attributes. Real-time information presents processing requirements that are different from the processing requirements of configuration data.
2. Including synchronous and asynchronous constraints as part of the model. The OSI management model treats the sources of events as external to the information model. Managed objects simply pass these external events to subscribers. In contrast, the constraint model used by ERC includes constraints to explicitly capture the sources of events. Explicit modeling of events permits event semantics to be represented in the model and used by applications.
3. Event correlation modeling. This area is informally hinted at by the "behavior" model of OSI. Event correlation is of central significance for network management. Unless events are correlated, a single event may generate a large number of alarms or trouble tickets, each initiating some corrective action. At best, this is a wasteful duplication of effort. At worst, different corrective actions may conflict with one another, causing an escalation of problems. In order to correlate events it is necessary to recognize the manner in which related events may propagate through a network system. Event correlation knowledge is therefore an integral component of the ERC model.

In spite of its additional expressive power, ERC is simpler than the GDMO. Simplification is accomplished through organizing objects within the generic class and relationships modeling framework. This generic class model can provide simple and uniform semantic models of networks of great complexity. In contrast, applying GDMO in various organizations has produced models of enormous complexity involving many hundreds of different classes and relationships. By keeping the set of generic relationships and generic classes small, this complexity can be reduced.

3. STATUS

The ERC model is implemented in C++ and provides the management information repository (MIR) for the SMARTS Management Server. The function of the Management Server is to encapsulate management responsibilities associated with a particular network domain. A domain can consist of a collection of related equipment (e.g., a backbone) or a complex device (e.g., a switch, router). A domain can also consist of vertical resources involved in the delivery of a particular application/service. For example, in managing a banking application, a domain may include clients, servers, transport connections among them, underlying routers and links.

The Management Server automates management through respective applications. Distribution of management applications to Management Servers utilizes the management by delegation (MBD) mechanism reported in [8]. Since Management Servers function as local managers with respect to domain resources and as agents with respect to higher level managers in a management platform, their role in network management is that of "Middle Managers". This organization of distributed management is depicted in figure 5 above. Agents are depicted as rounded rectangles, Management Servers are depicted as triangles, and management platforms are depicted as rectangles. This distributed organization of management should be contrasted with the centralized management depicted in figure 1.

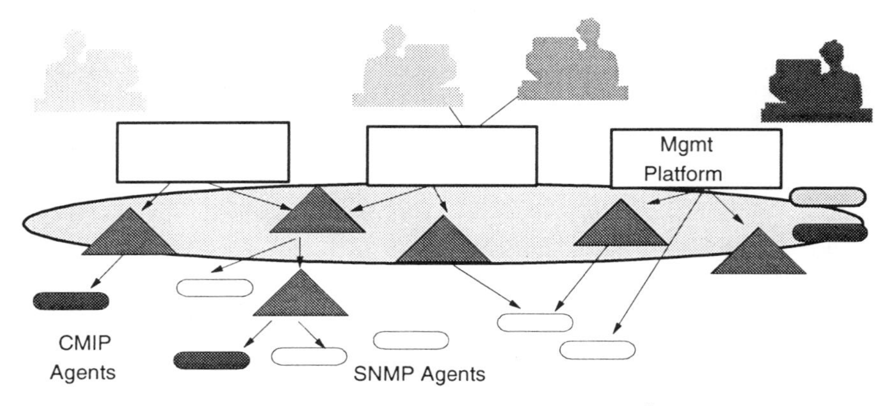

Figure 5. Distributed Management Using Management Servers

The MIR maintains a rich semantic information model of the domain managed by a Management Server. Management applications in Management Servers use the MIR to interact with entities they manage, as well as to provide them with their computational data. Platform managers (or other Middle Managers) can access the MIR information remotely using RPC, SNMP or CMIP. The Management Server provides an environment where different protocol agent interfaces can export different views of the MIR, subject to the limitations of the information models that they support. For example, SNMP does not provide representation of relationships. Thus, an SNMP manager can access the attributes of MIR entities but would require substantial extensions to access relationship information.

4. APPLICATIONS

This section provides a brief discussion of applying the ERC model to support some example management applications.

Applications to configuration management

Frequent changes in a network can result in configuration inconsistencies from time to time. Inconsistencies can lead to operation problems. For example, a port-switching LAN hub may switch a port of a domain name server to a different network, leading to loss of name services in a domain and the inevitable accompanying problems. A compressed voice circuit must be routed only through equipment that can support it. Upon failure in the network, automatic rerouting mechanisms may route the circuit through the wrong equipment leading to its loss. Maintaining consistency through changes is a central problem of configuration management. Often the task of ensuring consistency depends on the expertise of operations staff and their experience in handling the variety of network elements and their configuration.

The ERC model can simplify some of these management problems. Consistency constraints can be attached to the respective configuration-attributes and relationships of a managed object as part of its definition. Changes in these attributes will then cause automatic evaluation of the constraints as part of the attempt to modify configuration or relationship attributes. Attempted violations of the constraints will generate events that can trigger local (or remote) handlers. Local handlers can either prevent the change or reconfigure other parameters to allow the change.

This should be contrasted with the complexity of accomplishing similar function through a managed information model that does not support constraints. Consider, for example, an SNMP management system as depicted in figure 1. Configuration consistency must be ascertained by the centralized platform. The platform must evaluate the constraint upon any change that may cause its violation. However, certain changes may be manual (e.g., cables may be plugged into ports) and other changes are automated but are not performed through an application that is aware of the constraints. Synchronization of changes with platform processing may thus be very difficult to accomplish.

Applications to fault management

A fault in a given entity can manifest itself through anomalies in the various activity attributes of that entity. Moreover, these anomalies may propagate to other related entities, via relationships. For example, a bad connection at an ethernet port can cause an increase in the collision rate of the segment to which it is connected. Increased collision rates could lead to failures of applications using the LAN. As the problem propagates through multiple managed entities, a large number of correlated alerts may be raised through other ports. These alerts can cause a flood of trouble tickets generated at the network operations center.

The alert correlation model of ERC associates with each entity the events that it may cause (via constraints) and the manner in which they may propagate along relationships. This information can be used to correlate events and then resolve their common causes.

Policy-based management

A policy consists of a set of rules. A policy rule describes actions to be pursued upon occurrence of certain conditions. A policy rule can be described using a constraint to capture the condition and a handler routine to prescribe the respective action upon violation of the constraint. For example a policy governing priorities may prevent certain classes of users from accessing the network when resources are depleted. The constraints expressing resource depletion may be evaluated as a result of attempts to gain access to the network by a user. A handler can check the class of service to which the user belong and admit or deny access.

5. SUMMARY AND CONCLUSIONS

In order to perform complex management tasks, management applications require more information about their domain than that provided by standard MIBs. The ERC model provides a rich object-oriented management information model that expresses inter-object relationships, constraints, and events within a single uniform framework. Explicit modeling of these properties in the information model concentrates knowledge in a single place and thus simplifies the development of applications for fault, performance and configuration management. Individual applications no longer have to embed this knowledge and are thus shorter and less error prone. The additional information in the model also enables a whole new class of automated event management applications.

6. ABOUT THE AUTHORS

Professor Yechiam Yemini (yy@cs.columbia.edu) is the Director of the Distributed Computing and Communication Laboratory at Columbia University. Professor Yemini has over 20 years of R&D experience in a broad spectrum of distributed networked systems technologies including network and systems management, protocol design, high speed networks, distributed multi-media message systems, and performance analysis of distributed systems. He has published over 100 articles and edited 3 books in these areas. He was a co-founder and a Director of Comverse Technology and is a founder and Chief Scientific Advisor of System Management Arts. Professor Yemini received his PhD in Computer Science from UCLA.

Alexander Dupuy (dupuy@smarts.com) is a Principal Software Engineer at System Management Arts, a startup company developing network and systems management technology. Previously he was a Senior Research Staff Associate for the Distributed Computing and Communications Lab in the Computer Science Department at Columbia University.

Dr. Shmuel Kliger (kliger@smarts.com) is a Senior Scientist at System Management Arts. Before joining SMARTS, Dr. Kliger was a research scientist at the IBM Watson Lab and at the Weizmann Institute of Science. His research experience includes developing RPC in heterogeneous neworks, distributed concurrent logic programming languages and programming environments, distributed databases and operating systems. Dr. Kliger received hist PhD in Computer Science from the Weizmann Institue of Science.

Dr. Shaula Yemini (yemini@smarts.com) is President and a founder of System Management Arts, Inc. Before joining SMARTS, she was the Senior Manager for Distributed Systems Software Technology at the IBM Watson Research Center. At IBM, Dr. Yemini led research on communications, distributed systems programming languages, RPC in heterogeneous systems, and distributed systems fault tolerance. She has published over 50 articles, a book and edited several books on these areas. Her work on Optimistic Recovery in Distributed Systems received IBM's Outstanding Innovation Award and was granted a US patent. Dr Yemini received her PhD in Computer Science from UCLA.

7. REFERENCES

[1] Dupuy,A., Sengupta,S., Wolfson,O., Yemini,Y. *NetMate: A Network Management Environment,* IEEE Network (special issue on network operations and management), 1991.

[2] OSI, ISO., 10165-4 *Information Technology, Open Systems Interconnection, Guidelines for the Definitions of Managed Objects, 1991.*

[3] Rose,M., McCloghrie,K. RFC1155, *Structure and Identification of Management Information for TCP/IP-based Internets.* May 1990.

[4] Sengupta,S., Dupuy,A., Schwartz,J., Yemini,Y. *An Object-Oriented Model for Network Management.* in Object Oriented Databases with Applications to CASE Networks and VLSI CAD. Prentice-Hall, Englewood Cliffs, NJ, 1991.

[5] Stallings,W., *Networking Standards A Guide to OSI, ISDN, LAN, and MAN Standards.* Addison Wesley 1993.

[6] Ullman,J. *Principles of Database & Knowledge Base Systems,* Vol. I & II, Computer Science Press, 3rd Ed., 1988.

[7] Wolfson,O., Sengupta,S., Yemini,Y. *Managing Communications Networks by Monitoring Databases.* in IEEE Transactions on Software Engineering Vol 17, No 9 September 1991.

[8] Yemini,Y., Goldszmidt,G., Yemini,S. *Network Management by Delegation.* in Second International Symposium on Distributed Computing Systems pages 162-171 IEEE Computer Society June 1989.

DISTRIBUTED INFORMATION REPOSITORY FOR SUPPORTING INTEGRATED NETWORK MANAGEMENT [1]

James W. Hong, Michael A. Bauer and Andrew D. Marshall

Department of Computer Science
University of Western Ontario
London, Ontario, Canada N6A 5B7

ABSTRACT

In our previous work, we have investigated the potential role and feasibility of using the X.500 Directory Service to store information collected from network management tools. The results suggested that the Directory could be used to store various types of management information, without it being the bottleneck in most management operations. This paper reports on an extension of that work. We introduce an architecture that centers the Directory Service as a management information repository in integrated network management. Two prototype implementations of the integrated network management architecture and the performance of the Directory Service are also described.

INTRODUCTION

The management of networks requires information gathered from many network devices, such as bridges, routers, workstations, and computing hosts. Monitoring these network devices involves collecting a vast amount of information about a variety of managed objects throughout the network. Information gathered may relate to network devices, their configurations, security and performance and even the activities and relationships among network devices.

This information must be stored and updated somewhere in the network. In our previous work, we have examined the potential role and feasibility of using the X.500 Directory Service [3, 4, 5] for providing the storage and retrieval of managed information in networks. Our rationale for examining the use of the Directory Service in this context stemmed from several requirements of network management. First, in a heterogeneous network environment, a global naming scheme is central in order to uniquely identify entities throughout the entire network. The X.500 Directory Service provides such a name service and the means to use and manage it. There was little motivation for developing our own if the X.500 Directory Service proved satisfactory.

Second, the Directory Service Agents provide a distributed and cooperative mechanism for storing and accessing information distributed throughout a system. This distribution of information among agents and their interoperation are transparent to

[1]This research was supported by the IBM Center for Advanced Studies and the Natural Sciences and Engineering Research Council of Canada.

Network Management and Control, Vol.2, Edited by
I.T. Frisch *et al.*, Plenum Press, New York, 1994

users, including other applications, of the Directory Service. Moreover, the Directory defines a standard interface (via the Directory Access Protocol) for storing, updating and retrieving information from the Directory, regardless of its physical location. This meant that network management information could be gathered from a variety of sources and stored independently within the Directory. This also meant that information about new devices or systems could be incorporated by extending the information held in the Directory and by defining specific monitoring tools – existing tools would not have to be modified.

Further, the 1992 version of the Directory Service defines a mechanism for the replication of portions of the Directory information [2]. In network systems spanning large geographical regions, replication of information could be useful in ensuring availability or enhanced access to certain management information.

Although the Directory Service seemed attractive, several questions existed regarding its potential use in network management. What kinds of information could be usefully stored and updated within the Directory Service? Could it handle frequent updates? How would it perform? Our previous work explored these questions by building and evaluating a management environment which used a prototype X.500 implementation to store information collected from management tools. The results suggested that the Directory could be used to store such information as well as the management information needed to identify system components and their management attributes, such as the tools to perform the information gathering. Our prototype implementation also illustrated that certain information (*e.g.*, highly dynamic information), such as counts of packet transmissions, were not suitable for storage in the Directory. That work also outlined an architecture defining the relationships between monitoring tools, their start-up, and the collection and storage of information in the Directory.

This paper extends that architecture and builds on that work in the *integrated network management framework* [7]. A critical aspect of any collection of management information is its accessibility to management applications. Here, we use the term "management application" to refer to applications which collect, analyze, visualize, *etc.*, data gathered about managed objects. Such applications may, in turn, also produce "management information". In particular, we address the issue of whether the Directory Service can support such management applications. Again, the Directory offers several advantages: (1) the Directory User Agent protocol provides a generic interface for querying the Directory for information and updating the Directory about managed objects or other management information, (2) the generic interface means that new applications can be added to the management environment or existing applications modified to analyze different groups of information, (3) management applications depend less on specific hardware or devices, that is, the monitoring aspects have been separated from the analysis aspects, and (4) access to information from multiple monitors, geographically separated or even heterogeneous, can be achieved via a single interface.

We introduce an architecture which views the Directory Service as a management information repository. Monitoring tools can update information within the repository and other management applications can extract it – both using the same protocol. In particular, we report on the development of a simple management tool for visualizing information collected about network load in an interconnected network environment. It is particularly interesting to note that the visualization tool was not designed specifically for this application. By relying on the Directory Service it was possible to quickly adapt the tool to function in this capacity. We also report on a prototype implementation of a performance analysis tool which retrieves network load data being stored in a

Directory located in a geographically distant site. Finally, the performance of the Directory Service for supporting network management tools is also discussed.

OVERVIEW OF X.500 DIRECTORY SERVICE

The X.500 Standards [1, 2] specify a Directory Service which provides and manages information about entities. The X.500 Directory Service (or simply the Directory Service) and its Directory information is distributed physically, yet it provides a view to the user of a centralized system.

The information held by the Directory is called the Directory Information Base (DIB). The DIB consists of entries (or objects) which contain information about entities. Each entry consists of a set of *attributes* each with a type and one or more values. The types of attributes which are present at each entry are dependent on the *class* of entities which the entry describes. The Standards define a number of generic classes; classes can also be user-defined, and thus various types of information can be stored in the Directory. Entities in the DIB are represented by entries in a global, hierarchical name space called the Directory Information Tree (DIT). Entries are placed in the DIT according to the organizational relationships between the entities which they represent.

The X.500 Abstract Service Definition [1, 2] defines abstract ports and operations which provide the user with the functionality for retrieving, searching and modifying directory information. These operations form a simple interface protocol called the Directory Access Protocol (DAP). The current X.500 Standards support the read, compare, list, search and abandon interrogation functions and the basic add, remove and modify entry manipulation functions.

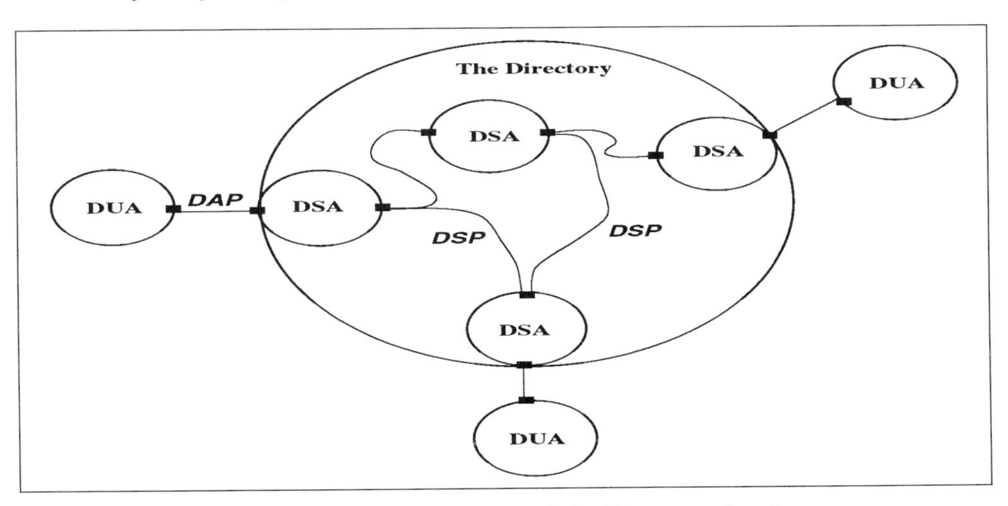

Figure 1. Functional Model of the Directory Service

The Directory Service (including the DIT) is distributed over physically separated entities (such as computer hosts) called Directory Service Agents (DSAs). This distribution is transparent to the user through the use of Directory Service Protocol (DSP) operations between DSAs. Each user or user-process is represented by a Directory User Agent (DUA) which is responsible for interacting with the Directory. The functional model of the Directory Service is illustrated in Figure 1.

It is assumed that some service will be provided regardless of network partitioning resulting from events such as non-local site failures. This is achieved by the distributed nature of the DSAs and replication of Directory information [2]. X.500 also provides a security mechanism so that directory information access by unauthorized users can be prevented. Caching is also recommended to increase the performance of the Directory Service but is considered to be a local issue, that is, the X.500 Standards do not provide caching algorithms and it is up to the implementor of the local DSA to do so.

INTEGRATED NETWORK MANAGEMENT ARCHITECTURE

In this section, we introduce an architecture that integrates the Directory Service as a management information repository. The architecture, shown in Figure 2, uses the Directory Service as a repository for storing management information collected by monitoring tools and for providing stored information requested by other management tools for other uses (such as analysis, visualization or report generation). These tools may be developed independently or as parts of an integrated management package. An important aspect that needs to be emphasized is that all management tools use a single standard interface for accessing the Directory. Another important aspect is that the management tools and the Directory can be geographically dispersed.

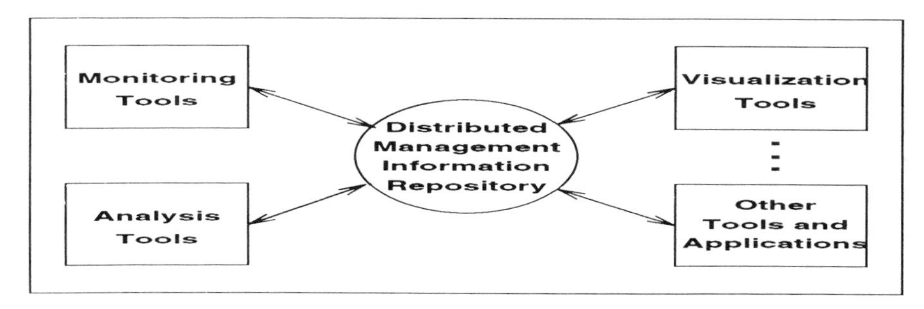

Figure 2. An Integrated Network Management Architecture

That single standard Directory interface is a DUA (Directory User Agent) interface. Each management application must be equipped with this interface in order to communicate with the DSAs in the Directory to perform storage or retrieval operations. This DUA interface should be identical for all applications as long as they conform and support all the operations defined in the X.500 DAP specification [2]. One can, however, have a DUA interface that supports a subset of the DAP operations. For example, if an application will only retrieve information from the Directory then it does not need to provide update (*e.g.*, add, modify) operations in the DUA interface.

Although the interface is uniform, the types of information that are stored and retrieved may be different from application to application. If an application needs to store information that is not defined in the generic object and attribute definitions in the Directory Service implementation, then they have to be defined and added to the existing definitions. Currently, most prototype Directory Service implementations provide object and attribute definitions related to human information. Fortunately, the X.500 Directory Service allows the user to define the objects and attributes of information which is to be stored in the Directory.

PROTOTYPE IMPLEMENTATIONS

In this section, we present two prototype implementations of the integrated network management architecture. The components that are common to both prototypes are the network load monitoring tool and the Directory. We first describe the network load monitoring tool and then the two prototype implementations.

Network Load Monitoring Tool

A management application used in both prototype implementations is a simple network management tool that monitors the network loads of subnets in a heterogeneous internetwork environment. A network load monitoring tool manages a number of subnets, each of which consists of one or more network devices. Each network device runs a small agent program for monitoring network traffic information from its interface(s). The number of packets that have been sent to the network by the device and the number of packets that have been read from the network from the device are examples of such traffic information. A network load monitor periodically polls these agents and collects the traffic information. The collected traffic information is then averaged out per subnet (per second, for example) and then stored in the Directory. When many subnets are present in a network, multiple network load monitors can be used to share the work of monitoring and storing.

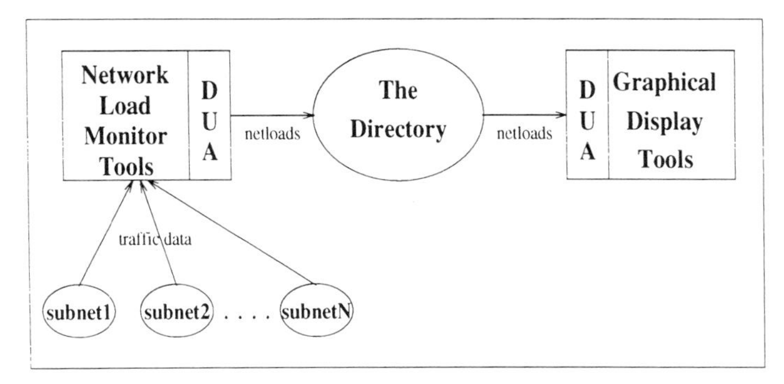

Figure 3. Integration of Network Load Monitoring and Visualzation Tools with the Directory

Prototype Implementation I

The first prototype implementation of the integrated network management architecture consists of three major components: a network load monitoring tool, the Directory service, and a graphical visualization tool. The network load monitoring tool periodically monitors the network load of subnets in an interconnected network environment and stores the values in the Directory. The graphical visualization tool, on the other hand, retrieves the network load information from the Directory and generates X Window displays for the user (*e.g.*, a network administrator). In this section, we describe the components involved and the additional object classes and attributes that were defined. Figure 3 illustrates the architecture of the prototype implementation.

The graphical visualization tool we have used to display network loads is **xmconsole** [10]. Xmconsole is an OSF/Motif based graphical visualization tool that can display

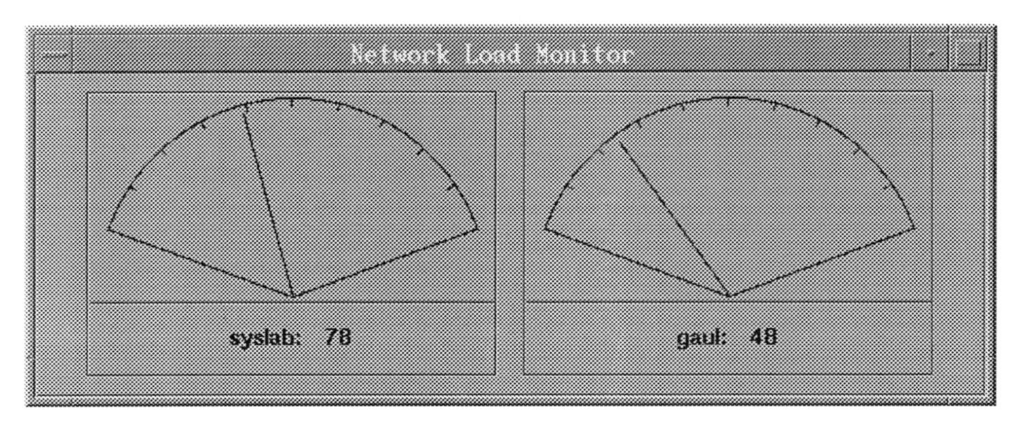

Figure 4. A Graphical Display of Monitored Network Loads

various types of statistics using windows. In our prototype implementation, xmconsole invokes an information-retrieval process which retrieves network load information on a set of subnets from the Directory. As a consequence of the design of the xmconsole program, we had to create an intermediate process that was equipped with the DUA interface that would retrieve the network loads from the Directory and feed them to xmconsole. A sample display of network loads of two subnets (called `syslab` and `gaul`) is shown in Figure 4. In this example, the numerical value beside the subnet name (*e.g.*, 78) denotes the number of Ethernet packets that have been transmitted on the subnet in the preceding 10 second interval.

Prototype Implementation II

The second prototype implementation of the integrated network management architecture also consists of three major components: a network traffic load monitoring tool, the Directory service, and a performance analysis tool (see Figure 5). In this case, however, the performance analysis tool is running from a remote site which is located hundreds of miles away from the rest. In this prototype, the network load monitoring tool also periodically monitors the network load and stores the values in the Directory. The monitoring component of the management service at the University of Michigan retrieves network load data stored in the Directory at the University of Western Ontario and feeds it to its performance analysis tool. The details of the setup and experimental results are reported in [12].

The Directory and Directory Interface

The implementation of the X.500 Directory Service we are using is Quipu version 7.0 [8, 11]. In order to interact with the Directory, a DUA interface is required in any application that accesses the Directory. In our earlier work, a DUA interface was developed that supported the *add*, *search*, and *modify* DAP operations [3, 4]. These operations are sufficient to store new information entries, retrieve or update existing information entries in the Directory. For the graphical visualization and performance analysis tools incorporated in our prototype implementations, we used the same DUA that we used for the monitoring tool. This shows that the interface to the Directory is simple and general and thus facilitates adaptation of various applications to use the Directory.

Two of the most important and relevant interface routines are called `queryObject` and `modify_object`. Figure 6 is a DUA code segment in C for the search operation. The `queryObject` routine is invoked when an object in the Directory is being searched. The routine sets up all the required parameters and options before a search query is actually sent to a DSA (via the `MonSearch` routine). Similarly, the `modify_object` routine (shown in Figure 7) sets up all the required parameters and options before a modify request is sent to a DSA (via the `MonModify` routine).

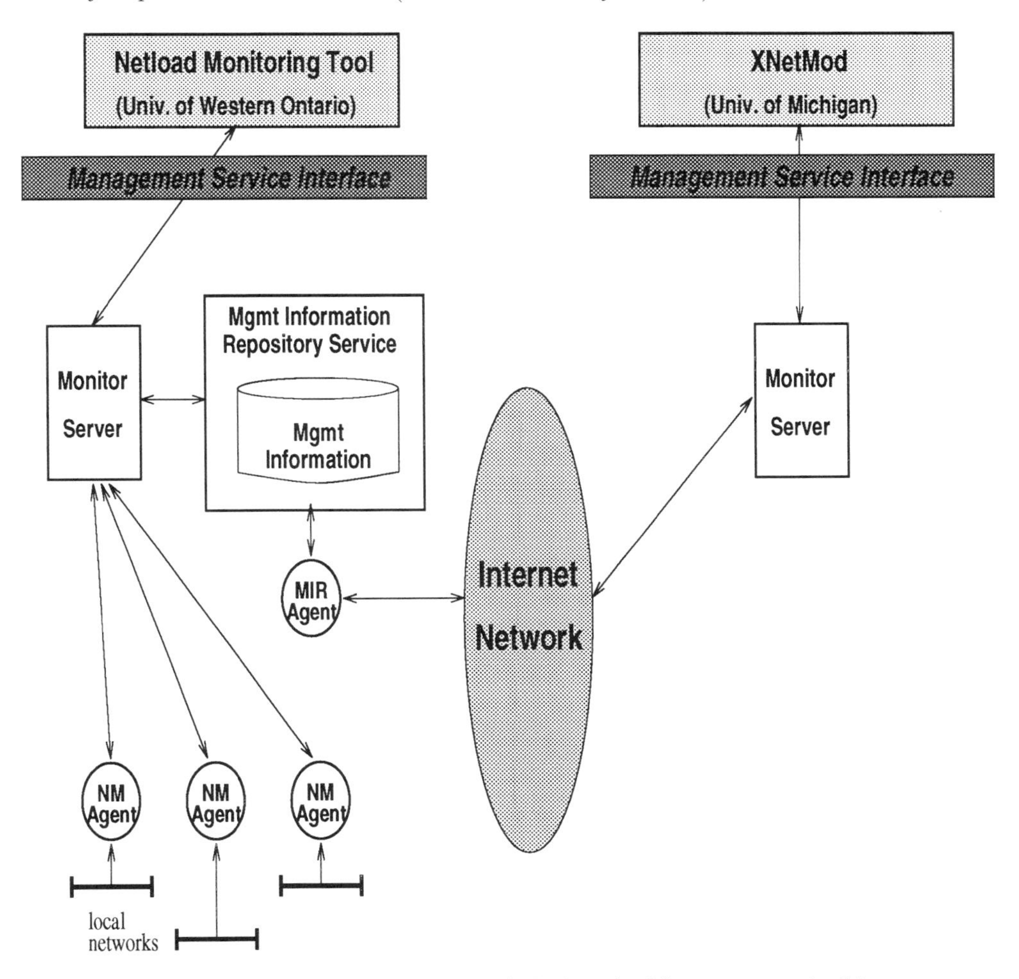

Figure 5. An Interoperability Example Using the Management Architecture

Managed Object Class Definitions

Since generic definitions of object classes and attributes in most X.500 implementations (including Quipu) are for information about people, they do not meet the requirements for the kinds of information we wish to store in the Directory, namely the domain, host and network load information for our prototypes. Thus, we defined a new set of object classes and attributes and added them to the existing Quipu definitions. Below, we present two of the most important object class definitions (in ASN.1 [6]) required for our prototype implementation.

```
int queryObject(target, attr_sequence)
char *target;
Attr_Sequence  attr_sequence;
{
  struct ds_search_arg    search_arg;
  char                   *str;
  Filter                 cf;
  ServiceControl         *sc;
  CommonArgs             *ca;
  char                   ftype;
  struct DAPindication   di;

  TidyString(target);
  if (*target == '~') {
      ftype = FILTERITEM_APPROX;
      target++;
  } else ftype = FILTERITEM_EQUALITY;

  sc = &search_arg.sra_common.ca_servicecontrol;
  sc->svc_options = SVC_OPT_PREFERCHAIN | SVC_OPT_LOCALSCOPE |
                    SVC_OPT_DONTDEREFERENCEALIAS;
  sc->svc_prio = SVC_PRIO_HIGH;      sc->svc_scopeofreferral = SVC_REFSCOPE_NONE;
  sc->svc_sizelimit = S_SIZELIMIT;  sc->svc_timelimit = SVC_NOTIMELIMIT;

  ca = &search_arg.sra_common;
  ca->ca_requestor = 0;  ca->ca_security = 0;
  ca->ca_sig = 0;        ca->ca_extensions = 0;

  search_arg.sra_baseobject = search_base;
  search_arg.sra_subset = SRA_WHOLESUBTREE;
  search_arg.sra_searchaliases = TRUE;

  /* set the attributes being requested */
  search_arg.sra_eis.eis_allattributes = FALSE;
  search_arg.sra_eis.eis_select = attr_sequence;
  search_arg.sra_eis.eis_infotypes = EIS_ATTRIBUTESANDVALUES;

  /* Root of filter specifies that the only leaf will be one item. */
  search_arg.sra_filter = filter_alloc();
  search_arg.sra_filter->flt_type = FILTER_OR;
  search_arg.sra_filter->flt_next = NULLFILTER;

  /* First leaf of filter specifies machine object. */
  cf = filter_alloc();
  search_arg.sra_filter->flt_un.flt_un_filter = cf;
  cf->flt_type = FILTER_ITEM;
  cf->flt_un.flt_un_item.fi_type = ftype;
  cf->flt_un.flt_un_item.fi_un.fi_un_ava.ava_type = str2AttrT("machine");
  cf->flt_un.flt_un_item.fi_un.fi_un_ava.ava_value =
    str2AttrV(target, str2AttrT("machine")->oa_syntax);
  cf->flt_next = NULLFILTER;

  /* Search the DIT for the objects. */
  if (MonSearch(&search_arg) == NOTOK)
      return(NOTOK);
  return(OK);
}
```

Figure 6. A Code Segment for a DUA Interface Routine: Search

```
int modify_object(name, attribute, new_value)
char  *name;                /* the name of the object to update */
char  *attribute;           /* the attribute to update          */
char  *new_value;           /* the new attribute value          */
{
  struct ds_modifyentry_arg modify_arg;
  ServiceControl   *sc;
  AttributeType    at;
  AttributeValue   av;
  AV_Sequence      as;

  at = str2AttrT(attribute);
  if ((av = str2AttrV(new_value, at->oa_syntax)) == NULLAttrV)
        return(NOTOK);
  as = avs_comp_new(av);

  /* First come the service control parameters. */
  sc = &modify_arg.mea_common.ca_servicecontrol;
  sc->svc_options = SVC_OPT_PREFERCHAIN | SVC_OPT_LOCALSCOPE
    | SVC_OPT_DONTDEREFERENCEALIAS;
  sc->svc_prio = SVC_PRIO_HIGH;
  sc->svc_timelimit = SVC_NOTIMELIMIT;
  sc->svc_sizelimit = S_SIZELIMIT;
  sc->svc_scopeofreferral = SVC_REFSCOPE_NONE;

  /* Then the remainder of the Common arguments. */
  modify_arg.mea_common.ca_requestor = 0;
  modify_arg.mea_common.ca_security = 0;
  modify_arg.mea_common.ca_sig = 0;
  modify_arg.mea_common.ca_extensions = 0;

  /* Distinguished Name of object for which to look. */
  modify_arg.mea_object = str2dn(name);

  /* Entry Modifications. */
  /* We first remove the old attribute value. */
  modify_arg.mea_changes = em_alloc();
  modify_arg.mea_changes->em_type = EM_REMOVEATTRIBUTE;
  modify_arg.mea_changes->em_what = as_comp_new(at, NULLAV, NULLACL_INFO);

  /* Then we put in the new one. */
  modify_arg.mea_changes->em_next = em_alloc();
  modify_arg.mea_changes->em_next->em_type = EM_ADDATTRIBUTE;
  modify_arg.mea_changes->em_next->em_what = as_comp_new(at, as, NULLACL_INFO);
  modify_arg.mea_changes->em_next->em_next = NULLMOD;

  /* Now we call MonModify to call DapModifyEntry. */
  if (MonModify(&modify_arg) != DS_OK)
      return (OK);
}
```

Figure 7. A Code Segment for a DUA Interface Routine: Modify

Domain Object. A *domain* is a logical entity within an interconnected network, and may represent a subnet consisting of a set of network elements. The definition for the domain object class is as follows.

```
domain OBJECT-CLASS
    SUBCLASS OF top
    MUST CONTAIN {
        domainName}
    MAY CONTAIN {
        description, hostList, subnetAddress, subnetMask, adminContact,
        techContact, netload}
    ::= {csdObjectClass 6}
```

Domain is the name of the *domain* object class. The domain object inherits attributes of the top object class and has a single mandatory attribute called domainName. The description, hostList, subnetAddress, subnetMask, adminContact, techContact, and netload attributes are optional. The hostList attribute contains the names of network devices in the domain. The network load information for this particular domain for some time interval is stored in the netload attribute, which can be retrieved by other applications for various uses. For example, overall network load of an entire departmental (possibly consisting of one or more domains) or campus-wide network can be analyzed. A sample directory entry of a domain object is given below:

```
domainName=syslab
objectClass= top & quipuObject & quipuNonLeafObject & csdLocalTop & csdDomain
description= Systems Laboratory, Dept. of Computer Science, UWO
subnetAddress= 129.100.14.0
subnetMask= 255.255.255.0
masterDSA= c=CA@o=University of Western Ontario@cn=Homer
netload= 107
hostList= rubble ford grabel crocker boop
adminContact= Mike Bauer
techContact= David Wiseman
```

Network Device Object. A *network device* object (*e.g.*, a host) is a physical or logical computing device on a network. It appears to the user as a distinct entity that can be used for running user applications (*e.g.*, management agent programs). The definition for the network device object class is as follows.

```
machine OBJECT-CLASS
    SUBCLASS OF top
    MUST CONTAIN {
        machineName}
    MAY CONTAIN {
        description, machineType, mailServer, operatingSystem, ipAddress,
        location, application, adminContact, techContact, cpuload}
    ::= {csdObjectClass 7}
```

Machine is the name of the network device object class. Objects of this class also inherit attributes of the top object class and has a single mandatory attribute called machineName. The description, machineType, mailServer, operatingSystem, ipAddress, location, application, adminContact, techContact and cpuload attributes are optional.

The `cpuload` attribute of a network element object contains its CPU load, which is monitored and stored by a CPU load monitoring tool [3]. It is interesting to note that the `application` attribute contains the names of management applications that a particular network element can run. It also tells where the application programs can be found. For example, it contains the file path of the agent program that collects network load information. This is where the manager program fetches the file paths of the agents when activating them. campus-wide network can be analyzed. A sample directory entry of a domain object is given below:

```
machineName=rubble
objectClass= top & quipuObject & csdLocalTop & csdAdmin & csdMachine
description= a SPARCstation in Systems Laboratory
machineType= SPARCstation EL
mailServer= rubble.syslab.csd.uwo.ca
networkAddress= 129.100.14.14
operatingSystem= SunOS 4.1.1
application= /sl/s0/research/jwkhong/AA/agent/agent
application= /sl/s0/research/jwkhong/CPUMON/Agent/cpumon-agent
adminContact= Mike Bauer
techContact= Dave Wiseman
location= Systems Laboratory, MC 325
cpuload= 13
```

In this section, we have described our prototype implementations of integrating the Directory Service in an integrated network management environment. In the next section, we examine the performance of the Directory Service in the prototypes.

PERFORMANCE

In order to determine the performance of the Directory Service for supporting the network management tools, we have measured the times taken for accessing the Directory by two management applications used in the first prototype implementation[2]. The use of the Directory basically involves two phases as follows.

1. management tool (DUA) connects to DSA, and
2. accesses the Directory.

The first phase involves the DUA representing the management application making a connection to a DSA before accessing the Directory. The time taken for this phase was identical for both applications. The second phase involves sending appropriate Directory request operations to the DSA. For the network load monitoring tool, this means sending *update* requests of the network loads of domains to the DSA. For the visualization tool, this means sending *retrieval* requests to the DSA for the domains of interest and receiving their network loads from the DSA.

Figure 8 shows the performance results for each operation. These measurements were taken from running our prototype implementation on our systems laboratory described earlier. The times were sampled once every 60 seconds. The times shown in the table are all in seconds.

The time taken for an update operation (0.045 second) is much less than the time taken for retrieval because update was done asynchronously and retrieval was done

[2]At the time of writing this paper, we did not have the performance figures for accessing our Directory from University of Michigan. Therefore, we only report the performance of accessing the Directory for the first prototype implementation.

Operation	Trials	Mean	Std. Dev.	Minimum	Maximum
DUA-DSA Connection	1000	0.64	0.22	0.5	1.6
Retrieval	4202	0.67	0.042	0.6	0.9
Update (asynchronous)	4152	0.045	0.037	0.011	0.44
Update (synchronous)	3981	1.1	0.23	0.85	9.4

Figure 8. Performance results for directory operations

synchronously. That is, the DUA sent update requests to the DSA and returned immediately to perform other operations (*e.g.*, next collection). For the sake of comparison, we measured the synchronous version of update operation; these results are also reported in Figure 8.

We have run some tests to determine the maximum rate of update frequency that can be handled by the Directory Service. The maximum rate that the Directory Service can handle without any problem was at every 2 seconds or less. We envisage that this performance by the Directory Service is sufficient to provide the storage and retrieval service to most (if not all) management applications.

CONCLUDING REMARKS

Our earlier work suggested that the X.500 Directory Service could be used for the storage of management information collected by network monitoring tools. In this paper, we extended that work to investigate the use of the Directory as an information repository in an integrated network management architecture. Our prototype implementation shows that the Directory interface is truly generic in that various management applications can be easily added to the management environment by simply adding the DUA interface to existing applications. The performance of the Directory Service in an integrated network management system is still acceptable for most management applications.

Although the X.500 Directory Service provides a rich storage functionality for storing all kinds of information, one of the main concerns the users and developers of applications that use the Directory currently have is its performance. An interesting future study is to investigate the use of Directory Services other than X.500 (such as the Cell Directory Service (CDS) which is part of DCE [9]) for the repository service of management information in integrated network management.

In our prototype implementation, we used two management tools accessing the Directory for monitoring and displaying the network loads of two subnets. Another interesting future study is to investigate the scalability in the performance of the Directory. That is, what would be the performance of the Directory if we have tens or hundreds of management applications accessing the Directory simultaneously? Also, if we increase the number of subnets to tens or hundreds in the management (and thus increase the number of update and search operations), what would the Directory's performance be? A carefully designed experimental or simulation study would be required to answer these questions.

ABOUT THE AUTHORS

James W. Hong (`jwkhong@csd.uwo.ca`) is a research associate and adjunct professor in the Department of Computer Science at the University of Western Ontario. He received his HBSc and MSc from the University of Western Ontario in 1983 and 1985, respectively. He received his doctorate from the University of Waterloo in 1991. He is a member of the ACM and IEEE. His research interests include distributed computing, operating systems, systems and network management.

Michael A. Bauer (`bauer@csd.uwo.ca`) is Chairman of the Department of Computer Science at the University of Western Ontario. He received his doctorate from the University of Toronto in 1978. He has been active in the Canadian and International groups working on the X.500 Standard. He is a member of the ACM and IEEE and is a member of the ACM Special Interest Group Board. His research interests include distributed computing, software engineering and computer system performance.

Andrew D. Marshall (`flash@csd.uwo.ca`) is a Ph.D. candidate in the Department of Computer Science at the University of Western Ontario. He completed his M.Sc. in 1989. He is a member of the ACM and IEEE. His research interests are in distributed computing, high-speed networks, and software re-engineering.

REFERENCES

[1] CCITT. *The Directory - Overview of Concepts, Models and Services, CCITT X.500 Series Recommendations.* CCITT, December 1988.

[2] CCITT. *The Directory - Overview of Concepts, Models and Services, Draft CCITT X.500 Series Recommendations.* CCITT, December 1991.

[3] J. W. Hong, M. A. Bauer, and J. M. Bennett. Integration of the Directory Service in Distributed Systems Management. *1992 International Conference on Parallel and Distributed Systems*, pages 142–149, Hsin Chu, Taiwan, December 1992.

[4] J. W. Hong, M. A. Bauer, and J. M. Bennett. Integration of the Directory Service in the Network Management Framework. *Proc. of the Third International Symposium on Integrated Network Management*, pages 149–160, San Francisco CA, April 1993.

[5] J. W. Hong, M. A. Bauer, and J. M. Bennett. The Role of Directory Services in Network Management. *Proc. of the 1992 CAS Conference*, pages 175–187, Toronto Canada, November 1992.

[6] ISO. *Information Processing Systems - Open Systems Interconnection - Specification of Abstract Syntax Notation One (ASN.1).* International Organization for Standardization, International Standard 8824, 1987.

[7] C. A. Joseph and K. H. Muralidhar. Integrated Network Management in an Enterprise Environment. *IEEE Network*, 4(4):7–13, July 1990.

[8] Stephen E. Kille. *Implementing X.400 and X.500: The PP and QUIPU Systems.* Artech House, Boston MA, 1991.

[9] OSF. *The OSF Distributed Computing Environment Rationale.* Open Software Foundation, Cambridge MA, May 1990.

[10] Marc Pawlinger. *The xmconsole Graphical System Monitor for IBM Risc System/6000.* Personal Systems Programming, IBM, Palo Alto, CA, September 1992.

[11] C. J. Robbins and S. E. Kille. The ISO Development Environment: User's Manual Volume 5: QUIPU, July 1991.

[12] G. Winters and T. Teorey. Managing Heterogeneous Distributed Computing Systems: An Analysis of Two Data Repositories. *Proc. of the 1993 CAS Conference*, pages 691–706, Toronto, Canada, October 1993.

TREE PROTOCOL FOR SUPPORTING BROADBAND MULTIPOINT, MULTICHANNEL CONNECTIONS

Lyndon Y. Ong

Bell Communications Research
331 Newman Spring Road
Red Bank, NJ 07701

Mischa Schwartz

Columbia University
Center for Telecommunications Research
New York, NY 10027

ABSTRACT

Broadband ATM networks will support multimedia and multipoint connections requiring new network control functionality and protocol development. This paper examines the functionality of a signaling protocol that configures network resources to create calls with multiple channels and multiple destination users, and also supports real time reconfiguration of the call while in the active state. Protocol design issues such as distributed vs. centralized control, call topology, and special procedures for multipoint and multimedia are considered. The protocol designed uses distributed control to create a spanning tree connecting destination users, and provides new functionality for handling multiple call components and multipoint handling procedures. The protocol is specified using SDL, and tested using simulation. Its performance and complexity is compared with existing point-to-point call control protocols.

INTRODUCTION

Broadband networks using Asynchronous Transfer Mode (ATM) multiplexing and switching can support connections with a variety of different transport characteristics, making the network well suited to multimedia services, especially where multiple connections are coordinated as part of the same call. ATM is also well suited to providing multipoint configurations by using the internal capability of ATM switch designs to multicast cells and merge cells from different locations, or using bridging within specialized network service nodes.

This paper looks at the design of a network protocol for multimedia and multipoint services. A protocol is designed to create minimal spanning trees within a

Network Management and Control, Vol.2, Edited by
I.T. Frisch *et al.*, Plenum Press, New York, 1994

network that connect the parties in a multipoint call, using current work on Broadband protocols as a basis. The protocol is formally defined using CCITT Specification and Description Language, and its operation is verified by simulation. Its performance and complexity are compared against current point-to-point call control protocols.

The inclusion of multiple media has been identified as a key improvement in the provision of group collaborative communications for problem solving and decision making. Supporting separate channels for different media allows greater control over network resource usage than combining multiple media into one network channel. Many prototype systems have used the latter approach, due to network limitations that should be eliminated in Broadband. There appear to be advantages to segregating different information streams into separate network connections, if the network can provide this functionality.

The use of separate channels does introduce problems of synchronization between different media in a multimedia signal. The most synchronization that can be provided by the network, short of being involved in the coding of the signal itself, is coordinating the routing of associated channels carrying different media, so that the transport delay over the network has very limited variability between different streams of information. Common route connection groups will be a capability of future Broadband releases.

Studies have also identified the potential benefits of functionality in the network that allows the user to control information from multiple sources, and that supports network-based bridging functions These functions will require new developments in network control. The network will need to provide the functionality to add and delete parties from a conference freely, and add and drop new media to the conference (e.g. adding a graphics capability such as an electronic blackboard.) Network support for control of the merging and distribution of signals will be an important factor to allow users to specify the characteristics of a multimedia conference, e.g., whether all conferees should be seen, or only one at a time.

Subconferencing is another possible network function. In subconferencing, an interaction may include only a subset of the parties in the call, and multiple subconferences may be taking place simultaneously within the call. To use network resources most efficiently, the network call control protocol should be able to specify the subsets of parties that are involved in each subconference, and to allocate only those network channels that are needed to connect the parties in each subconference.

When a call is divided into multiple subconferences, the call processing software must be capable of auditing the use of each network channel, and deciding based on this audit whether or not it is still needed. As subconferences are reconfigured, the network will also have to allocate new channels when previously non-interacting parties are joined into a new subconference.

An example configuration is shown in Figure 1. This shows a call that is divided into two subconferences, one involving parties A, B and C, and one involving parties C and D. In this configuration, party D cannot exchange information with parties A and B, only with party C. In the example, channels are not needed between parties A and D, since they do not communicate directly.

BASIC MODEL

The protocol defined here provides for control of multipoint, multimedia calls, creating and reconfiguring minimal spanning trees in the network to connect all parties. The following model for communication services is used:

call: defined to be a logical relationship between network users that are sources and sinks of information. The flow of information between users is determined by the connections making up the call.

party: network user participating in the call. The protocol provides the calling party with sole control over the configuration of the call, with the exception that called parties can drop themselves off from the call (however, they cannot reconnect without first asking permission from the calling party.) This is based on the assumption that the calling user is responsible for paying for the use of network resources, and therefore should have the final authority for use of those resources. Other service arrangements are possible and may be explored in the future.

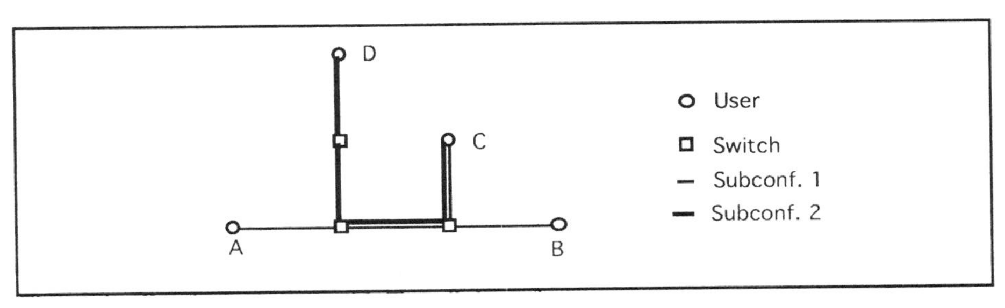

Figure 1 - Conference Call with Two Subconferences

connection: defined in a multimedia/multipoint context as a logical call element that specifies the sharing of information between parties in a call. A connection is characterized by a set of involved parties, a transport capability provided by the network, and a merging function that defines how input from these parties is combined and sent back to the parties. A connection, for example, might be defined as "connecting parties A, B and C with transport characteristics T, and providing a merging function F". A merging function might be defined, for example, as combining multiple video inputs into a multiwindow output.

This characterization is based on the EXPANSE call model by Minzer [1], particularly in the concept of the multipoint connection (similar to the "t-connection" in EXPANSE) as a collection of parties that share a certain type of information using some specified bridging or distribution algorithm.

The basic connection establishment procedure is shown in Figure 2, and is based on the Broadband ISDN User Part work [2] in international standards. Starting with the originating switch, the connection is established forward, link-by-link, to the destination switch. At each switch in the path, the request is received from a preceding neighbor switch, local resource allocation decisions take place, and a modified request is passed to the succeeding switch.

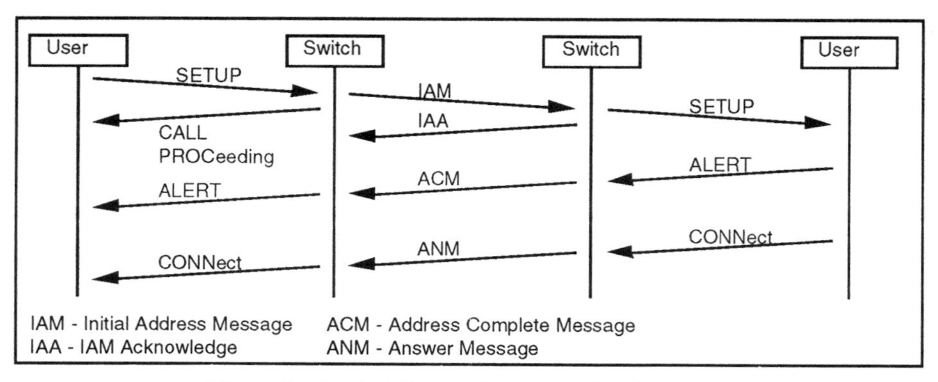

Figure 2 - Basic Message Sequence for Call Setup

PROTOCOL DESIGN ISSUES

Centralized vs. Distributed Control: The protocol incorporates distributed control, as currently used in call control protocols, rather than a centralized control as in network management. There were a number of reasons for this decision:

- As an evolution from the current protocols, the protocol for Broadband developed in standards will most likely use distributed control.
- The performance of fully centralized control in a large network needs to be investigated further, as the central connection manager may become a bottleneck for processing large numbers of user connection requests in real time.
- Fully centralized control is probably not possible in a multi-domain environment, where a connection may have to traverse networks belonging to different business entities.

However, centralized control has advantages due to having a single controller with a full view of network resources, allowing potentially more efficient routing of a call, for example in the case that scarce resources such as a conference bridge in a remote location are required--distributed control does not allow reservation of the scarce resource before proceeding with the call. Hybrid centralized/decentralized methods are also possible, and are explored in another paper [3].

Connection Topology: The protocol was designed to create a minimal spanning tree as the basic connection topology. The tree reduces the use of network transport resources by having information merged and distributed within branching points of the tree. The tree topology has been proposed for point-to-multipoint distribution [4] of data in packet networks, but can also be efficient for multipoint-to-multipoint interactions.

Other candidate topologies are star, ring and mesh. The star is only a subclass of the tree where there is only one branching point. Both the ring structure [5] and the mesh structure have also been proposed for voice and data conferencing However, the ring and mesh structures are geared towards provision of functions at the customer premises, rather than within the network, as there are no concentration points in the network where streams of information from different users are brought together. They require greater terminal complexity, and less efficient use of channels, and do not take maximum advantage of network functionality.

A number of methods exist for multipoint connection routing, although the problem of generating minimum spanning trees is an instance of the Steiner tree problem, and is NP-complete. No assumptions are made here as to the tree generating algorithm. For the protocol, it was assumed that correct routing information for generating the tree is given to each switch by a separate traffic process prior to connection establishment. For further information, the reader is referred to heuristic algorithms proposed in [6] and other papers.

Protocol Structure: A simple, monolithic structure was used for the protocol, i.e., no sublayering within the protocol. Since the tree protocol requires bridging to be considered at every node in the call, it is not possible layer the protocol and hide the complexity of multipoint aspects of the call from some nodes. Instead a monolithic structure is used for the protocol, and every node that is involved in processing the call must be capable of understanding multipoint and multichannel operation, even if it is not providing all of the functionality, e.g., it is not serving as a bridge for this particular call.

Multipoint Procedures: In the protocol, actions involving multiple adjacent nodes are implemented by *multicasting* control messages, i.e., sending the same message to several points simultaneously when an action involves a multipoint reconfiguration of the call, so that simultaneous processing occurs at each node. Each switch analyzes the set of adjacent switches that will be affected by an action such as adding or dropping a new connection, and multicasts the appropriate message to the affected switches. This results in a more complex protocol state machine, since the protocol must be capable of creating a multicast, and reacting to responses from several different switches.

Handling of Multiple Components: Since each request in a multimedia, multipoint environment may involve several components (individual connections or parties), and the response (or responses, in the case of a multicast request) may be a combination of confirmations and rejections, the handling of rejections is a major aspect of the protocol. When not all of the components requested in the call can be supported, the protocol allows call establishment to continue with fallback to a lower grade service, rather than automatically terminating the call. In the protocol, the user specifies the essential components of its request, and this information is carried into the network. If an essential component cannot be allocated to the call, the information is sent back, and a rollback of the action takes place. The rollback is handled by sending a message that results in the opposite action as the original message requested, e.g., sending a message to drops a connection if the procedure being rolled back was the addition of that connection.

Another alternative is to use a full two-phase commit procedure, as used for database transactions. A proposal using the two-phase procedure was defined by Bubenik [7], and some of the disadvantages in performance are described further in [3].

Procedure for Connection Establishment: The method of establishing a path between non-adjacent points is adapted from current protocol procedures, where a path is created link by link in a forward direction.

Existing point-to-point call protocols have evolved a consistent method of creating and controlling calls: each switch acts independently based on local knowledge of the most desirable route for a particular address, and states of adjacent switches with regard to a particular call. A call is established from the originating point to the destination point on a "link-by-link" basis, with the originating point sending a call request to an adjacent switch, then that switch establishing a link in the call to the next adjacent switch, and so on until the destination is reached. This same procedure is adopted here, providing an evolutionary approach from current Broadband control protocols, and ensuring that there is always a path available through the network before the called party is contacted.

Table 1 - Protocol Functions (without subconferencing)

Function	Message types	Description
add_call	req/conf/rej	initiate new call
drop_call	req/conf	release existing call
progress	indication	intermediate progress indication
add_party	req/conf/rej	add party to existing call
drop_party	req/conf	drop party from existing call
add_conn	req/conf/rej	add connection to existing call
drop_conn	req/conf	drop connection from existing call

PROTOCOL OPERATION

The functions provided in the protocol are listed in Table 1. The call establishment procedure allows the user to create a multipoint, multimedia call that can involve several called parties at once. In the example configuration in Figure 3, party A sends out an ***add_call_request*** message (1) to its serving switch S1, containing a call reference ID, a list of called parties, a list of desired connections, a list of local channels to be used, and a list of those components which are essential for successful call completion. At switch S1, the message is analyzed to determine the best routes to called parties B, C and D.

S1 will in this case initiate separate branches to S2 and S3, multicasting ***add_call_request*** messages on each branch (2), containing only the subset of called parties to be reached by that branch. In this way, race conditions are avoided where two branches attempt to reach the same called party. S1 also provides any bridging functions that are necessary to support the connections in the call.

Upon receiving the ***add_call_request*** messages, switch S3 sends an ***add_call_request*** message to party D, while switch S2 multicasts messages to parties B and C (3).

After any intermediate call progress indications are given, such as ***alerting***, then each party responds with a ***confirm*** or ***reject*** (4). Intermediate changes, such as when one party confirms before the other party has responded, are indicated to the calling party by sending a progress_indication (5) back towards the calling party. As the responses are received by S2 and S3, the call data are analyzed to determine if the essential components of the call have been accepted. S1 relays call progress indications back to the calling party, ultimately sending a ***confirm*** or ***reject*** to the calling party (6) after all responses have been received.

If the switch receives a ***reject*** for an essential component of the call, then the ***reject*** is sent back to the preceding switch, and a rollback is done by sending out ***drop_call_request*** messages to clear any remaining branches.

For the subconferencing feature, the procedures above must be modified to support connections that involve only a subset of the called parties in the call. The additional messages required are shown in Table 2. The main impact is that when the call is created, and for any reconfiguration of the call, a switch will have to analyze each connection to determine whether any channels or internal bridging needs to be allocated to the connection. In some cases, a switch may not need to allocate any local resources to a connection, because the connection only needs to be supported by remote switches. For example, the connection in the call of {B, C} does not require the use of any local resources by switch S1. When the call is initially created, and switch S1 receives an *add_call_request* message, it must determine from the message that plays no role in that connection, and can just pass the information on to switch S2.

New procedures are required to form a link to a subconference, if the subconference is on a different branch of the tree from the party to be added. For example, Figure 4 shows the case where party D is added to the subconference of {B, C}. An *add_party_req* message (1) is sent to add a branch to the tree along the path S1-S3-D, in order to add party D. At the same time, another branch is added by sending an *add_link_req* message (2) along the path S1-S2, in order to link party D with the subconference {B, C}. Different messages are used, since one procedure requires reaching a network user, and the other procedure will stop once it reaches a switch that is part of the subconference.

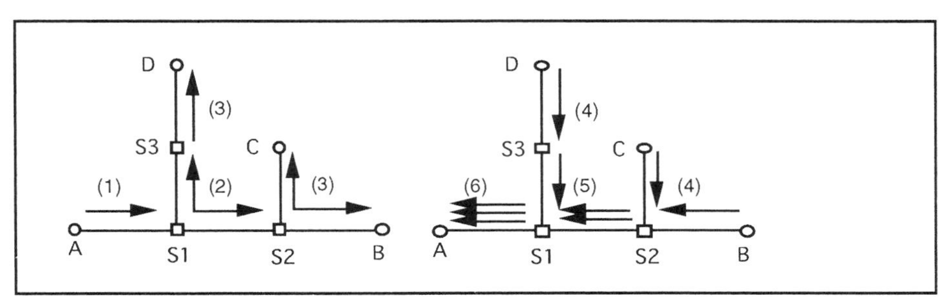

Figure 3 - Example of Protocol Operation

Table 2 - Additional Functions Required for Subconferencing

Function	Message Types	Description
add_sub	req/conf/rej	add existing party to subconference
drop_sub	req/conf	drop party from subconference
add_link	req/conf/rej	link party with subconference
drop_link	req/conf	drop link between party and subconference

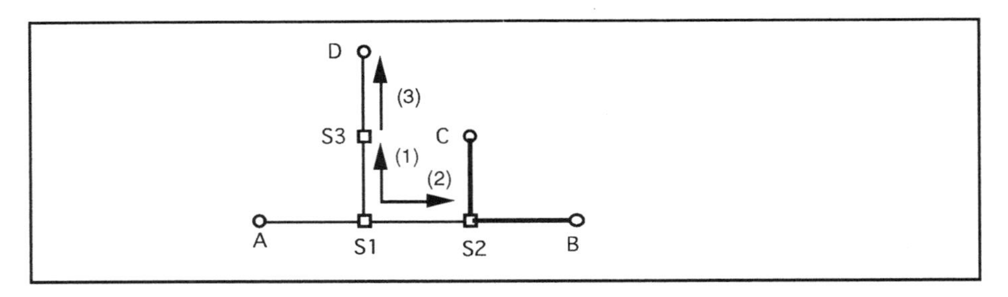

Figure 4 - Example of Subconferencing Operation

PROTOCOL SPECIFICATION AND VERIFICATION - SDL

CCITT Specification and Description Language (SDL) [8], which is widely used as a formal method of protocol specification in the telecommunications arena, was used to specify the tree protocol and compare its complexity with point-to-point protocols. An SDL diagram is essentially an expanded version of a state transition diagram, showing greater detail about the incoming and outgoing messages in a particular transition and showing processes that may take place during a transition. Each branch represents a different set of actions that must be taken in response to an external message. Each branch, then, may require a separate routine to perform this set of actions. While there are minor differences in complexity between branches, the number of branches overall provides a measure of the relative complexity of different protocols. An example of the SDL specification of the protocol is shown in Figure 5. This SDL diagram defines the actions of the protocol machine during the setup of a multipoint call.

There is no guarantee that a particular specification in SDL is the most compact possible, but comparison of SDL specifications for different protocols, especially if done at the same level of detail and with similar assumptions, provides a useful measure of the differences in complexity that would be required to implement alternative protocols. One advantage of an SDL-based comparison is that the SDL are independent of the programming language used for implementation, resulting in greater generality of results.

SIMULATION FOR VERIFICATION AND ANALYSIS

The operation of the protocol was verified by simulation using the L.0 simulation language [9]. The completed simulation then provided additional information on complexity by allowing a comparison of the number of lines of code needed to simulate different protocols, and comparing the performance of the simulations under different conditions. Since this form of comparison is implementation dependent, however, it is biased towards the specific command set of the L.0 language.

Two different network topologies were used for testing. The network in each case included transit and end switches, and several possible branching points on any given

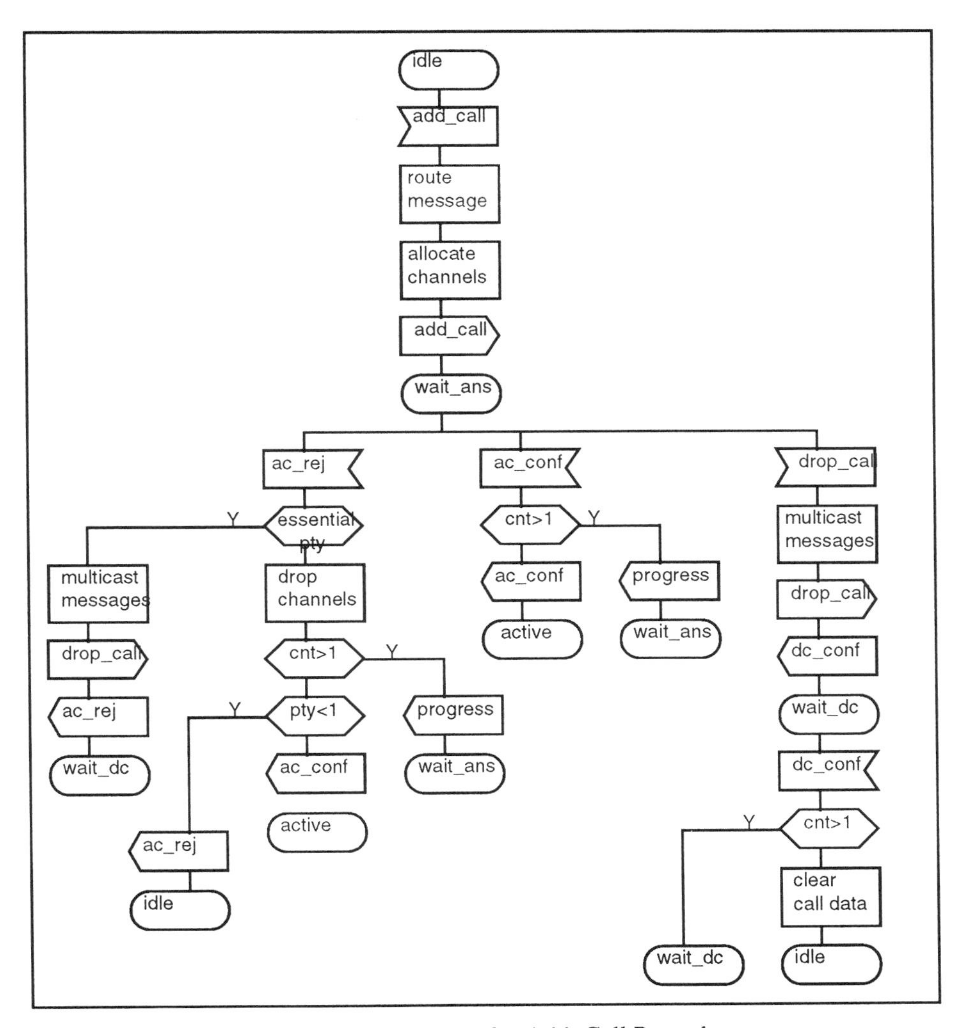

Figure 5 - SDL Example for Add_Call Procedure

call. The two configurations included one case where called parties were distributed evenly over the access switches, and one where called parties were clustered together, to look at the impact of clustering of parties on the number of messages and the time delay associated with operation of the tree protocol and its alternatives.

The protocol was tested by running a basic call script consisting of a sequence of all of the basic commands of the protocol, with calls of random sizes, and with each called party making a random decision on whether to accept or reject the call. The operation of the protocol was also verified by testing each of its component procedures independently, using requests with a random selection of parties. These random simulation tests were run for a long interval of simulation time, and the protocol tested for dead-end states or incorrect operation.

COMPARISON OF COMPLEXITY WITH POINT-TO-POINT PROTOCOLS

An SDL specification of ISDN point-to-point call control was developed using the standards for ISDN call control. These SDLs covered only basic operation, and did not include more specialized features such as overlap address, or channel negotiation. However, they provided a basis for comparing the complexity of the tree protocol with a basic point-to-point call control protocol.

As shown in Figure 6, the SDL for the tree protocol is considerably more complex than the SDL for point-to-point ISDN call control, due to the requirement to handle multicasting of control messages, and to handle reconfiguration of an active call such as party addition and connection addition. The tree protocol needed to be enhanced significantly to support subconferencing capability.

The number of lines of code needed for simulation of call processing increased similarly for the tree protocol as opposed to the point-to-point protocol. As shown in Figure 7, the complexity of the software required for simulation of the point-to-point call control protocol was considerably less than was required for the multipoint tree protocol, and in turn, the simulation of the tree protocol with the subconferencing feature required significantly more lines of code than the basic tree protocol.

The comparison shows a substantial increase in complexity required to provide subconferencing capability. This increase stems from the additional functions needed, such as linking, and also from the need to analyze each connection to determine where channels are needed in the case of a subconference that only includes a subset of the parties in the call. The increased complexity implies that the protocol will require more memory and processing at a switch.

PERFORMANCE

The messages required and time delay required for call setup and other commands represent load on the network. Each message causes some fixed delay when passed from one module to another (i.e., one node to another), as well as a variable queuing delay when a particular switch must process several messages received at the same time or very close together. Both the load and time delay are useful as gauges of the efficiency with which a particular protocol can carry out the functions required for multipoint and multimedia control.

Figure 8 shows the performance of the protocol during simulation of call setups with calls of different sizes, using randomly assigned called parties in each call. The messaging results shown in Figure 8 are counts of the total number of messages exchanged between switches during each call setup. The time delays shown in Figure 9 are in units of simulation time, and are used for comparison rather than as an indication of the actual setup time that would occur in an implementation. This includes processing delay and queuing delay that occurs when multiple messages are received simultaneously at a switch.

As shown in Figure 8, the number of messages increases with the number of parties, but at a slower rate than if separate calls were established to each party using ISDN call control protocols. The delay time shown in Figure 9 also increases with the number of parties in the call, but at a much slower rate than for separate point-to-point calls, due to the use of multicasting of control messages that allow the call to be processed in parallel by multiple switches or parties.

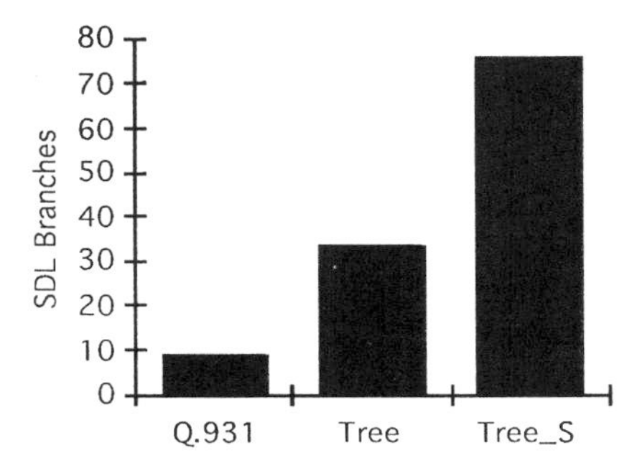

Figure 6 - SDL Complexity Comparison

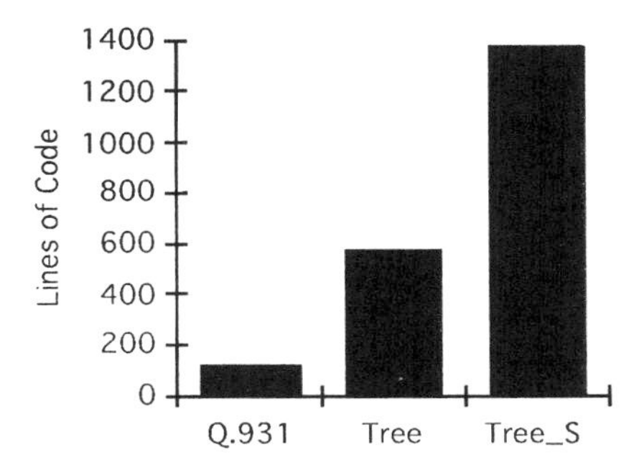

Figure 7 - Simulation Code Comparison

CONCLUSIONS

Potential network requirements for supporting multimedia, multipoint conferencing services have been explored, resulting in a proposal for a network call control protocol. The protocol uses distributed control to create a minimal spanning tree connecting all parties in a call, and supports a number of new features, such as multicasting of control messages, and special handling of multiple call components.

Comparison of its performance and complexity with point-to-point protocols indicates a significantly greater complexity required to handle message multicasting and subconferencing functions. However, due to multicasting, the procedure is significantly more efficient in messages and delay time than creating separate point-to-point connections to each party. The method of specification using SDL, and comparison of SDL complexity can be applied in future to other protocol designs, such as a centralized design based on management protocols, or hybrid approaches.

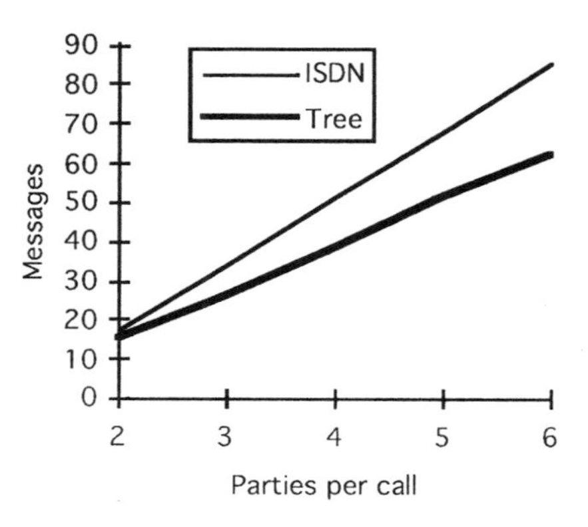

Figure 8 - Message Traffic Results

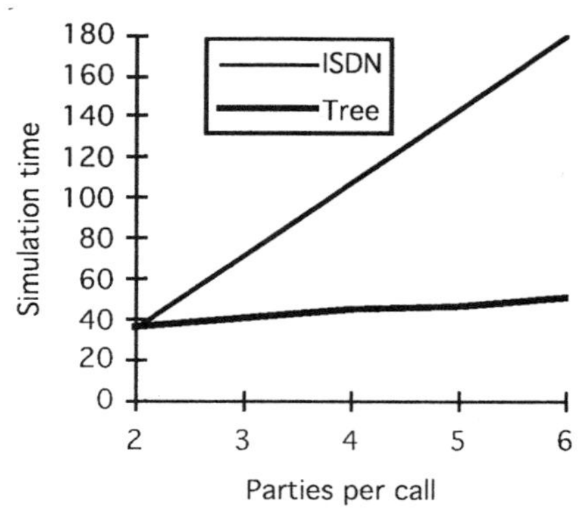

Figure 9 - Setup Delay Results

ABOUT THE AUTHORS

Lyndon Ong (lyo@cc.bellcore.com) is a Member of Technical Staff in the Protocol Architecture District of Bellcore. He received his doctoral degree in electrical engineering from Columbia University in 1992. His area of interest is the development of control protocols for ATM Broadband networks. Dr. Ong is currently chairing the international standards group in the ITU that specifies the network node signaling interface for Broadband ISDN.

Mischa Schwartz (schwartz@ctr.columbia.edu) - for biographical information, please see *Fault Diagnosis of Network Connectivity Problems by Probabilistic Reasoning*, also in this volume.

REFERENCES

[1] S. E. Minzer.
Signaling and Control for Multimedia Services.
Proceedings of IEEE Multimedia '90, November 1990.

[2] ITU Recommendations Q.2761-4.
Broadband ISDN - Signalling System No. 7 - B-ISDN User Part
Draft of December, 1993.

[3] L. Ong.
Design of Network Control Protocols for Broadband Multimedia, Multipoint Services.
PhD thesis, Columbia University, August, 1992.

[4] A. Segall.
Reliable Multiuser Tree Setup with Local Identifiers.
IEEE Journal of Selected Areas in Communications, 9(9):1427-39,1990

[5] C. Ziegler, et al.
Implementation Mechanisms for Packet Switched Voice Conferencing.
IEEE Journal of Selected Areas in Communications, 7(5):698-705, June 1989.

[6] C. H. Chow.
On Multicast Path Finding Algorithms.
Proceedings of INFOCOM '91, pp. 1274-1283.

[7] R. Bubenik, J. DeHart, and M. Gaddis.
Multipoint Connection Management in High Speed Networks.
Proceedings of Infocom 1991.

[8] CCITT Recommendation Z.100.
Functional Specification and Description Language.
1988 CCITT Blue Book Recommendations, Vol. X, Fascicle X.1.

[9] D. Cohen, T. M. Guinther and L. Ness.
Rapid Prototyping of a Communication Protocol Using a New Parallel Language.
First International Conference on System Integration, April 1990.

BROADBAND UNI SIGNALING TECHNIQUES

Thomas F. La Porta and Malathi Veeraraghavan

AT&T Bell Laboratories
Holmdel, NJ

ABSTRACT

This paper evaluates the possibility of extending the Release 1 Broadband Integrated Services Digital Network (BISDN) User-Network Interface (UNI) signaling protocol, called Q.93B, to support services in a Release 2 BISDN. Two Q.93B-based protocols are considered: a *monolithic protocol* and *separated protocol.* The resulting protocols are compared to a set of modular protocols referred to as the Distributed Call Processing Architecture and Protocols (DCPA+P). While the Q.93B-based approaches may be backward compatible with existing signaling protocols, their limited structure greatly reduces their flexibility. The DCPA+P also meets the functional objectives of Release 2 BISDN, while providing a flexible platform for rapidly introducing new services.

INTRODUCTION

Signaling is used to manage communications resources and services so that user information may be transported across a network. For narrowband ISDN, standards exist for signaling protocols across the User-Network Interface (UNI) and the Network-Node Interface (NNI). The existing standard for ISDN message-based signaling across the UNI is Digital Subscriber Signaling System No. 1 (DSS1[1]). The upper layer protocols of DSS1, Q.931 and Q.932, specify the information elements, messages and procedures needed to control connections and services across the UNI. The existing standard for the NNI is Signaling System Number 7[2] (SS7). For Release 1 BISDN, which will support mainly point-to-point, single connection services, CCITT has defined Q.93B[3] as the UNI protocol, and drafted the Broadband ISDN User Part (BISUP)[4] as the NNI signaling protocol. The ATM Forum has defined its own version of Q.93B[5] which also supports limited broadcast connections.

Several long-term functional objectives have been identified for signaling protocols. These functions will support Release 2 B-ISDN services, which includes multipoint multimedia (i.e. multi-connection) services. To handle these Release 2 UNI signaling functions, several approaches have been proposed. One widely considered approach is to extend the Release 1 Q.93B signaling protocol. This would also entail extending the corresponding SS7 NNI protocol, BISUP. For obvious reasons, such as backward compatibility and the reuse of large embedded bases, the industry would like to extend Q.93B to handle all the functions required of Release 2 B-ISDN signaling. On the other hand the technological transition to B-ISDN may be an opportunity to simultaneously transition to a new, modular

Network Management and Control, Vol.2, Edited by
I.T. Frisch *et al.*, Plenum Press, New York, 1994

signaling protocol and architecture. One such proposal is called the Distributed Call Processing Architecture and Protocols (DCPA+P)[6] [7]. This proposal is based on separating service control, call control, connection control, and channel control from a call processing and signaling protocol standpoint.

The purpose of this paper is to present two Q.93B-based approaches to BISDN UNI signaling, monolithic and separated, and to evaluate them in terms of performance and functionality. In addition, a comparison is made between the Q.93B-based approaches and DCPA+P.

The monolithic Q.93B is based on direct extensions to Release 1 Q.93B, providing call and connection control information elements in the same protocol messages. The separated Q.93B is based on dividing the current Q.93B into a call control protocol and a connection control protocol.

In addition to describing the *functionality* and *performance* characteristics of these approaches, the protocols are compared in terms of *the ease of introducing new services, common UNI/NNI signaling*, and *backward compatibility*.

Other proposals such an CMAP[8], GSP[9], and EXPANSE[10], also exist. ISCP[11] is a framework for defining B-ISDN signaling protocols based on the structured recommendation of OSI Application Layer Structure[12]. These protocols are not discussed here for the sake of brevity.

Section 2 describes functional objectives and performance considerations of B-ISDN signaling. Section 3 illustrates the two Q.93B-based protocols for establishing and modifying certain example calls. It also gives estimates of performance measures. Section 4 provides a short overview of DCPA+P. In Section 5, a comparison is made between DCPA+P and the Q.93B-based approaches based on additional criteria. Section 6 presents conclusions.

FUNCTIONAL OBJECTIVES AND PERFORMANCE CONSIDERATIONS

The primary concern with new signaling protocols being proposed for Release 2 B-ISDN is their ability to provide the *functionality* required to support advanced broadband services. Also of concern is the performance of the signaling protocols used.

Functional Objectives

The following subsection presents some of the functional objectives for Release 2 B-ISDN signaling and some possible approaches to provide this functionality. The CCITT B-ISDN service description [13] lists five classes of services.

Conversational services, such as multimedia conferencing, pose a number of functional requirements on signaling. Signaling should support establishment and release of *multiple connections* of differing quality within the framework of a call, and allow for bridged connections for *multicasting* information. It should also provide the ability to *modify* an active call, by either adding/dropping participants or by changing characteristics of the ATM connections within the call.

Messaging, data retrieval, and *distribution services* (both *with and without user control*) can be offered within an interexchange network, local exchange network, or customer premises network in entities called servers. An ATM connection is needed between any user and such a server in order to transfer information from the server to the user. Signaling for these services is handled by treating the servers as "users". Signaling must support the connections and associations between the users and servers.

A set of capabilities know as capability set 1 (CS1) has been defined[14] for Intelligent Networks. These support a set of services and associated features that include ISDN *supplementary services*, such as call forwarding, and services such as Freephone, Virtual Private Networks, etc. Signaling must support these capabilities.

To support multimedia, multipoint services, a single call may contain several connections, each with a different "quality of service". To support heterogeneous calls, not all parties on a call will have the same number of connections or same types of connections. In addition, connections within a call may require modification while a call is active. This has lead to the concept of separated call control and connection control. Call control is used to control associations between users and servers. It is responsible for negotiation between users, between users and servers, and between networks. Call control maintains state and configuration information pertaining to the call and is responsible for invoking and coordinating services in a call. The main attributes of a call include user IDs and server IDs.

Connection control is used to control the bearers which provide a path for exchange of user information between different parties in a call. Connection control is responsible for maintaining state information about each connection in the call and negotiating with the network for bearer attributes, such as end-to-end quality of service. The main attributes of a connection are user IDs and server IDs on which the connection is terminated (i.e. Adaptation Layer (AAL) termination), and transit switch IDs, quality of service attributes, AAL types, etc.

Negotiation may be carried out using *look-ahead* procedures within call establishment. Look-ahead allows end-to-end negotiation to take place before connections through the network are established. This "two-phased" type of connection establishment is referred to as *sequential call establishment.* The first phase, corresponding to call control, is dedicated to determining the status of the called parties and servers (e.g. busy) and performing end-to-end negotiation between users, servers, and networks. The second phase, corresponding to connection control, is dedicated to establishing bearers through the network. The first phase of the call establishment may gain importance with the proliferation of CPE with varying capabilities (e.g. car phones versus advanced work stations).

Alternatively, negotiation may be carried out using a single-phased call establishment procedure referred to as *simultaneous call establishment.* In this case bearers are allocated as signaling messages propagate through the network to the called party. If the called party modifies the call attributes affecting the allocated bearers, the bearers must be deallocated and then reallocated to match the final negotiated attributes.

Performance Considerations

Performance is another key aspect of a signaling protocol. Performance metrics such as call setup time must be considered when designing or adopting the signaling protocol for B-ISDN. While the *look-ahead* operation described above may assist in negotiation during complex multimedia calls, and allow a more efficient use of bandwidth within a network, it may increase call setup time. Therefore, for simple calls, and those that require fast setup time, traditional *simultaneous call establishment* should be provided in addition to sequential call establishment. In general, the performance of simple requests should not be penalized by the complexity in the protocol required to handle more complex requests.

In addition to sequential vs. simultaneous call establishment, other factors influencing performance include the number of messages that must traverse a network, the number of messages that must be processed at the various interfaces, and the processing complexity associated with each message, including protocol conversions.

In this paper, the performance measures of the signaling protocols techniques are characterized for the illustrated calls. The measures include the number of messages across the UNI, the number of end-to-end messages, and call setup delay.

Q.93B-BASED APPROACHES TO RELEASE 2 B-ISDN UNI SIGNALING

We examine two techniques for Q.93B-based broadband signaling. The *monolithic approach* uses a single protocol to perform both call and connection control operations. The *separated approach* uses one protocol to perform call control operations and a second protocol to perform connection control operations. The following sections compare the two Q.93B-based approaches.

Functionality

The following six subsections compare the two Q.93B-based techniques against the list of functions presented in Section 2. We illustrate various call establishment message flows. Call release and error handling are not addressed. In the discussion, we identify the origin of the information elements and messages as belonging to either Q.931 or Q.93B.

Support of Multipoint Calls. The message flow for establishing a simple point-to-point single media call using a monolithic Q.93B-based protocol is shown in Figure 1. The monolithic protocol examined in this paper is primarily based on using existing Q.93B message types with needed extensions for Release 2 B-ISDN. Any additional messages or information elements used in the analysis are presented in the discussion below.

The message flow for establishing a simple point-to-point single media call using a separated protocol is shown in Figure 2. The separated protocol examined in this paper is primarily based on using existing Q.93B messages that have been modified to be specific to either call control or connection control. For example, SETUP message is adapted to be a CALL_SETUP message used for call control, and a CONN_SETUP message used for connection control. The information elements used in the different messages are explained in the discussion below.

In the Release 1 Q.93B protocol, there are two information elements defined to specify a single called party. They are the *called party number* and the *called party subaddress*. In order to establish a multipoint call, several approaches may be taken. Several point-to-point calls may be established, and later conferenced together to form a single call. This is the approach taken in the first version of Q.93B as defined by the ATM Forum[5]. Alternatively, a single exchange of messages may be used to establish a multipoint call by extending the existing protocol messages. The first option is not desirable as a long term solution due to the additional overhead and time required to establish several individual calls and then forming the final call. Therefore, the second approach will be examined. Within the second option are further possible methods. Several addressing information elements may be carried in single messages, or group addressing may be used. We examine the options using multiple addressing IEs in single messages.

In a monolithic SETUP message which is responsible for initiating a call request, multiple *called party number* and *called party subaddress* IEs may be included in the message along with multiple Q.93B virtual channel identifier IEs. In a single-media call, one virtual channel ID would be assigned per called party. Depending on the service, in a multiparty call, multiple virtual channel IDs may be required. The network, upon receiving this message, must establish connections to the requested destinations and generate SETUP messages to each destination. The flow for establishing an N-way call using a monolithic protocol is shown in Figure 1. Likewise, both CALL_SETUP and CONN_SETUP messages in a separated protocol must be able to carry multiple destination address IEs.

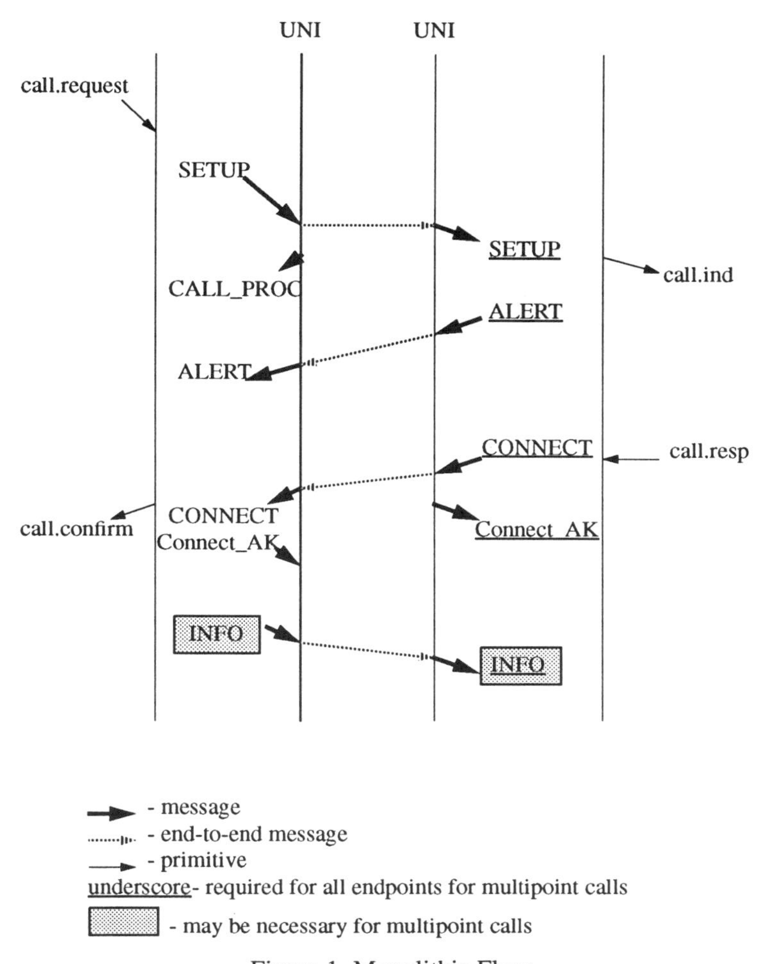

Figure 1. Monolithic Flow

The flow for establishing an N-way call using the separated protocol is shown in Figure 2. In general, any message that effects multiple parties in a call must be able to carry *multiple destination address IEs*. To support multiparty calls, the responses from called parties must be coordinated. For example, in Figure 1, each of the N-1 called parties generates a CONNECT message to accept the call. The network coordinates these responses and supplies the originating party with a single CONNECT message across the UNI. By designating the network instead of the calling party to coordinate responses from the called parties, the Release 2 Q.93B remains consistent with the current Q.93B. However, for

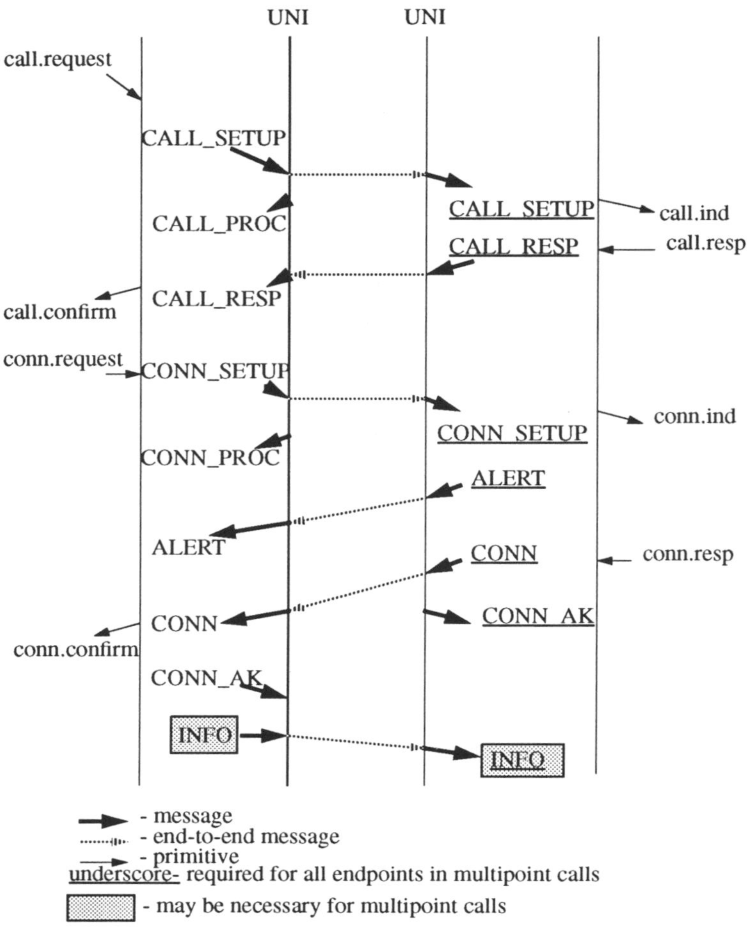

Figure 2. Separated Flow

certain services, such as voice telephony, it can be envisioned that a calling user may wish to coordinate multiple connect messages.

In addition, for many services, such as multimedia conferencing, each called party that accepts the call must be notified as to the final call configuration. As shown in Figures 1 and 2, an INFORMATION message is sent to each called party to carry this information. This message is needed so that the called parties know which other parties are participating in the call. The INFORMATION message defined in Q.93B must be extended to provide this functionality.

The extensions to the monolithic and separated Q.93B-based protocols for providing simple multipoint calls are similar. By specifying multiple destination addresses in the protocol messages, and extending INFORMATION messages to carry final call configuration information, both protocols may provide simple multipoint service.

In summary, in the monolithic protocol, the SETUP message must contain *multiple called party address and subaddress IEs and multiple Q.93B virtual channel identifier IEs.* The CALL_SETUP message of the separated protocol must carry *multiple called party address and subaddress IEs*, and the CONN_SETUP messages must carry *multiple Q.93B virtual channel identifier IEs.*

Support Multimedia Calls. The support of multimedia calls implies the need for multiple connections in a single call. It also implies the possibility of multiple protocols being used on an end-to-end basis. The current Q.93B protocol defines six information elements that must be examined: *broadband bearer capability (B-BC), user cell rate, low layer information (LLI), high layer information (HLI),* and *adaptation layer (AAL) parameters.* The B-BC, LLI, and HLI IEs are based on the Q.931 *bearer capability (BC), lower layer compatibility (LLC),* and *higher layer compatibility (HLC)* IEs respectively.

The B-BC IE is used to request bearer services from the network. The network may overwrite this IE if it cannot meet all of the requests. Currently, this IE specifies the traffic class, whether traffic is constant or variable bit rate, and end-to-end timing requirements. While additional quality of service parameters, such as delay variance, cell loss ratio, directionality, etc., must be defined to provide real-time B-ISDN services, the use of the B-BC IE will remain essentially the same in Release 2 B-ISDN.

The *low layer information* (LLI) IE is passed transparently through the network to the destination user. The LLI IE is used to supplement the B-BC IE by providing additional information required by the communicating endpoints. The LLI IE is checked by the endpoints in a call for compatibility while the B-BC IE is processed by the network. Again, although new parameters in the LLI IE may be necessary for Release 2 B-ISDN, its use will remain essentially the same.

The *high layer information* (HLI) IE is used to specify higher layer properties of the information transferred between users. The HLI IE is passed transparently through the network unless certain services are subscribed to, in which case the network may also examine the IE. The parameters for this IE must be expanded to support all of the services envisioned for B-ISDN, but its use will remain essentially the same in Release 2 B-ISDN.

The *adaptation layer parameters* IE is used to specify the adaptation layer (AAL) type, and other AAL-specific parameters. The *user cell rate* IE specifies the traffic characteristics or the ATM connection, for example peak and mean data transfer rates. Both of these IEs will have similar uses in Release 2 B-ISDN.

To support multimedia calls in which multiple connections will exist, both the monolithic and separated Q.93B-based protocols must allow for multiple B-BC, user cell rate, LLI, HLI, and AAL parameters IEs to be included in a single message. For example, in the monolithic protocol, a SETUP message requesting the establishment of a call using both voice and video must include a HLI IE denoting the teleservice (video-telephony), a LLI IE denoting the end-to-end coding scheme for the voice and video components, and the appropriate user cell rate, AAL parameters, and B-BC IEs for both the voice and video connections. These IEs would be present in all of the appropriate Q.93B messages.

In the separated protocol, the B-BC and user cell rate IEs would be present in the appropriate connection control messages, and the LLI, HLI and AAL parameters IEs would be present in the appropriate call control messages. In either case, by extending the defined IEs to include parameters germane to Release 2 B-ISDN such as directionality, permissions, call configurations, connection groupings requesting common path or diverse path routing, etc., and including multiple sets of IEs in a single message, either Q.93B-based technique can support multimedia calls.

Support Heterogeneous Calls. The B-ISDN signaling protocol must support heterogeneous calls in two ways. First, a user must be able to request a heterogeneous call. Using the same reasoning as presented in Section 3.1.1 for providing multiple addresses in a single message, we would like to provide this function in a single message. A second required functionality is to allow receiving parties of different capabilities, either due to CPE or network constraints, to accept different portions of a proposed homogeneous call resulting in a heterogeneous call. The latter capability will be addressed in the next section dealing with negotiation. This section will discuss requests for heterogeneous calls.

To request a heterogeneous call, the functionality discussed in the two preceding sections must be combined. In particular, it must be possible to associate LLI, HLI, and AAL parameters IE values with specific destination addresses, and B-BC and user cell rate IEs with specific connections. In this way, user transfer protocols and services may be requested for each party on the call. This must be done in the appropriate messages in the monolithic and separated Q.93B.

Support Active Call Modification. During a BISDN call, it is possible for a user to request a change in call or connection attributes. For a call, a user or service may be added or dropped. Likewise, connections within a call may be added or dropped. To support the request of new connection, new messages must be defined, perhaps ADD and DROP messages, that would be similar to existing CONNECT and RELEASE messages except that they would act on existing calls. These two new messages would be part of the monolithic Q.93B protocol. For the separated protocol, new messages are not required to add or drop connections. The CONN_SETUP and CONN_RELEASE messages may be used.

For a connection, a user may request a change in the current quality of service being received. To modify a connection, the current Q.93B allows the network or user to generate NOTIFY messages. The NOTIFY message carries a newly requested B-BC IE. To support full modification of existing connections, extensions must be made to support the modification of the LLI, HLI, AAL parameters and user cell rate IEs. One possibility to is extend the usage of the NOTIFY message. Likewise, the user may request that an endpoint be added or deleted to an existing connection. Again, new messages need to be defined to support this capability. Another possibility is to extend the NOTIFY message.

With modifications to the NOTIFY message, and possibly defining new messages, both the monolithic and separated protocols may be used to modify active calls and connections.

Negotiation. When establishing or modifying a call, many parameters may be negotiated. We focus on the bearer capability, low layer information, high layer information, user cell rate, and AAL parameters. These IEs convey both service and connection requests pertaining to the call. Because Q.931 is more developed than Q.93B with respect to negotiation features, we base our discussion below on Q.931.

Currently in Q.931, the bearer capabilities requested for a call may be negotiated with the network. Up to three BC IEs may be specified in a Q.931 SETUP message. The originator uses the *repeat indicator* IE to specify how the three IEs should be interpreted. Currently, only one value has been defined for the *repeat indicator* IE. This definition allows the user to inform the network that the repeated BC IEs should be used in priority order, i.e. if the first BC IE cannot be accommodated, the second BC IE should be tried next. In addition, if the network cannot accommodate any of the requested bearer capabilities, it may issue a NOTIFY message to the originating node to change the bearer capabilities.

The ability to negotiate in this fashion with the network should be provided in B-ISDN signaling. In the monolithic Q.93B-based protocol, this may be done as it is today, using multiple B-BC IEs with the repeat/indicator IE, and NOTIFY messages from the network. For the separated Q.93B-based protocol, bearer negotiation with the network may be performed in the same manner using connection control messages such as CONN_SETUP. A similar solution may be instituted to define negotiation procedures for the user cell rate.

For end-to-end negotiation, the LLI, HLI, and AAL parameters IEs must be negotiated. Currently in Q.931, LLC negotiation (which is analogous to LLI negotiation in B-ISDN) is allowed. The originator specifies the LLC IE in the SETUP message, including setting a bit to indicate negotiation is allowed. If the receiving party wishes, it may return an alternate LLC IE in the CONNECT message completing the call. In B-ISDN, it is desirable to also allow the originator of the call to propose alternative options to its preferred LLI. To do this, the repeat/indicator IE usage may be extended to be used with LLI as well as the B-BC IE. If this is done, then full negotiation will be allowed. Using the monolithic Q.93B-based protocol, the LLI negotiation may be performed using the SETUP and CONNECT messages. In the separated protocol, because LLI negotiation is performed end-to-end and is part of call control, the negotiation will take place using CALL_SETUP and CALL_RESP messages.

Currently in Q.931, HLC negotiation (which is analogous to HLI negotiation in B-ISDN) is not supported. While bits in the HLC are included to indicate how repeated *characteristic identifiers* within a single HLC IE should be interpreted, the only value defined supports a single request. When enhanced, these bits, know as *interpretation bits*, may specify relationships among multiple requests such as "use all", "prioritized list", etc. In addition, procedures for allowing the called party to propose alternate values for the HLI should be defined. If this is done, the monolithic Q.931-based protocol will perform the end-to-end negotiation using SETUP and CONNECT messages. The separated protocol will perform the end-to-end negotiation using CALL_SETUP and CALL_RESP messages. A similar solution may be instituted to facilitate negotiation of the AAL parameters.

With the above modifications to existing Q.931 IEs, namely expanding the use of the *repeat indicator* IE to support LLI negotiation, and fully defining the HLI *interpretation bits*, both the monolithic and separated Q.93B-based protocols may support end-to-end negotiation and negotiation with the network. Allowing negotiation in multiparty calls requires careful design of parameters such as timers in the procedures.

Support (Supplementary) Services. The Q.93B-based protocols may use the *facility IE* and FACILITY message of Q.932 to continue supporting existing supplementary services. For the monolithic protocol, the facility IE remains in the SETUP message to invoke services during call establishment. For the separated protocol, the facility IE is placed in the CALL_SETUP message. Based on the service, the facility IE can be placed in both the call and connection control messages. As new services are introduced, new values for the facility IE must be defined.

In summary, with the following modifications, a monolithic Q.93B-based protocol may support the B-ISDN functionality defined in Section 2:

1. allow multiple destination address IEs in single messages to support multipoint calls;

2. allow multiple B-BC, user cell rate, LLI, HLI, and AAL parameters IEs in a single message to support multiple connections in a single call;
3. extend the B-BC, LLI, and HLI IEs to include fields and values for fields germane to Release 2 B-ISDN;
4. associate sets of IEs with each destination address in a single message to allow for heterogeneous calls;
5. extend the use of the repeat/indicator IE, and provide procedures for full negotiation for LLI;
6. fully define the *interpretation bits* of the HLI IE and provide provisions for full HLI negotiation;
7. expand the use of the NOTIFY message to allow for active connection modifications;
8. define messages to add/drop users and services from active calls;
9. define new messages to allow the addition and deletion of connections to active calls.

In addition, consideration must be given to modifications of the procedures used to control connections and calls in light of the modifications to the messages and IEs in the protocol. For the separated protocol, items 1-8 must also be addressed. Item 9 is not required for the separated protocol because CONN_SETUP and CONN_RELEASE messages perform these functions. The changes to the separated protocol must be introduced into the appropriate call or connection control messages.

Performance

As stated in Section 2, the performance of the signaling protocol for call establishment is affected by the number of messages processed at each interface and the number of messages that must traverse the network. The following analysis concentrates on the number of messages passed across the UNI and message flows for each call establishment procedure. This may be used to approximate the call setup time for different call configurations based on the assumption that en bloc sending is used so that the entire called party number(s) is present in the SETUP or CALL_SETUP message.

Table 1 summarizes the number of messages that must be passed across the UNI and the number of end-to-end messages required to establish various call configurations for the Q.93B-based protocols. In the column *call*, the call configuration is given followed by the number of media included in the call. A "Pt-Pt" call refers to a two party call. A "Pt-Npt" call refers to a call in which a single party is broadcasting to N called parties for a total of N+1 participants in the call. An "N-way" call refers to a conference call between N total parties which may all communicate equally.

Entries in Table 1 for the number of messages generated across both the calling and called party UNI and the number of end-to-end messages for a point-to-point call using the monolithic protocol can be obtained from Figure 1. Similarly, entries in Table 1 for the number of messages generated across the UNI and the number of end-to-end messages for a point-to-point call using the separated protocol can be obtained from Figure 2. For this case it is assumed that all connection control information is transmitted in a single connection control message.

Table 1. Protocol Performance Summary

Call	Q.931-monolithic messages across UNI	Q.931-monolithic end-to-end messages	Q.931-separated messages across UNI	Q.931-separated end-to-end messages
Pt-Pt	9	3	14	5
Pt-Npt	5+4N	3N	8+6N	5N
N-way	5N+1	4(N-1)	7N+2	6(N-1)

Simultaneous or sequential call establishment can be used with either the monolithic or separated protocols. The option chosen does not impact the number of messages generated or the number of end-to-end messages for a point-to-point call, but it does affect the relative call setup time. The call setup time for both cases, assuming that called party answers the call immediately is:

$$\textit{call setup time (monolithic, simultaneous, point-to-point)} = R_1 \quad (1)$$
$$\textit{call setup time (separated, sequential, point-to-point)} = 2R_2 \quad (2)$$

R_1 and R_2 are equal to the processing time, queuing delay, and insertion time of the specific messages at each node, and the propagation delay across the network for the monolithic and separated messages respectively. $R_2 \leq R_1$ under identical assumptions because the the separated protocol messages generally contain fewer IEs than the monolithic protocol messages, thus simplifying processing. A multimedia call would use the same message flow; only additional IEs would be included in the messages.

Entries in Table 1 for the number of messages generated across both the calling and called party UNI and the number of end-to-end messages for a multipoint call using the monolithic and separated protocols can be obtained from Figures 1 and 2 respectively. Again, simultaneous or sequential call establishment can be used with either the monolithic or separated protocols affecting call setup time. The call setup time for both cases, assuming that called party answers the call immediately is:

$$\textit{call setup time (monolithic, simultaneous, N-way)} = 1.5R_3 \quad (3)$$
$$\textit{call setup time (separated, sequential, N-way)} = 2.5R_4 \quad (4)$$

R_3 and R_4 are equal to the processing time, queuing delay, and insertion time of the specific messages at each node, and the propagation delay across the network for the monolithic and separated protocol messages respectively. $R_4 \leq R_3$ for reasons similar to those for point-to-point calls presented above. A multimedia call would use the same message flow; only additional IEs would be included in the messages.

In Figures 1 and 2 the messages underlined must be sent to all destinations or are generated by all destinations. The final INFORMATION message is used to convey the final call configuration to the parties in the call. This is necessary so that each party is aware of which parties are involved in the call, and which capabilities (i.e. voice, video, etc.) each party is using. This final INFORMATION message is the cause for the additional 0.5 factor in the call setup times for the N-way calls. To establish a broadcast call to N destinations (Pt-NPt), the same flow as shown in Figures 1 and 2 is used except that the final INFORMATION message is not required. This is because the destinations do not communicate with each other and do not need knowledge of the final call configuration.

In general, for simple point-to-point calls, the monolithic protocol requires 1 round trip time to establish the call. The separated protocol may also achieve similar performance assuming simultaneous call establishment is used.

OVERVIEW OF THE DISTRIBUTED CALL PROCESSING ARCHITECTURE AND PROTOCOLS

The DCPA and its accompanying protocols are described in detail in [6] and [7]. DCPA separates the call control, connection control, channel control, and service control functionality of call processing. Each control function is instantiated as a separate application, perhaps executing on a physically separate processor. This separation was derived using a methodology known as Object Oriented Analysis (OOA)[15]. Call control and connection control are responsible for the functions outlined in Section 2. Channel control is primarily responsible for controlling the switch fabric and allocating switch resources, for example VCIs. Service control applications are written for each service that is being offered, for example 800-number services. This structure allows distinct portions of call processing to be developed and maintained separately, so that new services can be introduced in a modular fashion.

To support the DCPA, an interface, called an Application Service Element (ASE), has been defined for each portion of call processing. The ASE for each control function, i.e. call control, connection control, channel control, and service control, is a set of operations that may be invoked in the particular call processing application, along with the associated parameters. For example, a *setup connections* operation is defined in the ASE for connection control, along with parameters to specify the AAL type, connection endpoints, quality of service, etc. Therefore, if one portion of the distributed call processing application requires this service from the connection control application, it invokes the *setup connections* operation in the connection control process. By virtue of this separation, protocol modifications may be made in a modular fashion. Also, flexibility is provided to service developers that can use a combination of existing operations for the different control functions to provide a service.

The different call processing applications execute in different servers in the network. Users and other servers may access operations in these servers through the defined ASEs. The relationship between the entity invoking the operation and the servers providing the service is based on a client-server paradigm. Therefore, a transactional protocol, such as the Remote Operations Service Element (ROSE[16]) may be used to carry the operations. Within the current signaling protocols, Q.932 and the component sublayer of the Transactions Capability Application Part (TCAP) of SS7, use ROSE. In an experimental prototype of DCPA built at Bell Laboratories, TCAP is used to carry the operations. This demonstrates the possibility of re-using existing signaling software in the DCPA.

COMPARISON OF Q.93B-BASED APPROACHES AND DCPA

The section provides a brief comparison between the Q.93B-based approaches and DCPA. Table 2 summarizes a comparison of the protocols based on the additional criteria of *common UNI-NNI signaling, backward compatibility with existing signaling protocols*, and *ease of introducing new services*. For each criteria, positive (+) and negative (-) attributes for the protocols are listed along with an overall rating. Where not obvious, the following sections explain the goals and rating assigned to the various criteria.

Table 2. Protocol Summary

GOAL	MONOLITHIC Q.93B	SEPARATED Q.93B	DCPA+P
Common UNI/NNI Signaling	(–) perpetuates DSS1/SS7 split –	(–) perpetuates DSS1/SS7 split –	(+) ASEs used across UNI and NNI +
Backward operations Compatibility	(+) Built on Q.931 +	(+) Built on Q.931 +	(–)new (+)IEs defined as in Q.93B (+) may re-use TCAP and Q.932 =
Ease of introduction of new services	(+) facility IE for supp. services (–) service parameters embedded with call & connection control protocols –	(+) facility IE for supp. services (+) separation of call & connection control (–) service requests not separated into transactions =	(+)transactional (+)modular call processing software (+)modular protocols +

Common UNI-NNI Signaling

Commonality of the UNI and NNI protocols is desirable for the following reasons:

1. simplified protocol structure;
2. reduced complexity of equipment interworking across networks;
3. reduced development costs; and
4. single protocol access to all service providers regardless of their location in the network.

Q.93B, whether monolithic or separated, implies perpetuating the DSS1/SS7 split between UNI and NNI signaling. DCPA allows entities (users or servers) in any part of the network to invoke operations in servers. The interfaces are defined by control function, not network interface.

Backward Compatibility

Compatibility with existing signaling protocols, such as Q.931 and Q.932, is critical in providing ubiquitous service and controlling costs in implementing networks. Extensions to Q.93B clearly are advantageous in these terms from a protocol point of view. DCPA requires the development of the new ASEs. However, the IE coding of the DCPA ASEs is the same as in Q.93B, and therefore interworking is relatively straight-forward.

Easy Introduction of New Services

A signaling protocol should also allow the easy *introduction of new services.* The signaling protocol should provide flexibility and extensibility so that new services may be introduced without the need for extensive modifications to the signaling protocols and without constraint from other network providers.

Currently, in Q.93B, the *facility* IE is used to support supplementary services may still be used. To introduce a new service, further service specific parameters will have to be defined and implemented in the local exchange switches. This may potentially slow, or inhibit, service providers from offering new, non-standardized services. Also, due to the monolithic nature between a call and services in Q.93B-based protocols, enhancements made to the protocols to support new services may also require processing changes in switches and other affected network elements. For this reason, although it may be relatively easy to modify the protocol messages, it may not be easy to modify the protocol procedures or call processing software resident in the network nodes.

DCPA has many advantages of the Q.93B-based protocols in this area. The ASEs are transaction-oriented with operations defined on a relatively fine scale. Therefore, service developers can combine existing operations in novel ways to provide new services. Also, call processing software is arranged in a modular way. Therefore, new services may be introduced without affecting most of the call processing modules. Only the call control software must be modified. Finally, by separating protocols, any changes made to one protocol may be performed independently of the others. Therefore, fewer software changes must be made inside the network.

CONCLUSIONS

With appropriate extensions, either of the Q.93B-based protocols may provide the functionality to support B-ISDN services. Preliminary analysis also shows that the protocols should achieve similar performance in terms of call setup time based on the number of messages generated and the message flows. The major drawback of these approaches is that their monolithic nature may make it difficult to introduce new services rapidly. This characteristic, along with the complexity of new Release 2 BISDN services, will make call processing software extremely complex.

Two functions important for supporting BISDN services, negotiation between users and servers, and end-to-end quality of service calculations for connections, are particularly difficult to perform using the Q.93B-based protocols. Both of these functions are more efficiently performed in a centralized way on a per call basis. The Q.93B-based protocols perform these functions on a hop-by-hop, or distributed basis, in which each network node involved in the call or connection performs part of the function. Also, the existence of the UNI/NNI protocol split perpetuated by this architecture is no longer a valid model. As the transition to BISDN occurs, more users will be intelligent processors acting like nodes, or attached to local area networks. Therefore, a common set of protocols may be used between the user and network nodes and between network nodes. This is evidenced in the ATM Forum[5] which is extending the Q.93B protocol to be an interface between private networks as well as the UNI protocol.

DCPA provides an alternative solution. While new protocols are defined, they are easily made compatible with Release 1 BISDN protocols. Their major benefit is their modular structure, both of call processing software and protocols. This allows for easier software development and maintenance, and provides a platform that allows rapid service introduction. A single set of protocols is used across the UNI and NNI. DCPA also supports user-server negotiation and end-to-end quality of service calculation in a centralized fashion on a per call basis.

ABOUT THE AUTHORS

Thomas F. La Porta (tlp@research.att.com) is a Member of Technical Staff in the Wireless Networking Research Department at AT&T Bell Laboratories. Dr. La Porta received his B.S.E.E. and M.S.E.E. degrees from The Cooper Union, New York, NY, and his Ph.D. degree in electrical engineering from Columbia University, New York, NY. His research interests include signaling protocols for broadband and wireless networks, high-speed protocol design, and formal methods. He is a member of IEEE, a Technical Editor for *IEEE Personal Communications Magazine*, and is an adjunct member of faculty at Columbia University.

Malathi Veeraraghavan (mv@research.att.com) is a member of technical staff at AT&T Bell Laboratories. Her research interests include signaling and control of networks, fault-tolerant systems design, and reliability and performance modeling. Dr. Veeraraghavan received a BTech degree in Electrical Engineering from Indian Institute of Technology (Madras) and MS and PhD degrees in Electrical Engineering from Duke University, Durham. She is a member of the IEEE and an Associate Editor of the IEEE Transactions on Reliability.

REFERENCES

1. CCITT, "Recommendation Digital Subscriber Signaling System No. 1," *CCITT Blue Book*, Volume VI, Fascicles VI.10, VI.11, 1989.
2. CCITT Study Group XI, "Specifications of Signaling System No. 7," *CCITT Blue Book*, Geneva, 1989.
3. CCITT, "Recommendation Draft Q.93B," 1992.
4. CCITT SG XI, "Broadband Integrated Services Digital Network User Part (BISUP)," *BQ.761-764*, Sept.-Oct., 1992.
5. ATM Forum, "ATM User-Network Interface Specification, Version 2.4," Aug., 1993.
6. T. F. La Porta, M. Veeraraghavan, "A Functional Signaling Architecture for Broadband Networks," *Proc. of IEEE Globecom '93*, 1993.
7. M. Veeraraghavan, T. F. La Porta, "Call, Connection, and Service Control Protocols for Broadband Networks," *Proc. of ICCCN '93*, San Diego, CA., 1993.
8. J. DeHart, M. Gaddis, R. Bubenik, "Connection Management Access Protocol (CMAP) Specification," Department of Computer Science, Washington University, MO.
9. P. A. Miller, P. N. Turcu, "Generic Signaling Protocol: Architecture, Model, and Services," *IEEE Transaction on Communications*, May, 1992.
10. S. Minzer, "A Signaling Protocol for Complex Multimedia Services," *IEEE Journal On Selected Areas in Communications*, December, 1991.
11. A. Pagliulunga, A. Biocca, M. Siviero, "Signaling Protocol For B-ISDN," *Proc. ICC '92*, Chicago, IL, 1992.
12. ISO Recommendation DIS 9545 - ALS, August, 1989.
13. CCITT Recommendation I.211, 1991.
14. CCITT Recommendation Q.1200 Series, "Intelligent Network Recommendations," Working Document, March, 1992.
15. S. Shlaer, S. J. Mellor, "Object-Oriented Systems Analysis Modeling the World in Data," Yourdon Press, 1988.
16. CCITT, "Recommendation X.219," Blue Book, vol. VIII.5, Geneva, 1989.

CONNECTION CONTROL PROTOCOLS IN BROADBAND NETWORKS[1]

Minfa Fred Huang[a], Ivan T. Frisch[b], and C. Edward Chow[c]

[a]OpenCon Systems, Inc.
[b]Polytechnic University
[c]University of Colorado at Colorado

ABSTRACT

Driven by fiber/switching technologies and user service demands, the public telecommunications network has been evolving towards the universal Broadband ISDN (BISDN). The Asynchronous Transfer Mode (ATM) is an attractive communication transport technique for carrying future broadband services with a broad spectrum of different traffic characteristics. In our previous work, we presented two approaches for the multiparty connection establishment procedure, which are responsible for setting up end-to-end physical connections for data transfer, in BISDN signaling systems. In this paper we focus on the performance analysis of these connection establishment procedures under various traffic loads. Under the assumption of call tree structure and Poison call arriving rate, we use a *M/G/1* model to characterize the call processing within each node. The calculated results show the trade-offs between response time and number of outgoing control (or request) messages using a threading approach and a concurrent approach under various traffic loads.

INTRODUCTION

Broadband Integrated Services Digital Networks (BISDN) is an emerging international standard for the next generation telecommunications networks [CCIT 90], [Kawa 91]. The BISDN network, which uses Asynchronous Transfer Mode (ATM) technology [Thom 84], could provide higher speed and bandwidth on-demand solutions [Pryc 92] to satisfy future advanced applications and customer's needs. To support these applications, the high speed switch fabrics with high bandwidth capacity and multicast capability [Giac 90] [Lee 88] [Turn 86] have been designed.

Existing protocols based on a monolithic call can not readily provide such services. We need a protocol which is a service independent, and based on a paradigm that comprises a new object-oriented call model, a new protocol syntax, and a new message flow model [Minz 91]. We assume that future broadband networks have the intelligence and capability to support bridges [Adde 88] and gateways in a multimedia multiparty connection environment [Buss 89]. This requirement differentiates our work from the approach in [Bube 91] and [Gadd 92].

In our previous paper [Huan 92.1] we focus on the physical end-to-end connection management and propose two connection establishment procedures. These two procedures assume that the connection can be obtained from a centralized connection finding algorithm which is capable of

[1]This work was supported in part by the New York State Science and Technology Foundation and its Center for Advanced Technology in Telecommunications at Polytechnic University and by the National Science Foundation under Grant No. NCR-9003006 and by the Bell Communications Research.

finding point-to-point, multipoint-to-point (concast), point-to-multipoint (multicast), and multipoint-to-multipoint (e.g. video conference) connections [Chow 91]. The first procedure is called the threading approach (TA) where the resource reservation is done in serial by the involved nodes along the connections in a call. The assumptions are that every node is responsible for allocating all the resources within its domain. The second procedure is called the concurrent approach (CA) where the resource reservation is done in parallel.

This paper explores the comparisons between these two connection establishment procedures. A call structure is represented, in terms of the total number of nodes in a logical connection and the call splitting parameters. We use a *M/G/1* model, under certain assumptions, to find the processing time for the messages from a **user network interface** (UNI) and the messages from a **network node to a network node interface** (NNI). We derive the average number of outgoing control messages in terms of call structure parameters under various message blocking probabilities. Numerical results are provided.

TWO CONNECTION ESTABLISHMENT PROCEDURES

In this section we discuss the connection control protocols for managing a connection over an ATM-based transport network. We show how a complicated connection can be established, how an incremental change can be flexibly handled and how the round trip delay can be eliminated. Due to space limitation, discussion of this connection control protocol will not include the control messages formats, how these messages can be encoded, how these messages can be forwarded from an origination node to destination node(s) and the internal processor in each node to handle these messages. We refer the reader to the [Huan92.2] for a more thorough presentation.

Threading Approach

In this subsection we discuss a connection management procedure which sends request messages to downstream nodes along the connection and serially allocates the specified resources, node after node. We assume that there is a centralized connection finding algorithm that generates low cost connections and that each node in the connection is responsible for allocating the specified resources according to the given service requirements in terms of Basic Message Structure (BMS) and the internal resource allocation protocol.

Given a connection P (as in Figure 1), the threading connection management procedure proceeds as follows:

For the origination node

On receiving request messages, allocate the resources, then send the request messages with the partial connection information to all downstream nodes.

On receiving ACKs (positive acknowledgements) from all the downstream nodes, report to the higher layer that the connection has been established

On receiving a NAK (negative acknowledgement) from a downstream node, send CANCEL messages to the rest of the downstream nodes to ask for the cancellation of the previous request. Then wait for the CANCEL_CONFIRM message. On receiving all the CANCEL_CONFIRM messages, report to the higher layer why the connection is not established.

For the intermediate node i

On receiving a request message which contains the partial connection, P, for each downstream node in P, abstract the BMS of node i, reserve the resources that satisfies the requirement of the connection, including the quality and the bandwidth. If the resources are available, node i sends a request message which includes the partial connection of P with respect to the downstream node. If any of the resources are not available, node i sends a NAK message back to the upstream node.

On receiving a CANCEL message, node i releases previous reserved resource with respect to the call and sends a CANCEL_CONFIRM message back to the origination node.

For the destination node(s)

On receiving a request message, reserve resources for the destination user or users. Send ACK back to the upstream node.

On receiving a CANCEL message, release previous reserved resource with respect to the call and send a CANCEL_CONFIRM message back to the upstream node.

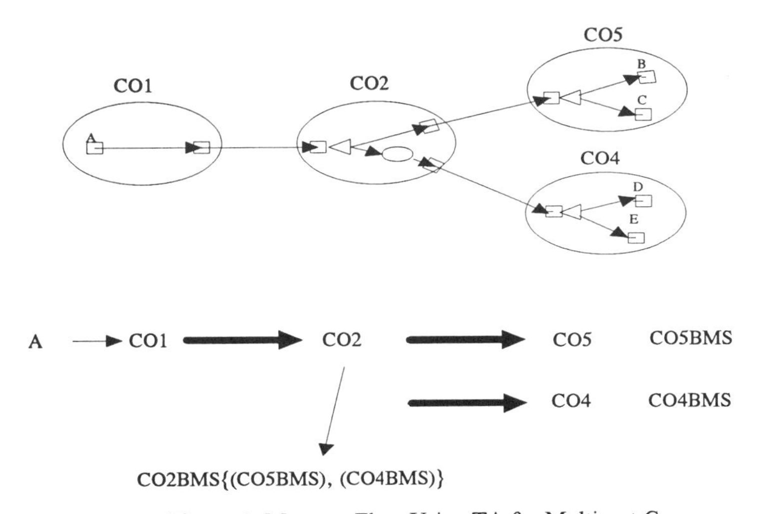

Figure 1. Message Flow Using TA for Multicast Case

Consider an example. User A wants to set up a multicast connection to users B, C, D and E in different locations. At node CO1, the connection generated by the connection finding algorithm. As shown in Figure 1, in Step 1, CO1 sends a request message with the partial connection representation CO2BMS {(CO5BMS), (CO4BMS)} to CO2. In Step 2, CO2 allocates a multicast edge and a signal code converter (from HDTV to NTSC) according to the local logical connection information derived from the CO2BMS and then sends a request message with the partial connection representation CO5BMS to CO5, and it also sends a request message with partial connection representation CO4BMS to the CO4. Even the bandwidth requirements are different, one being NTSC and the other being HDTV. When the CO2 receives ACKs from CO4 and CO5, it sends an ACK to the upstream node CO1 and the connection is established.

Concurrent Approach

In this subsection we discuss a connection management procedure that speeds up the connection procedure by sending request messages to all involved nodes and letting them reserve the required resources and establish the connection concurrently. As in the threading approach, we assume that the connection is given by running a centralized connection finding algorithm. Furthermore, it is based on the additional assumptions that out-of-band signaling channels exist among involved nodes in the connection and that each of the nodes is capable of setting up connections between any pair of its input ports and its output ports.

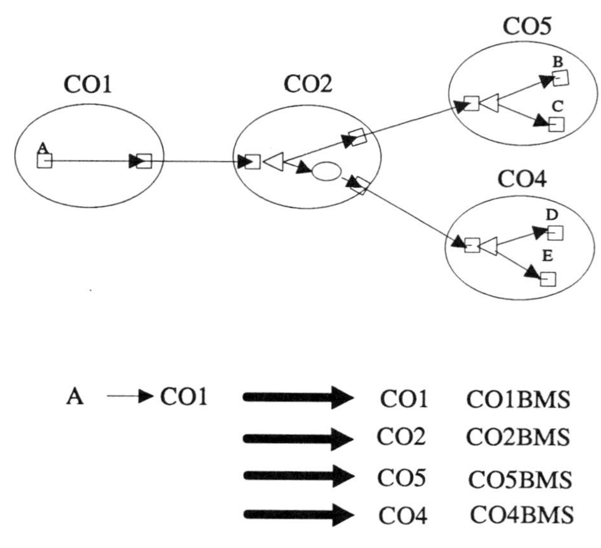

Figure 2. Message Flow Using CA for Multicast Case

Given a connection P, the concurrent connection management procedure proceeds as follows:

For the origination node

Send the individual basic message structure to all the involved nodes and wait for the acknowledgments.

On receiving ACKs from all the involved nodes, report to the higher layer that the connection has been established

On receiving a NAK from an involved node, send CANCEL message to the rest of the involved nodes to ask for the cancellation of the previous request. Then wait for the CANCEL_CONFIRM message. On receiving all the CANCEL_CONFIRM messages, report to the higher layer why the connection is not established.

For all the involved nodes

On receiving a request message which contains the BMS with respect to that node, reserve the specified resources that satisfy the service requirements of the connection, including the quality and the bandwidth. If the resource is available, then send an ACK message to the origination node. If the resource is not available, then send a NAK message back to the origination node.

On receiving a CANCEL message, release previous reserved resource with respect to the call and send a CANCEL_CONFIRM message back to the origination node.

Let us use the establishment of connection P1 in Figure 1 as an example of the concurrent approach. In Step 1, CO1 sends three messages to all the involved nodes (See Figure 2). It sends a request message with CO2BMS where the local logical connection specifies the bandwidth requirement to be 150 Mb/s for HDTV to the CO2. It sends a request message with CO5BMS where the LC specifies the multicast edge, the code converter and the connectivity of these and the bandwidth requirement on each link. It sends a request message with CO4BMS where the logical connection specifies the bandwidth requirement to be 44 Mb/s for HDTV to the CO4. Finally it sends one message with CO1BMS to its internal resource allocation module. In Step 2, on receiving the request message, each node sets up the connection among the incoming ports and the outgoing port and sends an ACK message back to the origination node. In Step 3, on receiving all the ACKs back from all the involved COs, the origination node can report that the connection has been established.

COMPARISONS BETWEEN THREADING AND CONCURRENT APPROACHES

To analyze the response time for call set-up is complicated since many factors are involved including call structure, network status, propagation delay and processing time. To simplify the problem we assume the central office call processing times are the same for all the control message either from UNI or NNI in the TA or CA. Let T_{proc} be the average total processing time for any control message spent in each node and T_{prop} be the maximum propagation delay between any two nodes for worst case consideration. We define T_{TA} as the time required to set up a call in TA and T_{CA} as the time required to set up a call in CA. Assuming all the acknowledgements are negligible, the response time is defined as the one way delay from the origination node to the destination node in our discussion. In a node-list call structure with n nodes involved, the time required to set up the call by using the threading approach and the concurrent approach is:

$$T_{TA} = n * T_{proc} + (n-1) * T_{prop} \quad (1)$$

$$T_{CA} = 2 * T_{proc} + T_{prop} \quad (2)$$

We assume the message transmission time is included in the T_{proc}. We need n processing time delays in threading approach since the control messages are sent in serial. In any intermediate node, the control message will be forwarded to downstream node(s) only if the specified service requirements are met in this particular node. Otherwise, it sends a NAK back to it upstream node. Thus $n * T_{proc}$ and $(n-1) * T_{prop}$ are needed to set up a call from the origination node to destination node. In the concurrent approach, all the control messages are sent from the origination node to all the involved nodes. Only two processing time delays, one in the source node and one in others, will contribute to the response time.

The total number of outgoing control messages from origination node to the destination node will be n-1 for the threading approach and the concurrent approach if the call is set up successfully. This is independent of how we forward these messages. Now we want to be more specific in finding out the processing time for different messages in the system. Notice that, the origination CO, always has to run the path finding algorithm to get the logical connection information for the new arriving call from UNI. This is very time consuming especially for a complicated connection call. We observe that, the total number of outgoing messages are different for TA and CA if the request fails in some node. These will impact the processing time for different kinds of control messages, either from UNI or NNI. To calculate the processing time within each CO for different control messages, we assume a **homogeneous system**; that is all nodes in the network are the same and each node has the same resources and the same call arrival rate with the same call structure. Under these assumptions, we derive a formula for the average number of outgoing control messages and the average time required to set up a call for both TA and CA.

Model for Call Processor

We use a model for the message processing procedure within the CO to show how different messages are invoked upon initiation of a call. As in the modeling of circuit switching, we model the call processing system within the CO as in Figure 3,based on a software architecture proposed in our previous work [Huan 92.2]. In this model, the messages arriving at each CO call processing module can be from end users through a UNI, or from another CO module through an NNI. After message application processing, each message is directed to a processor such as a UNI, NNI, router, switch controller and others. The UNI processor provides the interface between the end users and this central office. The NNI processor sends the control messages to other CO call processing modules in the network. The router will execute the path finding algorithm either in a centralized fashion or a distributed fashion. The switch controller has the interface to the switch fabric controllers within the central office. COs will carry out different tasks based upon the arriving control message. For example, only the origination CO is required to carry out the routing task which actually performs the path finding algorithm. We assume that once the processing of a message begins, it is carried to completion. Let λ_U be the aggregate call arrival rate from UNI to the input buffer 1 and λ_N be the aggregate control message arrival rate from NNI to the input buffer 2. It is assumed that all the queues in the following discussion are First Come First Serve (FCFS) service discipline. Each call arrival (from a UNI) at the origination node generates a sequence of control messages, contributing to the control message arrival λ_N at other nodes, according to it's call structure (the connectivity between origination node and the destination node) and the way we forward the control messages.

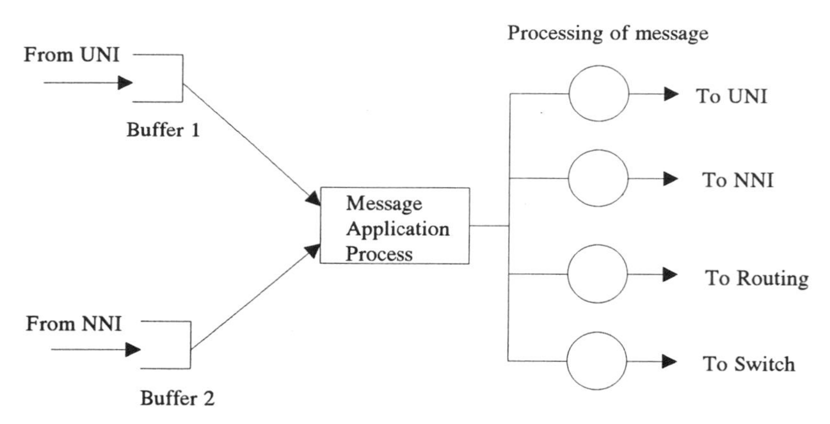

Figure 3. Simplified Model of Call Processing

Call Tree Structure

For simplicity to calculate the average time required to establish a call for the TA and the CA. We consider only the simple **call tree structure** in terms of number of nodes, links in a tree topology. Let n be the number of nodes involved in the logical connection for each call and f be the fan-out, or call-splitting, which is the number of out going links to the downstream nodes in the call tree structure. Let *Req(f,n)* be the call tree structure which consists of n nodes and f fan-out at each node of the logical connection except the leaf nodes. Let d be the depth or level of call structure in this particular tree structure of a logical connection. This is determined when *Req(f,n)* is given.

An example of *Req(f,n)* structure and the relation between n, f and d is depicted in Figure 4 for f=1, f=2 and f=n-1. For f=1, this is the node list case and d=n. For f=2, this is the binary tree structure and each node has two outgoing links to the downstream node except the leaf nodes. For any given *Req(f,n)*, we can determine d for the unique tree structure. In general, $d=i+1$ where $n_{i-1} < n \le n_i$ and $n_i = \sum_{k=0}^{i} f^k$. For $f=n-1$, this is the case of multicast from one source to n-1 destinations. In this case, d is equal to 2.

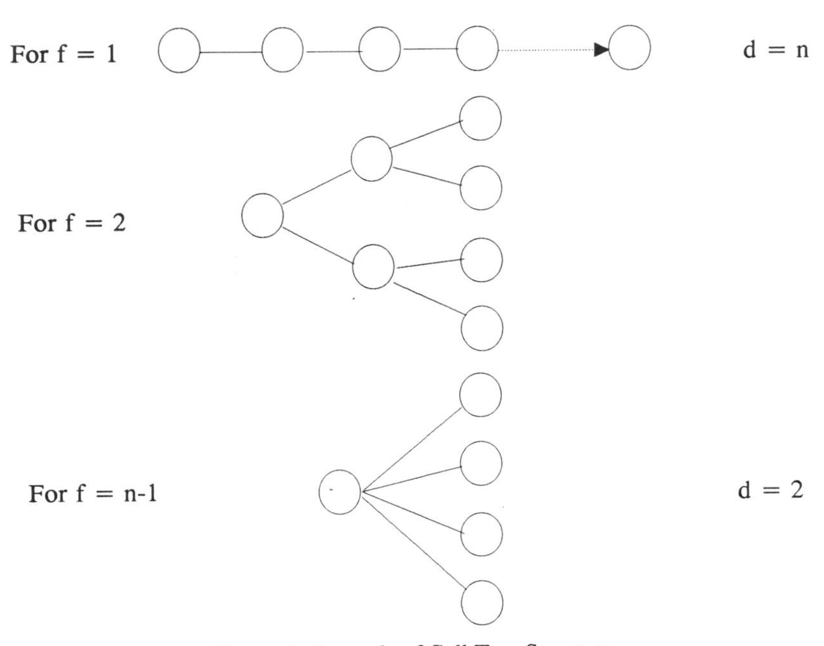

Figure 4. Example of Call Tree Structure

For any given call tree structure with *Req(f,n)*, our objective is to calculate the time required to set up that call, T_{TA} and T_{CA}, and the average number of outgoing control messages (for TA and CA) under various message blocking probabilities.

m1 : average service time for task in buffer 1

m2 : average service time for task in buffer 2

E(m^2) : second moment of composite service time

E(W) : average waiting time of the composite task

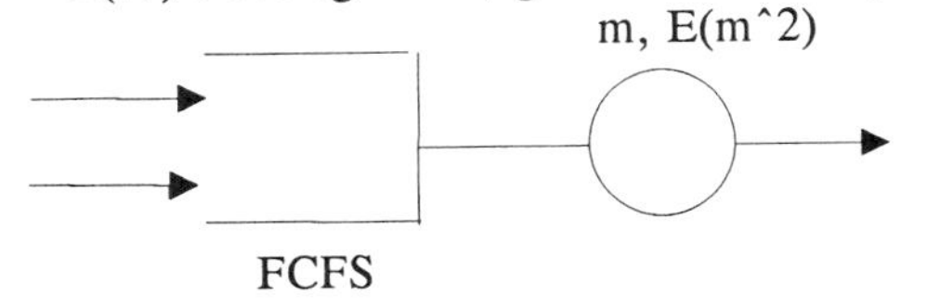

Figure 5. Equivalent Model of Call Processing

Response Time

We assume the aggregate calls arrive from UNI at a Poisson rate of λ_U calls/sec with a *Req(f,n)* tree structure and the aggregate messages arrive from NNI at a Poisson rate of λ_N. To calculate the T_{TA} and T_{CA} in the system, we further assume λ_U and λ_N traffic share the same message queue, an infinite buffer with first come first serve service discipline. Consider the no priority case; the equivalent model of call processing is depicted in Figure 5.

Let m_1 be the average service time for a request message in buffer 1 (from UNI) and m_2 be the average service time for a request in buffer 2 (from NNI). Since the composite service time process is a general distribution, we use a *M/G/1* model to characterize the system [Klei 75], [Schw 87]. By definition, the average service time m and the second moment $E(m^2)$ of composite request messages are

$$m = \frac{\lambda_U * m_1 + \lambda_N * m_2}{\lambda_U + \lambda_N} \tag{3}$$

$$E(m^2) = \frac{\lambda_U * E(m_1^2) + \lambda_N * E(m_2^2)}{\lambda_U + \lambda_N} \tag{4}$$

From the Pollaczek-Khinchine formula, the average waiting time (spent in the queue) of the composite request messages is

$$E(W) = \frac{(\lambda_U + \lambda_N) * E(m^2)}{2(1-\rho)} \tag{5}$$

where $\rho = (\lambda_U + \lambda_N)m$ is the traffic intensity. The total processing time for the message from UNI and NNI can now be calculated as $E(W) + m_1$ and $E(W) + m_2$ respectively. Once the processing time for origination node and intermediate and destination node is calculated, the time required to set up a call for TA and CA are:

$$T_{TA} = E(W) + m_1 + (d-1) * (E(W) + m_2) \tag{6}$$

$$T_{CA} = E(W) + m_1 + E(W) + m_2 \tag{7}$$

where the transmission time and the propagation delay T_{prop} are assumed to be negligible. To examine the effect of n and f in the T_{TA} and T_{CA} under various traffic loads, let us consider the case for $m_1 = 5$, $m_2 = 2$, $E(m_1^2)$=28, $E(m_2^2)$=5 and $\lambda_N = 2 * \lambda_U$. Plotting T_{TA} and T_{CA} versus traffic intensity ρ for n=7 and f=1, we get the result in Figure 6. Figure 7 shows the effect of call splitting f on the difference in response time, T_{TA} - T_{CA}, for n=12. In Figure 8, we demonstrate the variation in the difference of the response time (T_{TA} - T_{CA}) versus the traffic intensity when f=1 with n as a parameter.

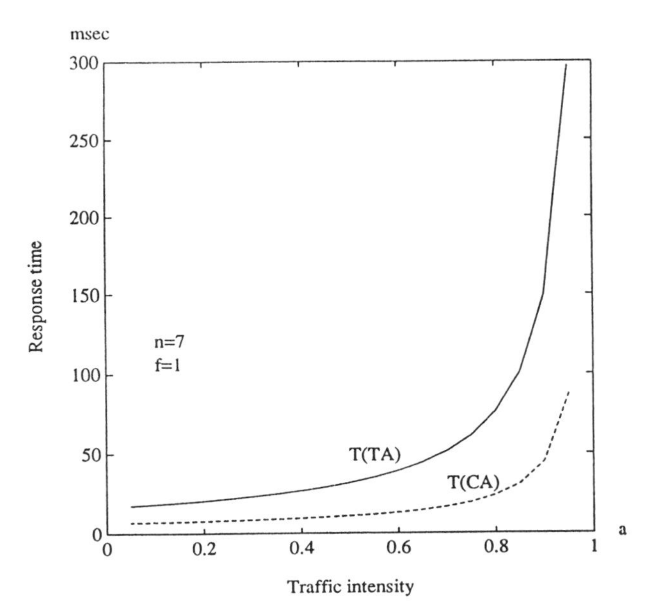

Figure 6. Response Time of Using TA and CA when n = 7 and f = 1

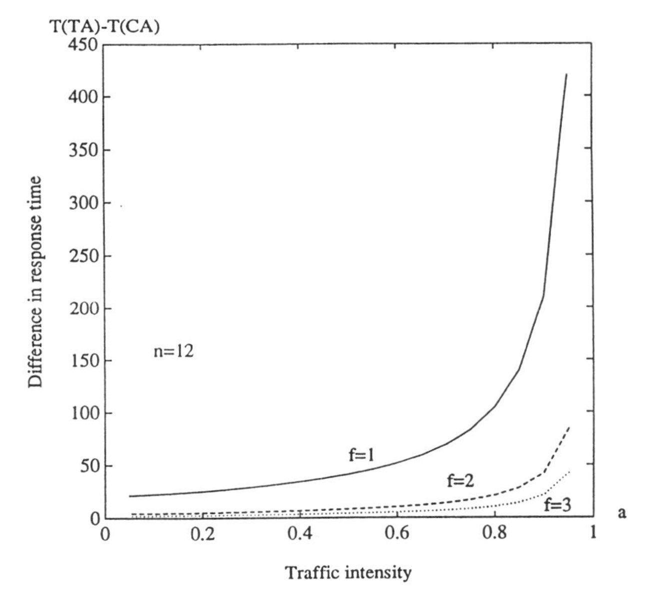

Figure 7. Effect of Call-Splitting f on T_{TA} - T_{CA}

Average Number of Outgoing Control Messages

Let p' be the probability that a control message can not be forwarded to downstream nodes due to unavailability of resources at the time the control message has been processed. We assume in a homogeneous system that every node has the same p' for any incoming control message from UNI and NNI. Let $E(N_c)$ is the average number of control messages, produced by each call, flowing into NNI. Then

$$E(N_c) = \sum_{n_c=1}^{n} n_c * P_{n_c} - 1 \tag{8}$$

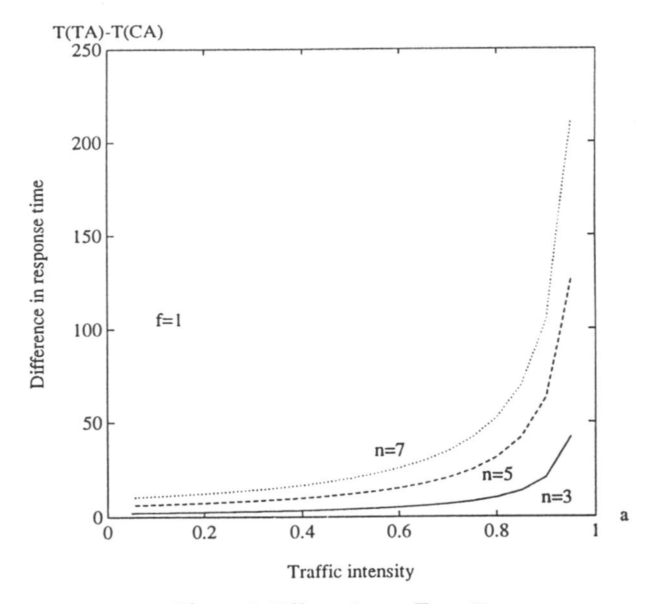

Figure 8. Effect of n on T_{TA} - T_{CA}

where n_c is the number of nodes been visited by control messages (including the origination node) in *Req(f,n)*. P_{n_c} is the probability of n_c nodes having been visited. $\sum_{n_c=1}^{n} n_c * p_{n_c}$ is the average number of nodes that have been visited (including origination node). The subtraction of 1 excludes the origination node according to the definition of $E(N_c)$.

Let us discuss two special cases for p'=0 and p'=1. When p'=0, all the control messages will eventually be forwarded to the destinations without blocking in any node. In this case, $E(N_c)$ is equal to n-1 for *Req(f,n)* in both TA and CA. Let us investigate another special case for p'=1. When p'=1, no control messages will be forwarded to any downstream node. As a result, no control message will flow into NNI and $E(N_c)$=0 in both TA and CA. In both special case, p'=0 and p'=1, $E(N_c)$ is independent of fan-out f and is independent of how we forward the control messages in TA and CA.

To discuss the general case for 0<p'<1 in TA and CA, we utilize the formula in Equation 8. For CA, we have

$$E(N_c) = 1 * p' + n * (1 - p') - 1 \tag{9}$$

The first term on the left, 1*p', corresponds to the case in which the call arrival fails at the origination node. Since it failed, the total number of nodes visited is only one (origination node) with the probability p'. The second term, n*(1-p'), is due to the success of the call at origination node at the time the arriving call message is processed. Since the origination node will generate n-1 messages simultaneously to the all other nodes in the LC according the description of CA, the total nodes visited is n (including the origination node) with the probability 1-p'.

Let us examine the general case 0<p'<1 for TA. Using Equation 8, similar procedures are carried out.

For f=1,

$$\begin{aligned} E(N_c) &= 1 * p' + 2 * p'(1-p') + \cdots + (n-1)(1-p')^{n-2} * p' + n(1-p')^{n-1} - 1 \\ &= \frac{(1-p') - (1-p')^n}{p'} \end{aligned} \tag{10}$$

For f=2,

$$E(N_c) = 1 * p' + 3 * p'^2 (1-p') + 5 * p'^3 (1-p')^2 * 2 + \cdots - 1 \quad \text{for } n \geq 7 \tag{11}$$

For f=n-1,

$$E(N_c) = 1 * p' + n * (1-p') - 1 \tag{12}$$

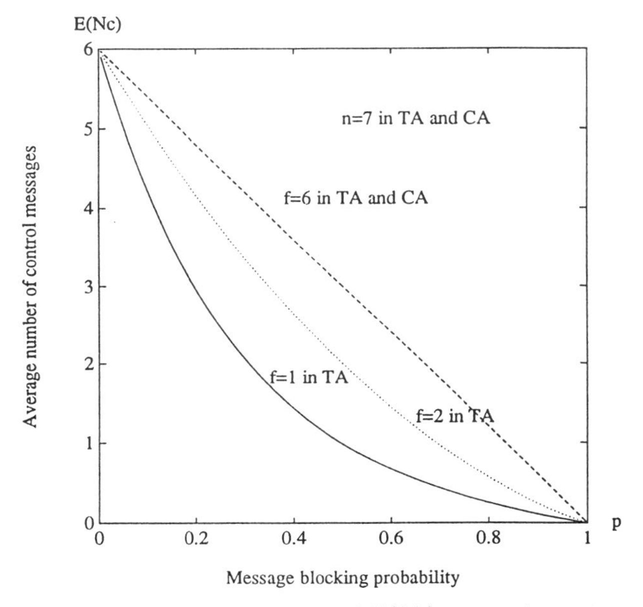

Figure 9. An Example of $E(N_c)$ versus p'

Notice that, for f=1, each time a message succeeds in any node, the number of nodes n_C visited is increased by 1. Similarly for f=2, the number of nodes visited is increased by 2. Only possible numbers of n_C (for example 1,3,5...) are allowed for f=2 (see Equation 11). In the case of f=n-1, the function of $E(N_c)$ in TA (see Equation 12) is the same as in CA (see Equation 9). Plotting $E(N_c)$ versus p' for f=1, f=2 and f=n-1 and n=7, we get the result in Figure 9. Notice that, for p'=0 and p'=1, $E(N_c)$ is equal to n-1 and 1 respectively. These are independent of whether we forward the messages as TA or CA and independent of fan-out f in TA. For the node list case (f=1) in TA, the function of $E(N_c)$ will serve as the lower bound for all kinds of call structures. Similarly for f=n-1, the function of $E(N_c)$ serves as the upper bound. For all other structures besides call tree structure *Req(f,n)*, the $E(N_c)$ will fall within the region between f=1 and f=n-1 as in Figure 9. For the case of

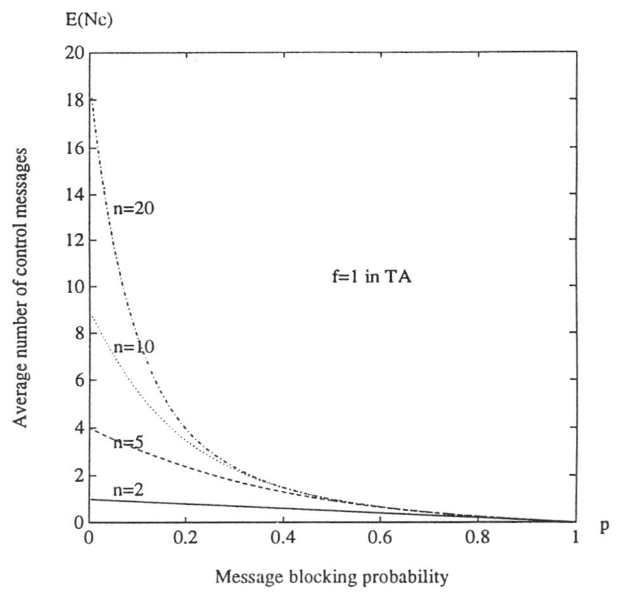

Figure 10. The Effect of n on the $E(N_c)$ in TA

f=n-1, TA and CA have the same $E(N_c)$ function. In practical situations, f is slightly greater than one and p' should be in the range of 10^{-3} to 10^{-1}. From Figure 9, we observe that while p' increase from 0 to 1/2, $E(N_c)$ drops dramatically when f=1. This is consistent with our previous observation that when the network is heavily loaded, TA will be better since fewer control messages are generated to aggravate congestion.

Figure 10 shows the function of $E(N_c)$ versus p' for n=2, n=5, n=10 and n=20 and f=1 (nodelist case) using TA according to Equation 10 Figure 11 describes the similar situation using CA according to Equation 9. These two figures show the impact on the $E(N_c)$ of a variation of n.

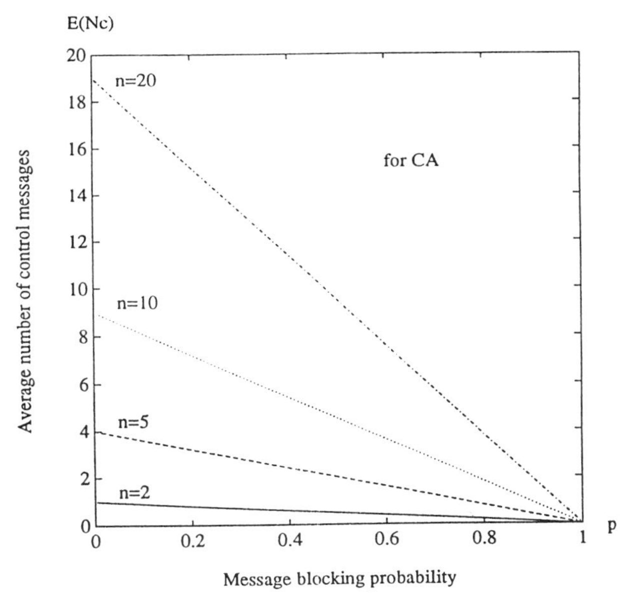

Figure 11. The Effect of n on the $E(N_c)$ in CA

CONCLUSION

We have discussed the procedures to compare the difference between TA and CA, in terms of the number of outgoing control messages and the time required to set up a call based on the assumption of Poisson call arrival rates with the call structure *Req(f,n)* in a homogeneous system. For any given *Req(f,n)*, λ_U, and λ_N, we can find T_{TA} and T_{CA} under a variation of traffic load conditions. We derived the average number of outgoing control messages for various message blocking probability. In general, the T_{TA} is always higher than the T_{CA}. But the threading approach is close to the ideal case in terms of average number of outgoing control messages. These are parts of the trade-offs in designing the connection control protocols for broadband networks.

ACKNOWLEDGMENTS

The authors wish to express grateful appreciation to Steve Minzer, F. D. Porter, Stanly Moyer and Rong Chang for their useful discussions and comments at Bell Communications Research. This work will not be as complete without the support and encouragement from Howard Bussey.

ABOUT THE AUTHORS

Minfa Fred Huang (mhuang@panther.opencon.com) is a Member of Technical Staff and project leader of the Software Download project in OpenCon Systems, Inc. He received his MS and Ph.D. in EE from Polytechnic University in Brooklyn. From 1989-93 he was a Teaching Fellow and then Research Fellow at Polytechnic University. During 1989-90 he worked for NYNEX (Department of Applied Research) and BELLCORE (Division of Broadband Systems Integration) as a summer intern. From 1990-91 he was a Resident Visitor in BELLCORE which involved in the design, analysis and implementations of routing protocols and resource allocation algorithm that support the signaling systems on the ATM-based broadband networks using object oriented techniques. He joined Research and Development Department of OpenCon Systems, Inc. in February of 1993 as a Member of Technical Staff. His research interests include network management and control, routing, resource allocation and flow control on the SONET and ATM based broadband networks. His current area of interest is in developing the software management applications for the SONET equipment over OSI seven-layer protocol stack.

Ivan Frisch has a BS in physics from Queens College, and a BS, MS and Ph.D. in EE from Columbia University. From 1962-69 he was an Assistant and then Associate Professor at the University of California at Berkeley. From 1969-1980 he was Sr. Vice President and a member of the Board of Directors at the Network Analysis Corporation (NAC), which was purchased by Contel in 1980. From 1980-86 he was General Manager of New York Operations and President of the Information Products Division at Contel. From 1987 to 1992 he was Technical Director, and then Director of the Center for Advanced Technology in Telecommunications at Polytechnic University in Brooklyn. He is currently Professor of EECS and Provost of the University.

He was a Founder of the Network Analysis Corporation, a developer of ARPANET, the world's first packet switching network, and NASDAQ, one of the first nationwide computerized financial transaction processing systems, and the world's first CATV design program. He was the founding Editor in Chief of Networks, published by Wiley, and he is co-author of "Communication, Transmission and Transportation Networks" Addison Welsley, 1971, which won an award from the Operations Research Society of America. He is a Fellow of the IEEE.

C. Edward Chow (Chow@quandary.uccs.edu) is an associate professor in the Department of Computer Science at the University of Colorado at Colorado Springs. He received his B.S.E.E. from National Taiwan University in 1977. He received his M.A. and doctorate from the University of Texas at Austin in 1982 and 1985, respectively. He is a member of IEEE and ACM. His research interests include network management, protocol engineering, distributed computing, and multimedia computing and communications.

REFERENCES

[Adde 88] E. J. Addeo, A. D. Gelman, A. B. Dayao, ``Personal Multi-media Multi-point Communication Services For Broadband Networks," *Proceeding of GLOBECOM '88*, vol 1, Hollywood, Florida, 1988, *pp*. 2.5.1-2.5.5.

[Bube 91] R. Bubenik, J. DeHart, and M. Gaddis, ``Multipoint Connection Management in High Speed Networks," *Proceeding of InforCom'91*, Miami, Florida, April 7-11, 1991, *pp*. 59-68.

[Buss 89] H. E. Bussey, F. D. Porter, and H. E. Raitz, ``A Second Generation BISDN Prototype," *International Journal of Digital and Analog Cabled Systems*, Vol. 1, No. 4, January 5, 1989.

[CCIT 90] CCITT Recommendation I.311, ``B-ISDN general network aspects," 1990.

[Chow 91] Ching-Hua Edward Chow, ``On Multicast Path Finding Algorithms," *Proceeding of InfoCom'91*, Miami, Florida, April 7-11, 1991, *pp*. 1274-1283.

[Chow 92] Ching-Hua Edward Chow, ``Resource Allocation for Multimedia Multiparty Connections in Broadband Networks," *Proceeding of 3rd International IFIP WG6.1/6.4 Workshop on Protocols for High Speed Networks*, Stockholm, May 13-15, 1992, *pp*. 121-132.

[Gadd 92] Michael E. Gaddis, Rick Bubenik and John D. DeHart, ``A Call Model For Multipoint Communication in Switched Networks," *Proceeding of ICC'92*, June 14-18, Chicago, IL, *pp.* 609-615.

[Giac 90] J. N. Giacopelli, M. Littlewood and W. D. Sincoskie, ``Sunshine: A High Performance Self-Routing Broadband Packet Switch Architecture, " Proceeding of Internati*onal Switching Symposium (ISS'90)*, Vol. 3, paper P21, May 27-June 1, Stockholm, Sweden, *pp.* 123-129.

[Huan 92.1] Minfa Huang, Ivan T. Frisch, C. Edward Chow and Howard E. Bussey, ``Two Multiparty Connection Establishment Procedures For Broadband ISDN," *Proceeding of the IEEE Network Operations and Management Symposium*, Memphis, TN, April 6-9, 1992, *pp.* 373-382.

[Huan 92.2] Minfa Huang, Ivan T. Frisch, C. Edward Chow and Howard E. Bussey, ``Nondistributed Multiparty Connection Establishment For Broadband Networks" *Proceeding of ICC'92*, Chicago, IL, June 14-18, 1992, *pp.* 1392-1396.

[Kawa 91] Masatoshi Kawarasaki and Bijan Jabbari, ``B-ISDN Architecture and Protocol," *IEEE Journal on Selected Areas in Communications*, Vol. 9, No. 9, December 1991, *pp.* 1405-1415.

[Klei 75] L. Kleinrock, *Queueing System. Volume 1: Theory*, John Wiley and Sons, New York, 1975.

[Lee 88] T. T. Lee, ``Nonblocking Copy Networks for Multicast Packet Switching," *IEEE Journal on Selected Areas in Communications*, Vol. 6, No. 9, December 1988, *pp.* 1455-1467.

[Minz 91] S. E. Minzer, ``A Signaling Protocol for Complex Multimedia Service," *IEEE Journal on Selected Areas in Communications*,, Vol. 9, No. 9, December 1991, *pp.* 1383-1394.

[Pryc 92] Martin De Prycker, ``ATM Switching on Demand: Either virtual channels or virtual paths can implement virtual connections in a B-ISDN ATM layer," *IEEE Network Magazine*, March 1992, *pp.* 25-28.

[Schw 87] M. F. Schwartz, *Telecommunication Networks: Protocol, Modeling and Analysis*, Addison Wesley, 1987.

[Thom 84] A. Thomas *et al.*, ``Asynchronous time division technique: An experimental packet network integrating video communication," *Proceeding of ISS'84*, May 1984.

[Turn 86] J. S. Turner, ``New directions in communications (or which way to the information age?)," *IEEE Communication Magazine*, Vol. 24, No.10, Oct. 1986, *pp.*8-15.

MANAGEMENT OF DISTRIBUTED SERVICES NETWORKS: SHARED TELE-PUBLISHING VIA ISDN - IBC IN EUROPE
The RACE Experience: From Electronic Data Interchange (EDI) to Electronic Process Integration (EPI)

Basil Maglaris*, Theodoros Karounos*, and Andreas Kindt**

Department of EE/CS
National Technical University of Athens (NTUA)
157 73 Zografou, Athens, Greece
e-mail: maglaris@theseas.ntua.gr

ABSTRACT

The work reported in this paper is part of RACE II Project 2037 (Distributed Documenting Services - DIDOS), partially funded by the Commission of the European Union (CEU) The project objectives are to demonstrate the feasibility of distributed service management, enhance the use of Pan-European ISDN and motivate the deployment of Integrated Broadband Communications (IBC or B-ISDN) in Europe. The concepts are being tested in three multinational "Application Pilots" within the business sector of electronic Technical Documentation (TD). All users that participate in the project (TD producers and service providers) are interconnected via ISDN in a "Distributed Service Centre." Participants are equipped with a Service Management kernel that performs basic "agent" functions and a user access shell, including directory and administration services. In the paper we outline a Reference Platform that evolved within the project, formalise user and functional requirements, and provide a vision for future use of public communications networks (ISDN and IBC) by other business communities in Europe. Our models provide an open, structured standard schema to all players (Clients, Service Providers and Public Network Operators - PNO's) to create, share and manage services taking advantage of the evolving Pan-European ISDN - IBC infrastructure.

INTRODUCTION

We present a Reference Platform for Service Management in a shared business community, as applied to electronic document composition, production and distribution. The concept is based on the notion of **Distributed Service Centre** [1]. All business participants (users), including customers, service providers and brokerage service providers, are interconnected via ISDN and can access each other on demand. Every user is equipped with a Service Management kernel that performs basic "agent" functions and a user access shell, including directory and administration services for the shared community. Our model integrates service management tools, usage scenarios and functional specifications. It offers a standard reference scheme to all players in a specific shared business community (clients, service providers and Public Network Operators-PNO's) for a cost-effective way to create, share and manage services.

* National Technical University of Athens, EE/CS, 157 73 Zografou, Athens, Greece
** DeteBerkom, RACE Project Office, Voltastr. 5, D-13355 Berlin, Germany

Network Management and Control, Vol.2, Edited by
I.T. Frisch *et al.*, Plenum Press, New York, 1994

This work is supported by **RACE II Project 2037** (Distributed Documenting Services - DIDOS). The models are being tested in Application Pilots, that inter-connect Trans-European users of Technical Documentation (TD) [2]. Users, i.e. TD producers - clients and service providers, generate electronically, communicate, order and, to a large extend, produce technical manuals and CD-ROM documentation by sharing available production facilities and public ISDN. The overall aim of the project is to demonstrate the concept of shared production in an international market of service providers, and the paramount importance of the necessary communications infrastructure. The business sector of TD publishing is ideally served via broad-band Integrated Communications (IBC) but at this pilot phase, most of the demonstrators employ the Pan-European narrow-band ISDN. The theoretical models, prototype implementations and results of the trials point to the development of a Trans-European **Inter-Organisational System.** Communications and electronic ordering protocols (EDI/EDIFACT) are just first steps towards efficient integration of specialised business sectors. The DIDOS reference model and pilots lead to a further stage, whereby users and providers of competing services negotiate and share resources by in fact integrating work-flow of independent enterprises. The resulting model is referred to by the RACE Strategic Planners as **EPI, Electronic Process Integration**. Hence the subtitle "from EDI to EPI".

In what follows we elaborate on the main DIDOS service management feature, i.e. the Distributed Service Centre concept as applied to shared TD production; we then proceed in describing the Reference Platform and conclude with possible evolution (migration) paths.

THE DISTRIBUTED SERVICE CENTRE

A distributed Service Centre (SC) describes the interaction between users and service providers, defines the management components and specifies the communication requirements over a distributed environment using IBC networks. An illustration of a Distributed Service Centre is shown in Figure 1. As it can be seen in the figure, various users are connected to a Service network via their own SC components (interface). The users initiate requests for document composition and production in electronic ordering form (EDI/EDIFACT). The documents are generated at the user equipment in some initial form, with the possible help of composing services made available to them by a Service Provider after consultation and agreement on pricing. The document data file in preparation (e.g. in POSTSCRIPT, SGML and JPEG compressed images) is transferred between the user and the Service Provider. The final document lay-out (including style, fonts, and archived graphics/images) is finalised by the Producer and Service Provider, with the possible intervention of a Broker to help with scheduling and selection. In a subsequent step, the Producer may request document production and distribution services (e.g., printing on demand in fast high quality printers, direct computer to plate, and multimedia CD-ROM fabrication and shipping). Throughout this process, the services provided are ordering, negotiations, selection of appropriate Service Provider and final billing.

The whole approach is based on the existence of Service Providers throughout Europe, whose services can be employed by various Users through an IBC/ISDN network. It is important to incorporate in this scenario a variety of base services (directory, management - selection - retrieval functions) that can facilitate competition on Quality of Service (QoS) and pricing. In a distributed Service Centre different types of service can be offered. As an example, a large manufacturing corporation can operate its private Service Centre and Service Provider facilities, but occasionally request from an external service provider some specialised service (e.g. CD-ROM production).

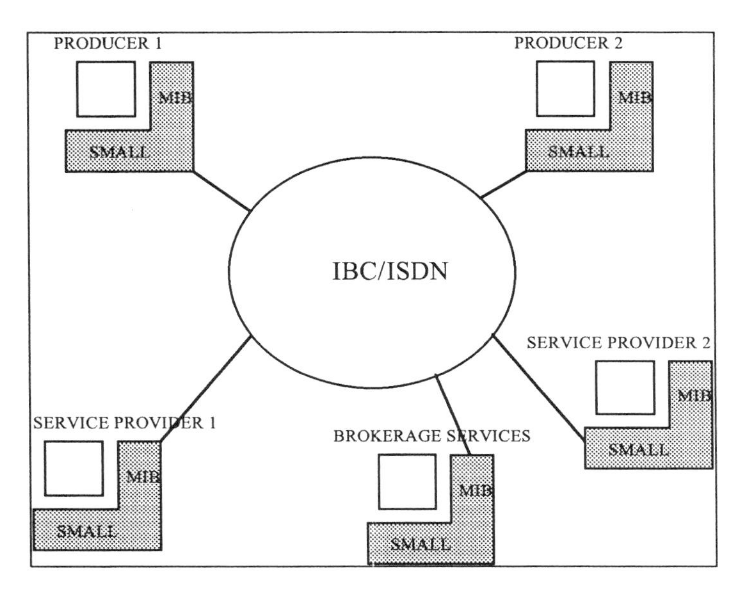

Figure 1. Service Centre

Summary Definition of Terms in Figure 1
User: The term ***user*** refers to end users depicted as squares in Figure 1 . Thus a user can be a Documentation Producer, a Service Provider or a Brokerage Service Provider.
Distributed Service Centre: The ***Service Centre*** concept refers to the distributed kernel, related user shell and information base that are necessary for a user to be part of a community. The distributed Service Centre is the collection of the shaded areas in Figure 1.
Service Management & Administration Logical Layer - SMALL: It refers to the interface of users that includes tools for communication and management, conversion, data representation and user interface.
Management Information Base - MIB: Contains objects maintained and used by SMALL, as well information pertaining to various layers of management (status, directory etc.)
Communications Network: The users are interconnected via public Pan-European ISDN/IBC (narrow-band ISDN or IBC) by means of standard generic interfaces.

REFERENCE PLATFORM

The Reference Platform provides SC users (producers, service providers, brokers) and implementation vendors with guidelines towards integration into a community. The Service Center concept is broad enough to allow for migration following maturing technological and business organizational aspects, yet tightly connected to present day user requirements.

A bottom-up approach reveals strong common threads, that can be summarised as: Computer Aided Composition, Processing and Production of Documentation in an Open, Shared Distributed Service Centre Environment via Pan-European IBC/ISDN. The commonalties are formalised in a top-down view, using the Open Distributed Processing (ODP) [3] model viewpoints and aspects. Hence, it is necessary to group user requirements and functional characteristics to specify a Reference Platform. The publishing User Community provides the necessary feedback and evaluates the different manifestations of the Platform in different applications.

The Reference Platform follows a manager-agent model. Information pertaining to functions in the Service Centre, Service Providers and Users will be stored in a **Management Information Base (MIB).** All entries will be in an object-oriented form maintained by and available to a local Manager referred to as **Service Management and Administration Logical Layer (SMALL).** These Managers (SMALL) will communicate local objects in their MIB's (in a prenegotiated manner). The SMALL-MIB relation in a local level and the SMALL interactions between different users follow the Manager-Agent model. The whole Management environment is shown in Figure 1.

A schematic view of the Reference Platform is shown in Figure 2. It consists of three layers as follows: At the low layer we specify all Common Technological Tools that users require (e.g. Documentation production tools, SMALL/MIB kernel, Communications - ISDN dialling etc.) In the middle level we specify Usage Scenarios for each category of players (Documentation producers, service providers, brokers). At the high level, the usage scenarios are related to Functional Specifications [4] (recommendations to Documentation producers, service providers, vendors and PNO's). We apply the object-oriented paradigm that is used by all standardisation organisations (OSI, CCITT, ETSI) to specify management functions and message sets. Viewing layers as objects (instances of classes) that communicate through messages, offers isolation that makes creation and management of services simpler and easier.

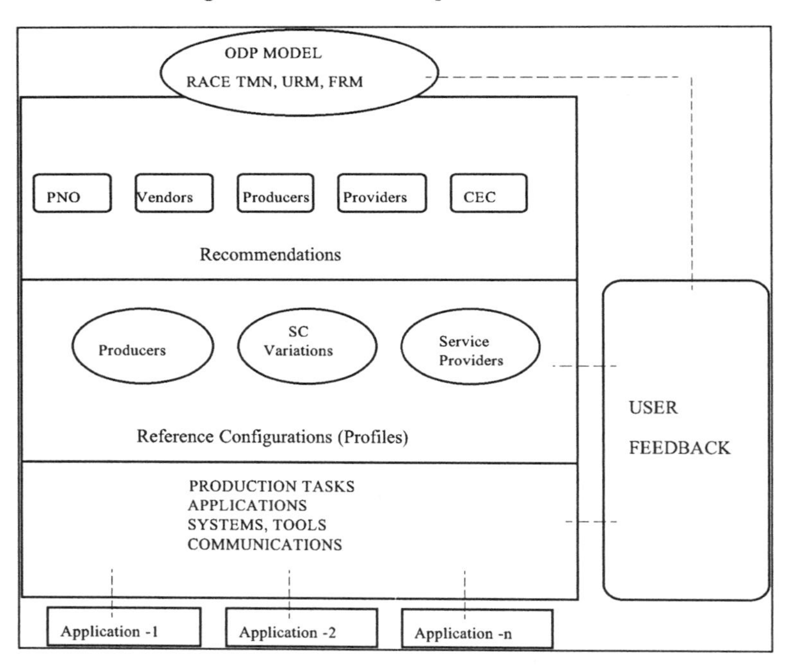

Figure 2. Reference Platform Framework

LOW LEVEL: COMMON TOOLS, SMALL - MIB

This level contains common tools for ordering, composing, producing, distributing and administering a user community. The innovation is the distributed SC concept, which at this level specifies the User Interface-SMALL and Management Information Base-MIB

[5,6,7] kernel for each participant. The SMALL/MIB interface allows Technical Documentation producers to share services provided by several providers aiming to:

1. Help producers to compose Documents using a Document Application Design Workbench and Multi-media Interactive Authoring Tools, combining text, image and/or multimedia files stored in distributed data-bases.

2. Negotiate, order and execute Just in Time Documentation Production, possibly allowing browsing of final documents prior to production, and implement invoicing and distribution of documentation to customers.

3. Provide CD-ROM Multi-media Production of Documentation; the interaction of service providers and producers may extend to the integration of workflow in all production phases, or in case of copy-right and confidentiality barriers may be limited to the final phases of mastering and replication.

MIDDLE LEVEL: USAGE REFERENCE MODEL - URM

The common tools, as co-ordinated via the SMALL/MIB concept, are used by specific users to accomplish various Usage Profiles. These are functionally grouped in the middle level of the Reference Platform, with hooks to the lower level (to draw from the SMALL/MIB and documentation production tool-bench) and the upper level (to link usage with functional specifications).

The middle level specifies Documentation usage in four stages, following the RACE URM (Usage Reference Model, **RACE Project R1077**):

i. **Definition of Reference for User Services (RUS):** This stage describes (1) goals, tasks, users and performance criteria and (2) minimal user attributes to satisfy the above criteria. The RUS model consists of communicating states and tasks (e.g. a possible state can be the dialogue phase between a Producer and Service Provider and associated tasks are price negotiations, service scheduling, ordering etc.)

ii. **Definition of Specific User Services (SUS):** It specifies infrastructural attributes needed to satisfy tasks in RUS. There are static attributes (e.g. type and speed of ISDN access supported by a SMALL - MIB interface) or dynamic (e.g. user control utilities for ISDN dialling and decision support algorithmic steps to allocate the best service provision).

iii. **Definition of Generic User Services (GUS):** They arrange and classify various SUS's into a number of SMALL concepts relevant to infrastructural implementation. A full description of each application (e.g. printing on demand, multi-media production etc.) is given by a pair of RUS and SUS. In all Publishing scenarios there are common functions such as ISDN dialling, ordering, security - authentication, billing etc.

iv. **Relating GUS Definitions to Common Functional Specifications (CFS):** The GUS requirements relate usage demands on infrastructure (Service Centre, IBC/ISDN). Sets of GUS's specify functional requirements, that are discussed in the Upper Level of the Platform.

HIGH LEVEL: FUNCTIONAL REFERENCE MODEL - FRM

This level groups Functional Specifications and provides input to the business users, as well as vendors, PNO's and standardisation bodies. It is anticipated that RACE projects will identify industry standards extending beyond the business sector of publishing, covering similar environments with customers requesting specialised services and providers that offer services in an open, distributed, Trans-European service centre environment.

In specifying this Level we follow the RACE FRM methodology (Functional Reference Model, **RACE Project R1044**). The objective of the FRM is to provide an implementation independent specification of the Reference Platform, a coherent functional model, with functions decomposed in a hierarchy of Levels of Abstraction. The model specifies (as formally as possible) the various functions and their relationship across levels.

Depending on their high level objective, functions are classified into Functional Groups. Following the **RACE Project R1024** NETMAN Cube Model for IBC TMN [8,9], we identified similar groups for the Publishing Reference Platform. We will refer to our model as the TDAM^3 Model (Technical Documentation Architectural Model - Cube) [10]. There are three orthogonal ways to group functions, as illustrated in the Cube Model of Figure 3. The three planes represent a Decision Making Model of functional grouping (Awareness, Decision, Implementation - ADI), a hierarchical grouping of Management Responsibilities and a grouping based on Applications.

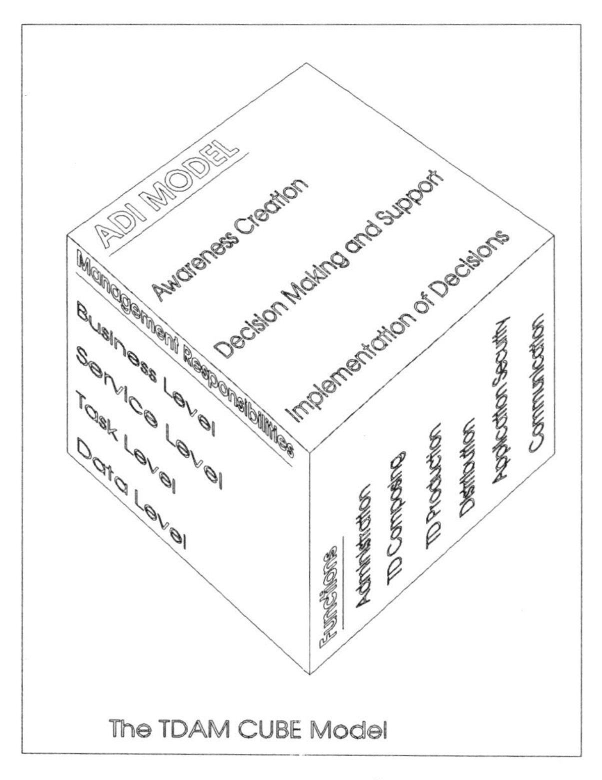

Figure 3. The TDAM^3 Model

i. **The ADI Plane** of the Cube decomposes the Publishing functions into Awareness, Decision and Implementation (ADI) groups. The ADI approach was introduced in the RACE Project NETMAN R1024 and fits all management environments, thus can be directly applied to the Service Management.

ii. **The MANAGEMENT RESPONSIBILITY PLANE** is a hierarchical grouping with four levels of management responsibilities. The interaction of the different levels is structured by the ADI Model.

iii. **The APPLICATIONS PLANE** functions are grouped into six Management Functional Areas that satisfy a set of Publishing and Telecommunication requirements.

THE ADI PLANE

The ADI plane consists of the following groups of functions:

A. **Awareness Creation** functions to trigger a decision at some management level. They are inititiated by the implementation part of a decision in the upper or lower management levels, or by a customer request. They are responsible for the activation of the appropriate decision making procedure.

D. **Decision Making and Support** functions are performed by the appropriate management personnel or software. The decision making is supported by information provided by upper or lower management levels and/or external sources. As the decisions flow from upper to lower management levels more implementation details are added.

I. **Decision Implementation** functions execute the decision. The decision implementation may include many activities. It may also be the awareness of the lower or upper management levels.

MANAGEMENT RESPONSIBILITY PLANE

This is a hierarchical grouping with four levels of management responsibilities. The interaction of the different levels is structured by the ADI Model. The four levels are:

Business Management

Functions in this level manage long-term interactions between the Service Centre and the external world. They are responsible for the long term planning and strategies, analyse the market trends and determine the tariff policies.

Service Management

This level contains functions that manage service interaction among SC users, i.e. TD Producers, TD Service Providers and (possily) Brokerage Service Providers. Functions in this level follow the client - server model. The SC advises the user about the possible offers (either directly from TD Service Providers or via a Brokerage Service Provider) and a service agreement is reached. This may include specifications on the appropriate document type, determination of service distribution and statement of the communications requirements. Finally, the total production cost is estimated before the production and is calculated after it. Part of the communication between users is done by EDI/EDIFACT.

Task Management

Functions are grouped according to the following management issues:

A. Control Management. Functions check whether all scheduled activities are performed according to an initial plan.

B. Administration Data Processing. They control the synchronisation between parallel steps.

C. Business Data Processing. Functions process business data like existed stock of needed materials, quantities of used materials, number of produced pages, etc.

D. Publication Production Workflow. Functions watch the state of the publication production and select workplans and deliverables. The gathered information is available on screen or paper lists.

E. Information Update Workflow. Functions control the updates of the database elements and are informed about all versions and variants. They can also provide different statistics about the contents of the database on screen or paper lists.

D. Publication Usage Control. Functions control the user access for on-line documentation and keep data for billing.

E. Review Management. Review management functions control the reviewing procedure. All parts are reviewed by appointed reviewers and their comments are attached to the reviewed elements.

F. Identification and Name Handling. Functions assign Identification codes to special elements so their access can be controlled.

G. Data and Document Routing. Functions define and control the production steps from data capturing to final production and usage.

Data Management

This level operates on a distributed object database that includes: Text, graphics, tables, images, audio, video, animation & references. It groups functions concerned with the following management issues:

A. Version Management. Database elements may be changed or modified and their new versions can be stored replacing the old ones. The data management functions control this procedure and keep statistics of all data transactions.

B. Variant Management. Modified elements can be stored as variants of the original ones. All variants are being controlled by the data management system. Statistics of all data transactions are available.

C. Retrieve Item. All database elements can be retrieved. The search is being done using content or attributes. References and pointers can also be used. The data management functions control and monitor all the retrieval procedures.

D. Store Item. The data management functions control the storing (with or without compression), the backup and recovery of data elements.

E. Assign Identification. Identification codes can be assigned to special data elements so that its access will be controlled.

F. Review Management. The revision procedure is being done by appointed reviewers and their reviews are attached to the data elements.

G. Item Inventory Management. Item Inventory Management functions provide statistical information about all the database elements.

APPLICATIONS PLANE

Functions are grouped into six Management Functional Areas that satisfy a set of Technical Documentation and Telecommunication requirements.

Administration. It includes Offer Processing, Order Processing, Accounting and Billing

Technical Document Composing. There are five functional groups:

A. Design and planning is performed by a special team of experts (product experts, legal experts, content experts, presentation experts, process experts, user experts, managers and end users). The planning is responsible for the cost effective implementation of the design.

B. Data Capturing. All data is transformed to electronic form and is stored in an internal electronic format, using the following subfunctions:

- Scanning
- Input Existing Data
- Create Content
- Manipulation
- Proofing
- Conversion to Internal Format
- Edit Content
- Translation in Other Languages

C. Document Element Preparation (Structuring). The specifications of the document structure and the database are analysed (usually in SGML). Then the structuring of the internally stored data is performed by assigning tags to data elements and building information units. The Document Element Preparation is integrated by the following subfunctions:

- Select Data
- Select TAGs
- Assign TAGs to Data

D. Document Composition (Selection). A revisable document is prepared by selecting the relevant information from the structured database. The document Composition is integrated by the following subfunctions:

- Create and Update Revisable Form Document
- Select and Insert
- Select and Delete

E. Final Layout Form. The layout format depends on the chosen output medium. The selected information will be converted to the corresponding layout format. For example, the layout of the paper based document will be page oriented, the layout of microfiche will be page oriented with additional information about the fiche format, the electronic media format (like CD-ROM) will optimise access including indexing. The Final Layout Form is integrated by the following subfunctions:

- Edit Layout
- Select Revisable Form Document

- Select Layout
- Format Revisable to Final Form

Technical Document Production. Using the electronic final layout format copies will be produced. The different subfunctions of the Technical Documentation Production are:

- Production Printing Copy (Film / Paper)
- Production Printing Plate
- Offset Lithographic Printing
- Laser Printing
- Binding
- Floppy Replication
- Mastering and Replication (Micrtofiche or CD-ROM)

Distribution. It includes Commissioning, Packaging and Shipping

Application Security. It includes Authorisation, Authentication, Verification and Encryption

Communication. This Functional Area includes the following sub-areas:

A. **Provisioning,** consisting of the following functions:

- Service negotiation and procurement phase
- Customer request for service
- Resource assignment phase
- Contractual phase
- Service activation phase
- Cessation or interruption of service
- Provision of Help information

B. **Maintenance** that retains the system at its nominal state and includes the following functions:

- Maintenance management
- Preventative maintenance procedures
- Testing and monitoring
- Trouble detection, notification, reporting, diagnosis and localisation
- System protection and recovery
- Put system resource out of service
- Repair scheduling, supervision and verification
- Return to service

C. **Performance** that includes the following functions:

- End system monitoring of connectivity and applications
- Analysis and problem alerting
- Diagnosis
- Optimisation and control
- Network configuration

D. **Security** that provides facilities for support of security services and includes the following functions:

- Authentication
- Authorisation

- Cryptographic support
- Security audit

E. **Accounting** that determines the cost of the use of IBCN resources and services by the customers. It includes the following functions:

- Cost accounting
- Charging and billing
- Inter-administration accounting

F. **Customer Query and Control** which comprises the management capabilities provided by the TMN to customers of IBC services, to value-added service providers, and to managers in other Telecommunication Administrations. It includes the following functions:

- Call control (based on Intelligent Network Services & ISDN)
- Virtual Private Network management
- Interadministration management communications

Implementation of some of the above functions requires the co-operation of Public Network Operarators (PNO's).

Table 1

MANAGEMENT RESPOSIBILITY	AWARENESS	DECISION	IMPLEMENTATION
Service	User A asks for offer	Select Service Providers X & Y	Order, Inform Task Levels of X & Y
Task	X & Y Informed	Capture text at X, images at Y	Inform Data Levels at X & Y
Data	X & Y Informed	Retrieve text at X & images at Y. Transmit images at X	Perform retrieval of images. Inform Task Level of X
Task	X is informed	Schedule production at X	Initiate production at Data Level of X
Data	X is informed of go-ahead for production	X initiates production	Document composition mastering, replication, inform Task Level of X
Task	X is informed of production completion	X decides to update workflow	X performs updates, informs Service Level
Service	X is informed of Job completion and statisitcs	X initiates billing functions	X transmits billing to A via EDI/EDIFACT

EXAMPLE

We give a simple example which explains how the TDAM [3] model is applied in Technical documentation. We suppose that **User A** requests a Printing on Demand (PoD) service via the Service Centre. The document includes text and images. We assume that the current version of the text is stored in the database of the **Service Provider X** and the images are stored in the database of the **Service Provider Y**. Text-image integration and production are assumed by **Service Provider X.**

EVOLUTION & MIGRATION PATHS

The Application Pilots are presently designed to employ narrow-band ISDN. This may suffice at this experimental phase, but in a production environment the limited circuit-switched nature of ISDN will be a bottleneck for the Service Porviders. Thus, a working scenario may be a combination of Nx64 Kbit/sec ISDN dial-up access for TD Producers and ATM (IBC) multiplexed access for Service Providers in an interworked ISDN - ATM Pan-European network, Figure 4.

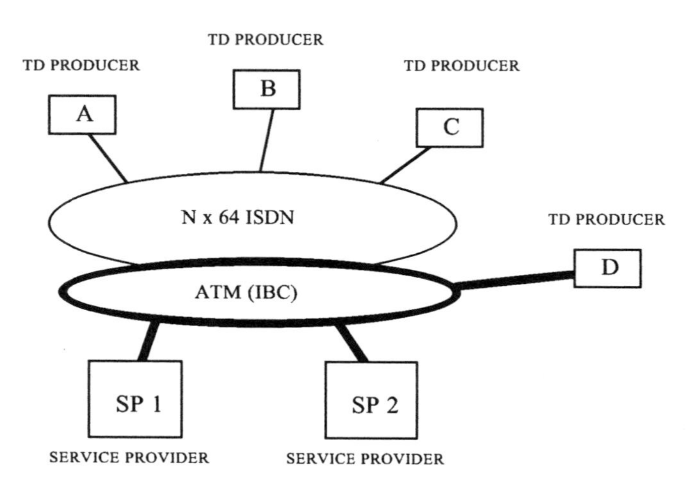

Figure 4. Networking Evolution Scenario

The current state (including short term goals) of the Service Centre implentation in DIDOS and migration paths are summarised in the Table below, as they relate to specific Platform Features.

Table 2

PLATFORM FEATURES	CURRENT STATE	FUTURE MIGRATION
Interactive Documentation Production Framework	EDI (Electronic Data Interchange) based	EPI (Electronic Process Integration) based, ODP (Open Distributed Processing) concepts
User Access	Interactive Composition, Batch Distribution (CD-ROM, Paper)	ON-LINE Multimedia Access
Communication Network	Pan-European ISDN	Pan-European IBC
User Terminal H/W	Fixed Location Graphics Terminal (MAC or PC)	Multimedia Personal Mobile Terminal
S/W	Object Oriented Analysis	Object Oriented Implementation
Network Protocols	TCP/IP	OSI, XTP, ATM AL
High Level Protocols	EDI, FTP, SGML, JPEG, MPEG	OSI, Integrated Multimedia Protocols (MHEG)

On the Interactive Framework and User Access of the Platform, the current state is to order and compose documents interactively, while production and distribution are performed in a batch mode (e.g. CD-ROM mastering and production, off-line printing on demand). The migration path is to move towards ON-LINE Multimedia access and close co-operation of Document Producers (customers) and Service Providers, using a Distributed SC via the underlined Pan-European IBC infrastructure. This conforms with the vision of European Inter-Organisation Co-operation, integrating shared resource management, ordering and workflow of a business sector in a dynamic, flexible way. The

DIDOS community, by sharing the distributed SC takes a step towards ***the transition from Electronic Data Interchange (EDI) to Electronic Process Integration (EPI)*** [11].

Publishing demonstrates the feasibility of the transition to a shared workflow environment using currently available Pan-European ISDN, off-the-shelf S/W and H/W and state of the art programming techniques. The success of pilot applications in RACE provides strong motivation for the deployment of Pan-European IBC and the establishment of Open Standards in all levels of the OSI model (including standards for multimedia communications such as MHEG).

The Reference Platform must conform with the evolving Open Distributed Processing (ODP) reference model, and allow for the migration of Documentation User H/W & S/W to technologically advanced solutions. Examples include provisioning of cost effective Multimedia Personal Terminals (possibly mobile) and the extension of Object Oriented Methodology to transcend all cycles of S/W development (analysis, design, prototyping, testing and production), including Object Oriented Multimedia Databases.

Professor Basil Maglaris (maglaris@phgasos.ntua.gr) was born in Athens, Greece in 1952. He received the Diploma in Mechanical & Electrical Engineering from the National Technical University of Athens (NTUA), Greece in 1974, the M.Sc. in EE from the Polytechnic Institute of Brooklyn (now Polytechnic University), Brooklyn, New York in 1975 and the Ph.D. in EE & CS from Columbia University, New York in 1979. From 1979 to 1981 he was with the Network Analysis Corporation, Great Neck, New York. In 1981 he joined the Department of EE at Polytechnic University, where he was promoted to Associate (tenured) Professor. Since October 1989 he joined the faculty of the EE & CE Department of NTUA, where he is Professor of CS. He founded and directs the Network Management and Optimal Design (NETMODE) Laboratory to conduct R&D in related topics. He is Vice - President of the board of the Research Institute of Communication and Computer Systems at NTUA, representative of the Department in the Research Committee of NTUA. He is the Scientific Director for building an integrated digital voice and data network at the University Campus in Zografou. His research interests includes communication networks, with focus on IBC protocols, queuing theory, performance models of protocols, optimal design and planning algorithms, and operations and management of communication networks.

Theodoros Karounos (karounos@phgasos.ntua.gr) was born in Mystras Greece in 1953, graduated from the Pre-engineering Programme, City University of New York in 1978, received his B.Sc. in CS in 1983 and the M.Sc. Degree in Information Systems, both from Polytechnic University, Brooklyn, New York in 1984. From 1984 to 1987 he was employed by IBM, New York, USA. From 1987 to 1990 he was a coordinator for systems support and R&D at the Informatics Development Agency of the Greek Ministry of Presidency and Special Advisor for informatics to the Greek Minister of the Interior. He is currently employed full time at the National Technical University of Athens (NTUA) as a Research Engineer. He is managing the R&D activities of the Network and Optimal Design Laboratory - NETMODE at the Department of EE & CE of NTUA. Mr. Karounos is Project Manager of the integrated digital telecommunications network under installation at the Zografou Campus NTUA. He represents NTUA at the National Committee for the development and operations of the Academic & Research internetwork. His research activities include design, planning and management of communication networks.

Andreas Kindt (kindt@deteberkom.detecon.d400.de) was born in German in 1964. He received the Diploma in Industrial Engineering and Management (IEM) from the Technische Universitat Berlin in 1987. From 1987 to 1989 he was a Product Manager in the Product Marketing Telecommunications at Nixdorf Computer AG, Product Design and Marketing for ISDN Endsystems and ISDN PBX, Project Manager for initial Prototyping of ISDN Videotex endsystems. He is currently Project Manager at the RACE Office of DETEBERCOM GmbH. He has been Project Manager for BERKOM project BILUS, Germany's largest broadband publishing project, BERKOM.

REFERENCES

[1] RACE Project DIDOS R2037, "Communication Requirements in a Distributed Technical Documentation Publishing Scenario," Deliverable 1, 1992.

[2] RACE Project DIDOS R2037 "Investigation of Technical Documentation Application Domains," Deliverable 2, 1992.

[3] ISO: "Recommendation X.9yy: Basic Reference Model of Open Distributed Processing, Part 2: Descriptive Model," ISO Proposal for Committee Draft, ISO/IEC/JTC/SC21 N6079, August 1991.

[4] R.Smith, E.H.Mamdami and J.Callaghan, "The Management of Telecommunications Networks," Ellis Horwood, 1993.

[5] G.Booch, "Object Oriented Design with Applications," The Benjamin Cummings Publishing Company, 1991.

[6] McCloghrie K., and M. Rose, "Management Information Base for Network Management of TCP/IP - based Internets," RFC 1156, May 1990.

[7] McCloghrie K., and M. Rose, "Management Information Base for Network Management of TCP/IP - based Internets, MIB II," RFC 1213, March 1991.

[8] "Telecommunications Management Conceptual Models," S. Plagemann and T. Turner, Broadcom, 1992.

[9] RACE Project NETMAN R1024, "Telecommunications Management Specifications," Deliverable 6, 1992.

[10] "TDAM Cube: A Conceptual Technical Documentation Architectural Model," D.Kalivas, T.Karounos & B.Maglaris, Proceedings of the BROADCOM - 92 RACE Conference, STG 4.2/92/55/2, Ireland, October 1992.

[11] DGXIII, RACE Office, "Advanced communications for economic development and social cohesion in Europe - A strategic audit of the development of advanced communications in Europe," April 1993.

V

IMPLEMENTATIONS OF ADVANCED TECHNOLOGY IN NETWORK MANAGEMENT - INTRODUCTION AND OVERVIEW

For some time now the field of Network Management has been a source both for new requirements in the area of computer system building, and one of the most strenuous testing grounds for ideas and methodologies in that realm. Indeed, the complexity of an effective management system may necessarily exceed the complexity of the system it manages, since it must first master the intricacies of the managed system and its operation, and then render them comprehensible to a human operator or manager. The topic of this session therefore generated considerable expectations among the participants in the Workshop.

In addition, a recurrent theme throughout this Workshop and other recent network management events has been that it is time to move past the ***protocol wars*** and begin to devote a substantial level of energy towards actually building network management applications that do more than just monitor raw MIB variables and generate dazzling graphics. While the papers presented in this session generally described prototypes or projects designed to establish feasibility of a given approach or technique, the questions from the audience often cut directly to practical issues such as scalability or applicability to a wider set of problems or situations. There was no choice but to conclude that there is substantial unsatisfied need in the marketplace for applications that can deliver on the promise to help synthesize or otherwise add value to the vast quantity of raw management data that is available in a modern network.

The papers in the session represent a broad cross section of current work in network management. The submission by Wu et al., ***On Implementing a Protocol Independent MIB*** focuses on a number of issues that arise in building flexible network management software. It draws on work in the field of software engineering in the way it deals with automatic generation of the code to realize a management agent, and it addresses the goal of rising above the limitations imposed by a particular network management data access protocol to achieve an implementation that can be readily reused in different contexts.

The trio of submissions from different operating units of NTT suggest that even within a single large company there is likely to be a microcosm of the network management community. The paper by Yata et al. deals specifically with the software environment needed to support network management applications. It attempts to rationalize and satisfy the demanding requirements of this field. ***Advanced Operations System for Mobile Communications Network*** by Fukushima et al. demonstrates how new technology offers both new requirements for network management, and new opportunities to rethink the implementation of management systems. It demonstrates the importance of structure in managing complex systems, and recommends more task orientation in the implementation of management systems. In their paper on ***IDSS-COME***, Inoue et al. describe a system that helps improve the quality of management for a network by enhancing operators' ability to understand and control the behavior of a complex network. It includes a range of simulation and analysis capabilities, knowledge-based features and decision support technology to help operators project the likely consequences of their actions.

Alarm correlation is one area where applications have been built that begin to deliver on the promise to produce a synthesis of raw input information and to contribute substantively to simplifying the task of diagnosing network problems. The papers by Jakobson et al., ***A Dedicated Expert System Shell for Telecommunications Network Alarm Correlation***, and Finkel et al., ***An Alarm Correlation System for Heterogeneous Networks***, deal specifically with this aspect of fault management. They demonstrate the variety of approaches that may be taken to any specific network management problem, and they also offer examples of how advanced software tools and techniques are being tailored to tackle problems in this domain.

Finally, in ***RREACT: A Distributed Protocol for Rapid Restoration of Active Communications Trunks***, Chow et al. describe efforts to embed intelligent behavior in the operational software of a network so as to achieve a higher level of network reliability and stability. This work lies on the boundary between areas that have traditionally been labelled network management and network control, and it gives a good example of how this traditional boundary has become blurred as network technology evolves.

Stephen Brady
IBM
T.J. Watson Research Center
P.O. Box 704
Yorktown Heights, NY 10598

ON IMPLEMENTING A PROTOCOL INDEPENDENT MIB

Shyhtsun F. Wu[1], Subrata Mazumdar[2],
Stephen Brady[2], and David Levine[2]

[1]*Columbia University, New York, NY*
[2]*IBM Research, Yorktown Heights, NY*

INTRODUCTION

The *Management Information Base (MIB)* plays a key role in a network management system (figure 1) simply because: first, network management applications need structured information about the managed network; second, the network itself usually does not provide a simple and structured interface for *information services*. Currently, when network management applications are developed, they are designed to work with either *Simple Network Management Protocol (SNMP)* [6, 5] or *Common Management Information Protocol (CMIP)* [2]. But, since a *management information base (MIB)* is only a repository of information about various network devices, the design of a MIB should not be constrained by the question, "what is the query access protocol." Instead, our major concerns in building a MIB should be the *data model* and the structure it imposes on network information. These concerns can be addressed in a manner which is *protocol independent*.

In figure 2, the concept of "protocol independence" as it relates to a MIB is illustrated in an object-oriented way. The MIB object encapsulates the structured information and provides a set of *generic* information services. For SNMP applications, a SNMP protocol object accepts a SNMP query, transfers that query into a generic form, and then sends it to the MIB object. In other words, the MIB object itself is also encapsulated by the SNMP protocol object. Furthermore, the protocol independent MIB object can be shared by several different *protocol objects*. For example, in figure 2, MIB objects are shared by both SNMP and CMIP protocol objects.

The mapping between protocol objects (both CMIP and SNMP) and the generic MIB is discussed in [4]. The design and implementation of the SNMP protocol object is described in [8]. Therefore, in this paper, we focus only on the design and implementation of the protocol independent MIB. We will first present a high level design of our MIB, where we show that three different levels of encapsulation are necessary. Then, the key component in the MIB, *CoreObject*, is described. We also show that *CoreObject* provides flexible choices in modeling management information, such as a *Counter*, of network devices.

Network Management and Control, Vol.2, Edited by
I.T. Frisch *et al.*, Plenum Press, New York, 1994

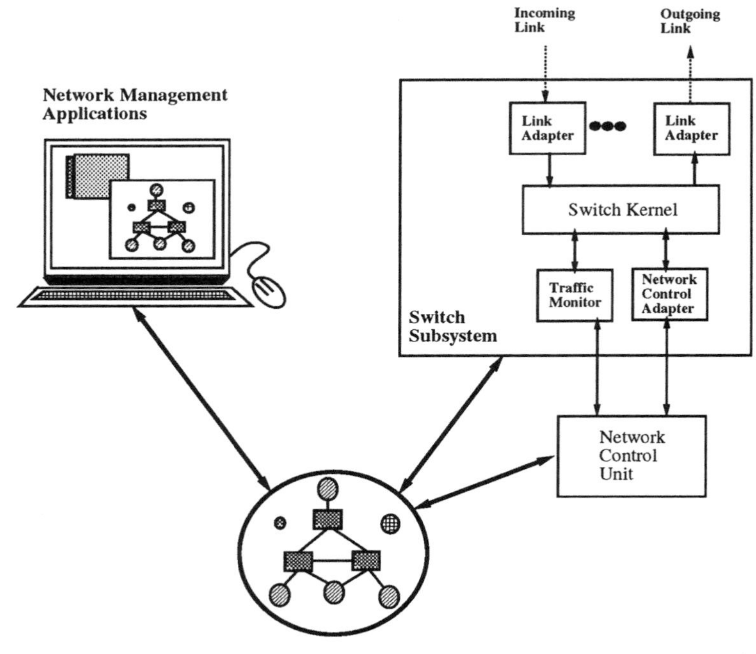

Figure 1. A Network Management System

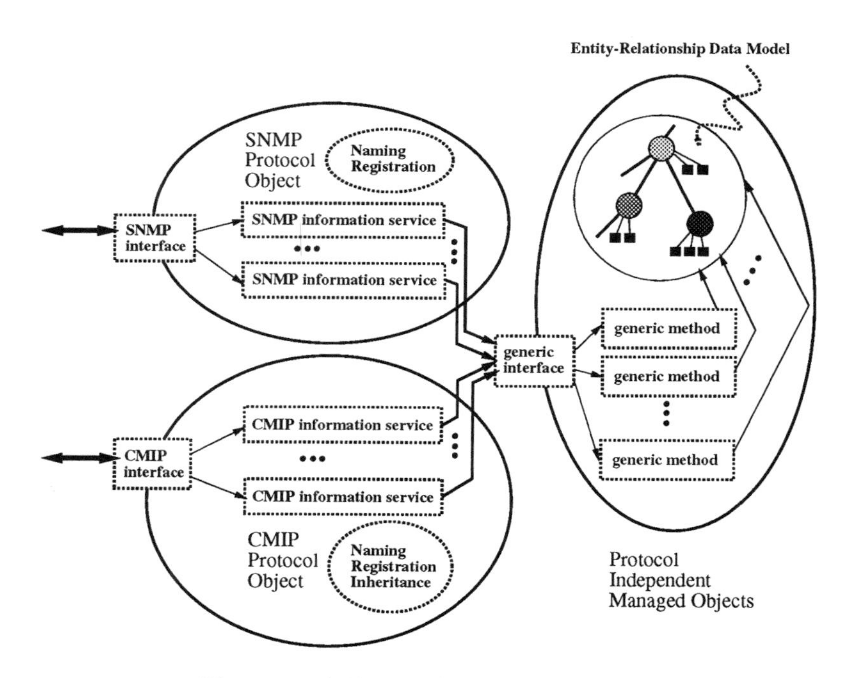

Figure 2. A Protocol Independent MIB

PIMIB HIGH LEVEL DESIGN

In this section, we introduce our approach to the design of the protocol independent MIB.

PIMIB Specification

For practical purposes, a MIB specification document is an ASN.1 description of management information [6]. This implies that all objects, all classes and all attributes (including key attributes) are clearly defined. Also, it is NOT necessary for the document to specify the query access protocol.

As shown in figure 1, both network management applications and the network itself are related to the MIB. Therefore, it is important to get a consensus about the MIB specification document among *application builders*, *network equipment designers*, and the *designers of the MIB* in order to prevent the following kinds of problems:

- the information needed by applications is not in the MIB;
- the information in the MIB can not be provided by the network.

Furthermore, depending on the approach we take to implementing the MIB agent, sometimes we need to concern the object structure in specifying the MIB. For instance, in [8], an SNMP MIB agent is generated by the EMOSY compiler, and one restriction in this EMOSY approach is that two SNMP MIB variables in the same column of a table must have the same type of information source. If one variable is from a shared memory region and the other is from UDP, then we have to either

- separate one table into two different tables in the MIB specification or
- provide a network device information objects to encapsulate both access methods.

The concept of network device information objects will be described in a later section. The integration of network device information objects and the MIB specification is illustrated in [8].

PIMIB Generic Interface and Protocol Objects

As shown in figure 2, the MIB consists of a protocol independent collection of managed objects, a set of protocol objects, and a generic interface in between. The advantage in having a generic interface is that the collection of managed objects can be developed once. With this collection specified, another programmer can implement the protocol objects focusing only on the *transformation* between the query access protocol and the generic interface, ignoring the details of any one MIB specification.

Traditionally, in specifying a MIB, one describes only what objects and attributes are available but not what the inheritance relations are among them. This is especially true for SNMP MIB design. However, when specifying an OSI compliant MIB, one would define an inheritance hierarchy. Thus, one issue that must be resolved is where to handle the inheritance relationship. In the current design, the inheritance hierarchy is imposed by the registration mechanism of the protocol objects. As a consequence, we can make the objects accessed through the generic interface *classless*.

The generic interface is the only way that protocol objects can access the classless objects in the PIMIB. This by no means implies that a PIMIB only contains one class of objects. Actually, in our implementation, a separate inheritance hierarchy is encapsulated inside the PIMIB. In other words, although an inheritance hierarchy of all the managed objects is present, protocol objects as well as network management applications have no knowledge about it through the generic interface. Furthermore, with a CMIP protocol object, two totally different class hierarchies co-exist in the system: one from the CMIP description of the MIB lives in CMIP protocol object (figure 2), and another, which will be described in a later section, lives in the PIMIB. Therefore, it is left totally to the person implementing the PIMIB to decide what the latter one should look like.

PIMIB and Network Device Information

At this point, the reader might ask, "Why do we have to worry about class hierarchy? Why don't we just write a compiler from ASN.1 to C++ or C (or whatever) since the MIB is so generic?" Unfortunately, at this moment, mechanical translation directly from the MIB specification to code is very difficult principally because (on the network inteface side of figure 1) there is no simple, well-defined standard interface between network devices and the MIB. This is especially true for multivendor networks with various types of devices.

The problem of integrating the MIB with various devices for network information is illustrated in figure 3. In principle, we have to deal with all the devices through various access methods, which means that while our MIB may be independent of management protocol, it is certainly not independent of the environment and resources it represents. However, in practice, we can add one or more *data access interfaces (DAI)* between devices and the MIB. Each DAI is responsible for communicating with network data sources using the appropriate protocols, and providing information to the MIB according to some set of rules. In our implementation, the protocol between the MIB and the DAI is *shared memory*. The shared data segment, called the *Controller Data Segment (CDS)*, contains the information about the devices and is updated by the DAI at regular time intervals. The relationships among the MIB, the DAI, and the CDS are depicted in figure 4.

3 Layers of Encapsulation

From various design concerns in previous subsections, we notice that there are three layers of encapsulation in a network management system containing a protocol independent MIB:

1. the query access protocol (SNMP/CMIP) layer provides a view for management applications;

2. the generic interface of MIB supports the kernel of information services for both SNMP and CMIP protocol objects;

3. the network information objects (i.e., network controllers or other devices) encapsulate various access methods to the network device information.

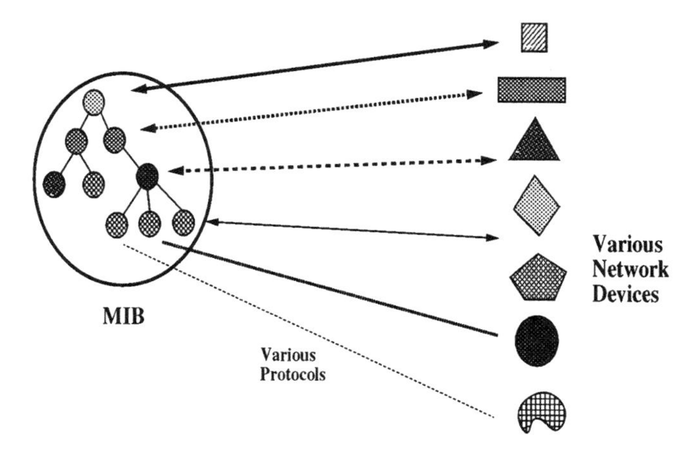

Figure 3. The Integrating Problem

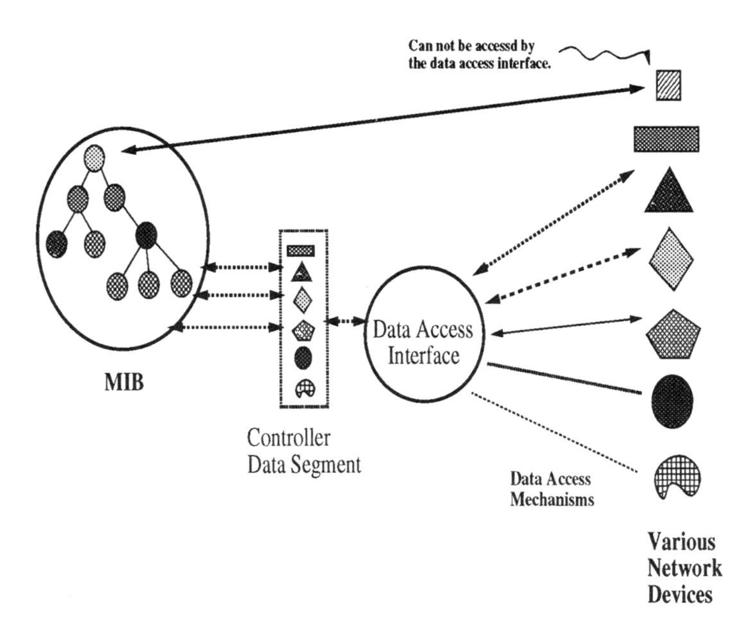

Figure 4. The Relationships among the MIB, the DAI, and the CDS

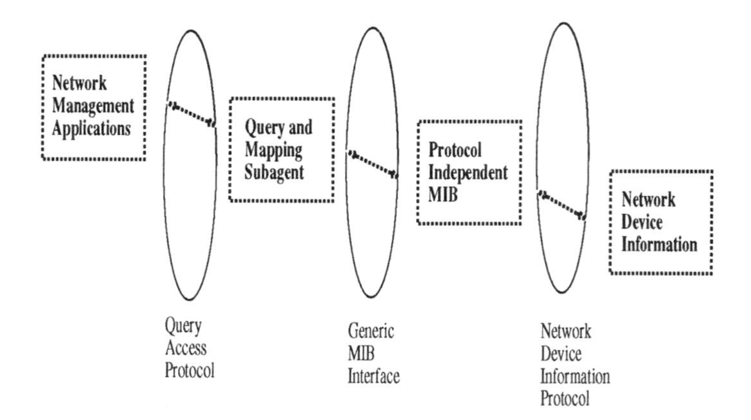

Figure 5. 3 Layers of Encapsulation

Thus, these three layers can separate the MIB development process into four parts (as in figure 5):

1. Network Management Applications (NMA).
2. Mapping between Query Access Protocols and MIB (QAP-MIB).
3. MIB.
4. Network Device Information Protocols (NDIP) and Data Access Interface (DAI).

This separation implies that we can have four different people with different kinds of background and knowledge implement these four different modules. We summarize the knowledge of the implementor for each of the four parts in figure 6. For example, the NMA builder only needs to know the MIB document and the SNMP/CMIP protocol, while the SNMP protocol object builder only knows about SNMP and the generic interface to PIMIB. Furthermore, since the MIB designer needs to know the source of the network information, it is necessary for him to interact with hardware devices. However, in our experience, this interaction is much simpler given a well-structured data access interface (DAI). This implies that, while implementing the DAI, the network equipment designers should consider what types of information will be needed by the MIB, and then provide an efficient and clean interface to access that information.

The knowledge separation concept can be applied not only to the network management system building process with human programmers and developers but also to the automatic process to build a SNMP management agent. In [8], the EMOSY compiler is introduced as a tool to mechanically generate the SNMP protocol object as well as the protocol independent MIB itself. To generate the MIB, the table in Figure 6 shows that merely the information in the MIB specification is not enough. Therefore, EMOSY needs to take an extended version of MIB specification (which contains some network device information) as its only input. The architecture of EMOSY/PIMIB is depicted in Figure 7.

Knowledge About \ Roles	*NMA*	*QAP-MIB*	*MIB*	*NDIP/DAI*
MIB Document (ASN.1)	Yes	No	Yes	Some
SNMP/CMIP Protocol Objects	Yes	Yes	No	No
Generic Interface to PIMIB	No	Yes	Yes	No
Inheritance (inside the PIMIB)	No	No	Yes	No
Devices	No	No	Some	Yes

Figure 6. Knowledge Separation

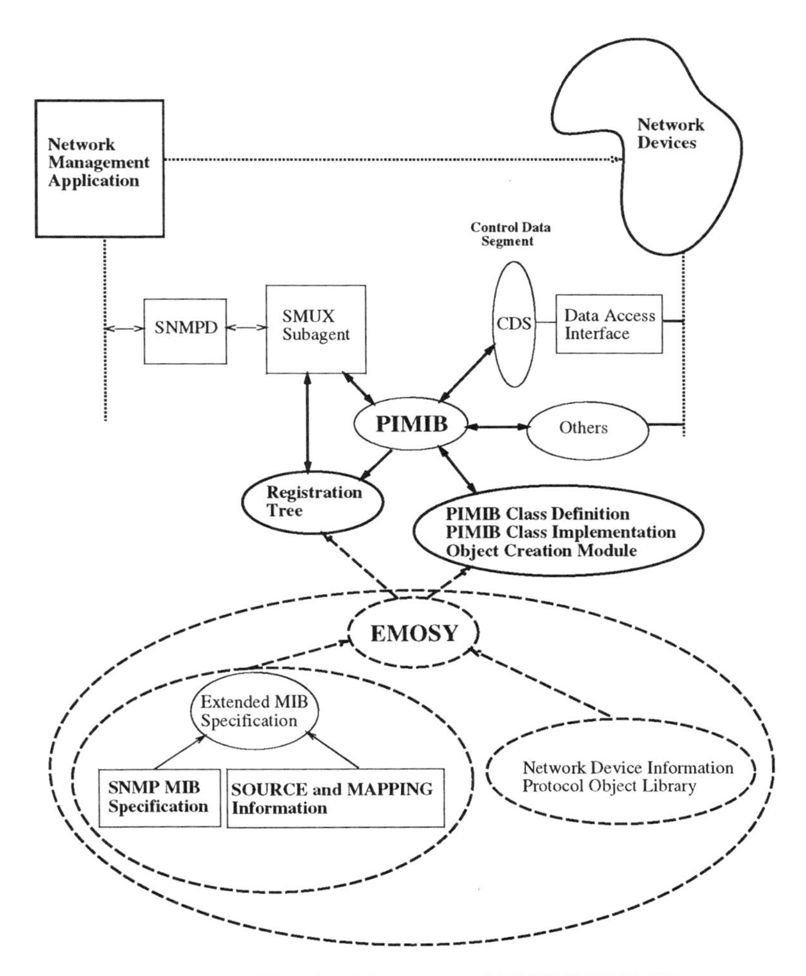

Figure 7. The Architecture of EMOSY/PIMIB

MIB GENERIC INTERFACE: *CoreObject*

The MIB generic interface in figure 2 and figure 5 is realized by way of an object class called *CoreObject*.[1] This class includes a collection of methods that protocol objects can call to access the encapsulated network information. Two generic methods are implemented:

- int *objid*.**getValue**(**attributeID**, **returnValue**)
- int *objid*.**setValue**(**attributeID**, **value**)

These methods are *overloaded* [7] (having more than one function signature) because **returnValue**, **value**, and **attributeID** could have many different types. Although the *CoreObject* class contains more than just these overloaded methods, these, combined with a name registration module, are enough for protocol objects to get all the management information.

The declaration of the class *CoreObject*, is contained in figure 8. In this section, we introduce the internal structure of a *CoreObject*. Then, the relationship among *CoreObjects* themselves is presented. The major goal for having such relations is to construct a *naming* and *containment hierarchy* for the MIB. Finally, a registration module for producing a mapping to be used by a protocol object to access the *CoreObjects* is described.

Internal Structure

A *CoreObject* contains a group of *CoreAttributes* and a set of methods to access those attributes. *CoreAttributes* are used to to model all kinds of management information. They are themselves *typed* objects kept in an *object hash table* owned by the *CoreObject* (in figure 8). There are two motivations of treating attributes as C++ objects in their own right rather than as attributes of C++ objects:

1. We want to associate more information with each attribute than is possible in C++. The information (in ASN.1 description) includes *attribute name*, *ASN.1 encoding*, *attribute type*, *access mode*, *attribute value*, and *default attribute value*.

2. We want to specify "where and how to get the attribute value" in the attribute level. For example, the *CoreAttribute* class provides a method to access network information locally, while the *CDSAttribute*, a derived class from *CoreAttribute*, provides an interface to extract data from a shared memory segment. Through this approach, a protocol object making a request of a *CoreObject* will not know how the information inside the object is stored and retrieved, but inside the PIMIB, the correct access method will automatically be invoked.

Because of the first motivation, in figure 9, all the ASN.1 attribute information is implemented as a collection of C++ attributes inside a *CoreAttribute* object. As a result of the second one, several subclasses of *CoreAttribute* are provided (*e.g.*, *CounterAttribute*, *StateAttribute*, and *CDSAttribute*).

Although different subclasses of *CoreAttribute* may have different method implementations, the data objects they keep are the all of the same type, *i.e.*, *CoreValue*. A *CoreValue* object includes the union of the following six primitive

[1]Our implementation language is C++.

```
// class_coreObject.h
#include <class_ADN.h>
#include <class_coreAttribute.h>
#include <class_hashTable.h>
class core_object
{
  private:
  protected:
  public:
    virtual char *className() { return STRDUP("core_object"); }
    // Naming and Containment Hierarchy.
    core_object  *parent;
    OI_class     *keyAttrID;
    ADN          *coreDN;
    char         *keyValues;
    virtual ADN *getDN(ADN *_tail = (ADN *) NULL);
    virtual OI_class *getKeyAttrID();
    virtual void buildDN();
    virtual void buildkeyValues();
    // constructors and destructor:
    core_object(core_object&);
    core_object(core_object *_parent, core_object&);
    core_object(core_object *_parent, core_attribute *_key);
    core_object(core_object *_parent, core_attribute *_key,
                attribute_list *_al);
    ~core_object();
    // Binary Hash: containing all the attributes.
    hash_table   *attrHashTable;
    // String Hash: containing the mapping from attrName to attrID.
    hash_table   *attrNameIDTable;
    // a message handling routine for messages from the EEP.
    virtual int reportFromEEP(char *_msg);
    // print routine: all the information about the core_object.
    virtual void dump(FILE *_fp = stderr);
    // attribute stuff.
    virtual core_attribute *getAttributeByAttrName(char *_name);
    virtual core_attribute *getAttributeByAttrID(OI_class *_id);
    virtual OI_class *getAttrIDByAttrName(char *_name);
    virtual char *getAttrNameByAttrID(OI_class *_id);
    virtual int assert(attribute_list *_list = NULL);
    virtual int assertNewAttribute(char *_name, OI_class *_id,
                                   int _mode, core_value *_val);
    virtual int assertNewAttribute(core_attribute *_attr_ptr);
```

Figure 8. The declaration of *CoreObject* in C++.

```
// generic set:
virtual int setValue(attribute_list *_list,
                     mod_operator _mod_op);
virtual int setValue(OI_class *_id, core_value *_value_ptr,
                     mod_operator _mod_op = REPLACE_VALUE);
virtual int setValue(core_attribute *_attr_ptr,
                     mod_operator _mod_op = REPLACE VALUE);
// generic get:
virtual int getValue(OI_class *_id, int *_attr_val);
virtual int getValue(OI_class *_id, char *_attr_val);
virtual int getValue(OI_class *_id, char **_attr_val);
virtual int getValue(OI_class *_id, float *_attr_val);
virtual int getValue(OI_class *_id, double *_attr_val);
virtual int getValue(OI_class *_id, void **_attr_val);
virtual int getValue(OI_class *_id, core_value **_attr_val);
virtual int getValue(attribute_list *);
```

Figure 8 (continued).

value types: *integer, float, double, character, string,* and *pointer*, plus the value type information. This makes *dynamic value-type checking* possible.

One example of a *CoreObjects* internal structure is depicted in figure 10. This *CoreObject* has two attributes, which are both integer-typed but which are obtained from different sources. The first one is modeled as a *CoreAttribute*, which means that the value can be obtained locally, *i.e.,* 100. The *CoreValue* type here is "**integer**." The second one is a *CDSAttribute*, which means the value is located in the control data segment. Therefore, the *CoreValue* type in this case is "**pointer**." Please note that although both attributes are integer-typed, the corresponding *CoreAttributes* have different value types at the *CoreValue* level. Since the knowledge of how to get the network information value is encapsulated in the attribute level, the *CDSAttribute* knows that the *CoreValue* type "**pointer**" doesn't mean a normal pointer but rather a pointer to an integer located in shared memory.

Naming and Containment Hierarchy

In this subsection, we consider how to identify each *CoreObject* in our MIB. We first introduce the containment hierarchy and the *key attribute* for a *CoreObject*. Then, the entity-relationship model as it relates to both SNMP and CMIP is briefly discussed. Finally, the naming schemes supported by each *CoreObject* are presented.

Containment Hierarchy and the Key Attribute

The *containment hierarchy* [2] presents a structural relationship among all the *managed objects*. It is also called the *management information tree* [3] because all the managed objects reside in the tree as either the root, the intermediate nodes, or the leaves. An example of a containment hierarchy is shown in figure 11, where a network node contains a few link adapters, and both the node and the link adapters contain some management information.

```
// class_coreAttribute.h
#include <class_OI.h>
#include <class_coreValue.h>
class core_attribute
{
  private:
  protected:
  public:
    char         *attributeName;
    OI_class     *attributeID;
    char         *attributeType;
    int          attributeMode;
    core_value   *attributeValue;
    core_value   *defaultAttributeValue;
    core_attribute(core_attribute&);
    core_attribute(core_attribute *);
    core_attribute(char *_n, OI_class *_id, int _m, char _c);
    core_attribute(char *_n, OI_class *_id, int _m, int _i = 0);
    core_attribute(char *_n, OI_class *_id, int _m, float _f);
    core_attribute(char *_n, OI_class *_id, int _m, double _d);
    core_attribute(char *_n, OI_class *_id, int _m, char *_s);
    core_attribute(char *_n, OI_class *_id, int _m, void *_p);
    core_attribute(char *_n, OI_class *_id, int _m, core_value);
    core_attribute(char *_n, OI_class *_id, int _m, core_value *);
    ~core_attribute();
    virtual core_attribute& operator=(core_attribute);
    virtual int setValue(core_value *,
                         mod_operator = REPLACE_VALUE);
    virtual int getValue(core_value **);
    virtual char *dumpValue();
};
```

Figure 9. the declaration of *CoreAttribute* in C++

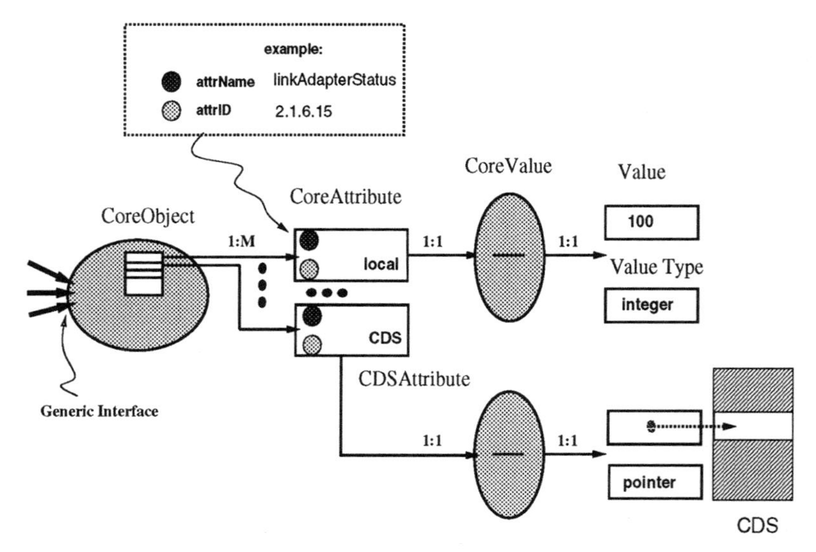

Figure 10. The Internal Structure of a *CoreObject*

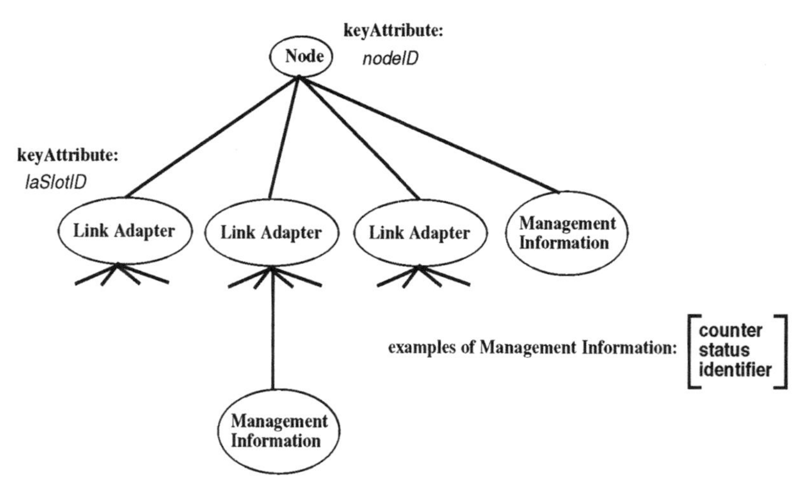

Figure 11. An Example about Containment Hierarchy

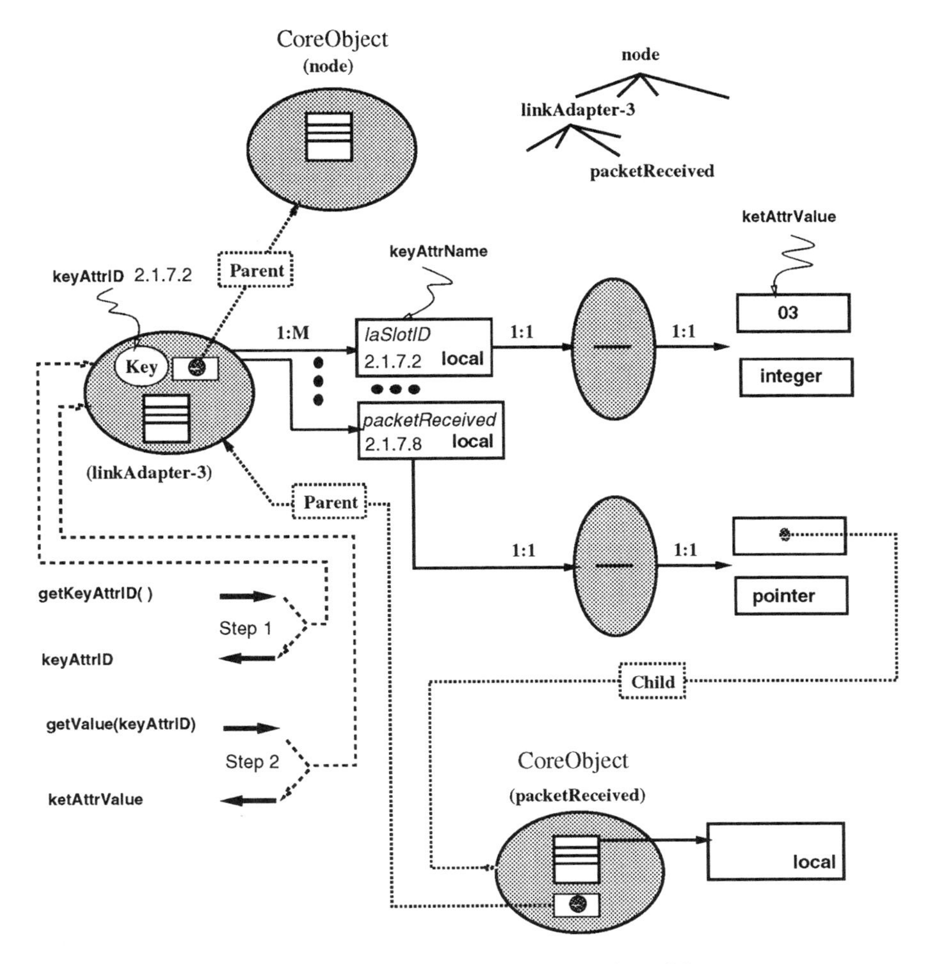

Figure 12. The *Key Attribute* in a *CoreObject*

A containment hierarchy can also be treated as a collection of *parent-children* relations. A *key attribute* is needed for a parent to distinguish each of its children. For example, *laSlotID* is the key attribute of a link adapter. Thus, as long as all the link adapters in one node have different **values** for their *laSlotID* attributes, the node can always distinguish one link adapter from the others.

In order to represent the containment hierarchy, two special attributes are added to the *CoreObject*: *parent* and *keyAttrID*. *Parent* is a C++ object pointer to another *CoreObject*, and *keyAttrID* is an object identifier for the key attribute in the hash table of a *CoreObject*. To determine the key attribute value for an object, we first get the *keyAttrID* (through a method called **getKeyAttrID**), and then, use the generic **getValue** interface with that ID as one parameter. This procedure and these two special attributes are illustrated in figure 12.

CoreObject Naming

Before considering the registration processes for SNMP and CMIP, we discuss the naming scheme provided by *CoreObject*. We briefly introduce the entity relationship model for the structured information. Then, we describe how the implementation of *CoreObject* supports both SNMP and CMIP.

Entity-Relationship Model As described in [4], both SNMP and CMIP structures can be mapped into an *entity-relationship (ER)* model [1]. From the implementation point of view, the difference between SNMP and CMIP is how to distinguish one particular entity (a variable or object instance) of a given entity type (a table entry or object class). In CMIP, we use a *distinguished name (coreDN)* to identify one particular entity (as in figure 13), while, in SNMP, the *values of key attributes (keyValues)* are used (as in figure 14). So, in order to identify one specific entity (variable or attribute), we first use the *entity type name* (*class name* in CMIP and *table/group name* in SNMP) to specify a set of typed entities. Then, we use either coreDN or keyValues to identify a single *CoreObject*. Finally, we can apply either **getValue** or **setValue** on that *CoreObject*.

In order to support both SNMP and CMIP, each *CoreObject* has two extra attributes: *coreDN* and *keyValues*.

coreDN: The key attribute in a *CoreObject* is equal to the *relative distinguished name (RDN)* [2] in the containment hierarchy. Within a superior node in the containment hierarchy, all subordinates are uniquely identified by the RDN. An RDN contains two parts: the identifier of an attribute and a value for that attribute. For those who are familiar with CMIP, the term *keyAttrID* in the subsequent text, this can be taken as equivalent to *RDNid*, and *keyAttrValue* can be interpreted as being equivalent to *RDNidValue*. For example, in figure 12, the RDNid (object-identifier) for link adapters is "2.1.7.2", and the RDNidValue for the particular link adapter is "3". Therefore, the RDN of **linkAdapter-3** is "[2.1.7.2 = 3]" (or "[laSlotID = 3]," where "laSlotID" is the text name for "2.1.7.2"). Furthermore, the parent relation in figure 12 is used to contruct the *distinguished name (DN)*. A special method **buildDN** recursively walks through the path to the root of the management information tree, and concatenates all the RDNs into one unique DN. The DN so constructed for **linkAdapter-3** would be "[2.1.5.2 = 5, 2.1.7.2 = 3]", where "2.1.5.2" is the object identifier for *nodeID* (as in figure 11) and the node's key value is "5" (node-5).

keyValues: A special method **buildKeyValues**, which is operationally similar to **buildDN**, is provided to construct a sequence of *keyAttrValues*. We can construct this sequence by concatenating the *RDNidValues* in the corresponding CMIP coreDN. For example, **linkAdapter-3**'s keyValues taken directly from the CMIP coreDN would be "5.3".

Some additional difficulties arise when trying to accommodate a pre-existing SNMP MIB description within this framework, because a natural CMIP containment hierarchy representing the same management information may require that a CMIP object identifier for a particular class have more DN components than there are indices for some SNMP MIB variable which corresponds to an attribute of this class . However, these difficulties can be readily overcome, following the discussion in [4].

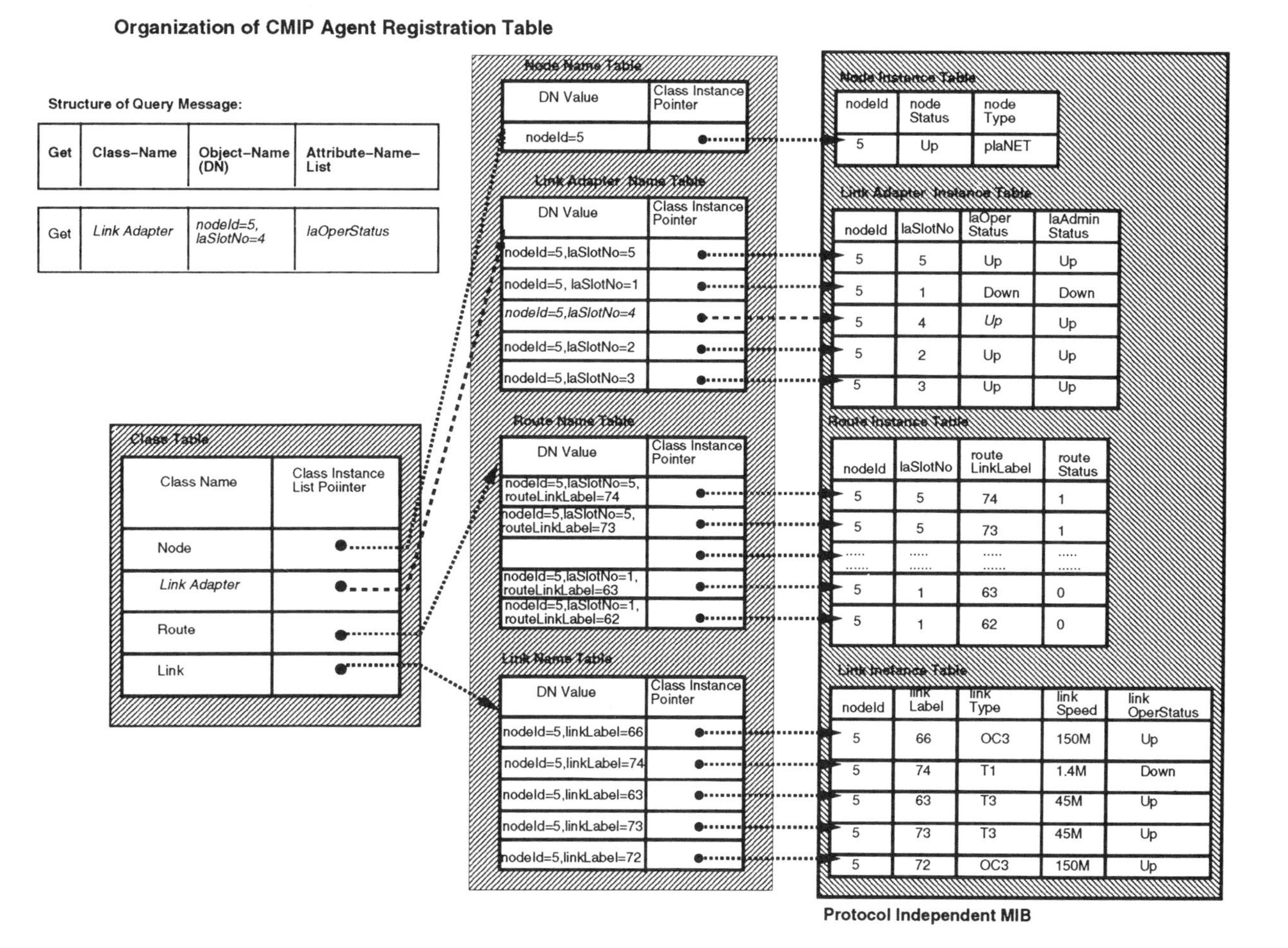

Figure 13. ER for CMIP.

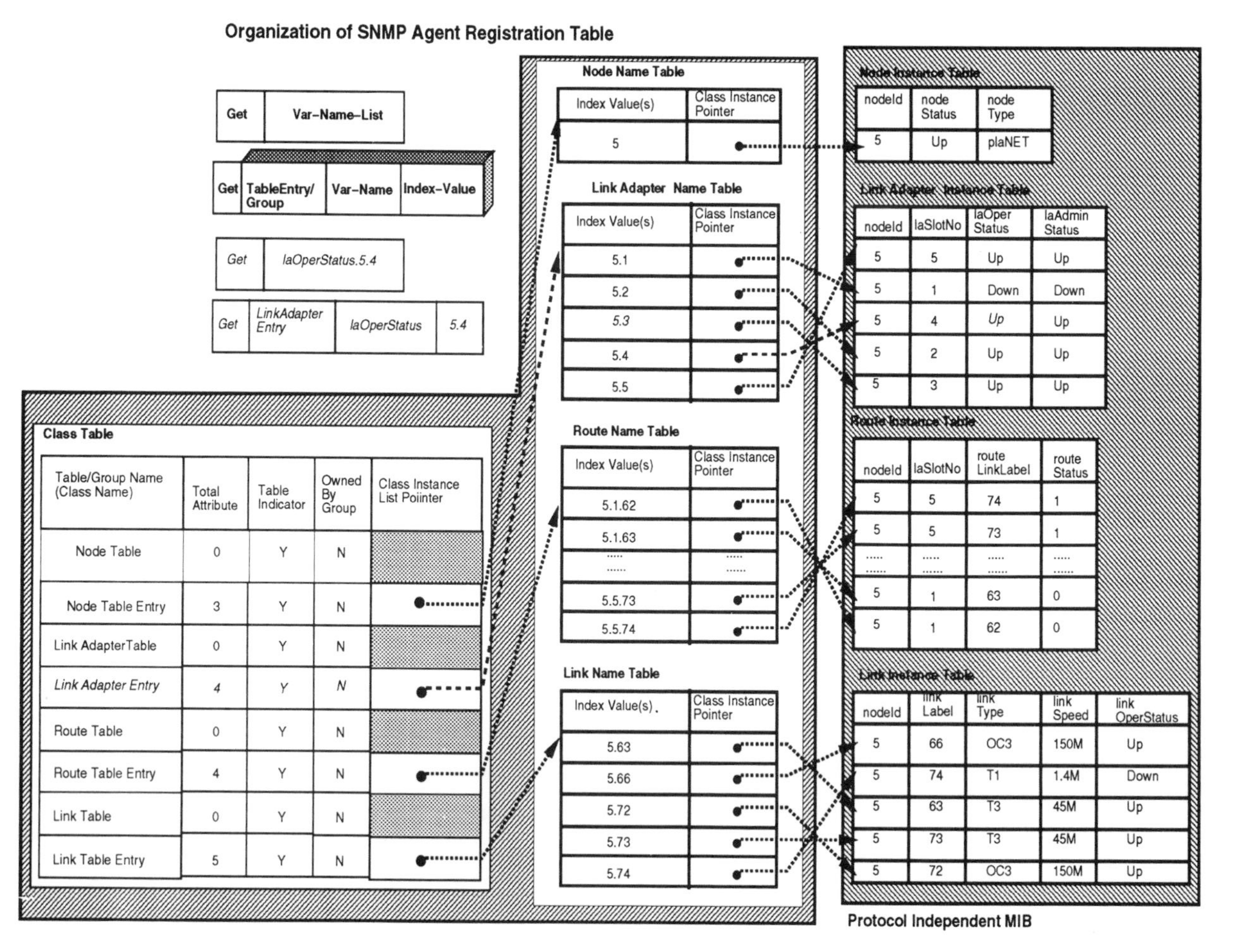

Figure 14. ER for SNMP.

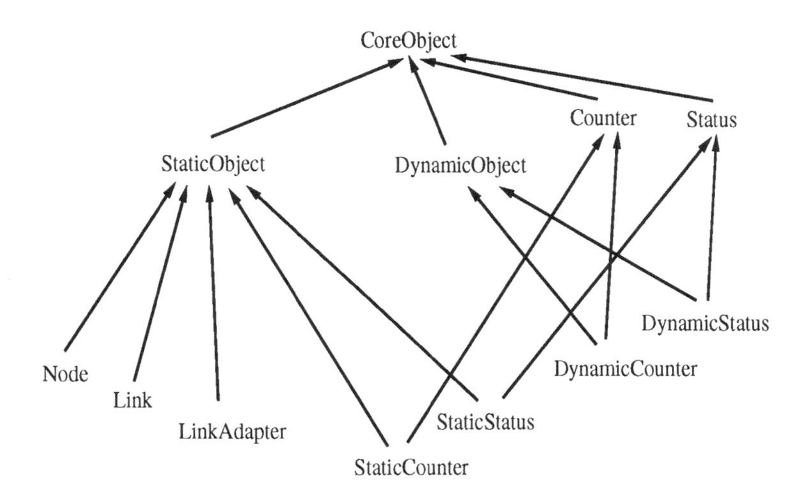

Figure 15. The Extended Inheritance Hierarchy

Registration

A registration process is used to build the mapping used by each protocol object to locate all of the *CoreObjects*. When the protocol object receives a request from a network management application, it can transform the object or variable identifier(s) in the request into a pointer(s) to a *CoreObject*(s), and forward the request to that *CoreObject* .

Not all the *CoreObjects* need to be registered, only those we wish to make visible to the various access protocols. We classify all the *CoreObjects* into two subclasses: *StaticObject* and *DynamicObject*. All objects that need to be registered inherit from *StaticObject*, and all other objects inherit from *DynamicObject*. Given an extended inheritance hierarchy as in figure 15, we can put a "registration" method call into the implementation of all the *StaticObjects*, and then, only *StaticObjects* would send a registration request to the CMIP/SNMP protocol objects. The "registration" functions are the following:

- int **DO_SNMP_Registration** (**className**, **keyValues**, **this**);
- int **DO_CMIP_Registration** (**className**, **coreDN**, **this**);

EXAMPLE: *Counter* and *CounterAttribute*

Counter is a very useful abstraction in modeling various kinds of network information. For example, number of packets received in one link adapter is conceptually a monotonically increasing counter. The easier way to implement a counter is to treat it as a *CDSAttribute* (figure 10). And, we need a five step procedure to get the value of the counter:

1. linkAdapter-3 receives a request: **getValue**(*packetsReceived*);
2. Using the attribute identity *packetsReceived* the object address of the *CDSAttribute* is obtained;
3. *CDSAttribute* performs a **getValue** on the *CoreValue*, and get back a pointer;

4. *CDSAttribute* uses that pointer to find the integer in the CDS;

5. *CDSAttribute* returns the integer to *CoreObject*;

However, if we want to attach other parameters to the *Counter* value, we need to treat the counter itself as an object. For instance, we might want to put a *threshold* or a *reset interval* together with the *Counter* value.

With the abstraction of *Counter* objects, the relation between a *CoreObject* (linkAdapter-3) and a *Counter* (packetsReceived) can be depicted as in figure 16. Now, the problem is that to satisfy a CMIP request, it is not sufficient to issue a "objid.**getValue**(*packetsReceived*)" request across the generic interface, because the *CoreValue* holds an object pointer to a *Counter*, not an integer. Our solution is to provide a class derived from *CoreAttribute*, *CounterAttribute*, which has the knowledge about the pointer held by the *CoreValue*. A typical CMIP query is then processed according to the following ten steps (in figure 17):

1. linkAdapter-3 receives a request: **getValue**(*packetsReceived*);

2. Using the attribute identity *packetsReceived* the object address of the *CounterAttribute* is obtained;

3. *CounterAttribute* performs a **getValue** on the *CoreValue*, and get back a pointer.

4. *CounterAttribute* uses that pointer to do a pointer.**getValue**(*value*); (*i.e.*, *CounterAttribute* makes an indirect call to the *Counter*.)

5. *Counter* forwards the request to *CDSAttribute*;

6. *CDSAttribute* gets the pointer stored in its *CoreValue*;

7. *CDSAttribute* uses that pointer to find the integer in the CDS;

8. *CDSAttribute* returns the integer to *Counter*;

9. *Counter* , after receiving the integer, sends it back to the *CounterAttribute*;

10. *CounterAttribute* returns that integer to the *CoreObject*.

Clearly, the first approach which needs fewer steps is more efficient in retrieving the information about a counter. The second approach, which treats a counter as a *CoreObject* , has two important advantages. First, if, in the MIB specification, a counter is directly addressable (i.e., we go through the registration process for all these counters), then we can simply use *StaticCounter* (figure 15) to model the counter directly and gain access to it more efficiently. Second, *Counter* introduces another layer of encapsulation. For example, there may be two different ways to get the counter values. One way might be through the *control data segment*, and the other way using some private access protocol (PAP). Now, if *Counter* is just an attribute, we need to have two different attribute types, *i.e., CDSAttribute* and *PAPAttribute*, and hence two different methods to get the values. However, this makes our MIB design harder because the designer has to worry about not only what attributes are contained in the object but also how to access the values. Thus, treating counters as objects, we separate concerns of "**what**" and "**how**" into two distinct design layers. In this way, the requirement for device knowledge in the "MIB" column of figure 6 can be reduced.

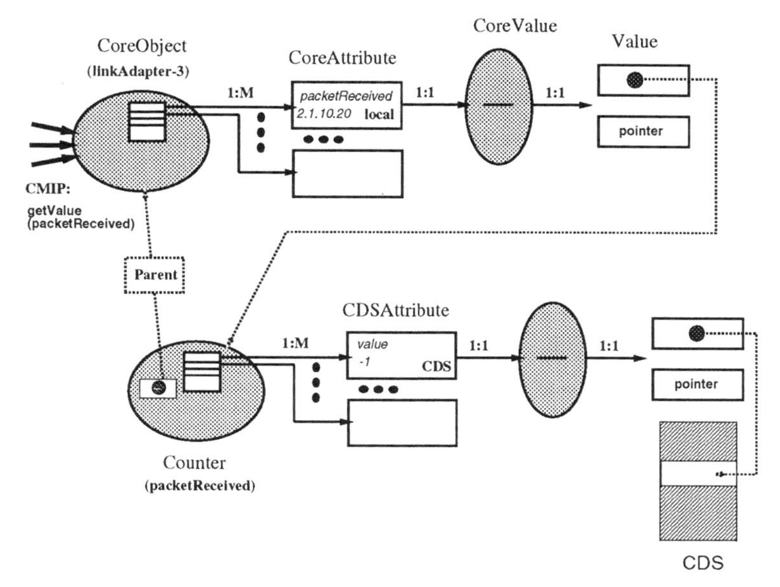

Figure 16. A *CoreObject*, a *CoreAttribute*, and a *Counter*

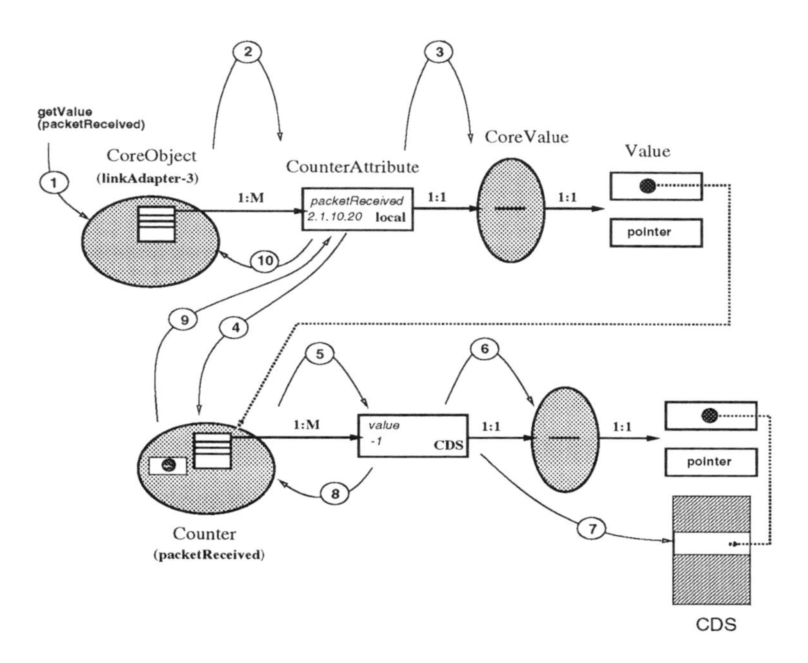

Figure 17. A *CoreObject*, a *CounterAttribute*, and a *Counter*

REMARKS

In this paper, we have analyzed the way different kinds of information can be encapsulated in layers of a network management system. We gave an object-oriented view of such a system and showed how various pieces of information and knowledge (*e.g.,* MIB specification, management protocols, and network devices) can be used by the layers. Then, the implementation of one particular layer, the MIB generic interface for all the protocol independent managed objects, was introduced. The key component of our implementation is the *CoreObject*, which contains various attributes, supports the naming and registration schemes for both SNMP and CMIP, and provides a flexible internal structure to model different types of network devices and information. Finally, we show that *CoreObject* provides various choices in modeling management information, such as a *Counter*, of network devices.

ABOUT THE AUTHORS

Shyhtsun F. Wu received the B.S degree in information science from Tunghai University, Taichung, Republic of China, in 1985, and the M.S. degree in computer science from Columbia University, NY, in 1989. He is currently a Ph.D. candidate in the Computer Science Department, Columbia University, NY, where he has held a IBM Graduate Fellowship since 1991. His research interests include real-time systems, mobile networking, network management, and object-oriented systems. Mr. Wu is a student member of ACM and IEEE.

Subrata Mazumdar has received the Ph.D. degree in electrical engineering from Columbia University, New York, NY, in 1990. Since 1990, he has been with IBM T.J. Watson Research Center as Research Staff Member. His current research interest are in the area of management of broadband networks and large-scale distributed systems.

Stephen Brady received the Ph.D. degree in Mathematical Logic from Cornell University in 1981. He joined IBM's Federal Systems Division in Owego, NY in 1982, as a member of the Advanced C^3 organization. In 1984 he became a member of the Network Management Systems project at the IBM T.J. Watson Research Laboratory in Yorktown Heights, NY. He currently manages that group as well as another project in the area of network design tools. His work in the Research Division has included investigation of expert systems and expert system - database interfaces, prototyping of network management systems, and definition of management architectures for future IBM networks.

David Levine is a Staff Programmer at the IBM Thomas J. Watson Research Center, where he is working on Mobile Computing and Personal Network Services. Prior work at IBM includes Several Network Management projects. He received a BA in Computer Science at Brandeis University in 1985, and is currently a Masters Candidate at Columbia University.

References

[1] P. P. Chen. The Entity-Relationship Model - Toward a Unified View of Data. *ACM Transactions on Database Systems*, 1(1):9–36, March 1976.

[2] J. Embry, P. Manson, and D. Milham. Interoperable Network Management. In I. Krishman and W. Zimmer, editors, *Second Integrated Network Management Symposium*, pages 29–44, Washington, D.C., April 1991.

[3] S. Mark Klerer. The OSI Management Architecture: an Overview. *IEEE Network*, 2(2):20–29, 1988.

[4] Subrata Mazumdar, Stephen Brady, and David Levine. Design of Protocol Independent Management Agent to Support SNMP and CMIP Queries. In *Third Integrated Network Management Symposium*, San Fransico, California, April 1993.

[5] M. Rose. Network Management is Simple. In I. Krishman and W. Zimmer, editors, *Second Integrated Network Management Symposium*, pages 9–25, Washington, D.C., April 1991.

[6] M. Rose. *The Simple Book*. Prentice Hall, 1991.

[7] Bjarne Stroustrup. *The C++ Programming Language*. Addison Wesley, 1986.

[8] Shyhtsun F. Wu, Subrata Mazumdar, and Stephen Brady. EMOSY: An SNMP Protocol Object Generator for the PIMIB. In *IEEE First International Workshop on System Management*, Los Angeles, CA, April 1993.

THE DESIGN AND IMPLEMENTATION OF AN EVENT HANDLING MECHANISM

Kouji Yata, Nobuo Fujii, and Tetsujiro Yasushi

NTT Transmission Systems Laboratories
1-2356 Take Yokosuka Kanagawa 238-03 Japan

ABSTRACT

This paper proposes an event handling mechanism and its application programming interfaces to the X Window System[1] for the purpose of telecommunication networks management application program development. The proposed mechanism is experimentally implemented on the UNIX[2] operating system using the C++ programming language based on object oriented concepts. The prototype of a telecommunication networks management system including OSI systems management protocol machines and sophisticated human-machine interfaces was implemented based on the proposed framework. It is confirmed that the mechanism provides sufficient environment for programmers without any knowledge of distributed application programming and window programming.

1 INTRODUCTION

The progress of standardization in ISO and CCITT of OSI systems management and Telecommunications Management Networks (TMN) provides inter-operability for network management [1][2][3][4][5].

One serious barrier to the creation of Telecommunication Networks Management (TNsM) Application Programs(APs) is the ever increasing complexity of the programs required. Distributed processing environments demand that the TNsM AP have a complex processing control mechanism to deal with messages received asynchronously from multiple APs.

---

[1] The X Window Sytem is a trademark of Massachusetts Institute of Technology.
[2] UNIX is a registered trademark of UNIX System Laboratories.

Network Management and Control, Vol.2, Edited by
I.T. Frisch *et al.*, Plenum Press, New York, 1994

These facts, coupled with the current lack of application programmers who have experience in TNsM APs, mean that a new and effective environment must be developed that enables even novice programmers to construct high performance TNsM APs.

This paper proposes an event handling mechanism and an Application Programming Interfaces(API) for the window system for a development environment for TNsM APs, based on the object oriented concepts. This paper also describes a method to implement the proposed framework on the UNIX operating system using the C++ programming language. The authors have built an experimental prototype of a TNsM AP with an efficient Human Machine Interface(HMI), that uses the standard OSI system management protocol.

Chapter 2 describes the implementation environment of the TNsM AP. In Chapter 3, the processing control model of the TNsM AP is discussed. In Chapter 4, the implementation method of the processing control mechanism is described. Chapter 5 describes the experimental TNsM AP prototype.

2 IMPLEMENTATION CONDITIONS OF THE TNsM AP

As the AP development environment, the following platform was selected.
(1) Operating System
UNIX OS : The UNIX OS has been implemented on a wide variety of computers so that UNIX OS-based APs have high portability. Many kind of development tools are available such as compilers, graphical window systems, and OSI protocol stack libraries. Inter-process communication is well supported with the interface function socket() that enables APs to communicate with other APs in a distributed processing environment [6][7][8].

(2) Window System
X Window System : The X Window System is the standard UNIX OS window system. It has been implemented on many kinds of UNIX work stations so that it is very portable. The X Window System is a network-oriented window system and depends upon the Server-Client processing model. The Server is a system program which draws graphics to the display and passes user input messages to the Clients. Each Client is an AP which sends drawing requests to the Server and receives user input messages from the Server using the Xlib. The Xlib is the function library that establishes communication with the Server through the socket interface. In this model, a TNsM AP is treated as a Client. User input messages are called X Events and occur asynchronously [9][10].

(3) Language
C++ : C++ is an object oriented programming language. Its specifications are a super-set of the C language, and it allows highly efficient access to the C language libraries such as the Xlib and OSI protocol stack libraries. Its object oriented framework allows application programmers to easily reuse modules written by other programmers. The key concepts are data encapsulization and class inheritance [11].

3 DESIGN OF AN EVENT HANDLING MECHANISM AND WINDOW SYSTEM INTERFACES

3.1 Processing Control Model in Distributed Processing Environment

The processing control model in the distributed processing environment is created based on the above AP implementation environment. We will now describe one example of a TNsM AP that uses the X Window System as HMI functions: the Manager AP in the

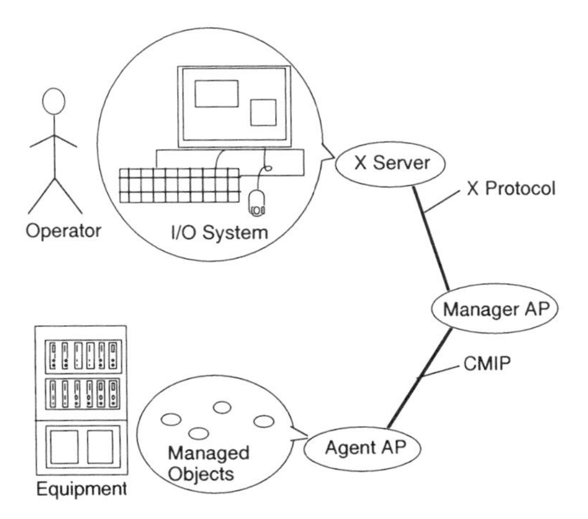

Fig. 1 Operation Architecture of the Telecommunication Network OpS

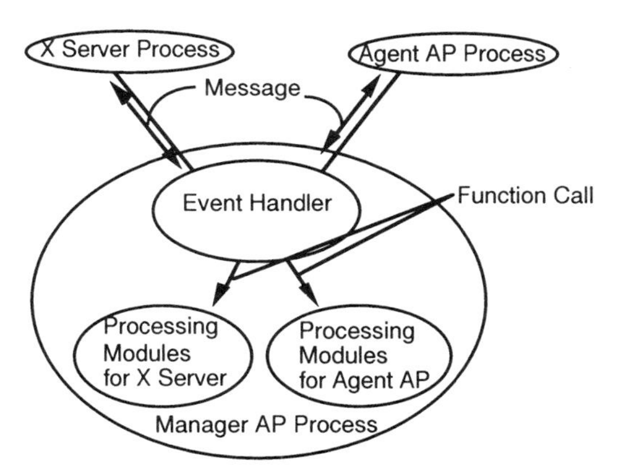

Fig. 2 Processing control model

OSI System Management model. Fig. 1 shows the operation architecture of the telecommunication network Operation System(OpS) using this Manager AP.

Fig. 2 shows the processing control model of the Manager AP which is realized as a process on the UNIX OS. The Manager AP process exchanges messages with the X Server process and the Agent AP process using the socket interface.

The ***event handler*** observes the messages from the other processes and dispatches it to the appropriate processing module. An ***application thread*** is established to group several processing modules, when they share the same socket. State transitions such as the arrivals of a message are called ***events***. An application thread has processing modules for each event type that it must respond. Each processing module, which is realized as a function, is called a ***task***. The event handler observes the occurrence of events, and calls the ***task function*** appropriate for the event that has occurred.

There are five event types described below.

(1) **Read event**: occurs when the descriptor of a socket is ready to read. This event means that some processing request data has been received by the descriptor of the socket from another process.

(2) **Write event**: occurs when the descriptor of a socket is ready to write. This event means that the matching process of the socket descriptor is ready to read.

(3) **Local event**: occurs when a processing request has been received but not directly through the descriptor. In the data communication framework of the X Window System, it is possible for one or more X events to have been already received and stored in the local queue.

(4) **Timeout event**: occurs when a specified time is exceeded. Using this event, a timeout function can be realized such as an acknowledgment timeout for communication.

(5) **IntervalTimeout event**: occurs on a fixed period if an interval timer is set. This event can realize blinking or animation on the display.

To observe event occurrence, the event handler must call functions to get descriptor values and to check the occurrence of the local event. These functions are called ***observation functions***.

3.2 Abstraction of Application Threads and the Object Oriented Concept

So that the event handler can call the task functions and observation functions, it must know the function names to be referenced. Therefore, when a new application thread is created, application programmers must update the event dispatching module of the event handler to call the new functions. The object oriented concepts of C++ allow these functions to be called in a unified way by the event handler. The basic technique is to abstract the application threads.

As the class of abstract application threads, ***AbstractApplicationThread*** class, which has the virtual functions of task functions and observation functions, is defined. The subclasses of the AbstractApplicationThread are defined as concrete application thread classes, and each virtual function is redefined according to the requirements of each application thread. Fig. 3 shows the class hierarchy of these classes.

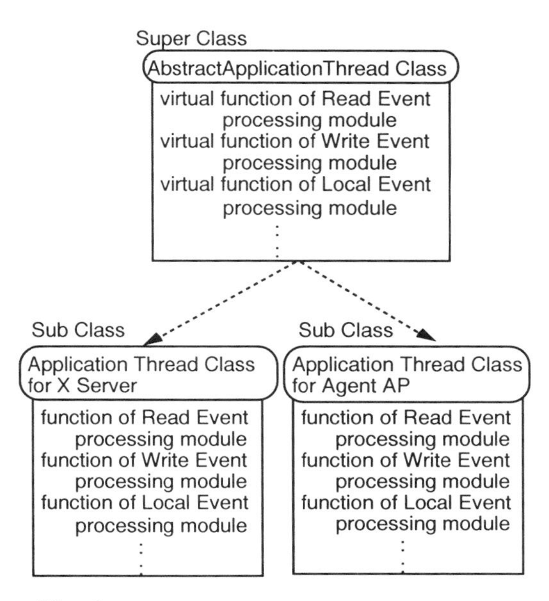

Fig. 3 Class hierarchy of Application Thread Objects

Fig. 4 shows the object oriented event handling mechanism. The event handler also acts as an object. The event handler object manages the various application thread objects. It treats any application thread object as an AbstractApplicationThread object and calls virtual functions for observe and dispatch events. As these virtual functions are redefined in the subclass, the event handling mechanism always works correctly.

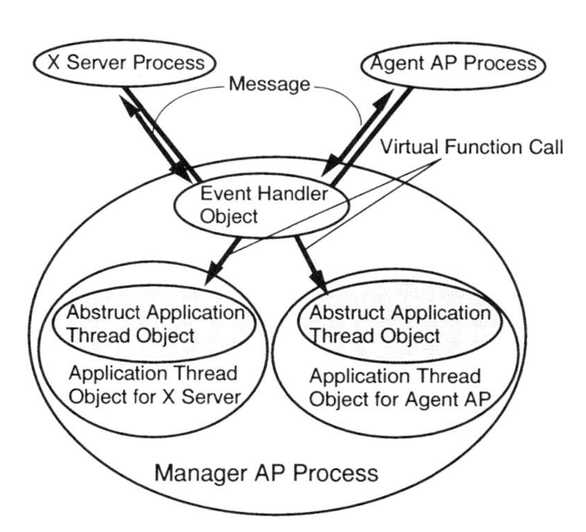

Fig. 4 Object Oriented Event Handling Mechanism

Using the technology described above, the event handler does not need to be changed, and the application programmers are freed from worrying about event observation or event dispatching.

As the AP development environment which considers the event handling mechanism, we offer class definitions of the ***EventHandler*** object and the AbstractApplicationThread object. This environment permits application programmers to easily make APs which work under the event handling mechanism; new application thread classes are defined as subclasses of the AbstractApplicationThread class.

3.3 Processing Control Model of the X Window System

This section describes the processing control model which constructs the API around the X Window System. Fig. 5 shows the processing control model for X events.

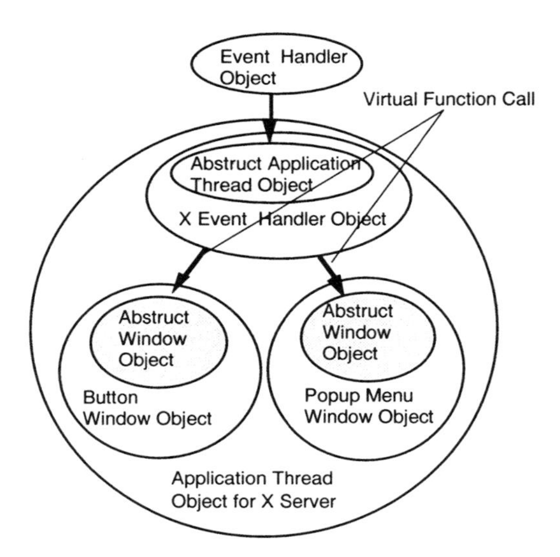

Fig. 5 X Event Handling Mechanism

As the window event handler, we designed the ***X_EventHandler*** class which is a subclass of the AbstractApplicationThread. The X_EventHandler object observes the occurrence of an X event and dispatches it to the window object. For the abstract window object, the ***AbstractWindow*** class is defined. The AbstractWindow object has virtual functions to process the X event. When an X event occurs, the X_EventHandler object investigates the window in which the X event occurred and calls the virtual function of the AbstractWindow object according to the type of the X event. The concrete window objects, such as button, pop-up menu and scroll bar window object, are defined as subclasses of the AbstractWindow class.

4 IMPLEMENTATION OF AN EVENT HANDLING MECHANISM AND WINDOW SYSTEM INTERFACES

This chapter describes the implementation method of object classes.

4.1 AbstractApplicationThread class

An application thread can be abstracted as an object which can be handled by the EventHandler object by implementing the AbstractApplicationThread class. The AbstractApplicationThread object has virtual functions which can be called by the EventHandler object to observe event occurrence and dispatch events.

The observation functions are follows.

get_read_fd() : returns the descriptor value of socket to read data.
get_write_fd() : returns the descriptor value of socket to write data.
check_lq() : returns a Boolean value whether there are Local events.
set_timeout() : sets the timeout value (1/1000 s). If the value is smaller than zero, the event will never occur.
get_timeout() : returns the timeout value (1/1000 s).
set_intervaltimeout() : sets the interval timeout value (1/1000 s). If the value is smaller than zero, the event will never occur.
get_intervaltimeout() : returns the interval timeout value (1/1000 s).

The task functions are follows.

read_fd() : called when the Read event is observed.
write_fd() : called when the Write event is observed.
process_lq() : called when the Local event is observed.
timeout() : called when the Timeout event occurs.
intervaltimeout() : called when the IntervalTimeout event occurs.

The method functions needed to determine priority are as follows.

set_read_fd_priority() : sets the processing priority value for the Read event.
get_read_fd_priority() : returns the processing priority value for the Read event.
set_write_fd_priority() : sets the processing priority value for the Write event.
get_write_fd_priority() : returns the processing priority value for the Write event.
set_lq_priority() : sets the processing priority value for the Local event.
get_lq_priority() : returns the processing priority value for the Local event.
set_timeout_priority() : sets the processing priority value for the Timeout event.
get_timeout_priority() : returns the processing priority value for the Timeout event.
set_intervaltimeout_priority() : sets the processing priority value for the IntervalTimeout event.
get_intervaltimeout_priority() : returns the processing priority value for the IntervalTimeout event.

The AbstractApplicationThread class has a facility for registering itself to the list which can be referenced by the EventHandler object. A specific application thread will be defined as a subclass of the AbstractApplicationThread class. Virtual functions must be redefined when defining a subclass.

4.2 EventHandler class

The EventHandler class is implemented as the object which observes event occurrence and dispatches events. Fig. 6 shows an object allocation model under the event handling mechanism. The EventHandler object manages AbstractApplicationThread objects using the list holding pointers to the AbstractApplicationThread objects. The ***MainLoop()*** module, which is the method function of the EventHandler, observes event occurrence and dispatches events.

To observe event occurrence, the ***select()*** function, which is a UNIX OS system call, can be used. This function examines the I/O descriptor sets to see if some of their descriptors are ready for reading or writing, and blocks the process until the descriptors are ready or timeout occurs. The usage of select() function allows the EventHandler to observe the following four sets of events:Read events: Write events: Timeout events: and IntervalTimeout events.

The MainLoop() module is constructed with following two stages.

At the first stage, the EventHandler makes a nonblock check of event occurrence in order of their priority. When an event is detected, the EventHandler calls the task function to process the event. Local events also can be processed at this stage.

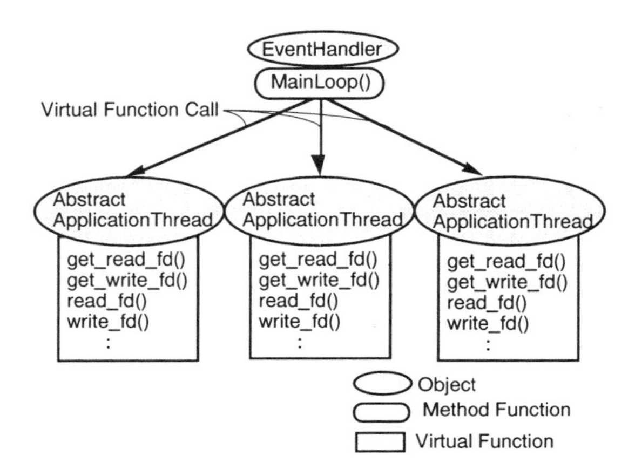

Fig. 6 Object Allocation Model under the Event Handling Mechanism

At the second stage, the EventHandler collects the read descriptor values, the write descriptor values, and the interval times until the Timeout event or the IntervalTimeout event occurs. Then, select() is called with the appropriate parameters. The timeout parameter of select() is the minimum value of the interval taken until the Timeout event or IntervalTimeout event occurs. When select() is returned, several events may be triggered, and the processing control return to the first stage.

When multiple events occur at the same time, they are processed in order of their priority. The priority values can be dynamically changed while the program is running. This allows application programmers to control the event processing priority in order to fit into the context of the AP.

Because the framework does not offer the preemptive processing control, the task functions must be carefully designed in order for the event processing to finish in a short period of time.

4.3 Execution of the Application Thread objects

In the AP implementation stage, the application programmers make a concrete application thread class as a subclass of the AbstractApplicationThread. The virtual functions must be redefined to match the requirements of concrete application thread. The AP works within the event handling mechanism by creating each object and the function calls of MainLoop().

4.4 X_EventHandler class

We define the X_EventHandler class as a subclass of the AbstractApplicationThread. An X_EventHandler object corresponds, one to one, to a ***Display Connection*** in the framework of the X Window System. The object opens the display using XOpenDisplay(), which is a function in Xlib, upon it's creation. The get_read_fd() function returns the descriptor of the socket to the X server process. The check_lq() function returns the number of X events in the local event queue using XPending() function in Xlib. X_EventHandler triggers X events using the XNextEvent() function in Xlib and dispatches X events to each window object by calling event processing functions of the window object.

4.5 AbstractWindow class

The AbstractWindow class was implemented as the abstract object that can be managed by the X_EventHandler object. Each AbstractWindow object corresponds to a ***window*** in the framework of the X Window System. At its creation, it registers itself to the list which can be referenced by the X_EventHandler object. The X event processing functions are virtual functions and are described as follows.

expose() : occurs when a window becomes visible on the screen, after being obscured or unmapped.
key() : occurs when the operator pushes or releases a key of the keyboard.
button() : occurs when the operator pushes or releases a mouse button.
enter() : occurs when the mouse pointer enters the window.
leave() : occurs when the mouse pointer leaves the window.
move() : occurs when the mouse pointer moves in the window.
resize() : occurs when the size of the window is changed.
client() : occurs when a message arrives from another client.

As the API to the X Window System, the AbstractWindow object has method functions which are coded using the Xlib. These functions were implemented by a programmer who has substantial experience in X window programming.

4.6 Execution of Window Objects

In the HMI implementation stage, application programmers make a concrete window as a subclass of AbstractWindow. The virtual functions must be redefined according to the concrete window. By creating each object and calling the MainLoop() function of the EventHandler object, the X_EventHandler object can be processed by the event handling mechanism so that every window object will fall under the X event handling mechanism.

5 EXPERIMENTAL IMPLEMENTATION OF A TNsM AP

To verify the proposed processing control model and the implementation method, we constructed an experimental TNsM AP. The prototype is a Manager AP in the OSI System Management model and has HMI functions.

5.1 Summary of Specification

The Manager AP manages the Managed Objects(MOs) in the Agent AP using Common Management Information Protocol (CMIP) [12][13]. For HMI functions, it uses windows that display the Managed Information Tree (MIT) and CMIP data editor. Using this AP, operators can display the MIT and send CMIP commands such as M-GET, M-SET and M-ACTION by mouse operations. The data from Agent APs, such as the response data of M-GET and M-SET, can also be displayed on the screen.

5.2 Class Implementation and Object Allocation

To implement the manager AP, we defined the subclasses of Abstract-ApplicationThread and AbstractWindow. Fig.7 shows the class hierarchy of the application threads. The ***CMIP*** class is defined as the object that hosts the API to achieve CMIP communication using ISODE and OSIMIS[14][15]. The ***SampleManager*** class is defined so that an instance of that class manages the telecommunication network using the API offered by the CMIP class.

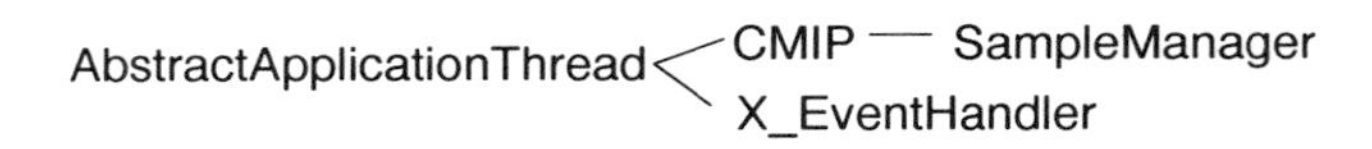

Fig. 7 Class Hierarchy of Application Threads

Fig. 8 shows the class hierarchy of windows. According to the HMI specification, the necessary windows were defined as window objects.

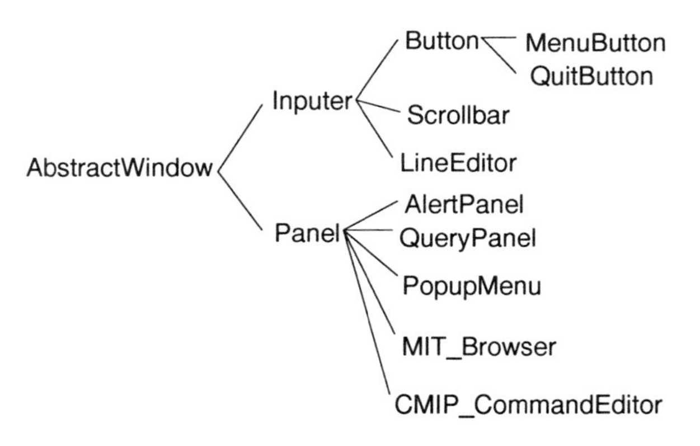

Fig. 8 Class Hierarchy of Windows

Fig. 9 shows object allocation. In the initialization routine, each object is created, and by calling the MainLoop() function of the EventHandler object, each object will perform its role.

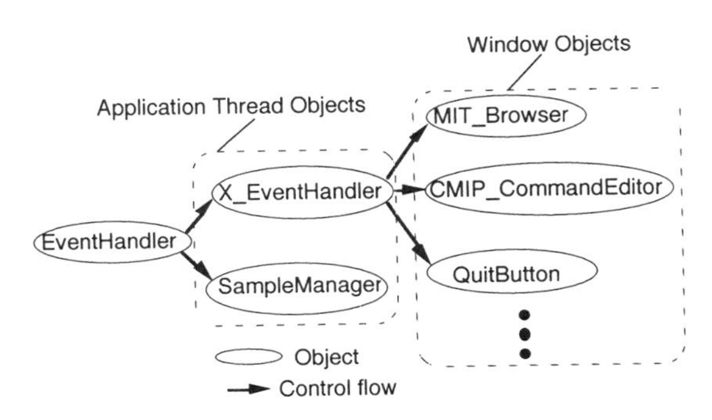

Fig. 9 Object Allocation of the Manager AP

6 DISCUSSION

By introducing a sophisticated event handling mechanism based on object oriented concepts, modules can be coded without considering the other modules' processing control. This eliminates the interdependency of modules. All application thread objects were found to work independently. By encapsulating internal processes and offering a well-defined interface, object reusability can be significantly increased. An object's behavior can be easily changed through the class inheritance mechanism. This means that application thread objects can be regarded as ***software parts***.

For example, the CMIP class, introduced in Chapter 5, can be thought of as the software parts that provides the CMIP communication interface. When implementing the experimental TNsM AP, the programmer who wrote the SampleManager class was different from the person who wrote the CMIP class. Although he did not know the details of the communication frameworks such as socket, ISODE and OSIMIS, he found it easy to construct the SampleManager class and implement the TNsM AP.

By offering software parts for TNsM APs, TNsM APs can be developed more easily. In the framework of the C++ language, the software parts are class definitions. Each class is a derived class of the AbstractApplicationThread or the AbstractWindow, and classes make hierarchical trees as shown in Chapter 5. A class near the leaf is a special class which has a complete specification for special processing. It can be easily used by novice programmers, but it does not have expandability. A class near the root of tree is a general class and is only partially specified; expert programmers can freely expand its behavior by defining subclasses. The hierarchical class trees allow programmers with varied levels of skill to develop TNsM APs.

By reusing the HMI software parts, the HMI operation environment is unified so that the operators can easily manage the telecommunication network in a consistent manner even for different APs. On the other hand, by offering multiple sets of HMI software parts, every set performing the same operations in a different manner, the HMI can be tailored to the operator's skills or preferences.

7 SUMMARY

In order for novice programmers to develop portable TNsM APs that offer high performance HMI functions, a TNsM AP development environment has been created on the UNIX OS using the X Window System. We have proposed an event handling mechanism and API to the X Window System, both based on object oriented concepts, as the development environment for TNsM APs. The method of implementing the proposed framework on the UNIX operating system using the C++ language was described. The framework was verified by the implementation of an experimental TNsM AP. The results clarified that the proposed environment, which offers reusable TNsM AP software parts, application programmers can easily construct highly effective TNsM APs even if they don't have much experience in distributed application programming or window programming.

ACKNOWLEDGMENTS

The authors would like to thank Dr. Tetsuya Miki, Dr. Haruo Yamaguchi, and Mr. Kazumitsu Maki of NTT Transmission Systems Laboratories for their continuous encouragement. We would also like to thank Mr. Yasumi Matsuyuki of NTT Telecommunications Software Headquarters, Mr. Ikuo Yoda, and Mr. Kenji Minato of NTT Transmission Systems Laboratories for instructive discussions and help in the implementation of the experimental TNsM AP.

ABOUT THE AUTHORS

Kouji Yata (kouji@ntttsd.NTT.JP) was born in Japan on 1961. He received the M.S. degree from the Osaka University in Information and Computer Sciences in 1987. Since then he has been working in NTT Transmission Systems Laboratories. His first research in NTT was the Human-Machine Interface for the Transport Networks Operation System. His current research field is development environment of the Telecommunication Networks Management Systems.

Nobuo Fujii (nobuo@ntttsd.NTT.JP) is a senior research engineer, supervisor in NTT Transmission Systems Laboratories. He received his B.Sc. and M.Sc. from Osaka University in 1977 and 1979, respectively. He has been active in the Japanese and International groups including ISO, ITU-T and NMForum on OSI Systems Management Standards. He is a member of IEEE. His research interests include distributed network management, information modeling and network operation systems engineering.

Tetsujiro Yasushi (yasu@ntttsd.NTT.JP) received the B.E. degrees from Waseda University, Faculty of Science and Engineering, Tokyo, in 1974 and 1976, respectively. After joining the Electrical Communication Laboratory of NTT in 1976, he worked on research and development of network synchronization, network control and asynchronous time division multiplex systems. He is presently working on the network operation and management system as a senior research engineer, supervisor. His current technical interests include distributed architectures of the Network Operation System. He is a member of the Institute of Electronics, Information and Communication Engineers of Japan.

REFERENCES

[1] OSI / NM Forum Release-1.
[2] ISO/IEC 10040:1991 : Systems Management overview.
[3] ISO/IEC 10165-1:1991 : Management Information Model.
[4] ISO/IEC 10165-2:1991 : Definition of Management Information.
[5] ISO/IEC 10165-4:1991 : Guideline for Definition of Managed Objects.
[6] B. W. Kernighan, R. Pike, "The UNIX Programming Environment", Prentice-Hall, 1984.
[7] M. J. Rochkind, "Advanced UNIX Programming", Prentice-Hall, 1985.
[8] S. J. Leffler, M. K. Mckusick, M. J. Karels, J. S. Quarterman, "The Design and Implementation of the 4.3 BSD UNIX Operating System", Addison-Wesley, 1991.
[9] Adrian Nye, "Xlib Programming Manual for version 11", O'Reilly & Associates, 1990.
[10] Adrian Nye, "Xlib Reference Manual for version 11", O'Reilly & Associates, 1990.
[11] Stanley B. Lippman, "C++ PRIMER", Addison Wesley, 1989.
[12] ISO/IEC 9595:1991 : Common Management Information Service Definition.
[13] ISO/IEC 9596:1991 : Common Management Information Protocol Specification.
[14] M. T. Rose, "The ISO Development Environment User's Manual", U. Delaware, 1990.
[15] G. Pavlou, G Knight and S. Walton, "Experience of Implementing OSI Management Facilities", Second IFIP Symposium on Integrated Network Management , Washington, April 1991.

ADVANCED OPERATIONS SYSTEM
FOR MOBILE COMMUNICATIONS NETWORK

Hironori Fukushima, Katsuhiro Tsujinaka, Motoshi Tamura, and Ichio Osano

NTT Mobile Communications Network Inc.
9-11 Midori-Cho 3-Chome
Musashino-Shi, Tokyo 180 Japan

ABSTRACT

This paper discusses operation problems and their causes in mobile communications network, and proposes advanced operation concepts that overcome these problems for a future mobile communications network. Network operation are becoming increasingly difficult as the number of users is increasing rapidly and the variety of services is also increasing. The functional architecture for these advanced operation concepts is described.

INTRODUCTION

Demand for mobile communications has grown rapidly, therefore mobile communications network will be expanded,and will be more intelligent and complex. Structure example of the mobile communications network, and operation systems for this network are shown in figure 1.

The network consists of the mobile gateway/transit stage (MGS) and the mobile local stage (MLS). Each mobile switching center (MSC) on the MLS has the base stations (BSs). A group consisting of both of them is called a subscriber access group in this paper.

There are three operation systems for this network; the operation system for the MGS-MSCs (OPS-MGS), the operation system for the MLS-MSCs (OPS-MLS) and the operation system for the base stations (OPS-BS).

Each operation system realizes the remote and centralized operation, but all three operation systems work independently.

The first part of this paper describes operation problems and their causes using the subscriber access group operational flow as an example.

PROBLEMS IN THE SUBSCRIBER ACCESS GROUP OPERATION

Figure 2 shows an example of the operational flow when trouble with a subscriber access group is reported by a customer.

In this example, an operator judges which system has trouble, MLS-MSC or BS, in the first period of the operational flow, and then searches the cause of trouble only in the system which is considered to have the trouble.

The operational flow may make trouble searching inefficient, because the causes of trouble may lie on several systems or elements and may related complicatedly. Therefore we must search the cause of trouble with testing related systems conjunctively and efficiently, and we must supervise the whole subscriber access group to grasp the condition of trouble correctly.

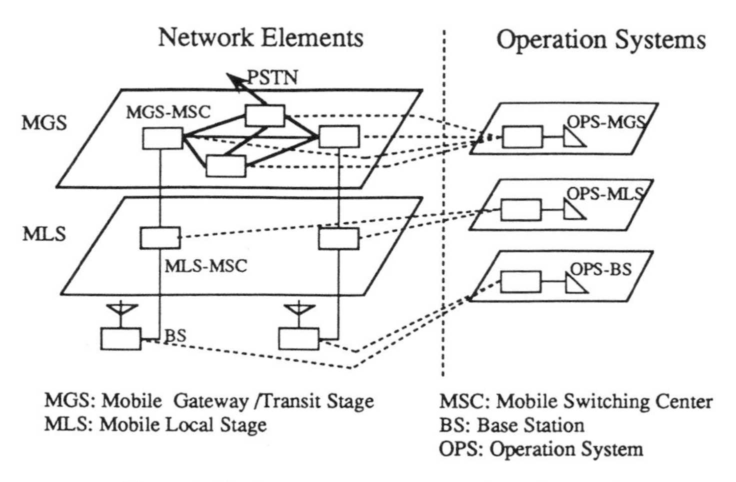

Figure 1. Mobile communications network and operation example

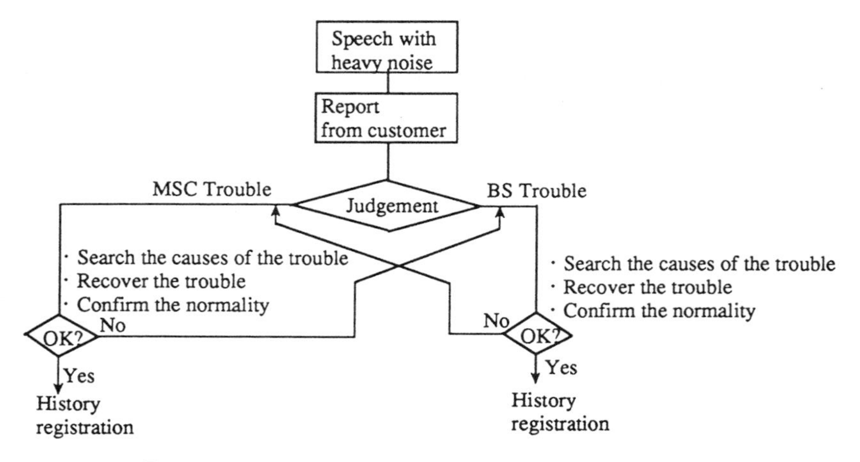

Figure 2. Task flow example in the case of subscriber access group trouble

NEW OPERATION CONCEPTS FOR THE NEXT STAGE

The fundamental concepts for efficient operation are considered to be available not only for the trouble shooting of the subscriber access group but all kinds of operation.

Mobile communications network will be expanded as the number of subscribers increases, and will be more intelligent and complex as variety of services increases. These factors make network operation more difficult. Two operation concepts are necessary for the mobile communications network operation to keep efficient operation under such network conditions. First, operators must control, test and recover a network following the efficient flow of each task with relating the referential network elements efficiently. We call this concept "Task Flow Oriented Operation" in this paper. Second, operators must supervise the whole network to grasp the network condition correctly, and then judge the most efficient flow to control, test and recover the network. We call this concept "Total Network Operation" in this paper. Figure 3 shows these concepts. New operation system is necessary to realize these concepts.

We use the word "task" in this paper. The "task" means each element of operation, for example network

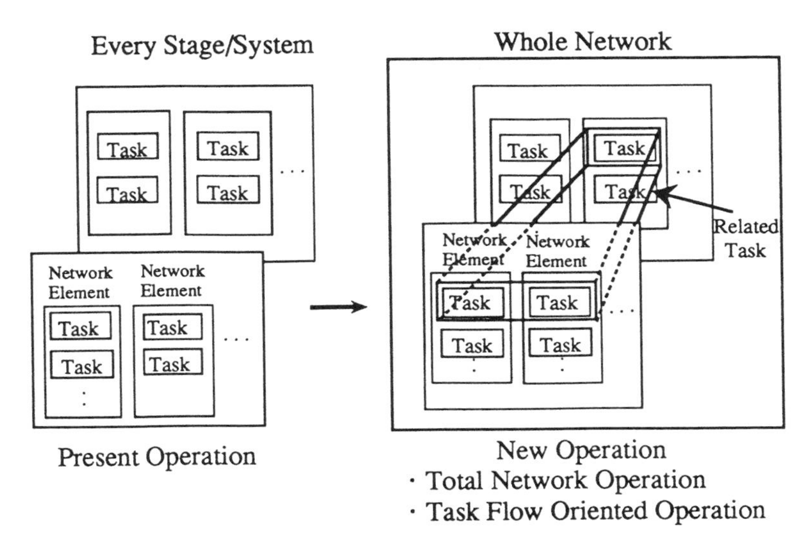

Figure 3. Image of new operation concepts

trouble shooting, traffic congestion control etc. and an operational flow in execution of the task is called "task flow" in this paper.

FUNCTIONAL ARCHITECTURE OF THE NEW OPERATION SYSTEM

Functional architecture of the new operation system is shown in figure 4.

In the present operation, network elements(MSC etc.) have several individual operation and maintenance functions, for example, signal path test and so on. These functions are carried out by operator's commands,

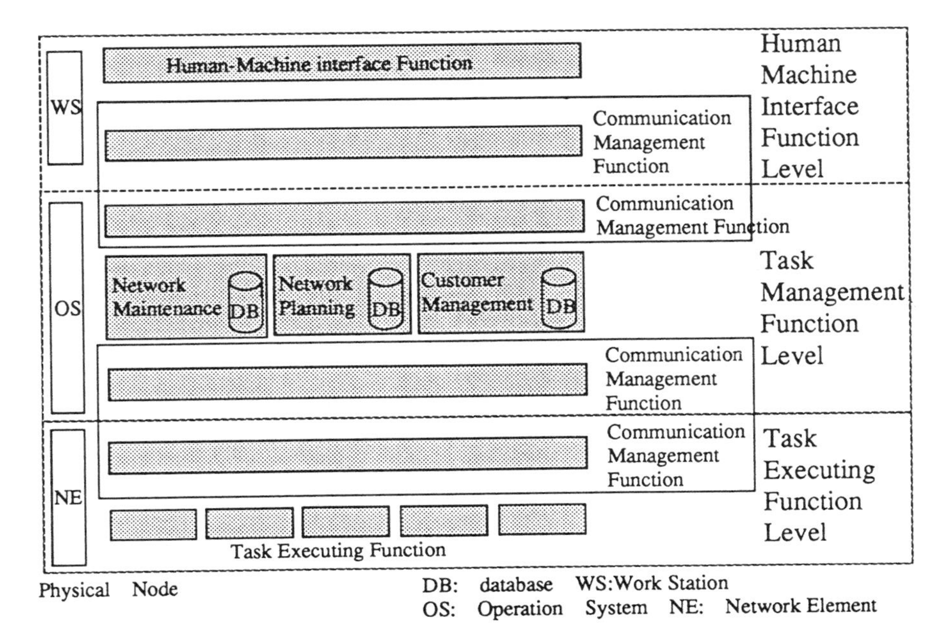

Figure 4. Functional architecture of the new operation system

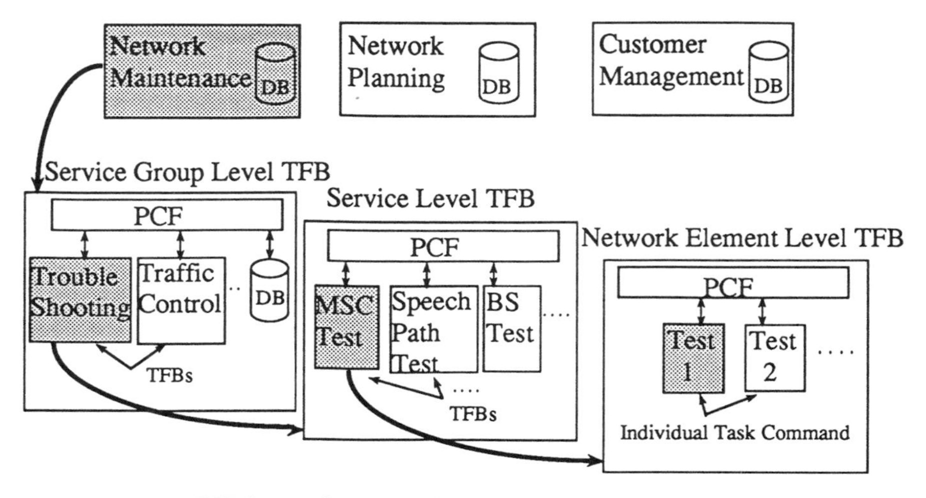

Figure 5. Task management function structure

and the operator manages the task flow with judging the answers of the commands. The new operation system should support the operator's task flow management for decreasing operator's misjudgment and for efficient task flow, and also should operates several kind of systems together for the total network operation and for more efficient task flow.

The task flow oriented operation and the total network operation are supported by a unified human-machine interface, and the operation system should be added the new task easily and each operation functions should have independency to the modification of other functions. The functional architecture must satisfy these conditions.

The architecture has three fundamental function levels. The upper level function links operators to operation functions. We call this level "Human-Machine Interface (HMI) Function Level". The middle level functions control and manage the task flows. We call this level "Task Management Function Level". This level is the new function which supports the operator's task flow management. The lower level functions execute the task under the control of the task management function. This level corresponds to the present individual operation and maintenance functions in the network elements. We call this level "Task Executing Function Level".

TASK MANAGEMENT FUNCTION LEVEL STRUCTURE

The task can be classified into three levels hierarchically. The general level task means fundamental group of operation elements, for example, network maintenance, network planning and customer management and so on. We call the task in this level "Service Group Task". The middle level task means main elemental unit of operation, for example, network trouble shooting and traffic congestion control and so on. We call the task in this level "Service Task". Each service task belongs to the service group task, for example, the network trouble shooting is an element of the network maintenance. The detailed level task means the network element operation in each service task, for example, the MSC test in the network trouble shooting. We call the task in this level "Network Element Task".

The task management function level will structured following this constitution as shown in figure 5. This level is classified into the service group level, the service level and the network element level, and the functions necessary for a task are grouped into unit called "Task Function Block (TFB)". An unit of the TFB corresponds to an unit of the task in each levels, so a service group level TFB has several service level TFBs in it, and one of the TFBs has several network element level TFBs in it. The network element level TFB has individual task commands in it, for example, common memory test command in the MSC test TFB, and so on. The individual task commands are the least unit of the task management function level. Each of the TFBs has a Process Control Function (PCF) in it. This is the main function in the TFB and manages the task flow.

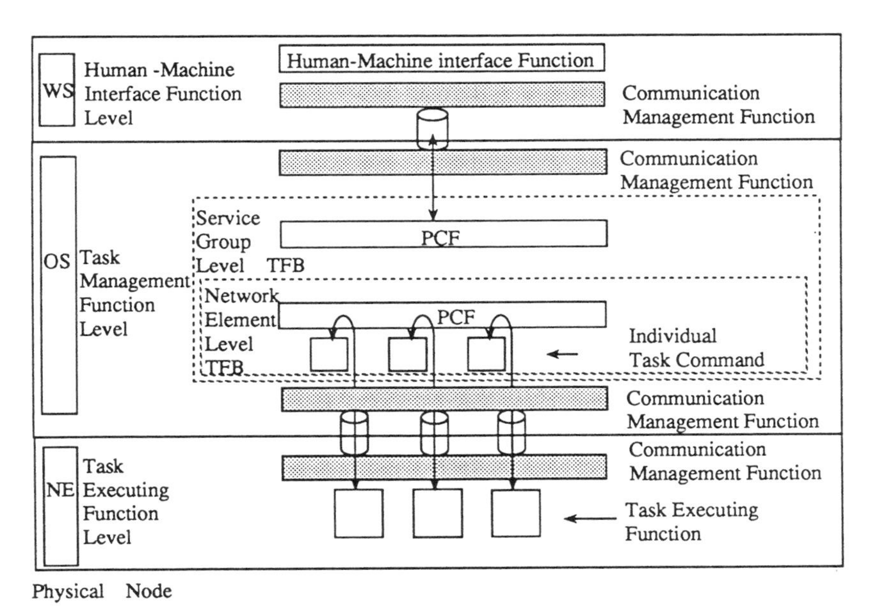

Figure 6. Relation and connections between function levels

RELATION BETWEEN THE HMI FUNCTION LEVEL, THE TASK EXECUTING FUNCTION LEVEL AND THE TASK MANAGEMENT FUNCTION LEVEL

Figure 6 shows the relation between the HMI function level, the task executing function level and the task management function level, and also shows how the TFB in the task management function communicates with other functions.

In the task management level, conversations with the HMI function level are managed by the PCF in the service group level TFB, so a logical connection will be established between them for conversations through a task operation.

On the other hand, each of task executing functions carries out the individual task, which is ordered by the PCF in the network element level TFB. Individual task commands are prepared in the network element level TFB in the task management function level, and each of the task executing functions corresponds to each of the individual task commands. There will be several logical connections, each of which is established between each task executing functions and the PCF in the network element level TFB.

TASK FLOW EXAMPLE THROUGH EACH FUNCTION LEVEL

Figure 7 shows an example of the whole task flow through each function level in the case of subscriber access group trouble shooting. In this example, network elements are restricted to only the MSC and the BS, and the flow is simplified as testing the MSC first and then testing the BS sequentially to make the example simple.

The task management function level is classified into three levels according to the task level, as described before, and PCFs in each level TFBs manage the task flows. The task management function level in this figure shows this concept. The horizontal arrows in the service group level shows the task flow managed by the PCF in the network service management TFB. The task flow managed by the PCF in the trouble shooting TFB is shown in the service level, and the task flows managed by the MSC test TFB and BS test TFB are shown in the network element level in the figure by horizontal arrows.

Operators do not need to think about the individual details of the task, which will decrease misjudgements and inefficient operation.

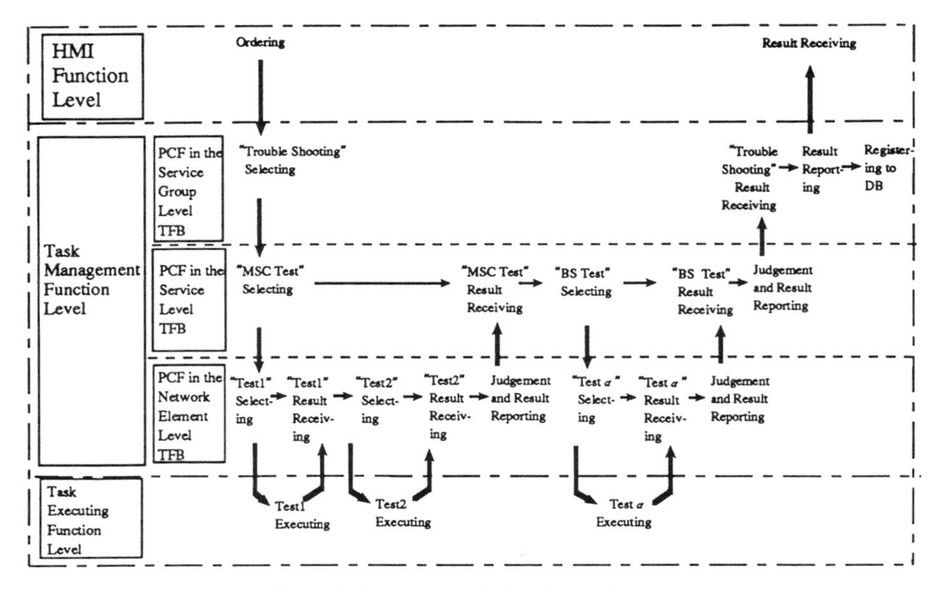

Figure 7. Task flow example through each functions

COMMUNICATION PATH TEST ON THE MOBILE COMMUNICATIONS NETWORK

One of the significant features of the mobile communications system is the mobility of mobile station (MS) during a call. While the operator tests the condition of a call through a subscriber access group communication path, the test may be stopped by the handover of the calling MS. Therefore, the operator must recognize a new communication path which is being newly used by the calling MS, and start to test the condition of the call again for continuous testing.

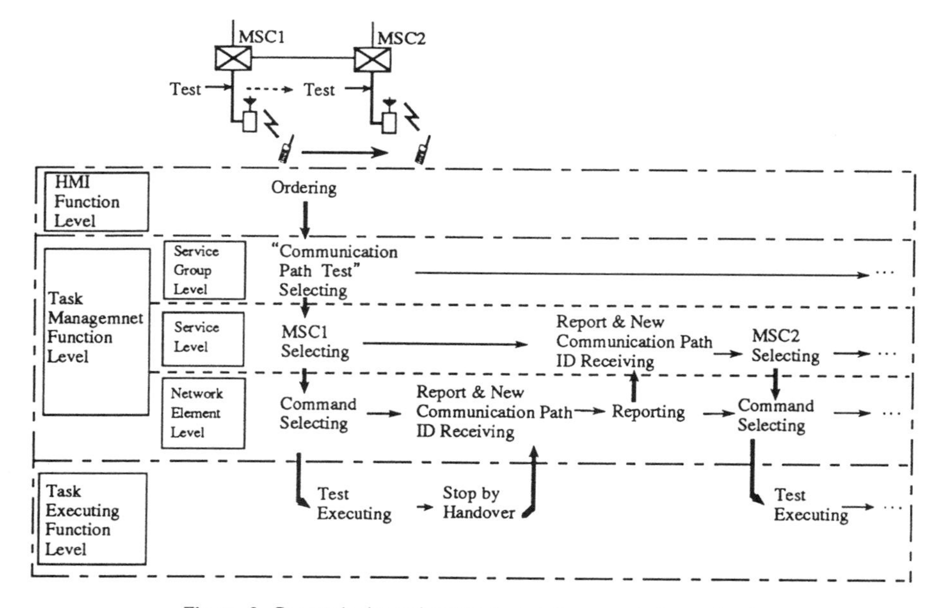

Figure 8. Communication path test on the mobile communications network

An operation system which operates the whole network will make such continuous testing easier, especially for inter-MSC continuous testing. The PCF in the service level TFB recognizes an MSC which has a new communication path, and orders the network element level TFB to send the command to the MSC. This command makes the MSC start the call condition test. The task flow example is shown in figure 8.

CONCLUSIONS

Problems in the present mobile communications network operation were investigated, and two operation concepts were proposed to overcome them. The functional architecture for such operation were discussed.

ABOUT THE AUTHORS

Hironori Fukushima is a engineer of Research and Development Department at NTT Mobile Communications network Inc.(NTT DoCoMo) where he has been engaged in research and deveropment of mobile communications switching systems, and operation systems for mobile communications network. He received his B.E. and M.E. from the Kyusyu Institute of Technology in 1983 and 1985, respectively.

Katsuhiro Tsujinaka is a engineer of Research and Development Department at NTT Mobile Communications Network Inc. (NTT DoCoMo). He joined the department in 1990, where he has been engaged in research and development of operation systems for mobile communications network.

Motoshi Tamura is a staff of Research and Development Department at NTT Mobile Communications Network Inc. (NTT DoCoMo). He joined the NTT DoCoMo in 1989, since he has been engaged in research and development of operation systems for mobile communications network.

Ichio Osano is a executive engineer of Research and Development Department at NTT Mobile Communications network Inc.(NTT DoCoMo) where he has been engaged in research and deveropment of mobile communications switching systems, and operation systems for mobile communications network. He received his B.E. from the Nippon University in 1965.

REFERENCES

[1] Narumi Takahashi, Masakazu Aso and Kisaku Fujimoto, "Telecommunication Network Operations System Architecture and Functions Towards Integration," Proc. NOMS'90, San Diego, California, February 11-14, 1990.

[2] Ikuo Naito, Akihiko Akaike and Makoto Yoshida, "System Interconnection Interface for Total Operation System," Proc. NOMS'90, San Diego, California, February 11-14, 1990.

[3] Soma Murthy, "ISDN Applications for operation Systems," Proc. ISS'92, Yokohama, Japan, October 25-30, 1992.

IDSS-COME: INTELLIGENT DECISION SUPPORT SYSTEM FOR NETWORK CONTROL, MANAGEMENT, AND ENGINEERING

Akiya Inoue, Haruhisa Hasegawa, and Hiro Ito

Traffic Management Research Group
Network Traffic Laboratory
NTT Telecommunication Networks Laboratories
9-11, Midori-Cho 3-Chome
Musashino-Shi, Tokyo 180 Japan

ABSTRACT

The Intelligent Decision Support System for network COntrol, Management, and Engineering (IDSS-COME) is being developed to better support network operators and engineers in running nation-wide telecommunication networks. The objective is to make and verify actions not only for network control and management, but also for network administration and engineering. This paper describes the concept and the functions of IDSS-COME, and the key techniques of each function. It introduces new approaches to parallel-processing simulation and a learning mechanism.

INTRODUCTION

The wide range of telecommunication services and their usage patterns, and the appearance of new common carriers have resulted in traffic variation that is larger and more complicated. Due to switching-system set-up and removal, trunk capacity increasing and decreasing, and network reconfiguration, various types of changes occur everyday in an actual telecommunication network. Under such circumstances, network providers should provide a high-survivability network while maintaining GOS (Grade Of Service). Telecommunication network capabilities are being improved to effectively handle unplanned or unforeseen traffic and to improve network survivability[1]-[6]. These improvements are achieved by advanced control techniques such as dynamic routing control [1]-[9], congestion control[6],[11] and network-facility assignment control [10], [12], [13].

Network operators and network management & administration centers require powerful decision support capabilities to make efficient use of advanced control capabilities and to manage telecommunication networks that involve an enormous number of factors concerning

Network Management and Control, Vol.2, Edited by
I.T. Frisch *et al.*, Plenum Press, New York, 1994

network structure, traffic patterns, and control parameters. The Intelligent Decision Support System for network COntrol, Management, and Engineering (**IDSS-COME**) [13] is being developed to meet these requirements.

This paper first describes the concept and functions of IDSS-COME. It then introduces new approaches to parallel-processing simulation and a learning mechanism as the key techniques of IDSS-COME.

IDSS-COME

Concept of IDSS-COME

IDSS-COME (Ver.1) achieves the first target shown in Figure 1. A network operator might meet a situation with which he has no experience. In such a case, he would like to test different control actions before the actions are actually performed. However, different control actions cannot be tested on an actual network. For the purpose of supporting these network management operations, a new environment for network management is being proposed[14]. This environment is referred to as the "virtual network" in this paper. In the virtual network environment, the operator can carry out the actions as if on an actual system. Then, on the basis of the experience in the virtual network environment, the operator can take the most appropriate action in the actual network.

The objectives of IDSS-COME are as follows:

- To evaluate the performance of each scenario under conditions similar to those of real networks, that is, with realistic network scale, network structure, traffic volume, control systems, etc.
- To evaluate the effects of multiple network controls, combining two or three of the main controls, several routing controls, congestion control and network facility assignment control.

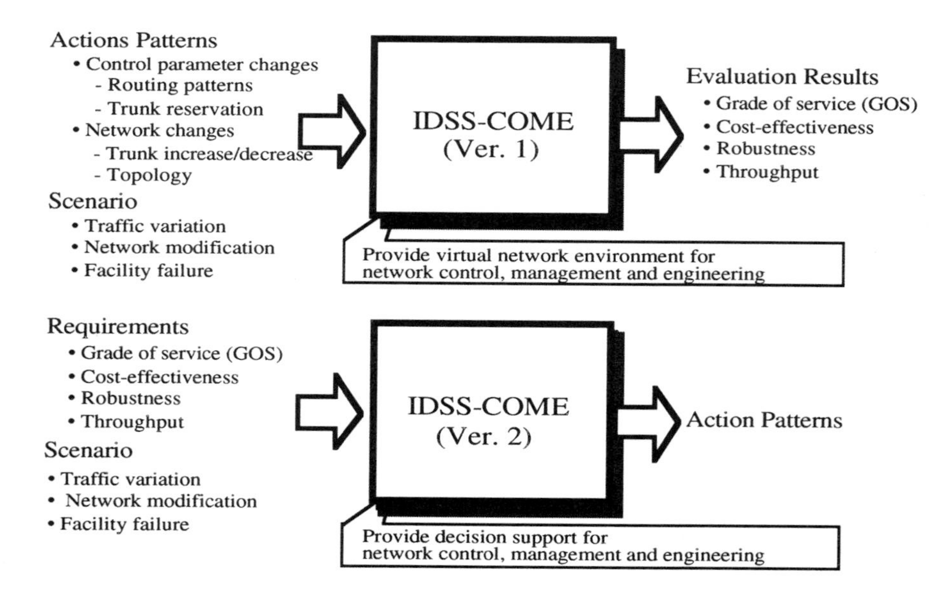

Figure 1. Concept of IDSS-COME

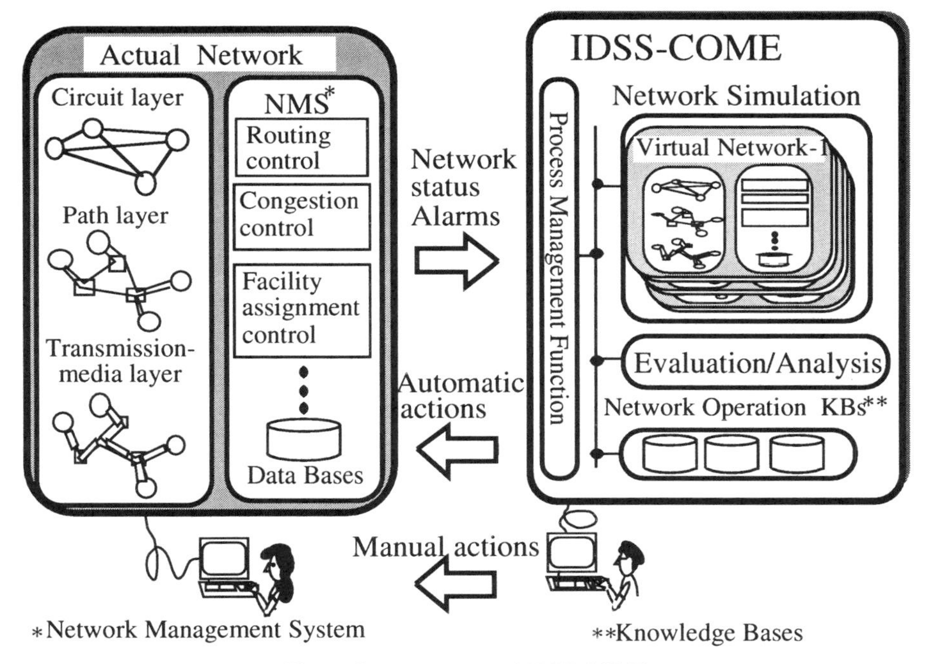

Figure 2. Architecture of IDSS-COME

IDSS-COME must support four functions:

- preparing pre-planned action patterns,
- making action patterns according to network conditions,
- verifying action patterns, and
- training network operators to appropriately deal with various situations.

The final target, IDSS-COME (Ver.2) shown in Figure 1, will implement the capability of choosing the most appropriate action plan for a specific network condition (a given scenario) from among many alternatives.

Functions of IDSS-COME

IDSS-COME consists of (a) a simulation module for telecommunication networks, (b) a network performance evaluation and analysis module, (c) network-operation knowledge-bases (KBs), and (d) a process management module. The system architecture is shown in Figure 2. The functions of each module are described below.

Network Simulation Module. Network simulation is the key to realizing the virtual telecommunication network environment. There are three network simulation modules, corresponding to the three layers of telecommunication networks: (1) a circuit-layer module that simulates call connection processes, (2) a path-layer module that simulates the conditions of circuit assignments in the path-layer network, and (3) a transmission-media-layer module that simulates the conditions of path assignments in the transmission-media-layer network. Each module can simulate, in the computer-based virtual network environment, various conditions such as the current conditions, future conditions, and ideal conditions. A telecommunication network model in the virtual environment is called a "virtual network" in this paper. A virtual network includes network management systems such as a routing control center and their related databases.

An operator, that is a decision maker, orders various control actions on these virtual networks, and then gets the results in the form of a performance evaluation and network conditions, as shown in Figure 3. The concept of this network simulator is similar to that of a flight simulator.

The first target is to develop the circuit-layer network simulation module. Its simulation techniques are described in Section 3.1.

Performance Evaluation and Analysis Module. Network throughput and GOS are calculated by this module. This function takes less time than the network simulation function, but is less accurate.

This module must support three functions:

- sensitivity analysis for link setting and removal, trunk capacity increasing and decreasing, routing-domain (a set of selectable alternate routes) changes, and so on,
- routing domain definition according to network conditions, and
- traffic flow analysis under a given network condition.

The flow-assignment method is the key technique for implementing these functions. For example, the flow between each origin-destination node pair is assigned to two-link alternate routes in a way that maximizes the total amount of flow under the given constraints of link capacity and the number of alternate routes in the routing-domain definition process[8].

Network Operation Knowledge Bases (KBs) . The KB module is one of the key factors enabling IDSS-COME to work as an expert system. Knowledge consisting of network conditions, control actions, and the results of a performance evaluation is stored for each event. The KB is implemented as the expert system platform. It is used to

retrieve related knowledge when a problem arises and

recommend action patterns under given network conditions.

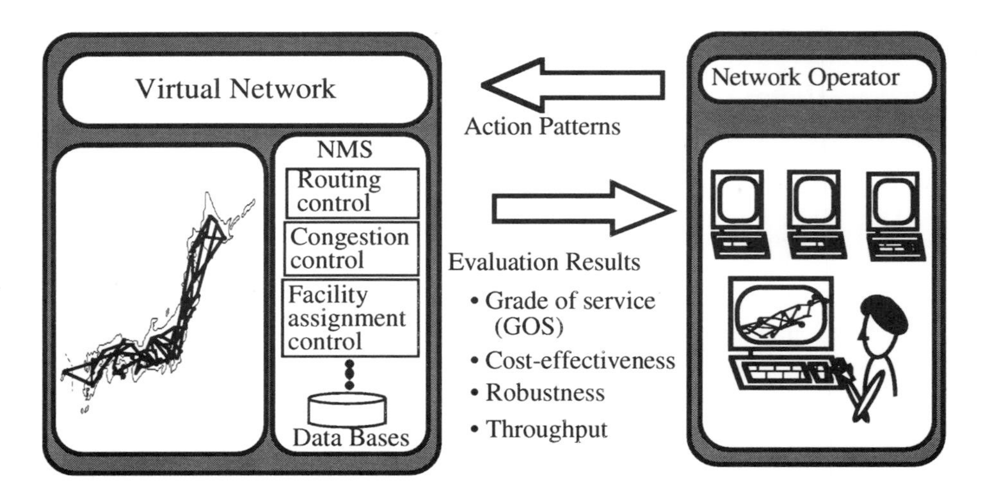

Figure 3. Virtual Network Concept

A learning mechanism is effective in choosing the appropriate action patterns for specific network conditions. A new learning mechanism based on success probabilities of action patterns is described in Section 3.2.

Process Management Module. This module controls various processes, such as scenario making, simulation, and statistical calculation, for managing module-to-module interfaces and graphical interfaces, and for maintaining and updating various types of data.

TECHNIQUES IN IDSS-COME

Parallel-Processing Simulation

Realistic-scale simulation of a nation-wide telecommunication network has been difficult because of the limitation of computer power (processing speed, memory capacity, etc.) and the cost of computer use. It is also difficult to introduce a super-computer system in each network management & administration center so as to be able to use such a new decision support system. However, the network simulator for IDSS-COME should satisfy the following conditions at low cost.

- **Realistic Network Conditions.** A network condition is defined by the network structure (number of nodes, number of links between nodes, topology), number of trunks in each link, the traffic volume, control parameters, and so on.

The scale of a realistic transit network in Japan is as listed below.
- Number of nodes: O (10^2)
- Number of trunks: O (10^5)
- Traffic volume: O (10^5) erl

- **High-speed Simulation.** The goal is that the required CPU time be made less than the simulation time by using a mini-computer with parallel processing.

- **Flexibility with Simulation Conditions.** The simulator must be able to handle many parameters for network models, traffic conditions, and control actions. It must also be able to evaluate the network performance under transient conditions and the condition when a new control method is introduced into a part of a network.

Parallel processing techniques[15] are effective for executing a realistic-scale simulation of a nation-wide telecommunication network. Network-facility dependent modeling in telecommunication networks has been developed for parallel processing. However, there are some difficulties regarding network simulation. One is that certain data is accessed very frequently. For example, each link between switching nodes can be used as an alternate route between any node-pair in a network with dynamic routing. Therefore, modeling in which each switching function is simulated in a separate CPU is not the best way to achieve high-speed simulation under realistic-scale network conditions. A new modeling method suitable for nation-wide network simulation is required to make good use of parallel processing capabilities. The simulation procedure in one such new modeling is divided into three layers to apply parallel processing techniques to network simulation (Figure 4)[14]. This method, which is being developed in NTT labs[14], is described below.

Parallel Execution of Simulation Units. Generally, there are several action patterns for solving problems. In layer 1, multiple simulations should be executed independently in parallel so that several action patterns can be evaluated quickly. Each

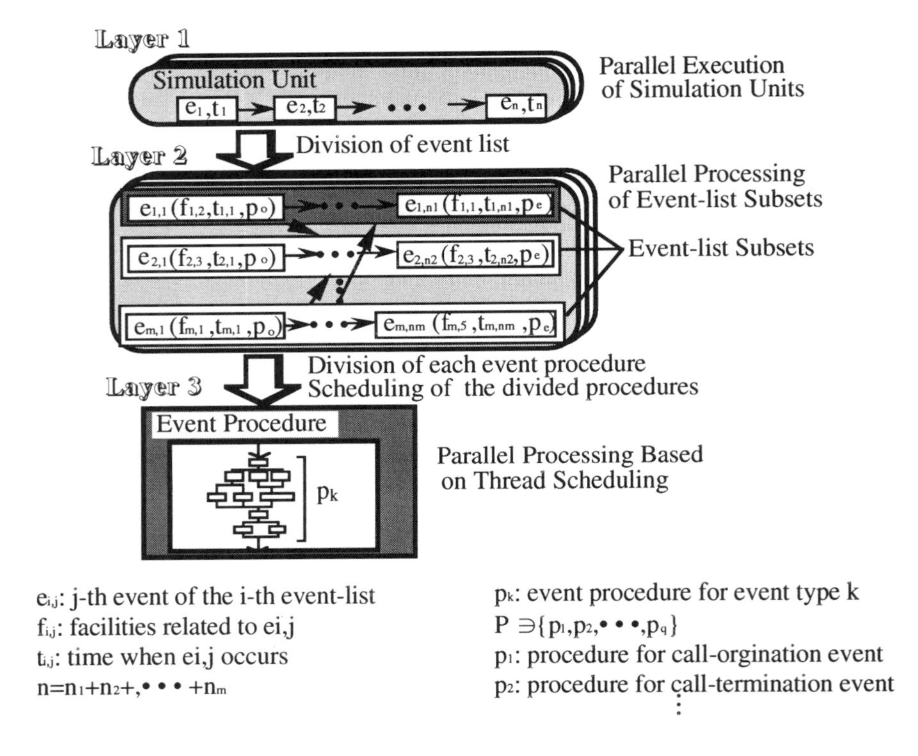

Figure 4. Hierarchical Modeling for Parallel Processing

simulation has its own scenario with a different action pattern. Each simulation is packaged as an independent simulation unit that includes everything needed to run the simulation, such as scenario files and an execution module. An execution module is generated for each simulation unit by copying and assembling the original files according to the purpose. Each simulation unit can be executed in parallel on a different set of CPUs, so it is easy to implement this layer 1 mechanism on a multi-CPU computer.

Parallel Processing of Event list Subsets. The conventional simulator executes an event-list sequentially by a time-true simulation mechanism[16]. The aim of layer 2 is to achieve high-speed simulation by dividing the event list into several subsets. Each subset can be executed in parallel on a different set of CPUs. Therefore, the simulation unit should be divided into subsets so as to minimize the total CPU time.

A simulation procedure executed in a simulation unit is expressed by an event list as shown in Figure 4. An event *e* is defined by event procedure *p* (for example, call origination or termination), facilities related to event, *f*, and the time at which the event occurs, *t*. In a telecommunication network with dynamic routing, an originating call might access any node and any link. This implies that almost all of the events may access the same data. A subset of an event list is made by dividing network facilities into several facility sets. Each subset consists of only the events related to the facilities in the same facility set. To minimize the occurrence of causality errors, each facility set is determined on the basis of traffic volume, the network structure, and the number of managed facilities.

In layer 2, event-list subsets made by the above-mentioned procedure are executed by a parallel simulation method. When a causality error is detected under simulation, the error is avoided by using a rollback mechanism.

Parallel Processing Based on Thread Scheduling. The aim of layer 3 is to apply parallel processing to the event-procedure level. The simulator has minimum-searching of the next event, searching and updating for available trunks, and other data manipulation procedures in an event procedure. Some of them do not have to run sequentially. Because the flows of all event procedures can be identified before running, each event procedure can be analyzed to identify parts suitable for parallel processing.

When scheduling the parts, the data dependency must be examined. In layer 3, thread scheduling is done so as to guarantee correct processing of data dependencies correctly. Therefore, each event procedure is carried out according to pre-scheduling by analyzing the data dependencies among all parts.

Learning Mechanism

The if-then-rule is the basic technique used in expert systems, such as an equipment failure diagnosis system. However, an actual telecommunication network has an enormous number of factors, some of which are uncertain, so it is difficult to achieve the ability to extract a single action pattern under a given set of network conditions. IDSS-COME uses a new technique, that is, a new learning mechanism, to indicate the success probability of each action pattern under given network conditions. This learning mechanism is described in this section.

Learning Mechanism using Discrete Choice Models (LMDCM). The LMDCM concept is shown in Fig. 5. The objective of LMDCM is to support network operators in choosing an appropriate action pattern from a set of alternatives. The feature of LMDCM is that the success probability of each action pattern under given network

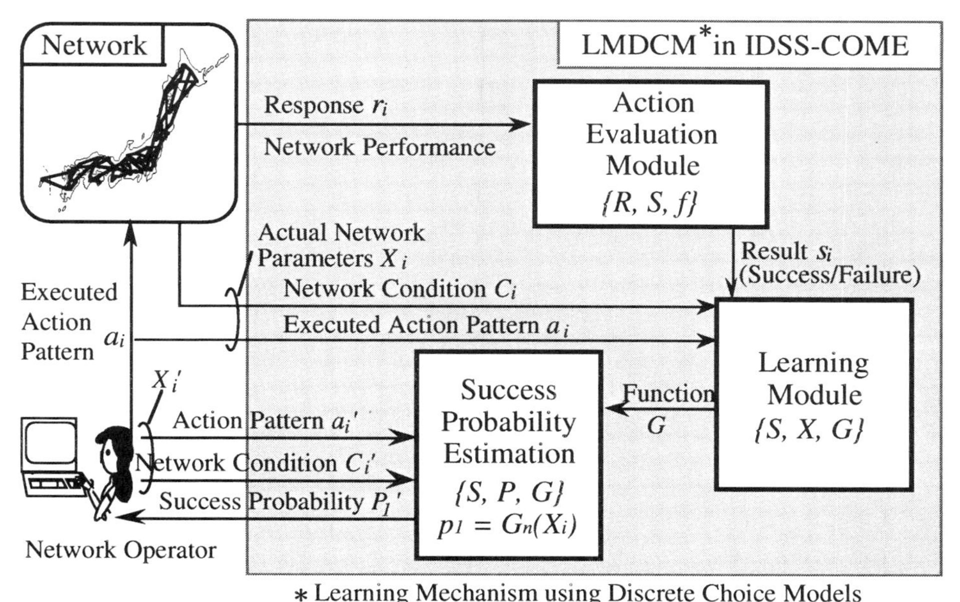

Figure 5. Architecture of LMDCM

conditions is indicated. A network operator can choose an action pattern on the basis of the success probabilities. For example, he may choose the action pattern whose success probability is highest among the alternatives.

The LMDCM consists of three modules: i) the learning module, ii) the action evaluation module, and iii) the success probability estimation module. The LMDCM is defined as $\{\boldsymbol{A}, \boldsymbol{R}, \boldsymbol{S}, f, \boldsymbol{X}, \boldsymbol{P}, G\}$, where

$\boldsymbol{A}=\{a_1, a_2, \ldots, a_m\}$: The set of alternatives of action patterns; a_i is an action pattern vector a_i expressed by several control parameters and m is the total number of alternatives

$\boldsymbol{R}=\{r_1, r_2, \ldots\}$: The set of responses (network performance); r_i is the response of the network when action pattern a_i is executed

$\boldsymbol{S}=\{0, 1\}$: The set of evaluation results of $\boldsymbol{R}$ and $s_i \in \boldsymbol{S}$ is an evaluation result of r_i, such that

$$s_i = \begin{cases} 1: \textit{success} \\ 0: \textit{failure} \end{cases} \tag{1}$$

f: An evaluation rule,

$$s_i = f(r_i) \tag{2}$$

$X_i=\{x_{i1}, x_{i2}, \ldots, x_{ik}\}$: The set of network parameters; X_i is the network vector that consists of two subsets; action pattern vector ai and network condition vector $\boldsymbol{C}_i$

$P_1=\{p_{11}, p_{21}, \ldots, p_{m1}\}$: The success probability vector p_{i1} is the success probability of action pattern a_i under network condition $\boldsymbol{C}_i$,

$$p_{i1} = Prob\{s_i = 1\} \tag{3}$$

$P_2=\{p_{12}, p_{22}, \ldots, p_{m2}\}$: The failure probability vector where p_{i2} is the failure Probability of action pattern a_i under network condition $\boldsymbol{C}_i$,

$$p_{i2} = Prob\{s_i = 0\} \tag{4}$$

and

$$p_{i1} + p_{i2} = 1, \text{ for all } t \tag{5}$$

G_n: The success probability function of action pattern a_i, where

$$p_{i1} = G_n(X_i). \tag{6}$$

The basic concept of the LMDCM is the same as that of learning automata [17]. That is, the LMDCM has the same features as conventional learning schemes[18]-[21] in which (1) an alternative is selected on the basis of estimated success probabilities, and (2) the evaluation result of whether the network response is a success or failure is input into the learning module. The difference between the LMDCM and conventional learning automata is the learning process. Conventional learning automata estimate the success probability $p_{i1}(t_{n+1})$ directly as[17]

$$p_{i1}(t_{n+1}) = T(p_{i1}(t_n), a(t_n), s_{i1}(t_n),), \quad (t_{n+1} > t_n, \quad n=1,2\ldots,), \tag{7}$$

and

$$\sum_{i=1}^{m} p_{il}(t_n) = 1 \tag{8}$$

where t_n is the n-th update time and T, which is called a reinforcement scheme, is an algorithm for updating probability $p_{in}(t_n)$[17],[18],[21]. On the other hand, the LMDCM, estimates the success probability function G_n, that is, the vector of the function's unknown coefficients. Success probability p_{i1} is thus obtained using the function G_n, which is based on the network vector X_i (see eq. (6)).

The functional form of G_n must be determined, as well as a method for estimating the function. In the LMDCM, discrete choice models are applied to determine the form of G_n and to estimate G_n, so as take into account many different network parameters. The discrete choice model[22] is a method for analyzing and predicting human-choice behavior based on multi-attribute utility theory and the principle of utility maximization. The choice set consists of discontinuous alternatives. This model consists of utility functions parameterized in terms of observable independent variables (alternative attributes) and unknown coefficients (weighted attribute vector). These unknown coefficients are estimated from a sample set of observed choices made by decision makers when confronted with a choice situation. The output of the model is the choice probability p_{ij} that alternative j will be chosen by decision maker i. This is defined as the probability that alternative j has the greatest utility among the available alternatives. Discrete choice models for special cases in which the choice set consists of exactly two alternatives are called "*binary choice models (BC models)*[22]". In this paper, BC models are applied to the learning module.

Estimation of Success Probability Function G_n. In IDSS-COME, a BC model is used as the learning module. The learning module substitutes action evaluation results for decision j. If the result is success, j =1, otherwise, j =0. Here, the utility U_j of result j is defined as the contribution to the improvement in network performance. In random utility theory, U_j is a random variable composed of two parts:

$$U_j = V_j + e_j \quad (j = 0,1). \tag{9}$$

V_j is called the *systematic (or representative)* component of utility U_j and is assumed to be deterministic; e_j is the random part and is called *disturbance*. Provided that the functional form for V_j has linear variables, the difference between the systematic component V_0 and V_1 can be defined as

$$V_1 - V_0 = b_0 + b_1 x_1 + b_2 x_2 + \ldots + b_K x_K, \tag{10}$$

where $b_K (k=1, 2, \ldots, K)$ is the coefficient denoting the weight for the value x_k of network parameter k, and b_0 is a constant related to the other attributes with the exception of $x_1, x_2 \ldots, x_K$. Under these assumptions, success probability p_1, can be described as

$$\begin{aligned} p_1 &= Prob(U_1 \geq U_0) \\ &= Prob(e_0 - e_1 \leq V_1 - V_0). \end{aligned} \tag{11}$$

Assuming that the disturbances are independent and identically Gumbel distributed (i.e., logistically distributed), the success probability, that is the function G_n, is given by[14]

$$p_1 = 1/(1 + exp(V_0 - V_1)). \tag{12}$$

Equation (12) is a function of the unknown coefficients $b_0, \ldots, b_K$. These coefficients are estimated from a set of N samples. In conventional discrete choice models, each sample used to estimate the unknown coefficients consists of a decision (an action pattern) and observed attribute values. In IDSS-COME, the attribute values are network vector X_i that consists of observed network condition vector $\boldsymbol{C}_i$ and action pattern vector a_i. A sample includes network vector $\boldsymbol{X}_i$ and evaluation result s_i.

After an action pattern has been executed N times, which is determined by the number of samples in the sample set, a vector is estimated for the unknown coefficients $b_0,\dots,b_K$ by maximum likelihood estimation[22]. The likelihood function L is defined as

$$L\,(b_0, b_1, \dots, b_K) = \prod_{n=1}^{N} p_1{}^{j}\, p_0{}^{\bar{j}}, \tag{13}$$

where

$j + \bar{j} = 1$ and N is the number of samples in a sample set.

Consequently, the success probability p_1 in IDSS-COME indicates the possibility that the action pattern a_i is the best for given network conditions.

CONCLUSION

IDSS-COME has a network simulation function, a network performance evaluation and analysis function, and network-operation knowledge bases (KBs). IDSS-COME can use both single functions and combinations of functions. In IDSS, various network simulation models, whose conditions are the same as or similar to actual network conditions, are installed in computer systems. Each scenario alternative can be simulated under various model network conditions. Operators can determine an appropriate action pattern on the basis of both evaluation results obtained from the network simulator and empirical knowledge stored in the network operation KBs.

Application examples are as follows:

- To obtain the best action pattern through repeated experiences under the same network conditions.
- To know future network conditions by putting the time forward.
- To determine the cause of the operational failure by tracing executed actions.

This paper presented new approaches to parallel-processing simulation and a learning mechanism as the key techniques of IDSS-COME. A prototype of IDSS-COME has been developed to evaluate the performance of routing control in circuit switched networks [16]. It can display network status data, including currently assigned alternate routes and end-to-end GOS data observed in a virtual model network simulated by computer. It can also evaluate routing performance on a model of a 60-node network with 1,500 links and a 400,000-erlang traffic volume.

IDSS-COME is under development. It is sure to be a powerful decision support system in the near future.

ACKNOWLEDGMENTS

The authors would like to thank Mr. Yuji Hirokawa of NTT Telecommunications Software Headquarters, Dr. Hisao Yamamoto of NTT Telecommunication Networks Laboratories, and Dr. Jun Matsuda of NTT Telecommunication Networks Laboratories for their useful comments and contributions to this work.

ABOUT THE AUTHORS

Akiya INOUE received the B.E. and M.E. degrees in electrical engineering from Nihon University, Tokyo, Japan, in 1979 and 1982, respectively. He joined the Musashino Electrical Communication Laboratory of the NTT Public Corporation in 1981. He has been engaged in basic research on network control and telecommunication-user behavior modeling. He is currently Senior Research Engineer of the Network Traffic Laboratory, NTT Telecommunication Networks Laboratories.

Haruhisa HASEGAWA has been engaged in research on telecommunication network control and the evaluation of telecommunication traffic at the Network Traffic Laboratory, NTT Telecommunication Networks Laboratories, since 1990. His research interests are focused on the simulation of telecommunication traffic and its applications. He received the B.E. and M.E. degree in electrical engineering from WASEDA University, Tokyo, JAPAN, in 1988 and 1990, respectively.

Hiro ITO received the B.S. and M.S. degrees in applied mathematics and physics from Kyoto University, Japan, in 1985 and 1987. Since 1987 he has been engaged in research on network control methods and mathematical programming, at NTT Telecommunication Networks Laboratories.

REFERENCES

[1] Hurley, B. R., Seidl, C. J. R., and Sewell, W. F., "A Survey of Dynamic Routing Methods for Circuit-Switched Traffic," *IEEE COM Magazine*, Vol.25, No.9, pp.13-21, 1987.

[2] Ash, G. R. and Schwartz, S. D., "Network Routing Evolution," *NETWORK MANAGEMENT & CONTROL WORKSHOP '89*, 1989.

[3] Pack, C. D. and Olson, D. W., "Advanced Routing Techniques Using Advanced Intelligent Network Functional Components and Data Base Controls," *NETWORK MANAGEMENT & CONTROL WORKSHOP '89*, pp.273-285, 1989.

[4] Regnier, J. and Cameron, W. H., "State-dependent Dynamic Traffic Management for Telephone Networks," *IEEE COM Magazine*, Vol.28, No.10, pp.42-53, 1990.

[5] Wolf, B. Richard, "Advanced Techniques for Managing Telecommunications Networks," *IEEE COM Magazine*, Vol.28, No.10, 1990.

[6] Mase, K. and Yamamoto, H., "Advanced Traffic Control Methods for Network Management," *IEEE COM Magazine*, Vol.28, No.10, 1990.

[7] Inoue, A., Mase, K., Yamamoto, H., and Suyama, M., "A State- and Time-dependent Dynamic Routing Scheme for Telephone Networks," *ITC13*, June 1991.

[8] Yamamoto, H., Mase, K., Inoue, A., and Suyama, M., "Dynamic Routing Schemes for Advanced Network Management," *IEICE Trans. on Commun.*, Vol.E74, No. 12, 1991.

[9] Inoue, A., Yamamoto, H., Ito, H., and Mase, K., "Advanced Call-level Routing Schemes for Hybrid Controlled Dynamic Routing," *IEICE Trans. on Commun.*, Vol.E74, No. 12, 1991.

[10] Yamada, J. and Inoue, A., "Intelligent Path Assignment Control for Network Survivability and Fairness," *ICC91*, June 1991.

[11] Nakajima, S. and Tokunaga, S., "A New Network Management Control System on a Telephone Network," *Network Oper. and Mngmt. Symp.*, 1988.

[12] Hasegawa, S., Kanemasa, A., Sakaguti, H., and Maruta, R., "Dynamic Reconfiguration of Digital Cross-connect Systems with Network Control and Management," *GLOBECOM'87*, 1987.
[13] Inoue, A. Ito, H. and Yamamoto, H., "Hybrid Control Strategies for Advanced Network Control and Their Intelligent Decision Support System.," *Institute of Electrical and Electronics Engineers Network Planning Symposium (NETWORKS 92)*, 2.2, 1992
[14] Hasegawa, H., and Inoue, A., "Approach to Nation-wide Network Simulation Making Virtual Reality for Telecommunication Network Management," *WSC92*, Dec. 1992.
[15] Fujimoto, R. M., "Parallel Discrete Event Simulation.," in *Communications of the Association for Computing Machinery.* 33 (10):30-53, 1990.
[16] Inoue, A., Yamamoto, H., and Harada, Y., "An Advanced Large-scale Simulation System for Telecommunication Networks with Dynamic Routing," *NETWORKS89*, pp.77-82, Sept. 1989.
[17] Narendra, K., and Thathacher, M. A. L., " Learning Automata," Prentice Hall, Inc. NJ, U.S.A., 1989.
[18] Narendra, K. S., Wright, E. A., and Mason, L.G., "Application of Learning Automata to Telephone Traffic Routing and Control," *IEEE Trans. Syst., Man, and Cybern.*, vol. SMC-7, pp. 785-792, Nov. 1977.
[19] Narendra, K. S., and Mars, P., "The Use of Learning Algorithms in Telephone Taffic Routing --- A Methodology," *Automatica,* vol. 19, pp. 495-502, 1983.
[20] Srikantakumar, P. R., and Narendra, K. S., "A Learning Model for Routing in Telephone Networks," *SIAM J. Contr. Opt*, vol.20, pp. 34-57, Jan. 1982.
[21] El-Hadidi, M. T., El-Sayed, H. M., and Bilal, A. Y., "Performance Evaluation of A New Learning Automata Based Routing Algorithm for Calls in Telephone Networks," *Proc. 11th Int. Teletraffic Congr.*, Kyoto, Sep. 1985, session 4.4A, paper 2.
[22] Ben-Akiva, M., and Lerman, S. R., "Discrete Choice Analysis," MIT Press, MA, U.S.A., 1987.

A DOMAIN-ORIENTED EXPERT SYSTEM SHELL FOR TELECOMMUNICATION NETWORK ALARM CORRELATION

Gabriel Jakobson, Robert Weihmayer, and Mark Weissman
GTE Laboratories Incorporated
40 Sylvan Road
Waltham, MA 02254

ABSTRACT

This paper focuses on domain-oriented expert systems dedicated to the tasks of real-time telecommunication network management. The knowledge acquisition and application development tools embedded in such shells are specialized to represent specific telecommunication domain models and network management practices. IMPACT (Intelligent Management Platform for Alarm Correlation Tasks), developed at GTE Laboratories, is such an expert system shell for telecommunication network alarm correlation. The IMPACT knowledge framework comprises the following interrelated elements: network configuration and alarm correlation knowledge components, namely, network element classes, network element instances, network element message classes, correlations, and correlation rules. This paper will provide an overview of IMPACT, along with some direct insights into the underlying alarm correlation model, knowledge representation, and the end-user-oriented knowledge-based editors for alarm correlation knowledge acquisition and application development. IMPACT has been used in two applications systems: AMES, for a land-based telecommunication network alarm correlation, and CORAL, for a cellular network alarm correlation. Many of the concepts and solutions presented in this paper are generic in nature and can be used beyond the domain of network alarm correlation.

INTRODUCTION

Expert Systems (ES) are now standard components in many telecommunication network management application systems [1,2]. As a rule, most ES fielded in the telecommunication area are built by system programmers, database specialists, knowledge engineers, and other people versed in the technology, using general-purpose expert system shells. Such expert systems are installed at network operations centers, where they are operated and maintained, in minor ways, by end-users and network operators (Figure 1). The application development process itself is a complex, error-prone, and time-consuming process. Practically, the ultimate user, the network operator, is out of the loop of application development, testing, and modification. To make required knowledge base modifications, for example, in order to

[1] Based on "A Dedicated Expert System Shell for Telecommunication Network Alarm Correlation" by Jakobson *et al.* which appeared in Second IEEE Network Management and Control Symposium, September 21-23, 1993, Tarrytown, New York, pp. 277-288. ©1993 IEEE.

Network Management and Control, Vol.2, Edited by
I.T. Frisch *et al.*, Plenum Press, New York, 1994

introduce a vendor change into the network element model, to update the network configuration model, or to change the message parsing algorithm, the network operator needs help from a knowledge engineer or a system programmer. As networks become larger by integrating multiple subnetworks, and as the tasks of network management become more knowledge intensive, the network operators are pushed more and more away from the application development and modification process. A fundamental benefit of our approach will be to bring network operators into the network monitoring application development process. In order to do that, we will focus on a domain-oriented expert system shell dedicated to the tasks of telecommunication network management.

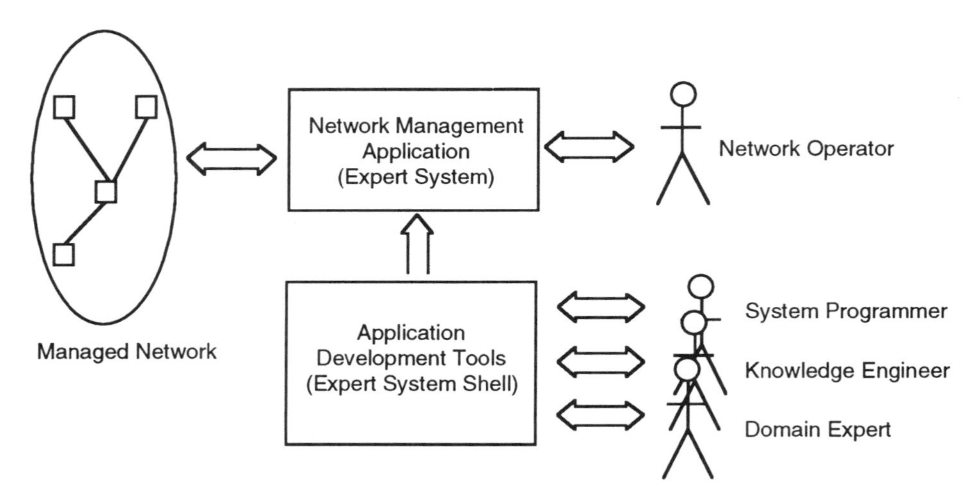

Figure 1. The use of general-purpose expert system shells to develop network management applications.

Central to the idea of dedicated expert systems is the availability of specialized knowledge structures, graphical user interfaces, and application development tools which are oriented toward end-users and reflect the idiosyncracies, constraints, and practices of the domain. Figure 2 represents an idealized situation in which the end-user is the sole developer of the network management application and is using the application development tools of a dedicated expert system. A realistic solution falls between the extremes represented in Figures 1 and 2, namely, where the end-user is able to participate in the development of particular network configuration models and network management rules, while domain experts and knowledge engineers continue to define the generic classes of network element models and classes of network management rules.

IMPACT, developed at GTE Laboratories, is a telecommunication network alarm correlation ES shell that exemplifies the notion of a dedicated ES shell. It supports end-users, as well as domain experts, in developing and maintaining real-time network surveillance and alarm correlation applications. Alarm correlation is a generic process which underlies different network management tasks, such as alarm filtering, network fault isolation and diagnosis, selection of corrective actions, pro-active maintenance, and trend analysis.

Different components of the IMPACT knowledge base, such as network element classes, network connectivity models, network element message classes, correlations, and correlation rules, are interrelated, forming the IMPACT knowledge framework. The structured application development environment contains a set of dedicated network configuration and alarm correlation editors, which help the users to define a network, the message generated by its network elements (NE), and the rules that link messages from those NE into more abstract network correlations. IMPACT guides the user through successive stages of knowledge elicitation, while enforcing local and global domain-specific constraints. Ultimately, the user is led to enter complex knowledge structures through a simple interface that ensures correctness, completeness, and consistency of the knowledge specifications.

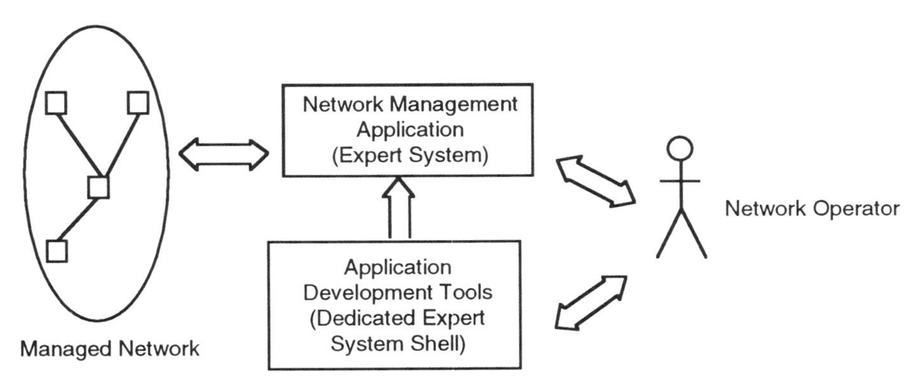

Figure 2. Idealized use of dedicated expert system shells to develop network management applications.

This paper provides an overview of IMPACT, along with some direct insights into the IMPACT alarm correlation model, the alarm correlation knowledge representation methods, and the end-user-oriented editors for alarm correlation knowledge acquisition and application development.

NETWORK ALARM CORRELATION DOMAIN

Modern telecommunication networks produce large numbers of alarms. Inadequate handling of these alarms can lead to incorrect interpretation of network events and delays in restoration of network services. Alarms are considered an external manifestation of faults occurring inside the managed network that affect its hardware and software components. Faults can be causally related, where they can be represented as an acyclic fault propagation graph, or they can be independent (causally unrelated). Not all faults have alarms associated with them. Such faults can be recognized indirectly by correlating available alarms (Figure 3).

In this paper, alarm correlation will be considered as a conceptual interpretation procedure in the sense that a new meaning is assigned to a set of alarms that happen during a predefined time interval. This interpretation process involves knowledge about underlying physical entities, network elements, and relations between events happening within these physical entities.

In general, any event affecting the network may be correlated with any other. Such events may be network alarms and status messages, environmental state parameters, or events created by the user or by external systems.

We will consider the following types of alarm (event) correlation:

1. $[A, A, \ldots\ldots, A] \Rightarrow A$	Compression
2. $[A, B, p(A) < p(B)] \Rightarrow \varnothing$	Suppression
3. $[n \times A] \Rightarrow B$	Count
4. $[A, A \subset B] \Rightarrow B$	Generalization
5. $[A, A \supset B] \Rightarrow B$	Specialization
6. $[A \rightarrow B] \Rightarrow C$	Temporal relation "after" ("before")
7. $[A, B, \ldots T, \wedge, \vee, \neg] \Rightarrow C$	Boolean pattern

Alarm compression (1) is thus the task of reducing multiple occurrences of an alarm into a single alarm. No number of occurrences of the alarm is taken into account. Alarm suppression (2) allows inhibiting a low priority alarm A in the presence of a higher priority alarm B. Alarm suppression is a context-sensitive process in which an alarm is suppressed depending on the dynamic operational context of the network management process. The context could be determined by the presence of some other alarm(s), as shown above, or of any other network event, level of network management task, or user command. This dependency makes the alarm suppression task different from the widely used alarm filtering procedure.

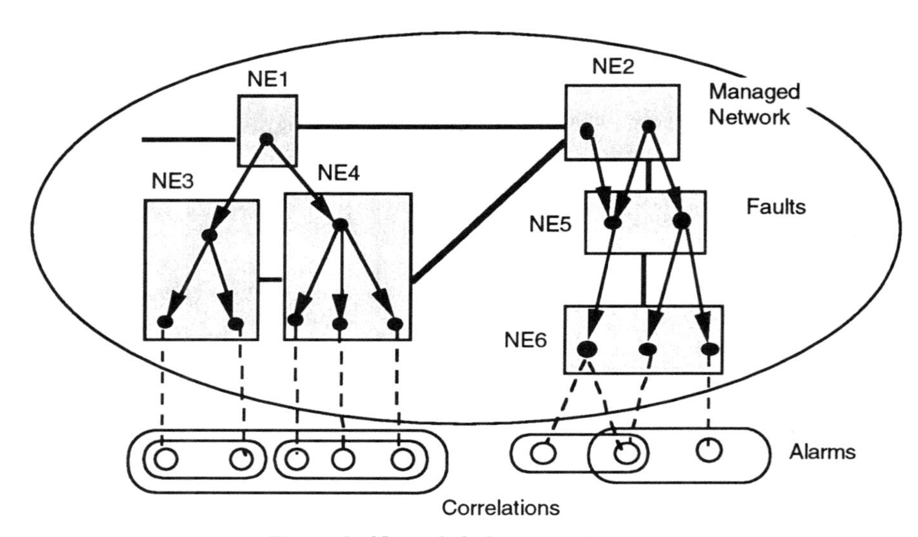

Figure 3. Network fault propagation model.

Another type of correlation would result from counting (3) and thresholding the number of repeated arrivals of the "same" alarm. Alarm generalization (4) is a correlation type in which an alarm A is replaced by its superclass B. Alarm generalization has a potentially high utility for network management. It allows one to deviate from a micro perspective of network events and view situations from a higher level. Alarm specialization (5) is the opposite to alarm generalization: it substitutes for an alarm a more specific subclass of the alarm. Temporal relations (6) allow alarms to be correlated depending on the time of their arrival. There are two basic types of temporal correlations: The first type takes into account only the fact that some alarm happened "after" or "before" some other alarm; the second type depends, in addition, on the length of time between the events, in either absolute or relative time. Complex temporal correlations could be created using the logic of temporal intervals described in [3].

There are two time intervals assigned to each correlation: a correlation time window and a correlation life span. The correlation time window defines a time interval during which the correlation process takes place. Correlation life span is counted from the moment a successful match of the correlation pattern is made and determines how long this correlation is valid. Boolean logic allows the creation of complex correlation patterns using logical operators over component terms. The terms in the pattern could be primary network events, previously defined correlations, or tests on network configuration.

THE CONCEPTUAL FRAMEWORK OF ALARM CORRELATION

The conceptual framework of the IMPACT Alarm Correlation Model is shown in Figure 4. It contains two major components: the structural component and the behavioral component. The structural component is the description of the managed network. It contains two parts: the Network Configuration Model and the Network Element Class Hierarchy. The Network Configuration Model describes the NEs (managed objects), and the connectivity and containment relations between them. The Network Element Class Hierarchy describes the NE types and the class/subclass relationships between the types. Each NE in the Network Configuration Model is an instance of a terminal NE class from the Network Element Class Hierarchy. The behavioral component describes the dynamics of alarm correlation. It contains three components: the Message Class Hierarchy, the Correlation Class Hierarchy, and the Correlation Rules. The Message Class Hierarchy describes the messages generated by NEs. The Message Class Hierarchy is used to control the alarm message parsing process.

The NE classes, message classes, correlation classes, and correlation rules are organized into hierarchies. These hierarchies are related by *producer/consumer* dependencies. NEs are *producers* of alarm messages, messages *produce* correlations, and rules are *consumers* of all the above.

As it appears from the model, IMPACT distinguishes between correlations and the correlation rules. A correlation is a statement about the events happening on the network, for example, Bad-Card-Correlation, which states that some port contains a faulty circuit card. A correlation rule defines the conditions under which correlations are asserted, for example, if there is a red carrier group alarm (CGA) from a digital cross-connect (DCS) and there is a yellow CGA from another DCS, and these DCSs are connected, then Bad-Card-Correlation should be asserted. It is important to note that many different correlation rules can lead to the assertion of one and the same correlation.

The conditional, or so-called left-hand side (LHS), part of a correlation rule uses (*consumes*) network elements, messages, and correlations as arguments to form the rule firing condition. The condition can contain Boolean patterns, sequences of alarms based on

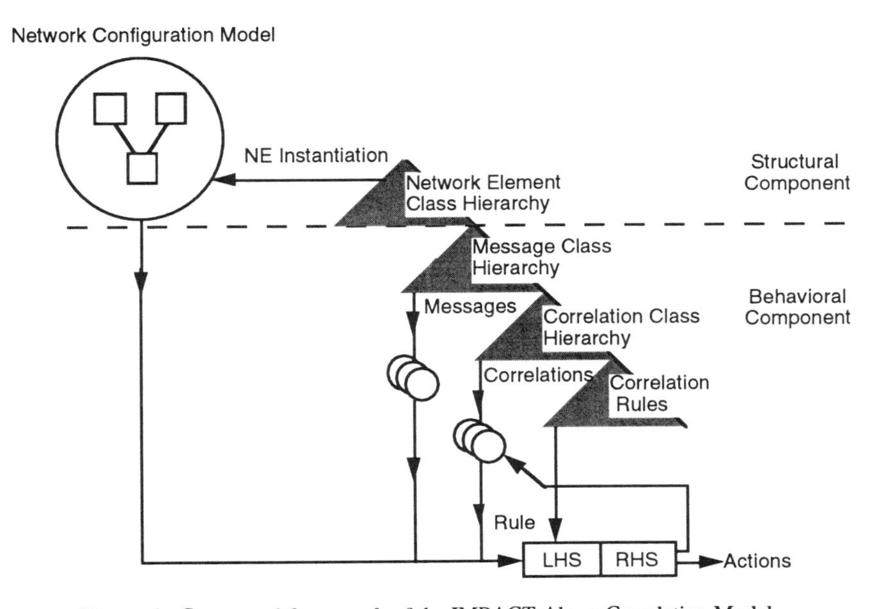

Figure 4. Conceptual framework of the IMPACT Alarm Correlation Model.

time relations, as well as alarm counters. The arguments for the Boolean patterns could be alarm messages (or their components), network connectivity statements, correlations, or test data stored in a database. The arguments for the sequences and counters could be alarm messages or correlations. The right-hand side (RHS) of the rule contains actions, including actions that assert (*produce*) correlations resulting from the firing of the rule. The subsequent application of correlation rules, *production* of correlations, and *consumption* of the *produced* correlations by the next rule describes the event (correlation) propagation process in IMPACT.

The *producer/consumer* relationships represent the deep domain-oriented dependencies between the entities of the alarm correlation model. For example, a message hierarchy is strictly related to a particular network element that generates these messages. This network element cannot be deleted from the knowledge base unless the related message hierarchy has first been deleted. The network element-message relationship also has an opposite effect: before selecting a message for a conditional part of the rule, the user has to identify the network element that produces this message. In a similar fashion, no messages or network elements that are *consumed* by some correlation can be removed.

The *producer/consumer* dependencies are enforced by IMPACT during the application development process. These dependencies (along with other domain-oriented constraints attached to network element and message classes) are used in two ways: to support the correctness, completeness, and consistency of the knowledge base; and to guide the user through the process.

IMPACT SYSTEM OVERVIEW

IMPACT was designed as a general-purpose telecommunication network alarm correlation system that can be used to develop a variety of network alarm correlation applications. Developing a new application involves defining network configuration, alarm correlation, and alarm message data specific to the particular application. The functional architecture of IMPACT is shown in Figure 5. It contains three major components: the Application Run-Time Component, the Application Development Component, and the Network Knowledge Base.

The Application Run-Time Component monitors the network events in real-time, correlates alarms, and responds to operator commands. In addition to those functions, it provides information on network status, explanations and help. The Application Run-Time Environment consists of four major modules: the Graphical User Interface, the Command/Message Processor, the Action Processor, and the Alarm Correlation Engine. The Command/Message Processor takes incoming alarm messages, analyzes them, and turns them into objects. It also processes the commands coming from the user. The Alarm Correlation Engine is a rule-based system that reasons about the messages and generates correlations. The Action Processor performs the functions determined by the correlation rules, such as displaying correlation messages, performing diagnostic procedures, storing data in a database, or executing external procedures. The Command/Message Processor implements a novel approach to message processing based on message class hierarchies. The essence of this method is to have a universal message parsing procedure that can be tuned to parse messages from different classes of NE using associated message class hierarchies [4].

The Graphical User Interface of the Application Run-Time Environment provides the network operator with several windows to perform the tasks of network surveillance and fault management (Appendix 1). In Appendix 1, the Map Window displays the managed network and two Bad-Card-Correlation icons. The references to the corresponding messages and correlations may be seen in the Message/Correlation Display Window. The BAD-CARD faults happened on ports #005 and #007 of the Los Angeles digital cross-connect, designated as LSANCASF. The Message Window displays the full text of the CGA-Red alarm message selected from the Message/Correlation Display Window.

The Application Development Environment provides powerful tools for building the Network Knowledge Base. The core of the environment consists of eight editors, with a common look and feel, which are grouped into three sets of tools: Network Configuration Tools, Alarm Correlation Tools, and Network Graphics Tools (Figure 5). The design of those editors is based on the general alarm correlation framework discussed in Section 3. The *producer/consumer* relationships of the framework are enforced by the editors. Tight integration between the editors allows simultaneous editing of conceptually related knowledge structures. Wherever a class or object is presented, either as text or iconically, a menu of common functions associated to that class or object is available. These menus offer choices such as "display that entity" or "access information about its relationship to others in the alarm correlation framework." The editors apply telecommunication domain knowledge by validating the correctness and completeness of entered data. If a physical port may only be connected to a T1 trunk, then only such trunks are offered to the user.

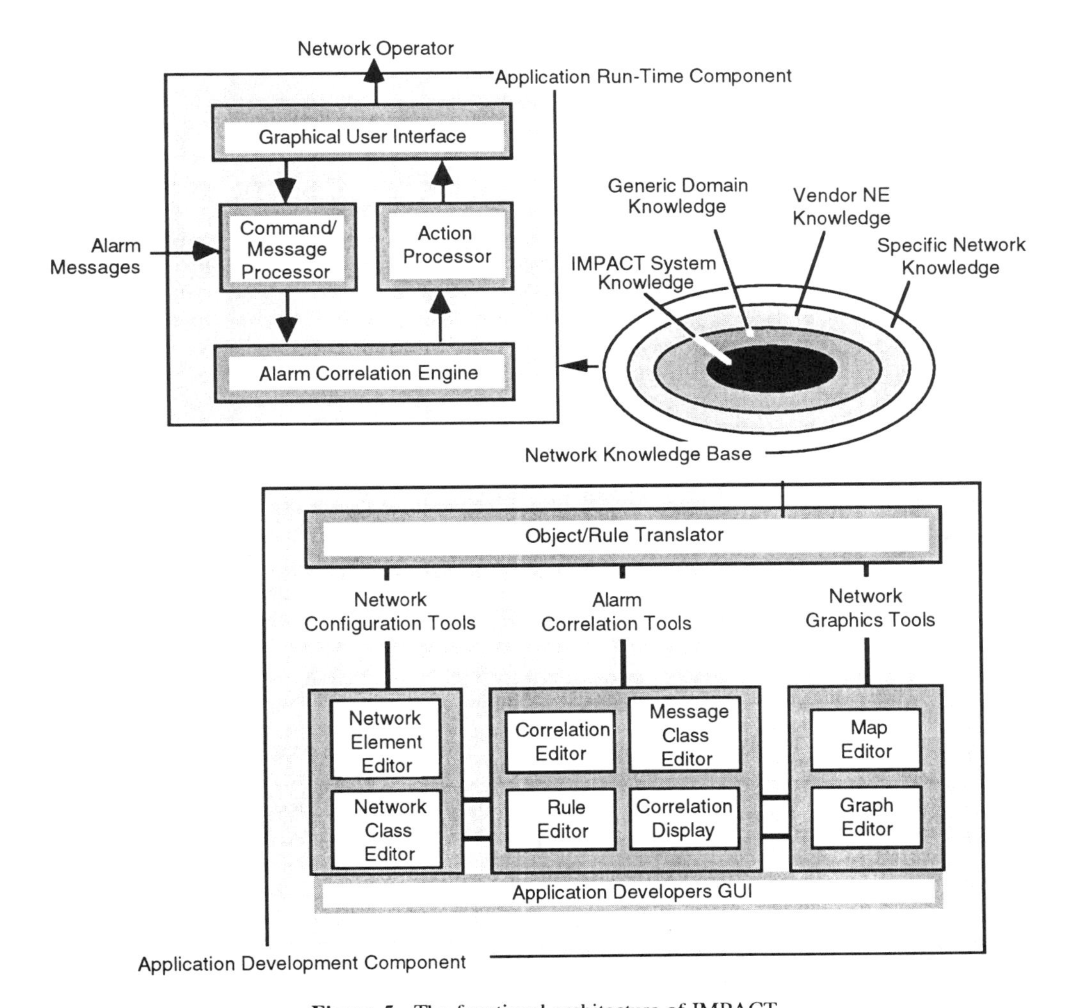

Figure 5. The functional architecture of IMPACT.

NETWORK KNOWLEDGE BASE

To perform the tasks of alarm correlation, the Application Run-Time Component has continuous access to the Network Knowledge Base. From the viewpoint of generality of the information, the Network Knowledge Base could be divided into four parts: IMPACT System Knowledge, Generic Domain Knowledge, Vendor NE Knowledge, and Specific Network Knowledge. The System Knowledge is the most general (and stable) part of IMPACT's knowledge. It defines the internal rules and objects of IMPACT's functioning. This knowledge is accessible only to the IMPACT developers and authorized personnel at the IMPACT deployment site.

The Generic Domain Knowledge defines the basic network specifications and management practices. It also describes the generic constraints and principles specific to the domain. For example, the Generic Domain Knowledge contains specifications of abstract network element classes, such as switch, digital cross-connect system, trunk, etc. It determines basic class, containment, and connectivity relations between the network elements. The Generic Domain Knowledge is the most general domain-specific knowledge, and as a consequence of that, the most static part of it. Development of the Generic Domain Knowledge is a one-time investment, and in most cases, the Generic Domain Knowledge is carried from one application to another with minor changes. Usually, the Generic Domain Knowledge is developed by the most experienced domain experts and knowledge engineers.

The Vendor NE Knowledge describes the specific types of vendor equipment which are used in managed networks, the environmental hardware, and the types of sites, buildings, etc. It also describes the classes of alarms, status messages, clearing messages, and general alarm correlation rules associated with the specific vendor equipment. The Vendor NE Knowledge is built either by the domain experts or by network operations personnel at the network management centers.

The Specific Network Knowledge is the least stable part of the Network Knowledge Base. It should be built for each managed network, and should be modified if changes are introduced into the network configuration model, or into the way the network is managed. The Specific Network Knowledge includes description of the managed network and the alarm correlations which are specific to the managed network. The network operations staff builds the Specific Network Knowledge.

The Network Knowledge Base could be, according to the alarm correlation knowledge framework (see The Conceptual Framework of Alarm Correlation section), conceptually divided into five components: Network Element (NE) Classes, the Network Connectivity Model, Message Classes, Correlations, and Correlation Rules.

NE Classes describe network elements, such as switches, digital cross-connects, multiplexers, etc., used to model specific networks. NE Classes are organized into class hierarchy using parent-child relations. The root of the hierarchy is a generic-NE-class which contains the most general information common to all NEs. The next level of the hierarchy describes the basic NE Classes, such as trunk-class, transmission-interface-class, switch-class, building-class, and others. Each of these classes refer to their own subhierarchy, for example, the trunk-class refers to the logical-trunk-class and the physical-trunk-class; and the physical-trunk-class to the super-link-class, the T1-trunk-class, and the T3-trunk-class, etc. Each subclass (child) inherits parameters, their values, attributes, and constraints from their corresponding superclass (parent). IMPACT permits multiple inheritance, that is, a child might have more than one parent.

Classes may have parameters, and may contain other classes as their components. The attributes of the parameters are used to describe specific information or requirements associated with the parameters, for example, the default values of the parameters, the default values, and incomplete specification conditions. IMPACT also allows the attachment of constraints to the values of the parameters. The constraints could be individual to each parameter or express interparameter relations. The components of the class are described using the containment relations WITHIN in the subcomponent, or CONTAINS in the main component: for example, switch can be contained in a building, and a logical trunk can contain physical trunks.

The NE Class Hierarchy, except the terminal nodes of the hierarchy, describe the mathematical abstractions of existing physical network components, while the terminal nodes describe the network element types (classes) produced by manufacturers. For example, specific digital cross-connect products, such as AT&T's DACS II, Rockwell's RDX-370, or Tellab's TCS-532, are terminal child nodes of the parent digital cross-connect class. The NE Class Hierarchy is specific to the applications, and could be modified by adding, deleting, or editing existing classes. The upper levels of the hierarchy are more general and, therefore, do not change much moving across different applications.

The Network Class Editor shown in Appendix 2 describes ROCKWELL-DEXCS, which is a subclass of the generic digital cross-connect class DEXCS-CLASS. Message Class refers to BASIC-DEXCS-MESSAGE, which is the root node of the associated Message Class Hierarchy. The Connected Filter specifies that ROCKWELL-DEXCS may be connected only to a digital cross-connect or a switch. The Within Filter is used to specify that ROCKWELL-DEXCS can be placed within a building or a network operations center, while the Contains Filter specifies that only physical and logical ports may be contained within a building or a network operations center.

The Network Configuration Model is a description of an actual physical or logical network. The connectivity model is built by instantiating terminal NE classes from the NE Class Hierarchy and connecting them according to the network configuration. This process could be performed by the network operating staff using the IMPACT Network Instance Editor. During the NE class instantiation process, IMPACT enforces the constraints defined in the class specification. For example, the user cannot make connections which violate the physical behavior of the connected elements, or leave the values for required parameters of the elements unspecified. The Network Element Editor shown in Appendix 2 describes LOS ANGELES-DEXCS, which is an instance of ROCKWELL-DEXCS. It is installed at a Los Angeles network operations center, it is connected to a DCS in Sacramento, and it contains four physical ports

All alarm messages produced by a specific NE are organized into a Message Class Hierarchy using the class-subclass relation, as with NE classes. Each class in the hierarchy contains a message parsing pattern and a translation schema, common to a subset of all messages of this network element. A trace from the root node to a some class node *n* in the hierarchy determines a sequence of patterns to be recognized by the IMPACT Run-Time Component to detect whether incoming messages belong to the Message Class determined by the node *n*. The translation schema in the Message Class determines how machine codes for this network element can be substituted with a common expression, or expressions in a more readable form to the network operator. In this capacity, the Message Class Hierarchy guides the alarm message parsing process. Development of the Message Class Hierarchies for all NE (which generate messages) allows the customization of the message parsing procedures. Message Class Hierarchies are developed in IMPACT using the Message Class Editor.

A Correlation Class is a generalized description of the state of the network based on interpretation of network events. The conditions under which correlations are asserted are described in the Correlation Rules. Each assertion creates an instance of a Correlation Class. A Correlation Class contains components, a message template, and parameters (slots). The components may be NEs, alarm messages, or other correlations. Correlation components are used to pass information from a Correlation Rule to the asserted correlation. Parameters provide information about a correlation to higher level correlations, of which it may be a component. Correlation BAD-CARD-CORRELATION, which is described in Appendix 3, contains two components, a digital cross-connect system DEXCS-CLASS, and a physical port PHYSICAL-PORT-CLASS. During assertion, a Correlation Rule assigns values to the slots CLLI (a universal code, which identifies the location of the equipment) and PORT-NUMBER. These values are used by the message template and asserted into the slots DEXCS-ID and PORT-NUMBER. Variable names are identified by a leading question mark.

A version of BAD-CARD-CORRELATION-RULE-1 is given in Appendix 3. Time is an important correlation criterion. Correlations are determined on a fixed length time interval. The correlation time interval may be absolute or relative. In the latter case, the time interval is considered to be a dynamic window where alarm correlation is performed continuously. This correlation rule is as follows: if physical ports *?near-port* and *?far-port* belong to two digital cross-connect systems, respectively, *?near-DEXCS* and *?far-DEXCS*, and these ports are connected by a T1 trunk, and Yellow Carrier Group Alarm *?yellow-msg* is reported from *?far-port*, and Red Carrier Group Alarm *?red-msg* is reported from *?near-port*, then assert BAD-CARD-CORRELATION. After matching the rule conditions, *?near-DEXCS* and *?far-DEXCS* are bound to particular NEs. These NEs are provided as components to BAD-CARD-CORRELATION.

AN EXAMPLE OF KNOWLEDGE ACQUISITION WITH IMPACT

Correlations in IMPACT thus represent complex events in a network. The prototypical network, for which an alarm correlation expert system was developed using IMPACT, is an integrated voice/data network with a mix of high bandwidth optical and low bandwidth metallic facilities. Digital cross-connect systems (DCS) create a flexible reconfigurable physical backbone transmission network. Voice, telemetry, and data overlay networks provide switched service connectivity. Any number of different instances of networks can be created from this mix. The following examples of alarm correlations demonstrate some variety of events that can be described in IMPACT:

- Simple hierarchical subsumption, for example, correlation of all DS0/DS1 contiguous alarms that accompany a DS3 (28 × DS1) facility failure.
- Attached alarms from different network elements, for example, combination of Red and Yellow Carrier Group Alarms from connected digital cross-connect systems in the transmission network that point to bad circuit card or facility failure problems.
- Causally correlated alarms from service/overlay layers and physical network layers, for example, call processing failures being reported over logical trunks, while physical alarms are reported from the physical facilities used to build those logical trunks.

We now describe an example of network alarm correlation with the goal of showing how an end-user can record knowledge using the IMPACT user interface. The situation described is real, but any direct references to actual locations have been removed.

The situation is as follows. Due to an administrative error at a primary network control center, a circuit disconnect order was issued in error to a common carrier, but withdrawn soon after. An additional human error at the common carrier site led to the disconnect order's being carried out in spite of the cancellation. This meant that a live circuit was disconnected, and a catastrophic failure occurred on a major DS-3 link between City A and City B (Figure 6). Usually, normal disconnects lead to loopback conditions on digital cross-connect (DCS) equipment at both ends of the circuit. In this particular case, the loopback conditions were not reported on the DCS equipment considered. Therefore, the DCSs did not issue any carrier group alarms. Nevertheless, the packet and voice switches that had logical trunks over the disconnected circuit sent large numbers of messages to the network control center. The operators were puzzled for a while before they realized what had happened. The task at hand was to correlate the call processing alarms from the switches with the fact that there were no carrier group alarms from the DCSs, and to recognize that the trunk was actually disconnected, despite the incorrect record in the database listing the circuit as live.

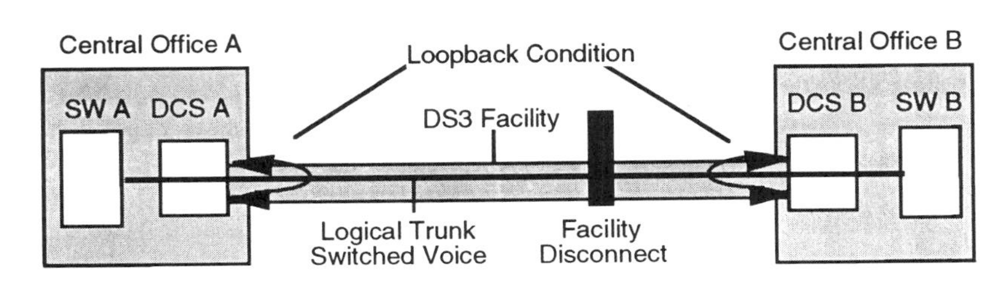

Figure 6. Facility disconnect.

We assume that the NE classes for switches, DCSs, ports, and trunks have been already described, as well as the message classes generated by the switches and the DCSs, including the voice switch messages for Call Processing Failures (CPF) on logical ports, and Carrier Group Alarms (CGA) on the DS-1 ports of the DCSs.

The Network Instance Editor allows the user to instantiate these classes into the following objects relevant to the example, along with their connection and containment properties:

- Physical trunks (supported by the DS-3 facility) connected to the ports on the DCSs.
- Physical ports contained in the DCSs and connected to the physical trunks.
- Logical ports contained in the voice switches and connected to the switched logical trunks.
- Logical trunks containing the underlying physical trunks and ports (within and between the central offices).
- DCSs and voice switches ("A" and "B") contained in their respective central offices and connected to the appropriate set of ports.

This configuration model will enable IMPACT to test the fact that DCSs from City A and City B are directly connected via the DS3 facility, that the switches and DCSs are connected accordingly, and that the logical trunks are built from the physical trunks in the DS-3 between City A and City B.

The analyst uses the Correlation Class Editor to define the set of correlations that will be activated by successive applications of Correlation Rules. The Live-Facility-Disconnect-Correlation (Figure 7) is activated when two conditions are met: the circuit is normally disconnected, and there are call processing failure (CPF) alarms from the corresponding switches. The Normal-Disconnect-Correlation is activated under two conditions: DCSs at both ends of the circuit should be in loopback. Finally, the two Loopback-Correlations are recognized by a pattern containing the CPF alarm from the switch, and no carrier group alarm from the DCS.

IMPACT was built using ART-IM [5], a general-purpose expert system shell. User interfaces for the run-time and application development components were developed in Tcl/Tk [6], a programming system for building X-based windowing applications. IMPACT has been used for developing two application systems: AMES, for telecommunication network alarm correlation in a voice/data network such as described above; and CORAL, for cellular network alarm correlation.

The structure of the Network Knowledge Base and availability of different application development editors determine the particular network knowledge acquisition to be done. As was discussed earlier, the Network Knowledge Base could be divided according to generality of the information and according to conceptual structure of the information (Figure 8). Personnel of different knowledge and skills take part in the knowledge acquisition process using different tools, as shown in the figure.

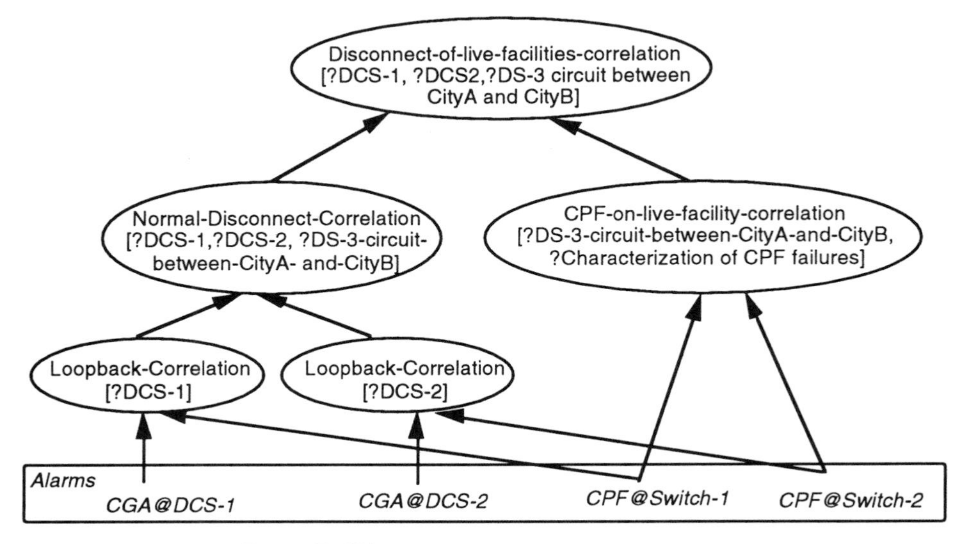

Figure 7. "Disconnect-of-live-facilities-correlation."

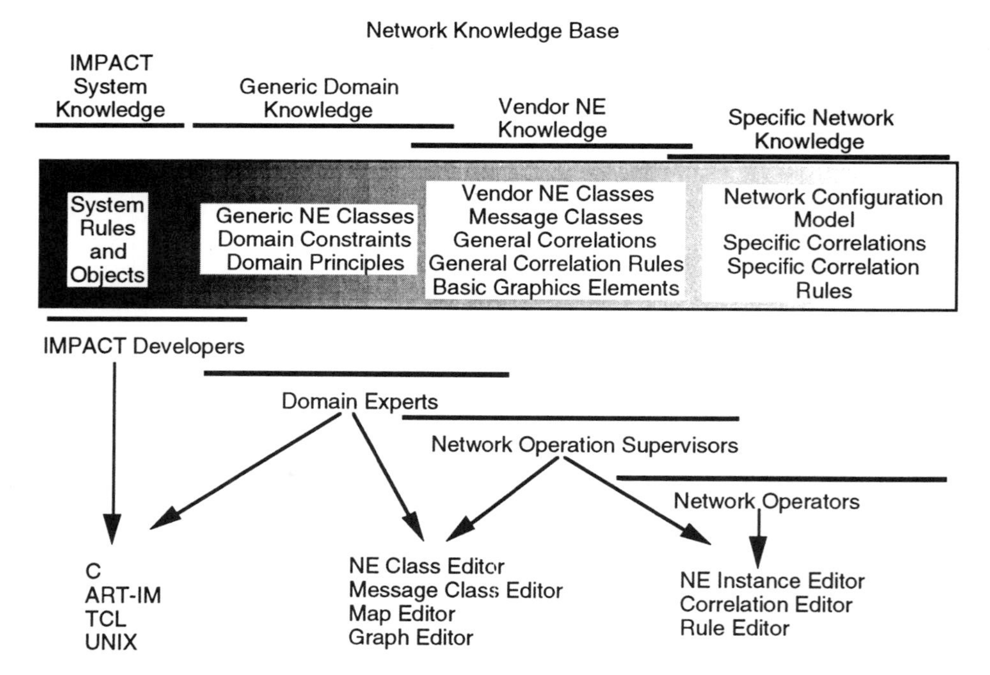

Figure 8. Technology of network knowledge acquisition with IMPACT.

CONCLUSIONS

Domain-oriented expert system shells offer significant potential for building communication network management applications. One of the most far-reaching benefits of such shells is that the end-users, network operations staff, can actively participate in the application development process. This paper presents a concrete example of a domain-oriented expert system shell IMPACT, which is currently being used at GTE for land-based and cellular network alarm correlation tasks.

This paper does not discuss the real-time performance of IMPACT, which was an important aspect of the system development. Preliminary tests of IMPACT on a SPARC 10 showed that the system was able to parse and correlate 12–15 alarms per second.

Many of the concepts and solutions presented in this paper are generic in nature, and can be used beyond the network alarm correlation tasks. For example, in the same way that correlations and correlation rules are described, one can define network faults and corresponding diagnostic rules. The network element classes and connectivity models can be used for a network configuration and planning expert system.

ABOUT THE AUTHORS

Gabriel Jakobson is a Principal Member of Technical Staff at GTE Laboratories, where he has been project leader of several expert systems and intelligent database systems development projects. He received a MSEE from the Tallinn Polytechnic Institute and a PhD in computer science from Estonian Academy of Sciences in 1964 and 1971, respectively. Prior to his current position, he was a Senior Research Staff Member at the Institute of Cybernetics, Tallinn, Estonia.

Dr. Jakobson is the author or co-author of more than 40 technical papers in the areas of automata theory, man-machine interfaces, artificial intelligence, expert systems, and telecommunication network management. He is a member of IEEE and AAAI.

Robert Weihmayer is a Senior Member of Technical Staff at GTE Laboratories. He received a BS in electrical engineering and an MS in computer science from McGill University and Boston University, respectively. Prior to his current position, he was a Member of Scientific Staff at Bell-Northern Research.

He has published over 20 papers in the areas of combinational optimization for communication network planning, data network design, and applications of distributed artificial intelligence and expert systems to telecommunication network operations and management. He is a member of IEEE and AAAI.

Mark D. Weissman is a Senior Member of Technical Staff at GTE Laboratories, where he has been a major contributor to the development of several expert systems for network management applications. He received a BS in chemical engineering and a BA in computer science from the State University of New York at Buffalo in 1983 and 1984, respectively.

He is the author or co-author of several technical papers in the area of language parsing and using expert systems for telecommunication network management. He is a member of IEEE and AAAI.

REFERENCES

[1] Liebowitz, J., *Expert System Applications to Telecommunications,* John Wiley & Sons, 1988.

[2] Ericson E., Ericson L., Minoli D., *Expert Systems Applications in Integrated Network Management,* Artech House, 1989.

[3] Allen, J.F., "Maintaining Knowledge About Temporal Intervals," *Communications of the ACM,* pp. 832–853, 1983.
[4] Jakobson, G., Weissman, M., "A New Approach to Message Processing in Distributed TMN," *Proceedings of the Fourth IFIP/IEEE International Workshop on Distributed Systems,* Long Branch, NJ, 1993.
[5] ART-IM Programming Language Reference, Inference Corporation, 1991.
[6] Ousterhout, J., "Tcl: An Embeddable Command Language," *Proceedings of the Winter USENIX Conference,* pp. 133–146, 1990.

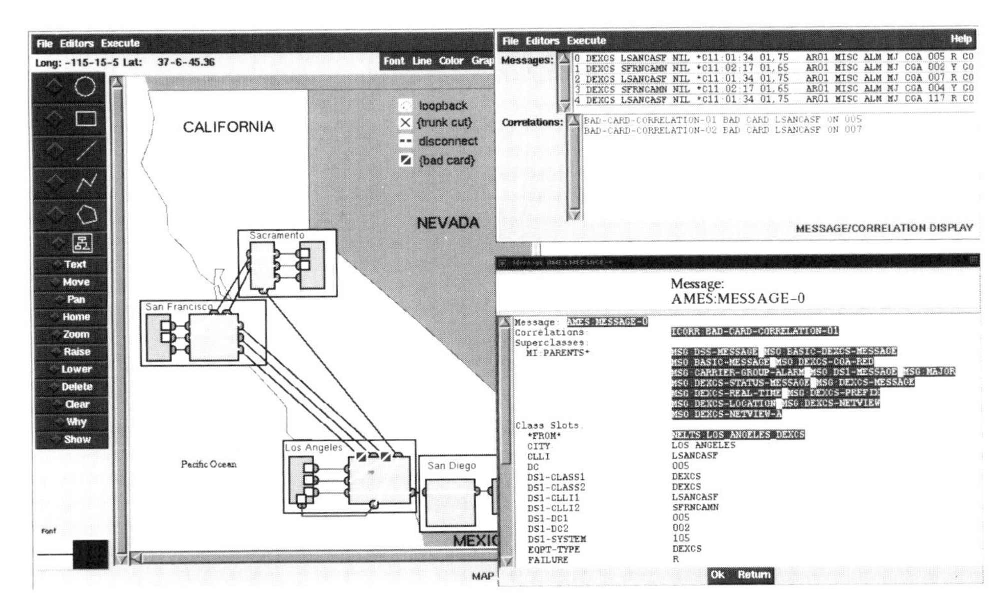

APPENDIX 1

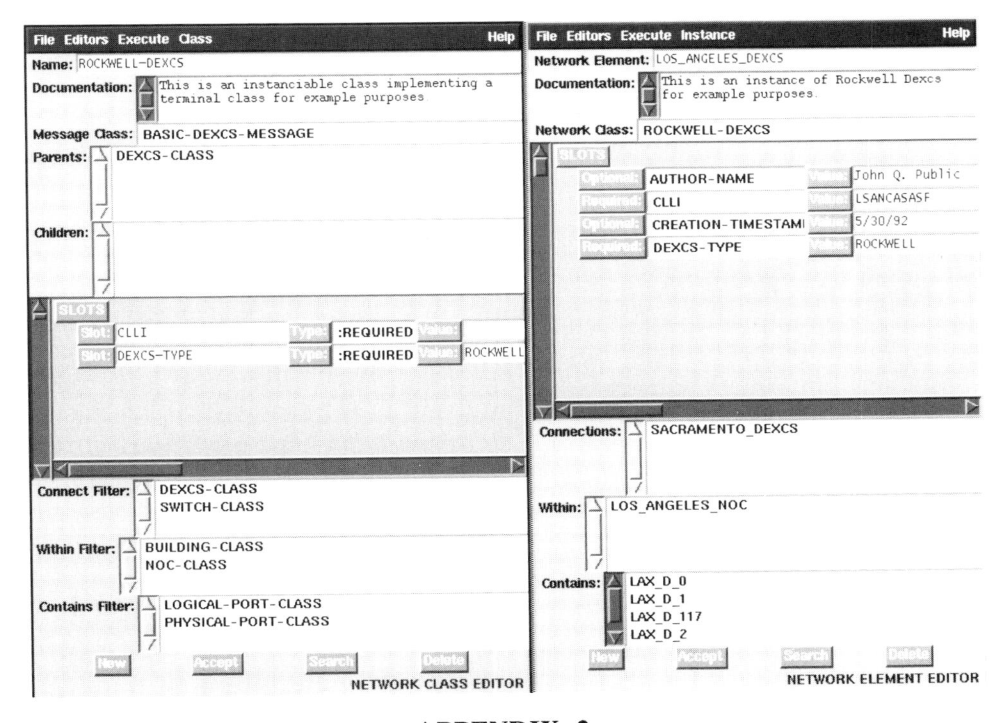

APPENDIX 2

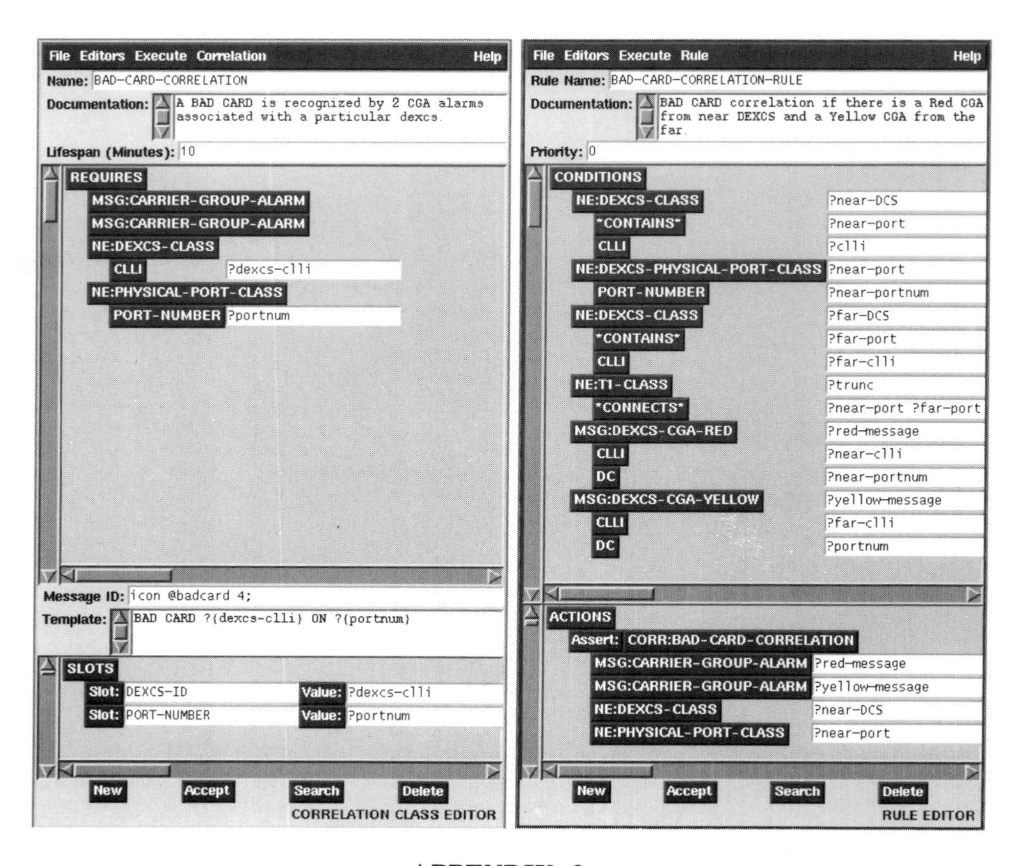
File Editors Execute Correlation Help
Name: BAD-CARD-CORRELATION
Documentation: A BAD CARD is recognized by 2 CGA alarms associated with a particular dexcs.
Lifespan (Minutes): 10
REQUIRES
MSG:CARRIER-GROUP-ALARM
MSG:CARRIER-GROUP-ALARM
NE:DEXCS-CLASS
CLLI ?dexcs-clli
NE:PHYSICAL-PORT-CLASS
PORT-NUMBER ?portnum
Message ID: icon @badcard 4;
Template: BAD CARD ?(dexcs-clli) ON ?(portnum)
SLOTS
Slot: DEXCS-ID Value: ?dexcs-clli
Slot: PORT-NUMBER Value: ?portnum
New
Accept
Search
Delete
CORRELATION CLASS EDITOR
File Editors Execute Rule Help
Rule Name: BAD-CARD-CORRELATION-RULE
Documentation: BAD CARD correlation if there is a Red CGA from near DEXCS and a Yellow CGA from the far.
Priority: 0
CONDITIONS
NE:DEXCS-CLASS ?near-DCS
CONTAINS ?near-port
CLLI ?clli
NE:DEXCS-PHYSICAL-PORT-CLASS ?near-port
PORT-NUMBER ?near-portnum
NE:DEXCS-CLASS ?far-DCS
CONTAINS ?far-port
CLLI ?far-clli
NE:T1-CLASS ?trunc
CONNECTS ?near-port ?far-port
MSG:DEXCS-CGA-RED ?red-message
CLLI ?near-clli
DC ?near-portnum
MSG:DEXCS-CGA-YELLOW ?yellow-message
CLLI ?far-clli
DC ?portnum
ACTIONS
Assert: CORR:BAD-CARD-CORRELATION
MSG:CARRIER-GROUP-ALARM ?red-message
MSG:CARRIER-GROUP-ALARM ?yellow-message
NE:DEXCS-CLASS ?near-DCS
NE:PHYSICAL-PORT-CLASS ?near-port
New
Accept
Search
Delete
RULE EDITOR

APPENDIX 3

AN ALARM CORRELATION SYSTEM FOR HETEROGENEOUS NETWORKS

Allan Finkel, Keith C. Houck, Seraphin B. Calo, and Anastasios Bouloutas

IBM Research Division
Thomas J. Watson Research Center
P.O Box 704
Yorktown Heights, NY 10598

ABSTRACT

Heterogeneous network management is a challenging area of increasing importance. The NetFACT system provides a platform for correlating alarms and localizing faults in a heterogeneous network. NetFACT uses management techniques that are independent of the underlying topology of the network, the communications protocols employed, and the alarm structures provided. In this paper, we give an overview of the architecture of NetFACT, and discuss the features which enable fault management in heterogeneous networks.

INTRODUCTION

The last few years have seen dramatic growth in both the size and scope of telecommunications networks. Large telecommunications networks may encompass dozens of nodes, and include network elements and software from numerous manufacturers. A large enterprise may find itself dealing with a heterogeneous environment that includes multiple communications protocols, and separate voice and data networks. To make matters worse, the element management systems supplied by vendors often only manage the network elements manufactured by that vendor. It is not uncommon for each element management system in an enterprise to require its own operator.

The increasing complexity of telecommunications networks poses serious challenges for the management and integration of these networks. An important aspect of network management is the fast identification and resolution of *faults* that may occur.

Unfortunately, individual network elements are rarely able to decisively identify faults in a network. In general, a network element can only detect problems with its own internal operation, with its peer device, or with its interface to the network. Network elements are unaware of the overall structure and configuration of the system. When a network element detects a problem, it emits an alarm. Alarms are usually text strings that describe the problem as perceived by the network element. These alarms are typically unarchitected and have no semantic regularity.

A single problem or fault in a network may result in dozens or hundreds of alarms being reported to management systems. These alarms may overwhelm network operators and even obscure the underlying problem. Automatic software to correlate alarms and localize and diagnose faults is of signal importance to network management.

Network Management and Control, Vol.2, Edited by
I.T. Frisch *et al.*, Plenum Press, New York, 1994

NetFACT, the Network Fault and Alert Correlator and Tester is a software system that receives alarms from element management systems, correlates these alarms, and localizes faults. Network operators are presented with a list of open INCIDENTS. Associated with each INCIDENT is a set of correlated alarms, and a list of network elements and communications lines that may be at fault.

NetFACT's alarm correlation capabilities are not tied to the underlying network topology, the structure of the alarms, or the underlying communications protocols used by the network. In contrast to other approaches, (see [7] for example), NetFACT does not rely on a rule set as the basis for its alarm correlation, since rule sets may not always prove robust to network changes. Instead, NetFACT employs a family of correlation algorithms fully described in [2].

In this paper we discuss the NetFACT system. In "NETFACT ARCHITECTURE" we discuss the overall organization and architecture of NetFACT; "DATA MODEL" covers the NetFACT data model, which is crucial to the underlying correlation algorithms; "ALARM CORRELATION" covers functional aspects of NetFACT; and, our conclusions are stated in "CONCLUSIONS."

NETFACT ARCHITECTURE

NetFACT is intended to enhance a network operator's productivity by providing timely and concise information. The organization of NetFACT was influenced by careful analysis of the tasks performed by network operators. Network operators:

1. monitor alarms that are received from network elements,
2. filter out unimportant or low priority alarms,
3. group or *correlate* related alarms that may indicate a problem situation,
4. use these correlated alarms as a guide for further testing and diagnosis, and
5. diagnose the underlying fault and take appropriate action.

NetFACT is divided into several components based on the observations noted above. Figure 1 provides an overview of the NetFACT architecture. In the most recent implementation of NetFACT all the components in Figure 1 except the Display Manager are implemented under IBM's MVS operating system. The Display Manager provides a graphical user interface which runs on a personal computer under OS/2. The key components of NetFACT described in Figure 1 are:

Management System Interface

Alarm messages are forwarded to NetFACT through the Management System Interface. NetFACT does not connect directly into individual element management systems. It assumes that operational connections are provided by some integrated network management package. In the current implementation these connections are provided by NetView ([5]), IBM's host based network management platform. The architecture will support connections to other systems however, such as NETMATE (see [3]).

A Driver component may be used in place of the Management System Interface to test NetFACT using various scripts.

Alarm Recognizer

Alarms are the basic input to NetFACT. The alarm recognizer translates alarms received through the Management System Interface into a canonical format. This format is device independent and also independent of the underlying network topology. The translated alarms are called *normalized alarms*. Further details may be found in "ALARM CORRELATION."

Configuration Manager

The Configuration Manager is a database which contains a model of the telecommunications network. Captured in this model are the characteristics of the devices in the network and their inter-relationships. The Configuration Manager is object-oriented. Further details are in "DATA MODEL."

Alarm Manager

The Alarm Manager component contains the bulk of the correlation code. The Alarm Manager uses network topology information contained in the Configuration Manager, and the normalized alarms to construct INCIDENTS.

Fault Recognizer

The Fault Recognizer tries to further narrow the cause of the fault by performing additional diagnosis on an INCIDENT. The Fault Recognizer has not been fully implemented in the current version of NetFACT.

Tester

An important part of network problem diagnosis is iterative testing to pinpoint failed components. The NetFACT Tester provides for iterative testing of network elements. The Tester is not fully implemented.

Interface Manager

The Interface Manager gathers display information and forwards it to workstations. The architecture provides for multiple concurrent operators.

Display Manager

The personal computer based code provides a windowed interface into NetFACT. Operators can examine a list of open INCIDENTS, and graphically view affected areas of the network.

Incident Manager

INCIDENTS may be opened, closed, or assigned to operators. The Incident Manager oversees these tasks.

Figure 2 shows how information flows through the NetFACT system. Alarms are forwarded by the Management System Interface to the Alarm Recognizer, where the alarms are normalized. The normalized alarms are forwarded to the Alarm Manager. The Alarm Manager correlates these normalized alarms and produces INCIDENTS, which are then sent to other components for processing.

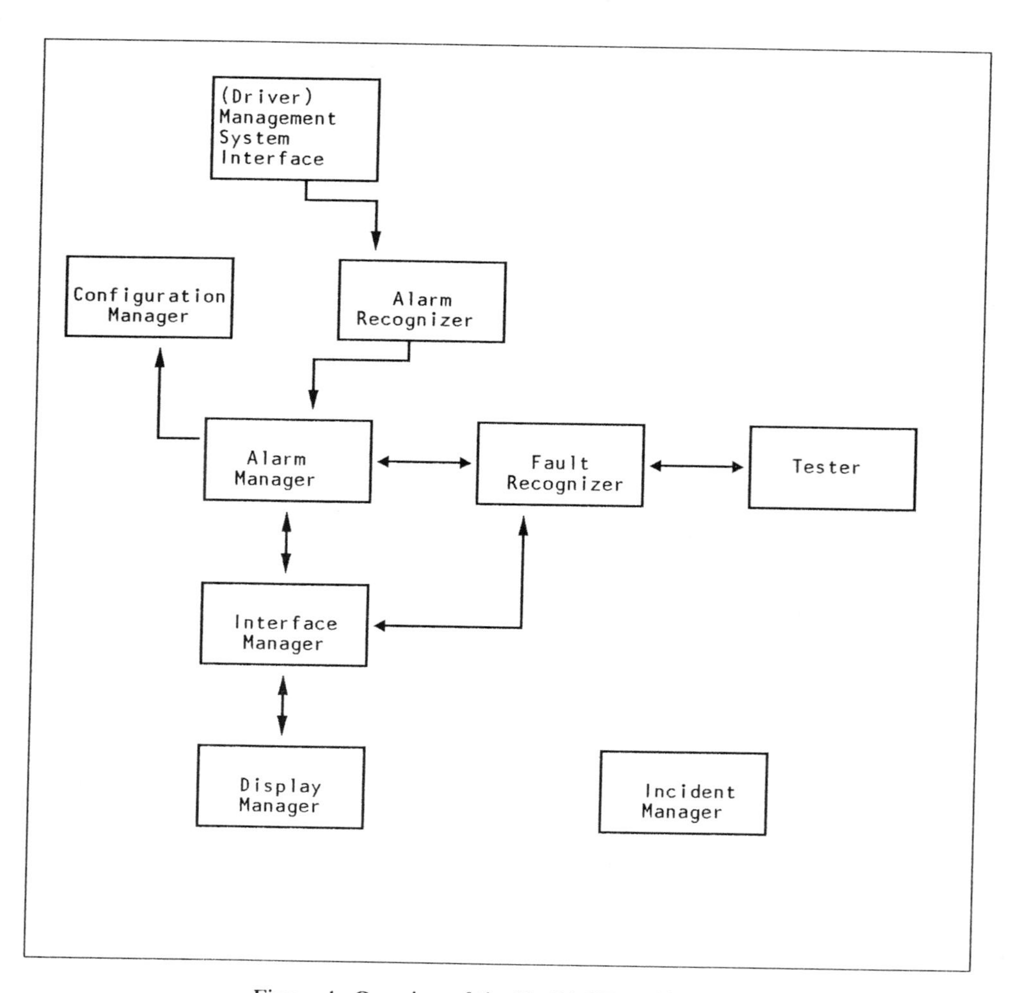

Figure 1. Overview of the NetFACT Architecture

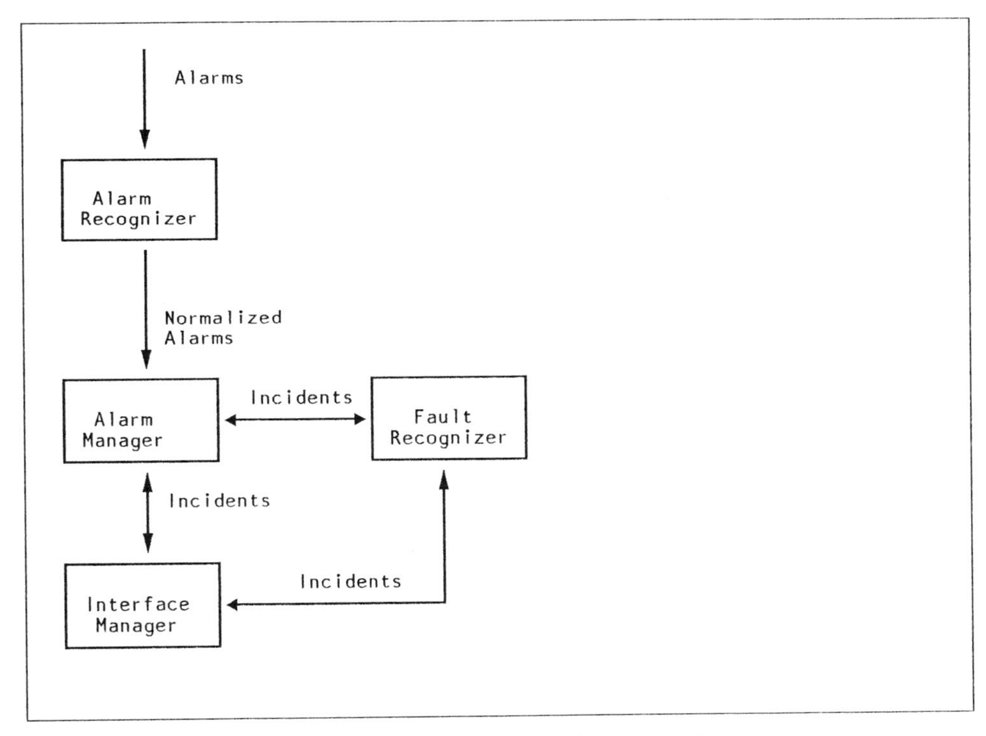

Figure 2. Information Flow in NetFACT

DATA MODEL

A suitable information base and data model are crucial to any network management system. An information base provides data and information management facilities, and serves as a repository for representations of the software and hardware configuration of the network. The information base must be able to capture the complex inter-relationships among network elements. Furthermore, it must be able to process, store and distribute the real-time status updates received by the network management system.

In many network management systems, object-oriented technology is the corner-stone of the information base (see [8], [6], for example). Although object-oriented, the information base sometimes uses a relational database as its foundation (see [3]).

The NetFACT Configuration Manager uses RODM ([4]) as its base. RODM is an object-oriented information manager and it contains facilities for establishing relation-ships between objects. Its capabilities make it easy to model network configuration. RODM supports multiple concurrent users and can handle a high volume of transactions. It can be used to model the status of numerous network elements.

Network elements may be modeled as RODM objects. The NetFACT data model contains:

1. an elaborate model of each device in the network,
2. connectivity information, and
3. facilities for storing real-time status information.

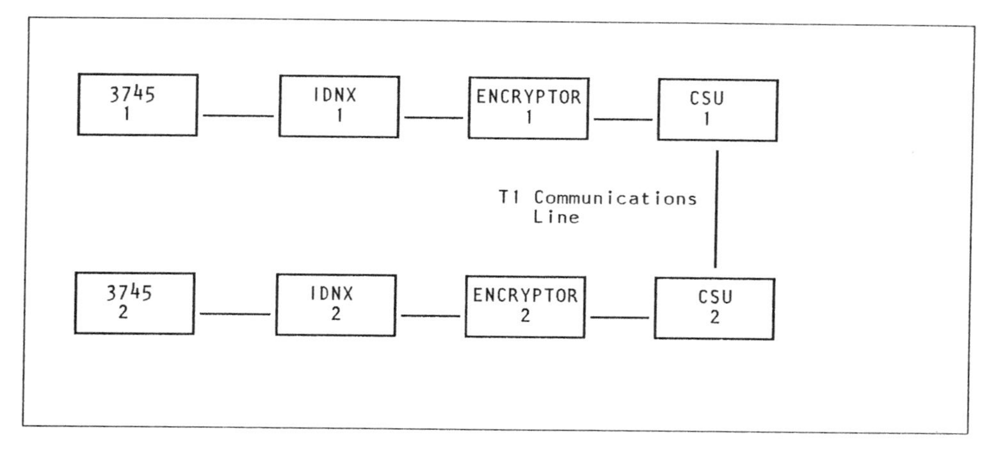

Figure 3. An Idealized T1 Circuit, and its Data Terminal Equipment

Of signal importance is the method with which NetFACT captures the interrelationships among network elements. The representation of network elements is crucial to the smooth functioning of the correlation algorithms. The modeling technique is adapted from [1] and is similar to the modeling technique used in [3]. Consider an idealized T1 circuit as depicted in Figure 3. The circuit establishes a connection between two IBM 3745 communication controllers. These 3745 communication controllers are connected to Integrated Digital Network Exchange (IDNX) switches, which are in turn connected to Encryptors. The encryptors are connected to Channel Services Units (CSUs), which maintain the integrity of the T1 line. The circuit described is an idealization, since IDNX ports and trunks have been omitted, and the wiring between devices has been simplified.

Figure 4 shows how this simple circuit would be represented inside the NetFACT Configuration Manager in a tree structured graph. A LINE is an abstraction that represents the connectivity between two devices. They do not necessarily correspond to actual physical connections. LINES are represented by nodes in the graph. The children of a LINE node represent the devices and connections that make up the LINE. A network can thus be hierarchically decomposed. This representation makes it easy to trace the connectivity between devices.

According to Figure 4 , then, the two 3745 communication controllers are connected by LINE 1, which consists of two IDNX switches and another LINE, LINE 2. LINE 2 in turn consists of two encryptors and another LINE, LINE 3, etc.

ALARM CORRELATION

Alarms emitted by network elements are often indicative of fault situations. However, a single fault in a network may result in alarms being generated by several devices. At present, telecommunications network operators monitor these alarms, and use them as a guide for determining network problems. Many integrated network management systems highlight and group alarms by severity, and present these groupings to network operators. Unfortunately, a single fault can result in the display of many alarms of differing severity. NetFACT attempts to ameliorate this problem by presenting a display of open INCIDENTS. Each INCIDENT contains correlated alarms and fault localization information.

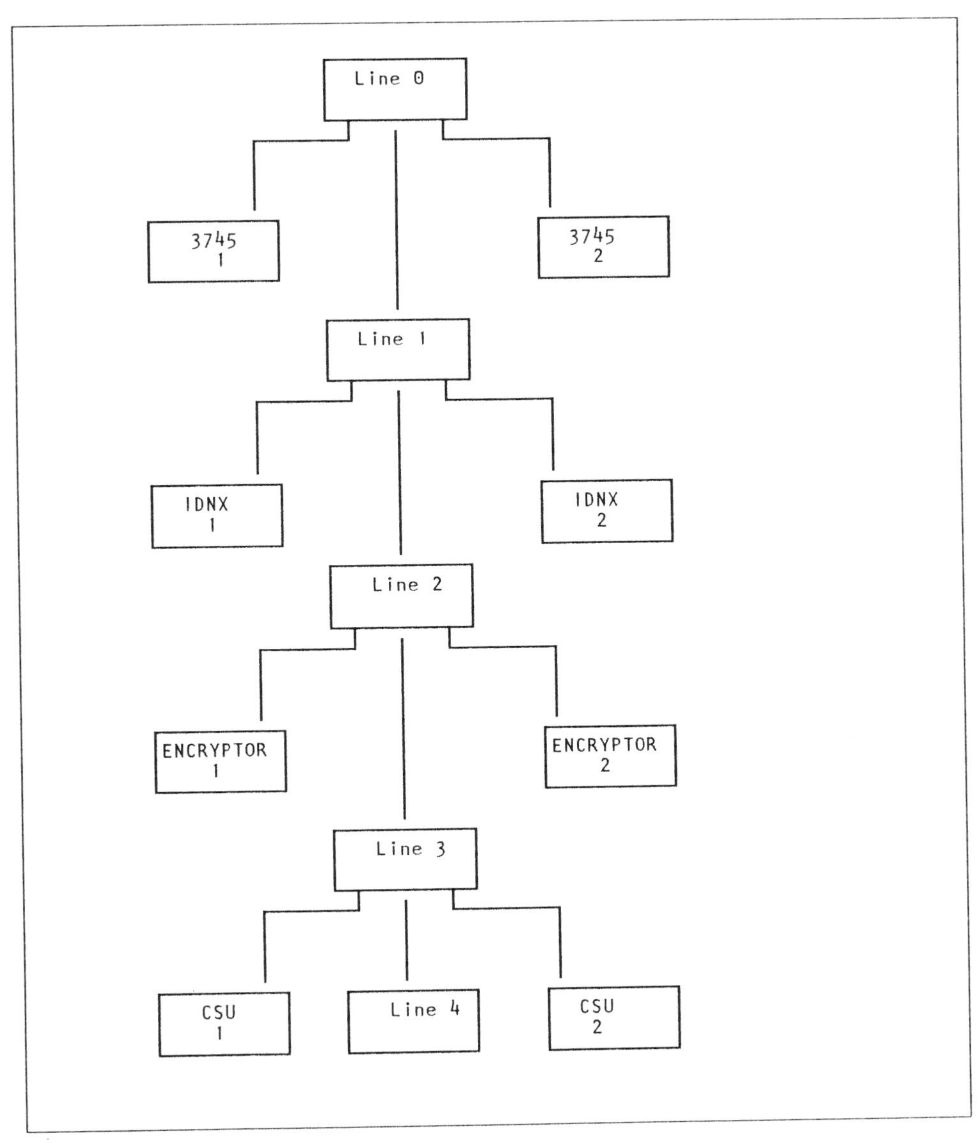

Figure 4. Model of a T1 Circuit

To see how NetFACT groups alarms into INCIDENTS, consider Figure 5 which shows correlated alarms for two possible problems in the circuit in Figure 3. INCIDENT 1 involves a problem in the connection between IDNX 1 and CSU 1, and INCIDENT 2 indicates a problem on the T1 line between CSU 1 and CSU 2.

At first glance it may seem that these diagnoses are highly contextual, and that an operator might arrive at the same conclusions by performing pattern recognition. Having recognized a particular combination of alarms in a given network configuration from a previous problem occurrence, an operator might more quickly reach a diagnosis the second time the same situation is encountered.

Rule based techniques thus appear to be a promising approach to alarm correlation. In fact, some network management systems (see [7]) provide rule based capabilities, and rules can play an important role in network management systems when used appropriately. However, any rule base that implements productions that search for particular combinations of alarms (such as the combinations found in Figure 5) cannot be robust. The absence or addition of alarms must be accounted for in a seamless manner. To do this by enumeration of all interesting combinations of alarms that might relate to a given problem is prohibitively costly in terms of information specification and maintainability for networks of any significant size.

NetFACT therefore does not use a rule based approach to correlation. Instead it recognizes that each alarm contains implicit localization information. The message Loss of Signal from DTE indicates a problem on the Data Terminal Equipment side of the CSU. The message Failed Signal indicates a problem in the direction of the T1 line.

The NetFACT Alarm Recognizer uses such information to associate fault localization information with each alarm. Through a table lookup, each alarm is appropriately normalized and associated with one of several primitives. These primitives can take such forms as: INTERNAL, EXTERNAL, UPSTREAM, or DOWNSTREAM. The exact verbs may vary. The important point is that an implicit (relative) location is given, rather than a set of suspect devices.

Once the alarm is normalized, it is transferred to the Alarm Manager. The Alarm Manager may then **dynamically** call the Configuration Manager to determine for each alarm an *explicit* set of devices which may be at fault. We call this set of devices the explicit fault localization field of the alarm.

Several methods, detailed in [2], may then be used to correlate the alarms. In one method, INCIDENTS are computed iteratively. Alarms are correlated when they share common devices in their explicit fault localization fields. The first INCIDENT consists of the largest set of alarms that share at least one common device in their explicit fault localization field. (In the event of a tie, an arbitration method is used). These alarms are removed from consideration and the process is repeated with any alarms that have not yet been correlated.

For example, consider INCIDENT 1 of Figure 5. The first alarm, IDNX Trunk Down, indicates a DOWNSTREAM problem and has a fault localization field consisting of IDNX 1, ENCRYPTOR 1, CSU 1, and the T1 line. The second alarm, Loss of Signal from DTE, indicates a problem with either IDNX 1 or ENCRYPTOR 1. IDNX 1 and ENCRYPTOR 1 appear in the fault localization fields of these two alarms. They can then be correlated, and the common devices indicate possible problem areas.

CONCLUSIONS

The integrated management of telecommunications networks is of increasing importance. An integrated management system must deal with heterogeneous devices and protocols. The first step in integrated network management is the unification of diverse element management systems, and the forwarding of management information to network control points. NetFACT provides another level of integration.

NetFACT integrates heterogeneous networks into a unified whole by merging seemingly disparate events into INCIDENTS. The correlation facilities of NetFACT allow network operators to manage the network as a single entity instead of a collection of disparate pieces. A network operator is presented with a set of open INCIDENTS which guide testing and diagnosis. INCIDENTS contain lists of correlated alarms, and fault localization

```
INCIDENT 1

 IDNX 1 --  IDNX Trunk Down
 CSU  1 --  Loss of Signal from DTE

INCIDENT 2

 IDNX 1 --  IDNX Trunk Down
 CSU  1 --  Failed Signal
```

Figure 5. Two NetFACT INCIDENTS

information. A graphical interface helps direct testing and diagnosis by displaying the network elements and telecommunications lines which may be affected by a particular INCIDENT.

NetFACT's capabilities do not depend on the underlying communications protocol or network topology. NetFACT can be easily adapted to new devices and configurations. As telecommunications networks grow in size and scope, automation software of this kind becomes increasingly important for their smooth operation.

ABOUT THE AUTHORS

Dr. Allan Finkel (afinkel@watson.ibm.com) is a Research Staff Member and project leader of the *Systems Management* project of the IBM Research Division. Dr. Finkel received a bachelor's degree from the State University of New York at Binghamtom in 1977 and a Ph. D. in mathematics in 1982 from New York University where he studied at the Courant Institute of Mathematical Sciences. During 1982-1983 he was a Member of the Institute for Advanced Study at Princeton, New Jersey. He joined the Mathematical Sciences Department of IBM Research in the fall of 1983 as a postdoctoral fellow and moved to the Computer Sciences Department in 1985. His research interests include expert systems, systems and network management and object-oriented databases.

Keith Houck (houck@watson.ibm.com) is currently a Senior Programmer in the network applications project at the IBM T.J. Watson Research Center. Prior to joining IBM Research he was a software designer in IBM's Network Management Products organization, where he worked on incorporating automation facilities in IBM's network management platform, NetView. His current area of interest is in developing applications to automate network management processes.

Dr. Seraphin B. Calo (calo@watson,ibm.com) received his Ph. D. degree in Electrical Engineering from Princeton University in 1976. Since 1977 he has been a Research Staff Member in Computer Sciences at the IBM T. J. Watson Research Center, Yorktown Heights, New York, and has worked and published in the areas of queueing theory, data communication networks, multi-access protocols, expert systems, and complex systems management. He has managed research projects in the communications and systems performance areas and has served on the staff of the IBM Research VP, Systems. He joined the Systems Analysis Department in 1987, and is currently Manager of Systems Applications. This research group is involved in studies of architectural issues in the design of complex software systems, and the application of advanced technologies to systems management problems. As part of his interest in systems management research, Dr. Calo was instrumental in establishing the IEEE International Workshop on Systems Management, and served as its first Chairman. Dr. Calo holds two United States patents.

Dr. Anastasios Bouloutas (atb@watson.ibm.com) was born in Lamia, Greece in 1959. He received his Ph.D. in 1989 from the Department of Electrical Engineering of Columbia University. For the year 1990 he worked as a Postdoctoral fellow at IBM Research. From 1991 to Sept. of 1992 he worked as a research scientist at the National Technical University of Athens Greece and he participated in RACE and ESPRIT projects. Since Sept. 1992 he has been a Research Staff Member at IBM Research. His research interests include systems and network management.

REFERENCES

[1] Anastasios Bouloutas.
Modeling Fault Management in Communication Networks.
PhD thesis, Columbia University, May 1991.

[2] Anastasios Bouloutas, Seraphin Calo and Allan Finkel.
Alarm Correlation and Fault Identification in Communication Networks.
IBM Research Division RC 17967, 1992.

[3] Alexander Dupuy, Soumitra Sengupta, Ouri Wolfson and Yechiam Yemini.
NETMATE: A Network Mangement Environment.
IEEE Network, 5(2):35-43, March 1991.

[4] Allan Finkel and Seraphin B. Calo.
RODM: A Control Information Base.
IBM Systems Journal, 31(2):252-269, May 1992.

[5] IBM Corporation, number GG24-3737.
NetView V2R2 Primer.
1992.

[6] Aurel A. Lazar.
Object-Oriented Modeling of the Archtecture of Integrated Networks.
Columbia University CTR 167-90-04, 1990.

[7] Gary Tjaden, Mark Wall, Jerry Goldman and Constantine N. Jeromnimon.
Integrated Network Management for Real-Time Operations.
IEEE Network, 5(2):10-15, March 1991.

[8] Ouri Wolfson, Soumitra Sengupta and Yechiam Yemini.
Active Databases for Communication Network Mangement.
Columbia University CUCS-059-90, 1990.

RREACT: A DISTRIBUTED NETWORK RESTORATION PROTOCOL FOR RAPID RESTORATION OF ACTIVE COMMUNICATION TRUNKS[1]

C. Edward Chow and Steve McCaughey

Department of Computer Science
University of Colorado at Colorado Springs
Colorado Springs, CO 80933-7150, USA

Sami Syed

MCI Telecommunications Corporation
4678 Alpine Meadows Lane
Colorado Springs, CO 80919, USA

ABSTRACT

Commercial telecommunications networks have tight real-time requirements for restoration after a failure. The problem of finding the available restoration paths and reassigning the interrupted traffic within such tight real-time requirements places difficult demands on the restoration protocol employed. This paper reviews several distributed network restoration protocols and presents a new distributed protocol called RREACT, for performing this function with a distributed algorithm which uses no prior network status or topography knowledge and which supports multiple simultaneous link restorations. Simulation results show that this protocol significantly outperforms other existing algorithms based on the Sender-Chooser approach, usually completing the total restoration in well under one second. In addition, enhancements to the basic algorithm are described which help to ensure near-optimal use of network spare channel resources and address such restoration situations where a complete restoration is possible but not achieved due to the poor choice of initial restoration path(s).

INTRODUCTION

To meet the increasing demands of telecommunications customers, both commercial and governmental, improvements must be made in the restoration capabilities of existing digital telecommunication networks. With the widespread deployment of fiber optic transmission systems and the alarming rate of outages due to fiber cuts [27], there is great interest in strategies for improving the process of restoring disrupted traffic from minutes to sub-seconds following a fiber cut [28]. Automatic protection switching probably is the fastest technique and can switch the disrupted traffic to dedicated spare links in under 50 milliseconds.

1. This research is supported by MCI with grant #01-80046.

However, it requires high dedicated spare capacity. With recent advances in digital cross-connect systems, DCS, there is increasing interest in using DCS in network restoration [2, 7, 9, 12, 14, 16, 21, 29]. The centralized DCS-based network restoration approach [9, 12, 21] requires reliable telemetric links between the DCS nodes and the network operation center. It is slower than distributed DCS-based network restoration, where the affected DCS nodes exchange messages directly to restore the disrupted traffic. The hybrid preplanned approach proposed in Bellcore's NETSPAR uses a distributed topology update protocol to identify the fault and then downloads a precomputed routing table according to the fault. The problem with this approach is that the memory required for storing the routing tables is too great [7]. In this paper, we will focus on distributed network restoration algorithms for fiber networks.

There are two basic approaches to reroute the disrupted traffic due to a fiber span cut. The link restoration approach replaces the affected link segment of a disrupted channel by a spare path between the two disrupted ends. The path restoration approach releases each disrupted channel and lets the source and destination end of the channel re-establish the connection. With the additional release phase the path restoration will take more time than the link restoration. However, the path restoration can find more efficient spare paths with fewer link segments and can handle the node failure situation with the same logic. The network restoration techniques described in [9, 12] fall in the path restoration category. In order to achieve fast network restoration, we focus on the link restoration approach in this paper.

The existing shortest path algorithms provide the basic building blocks or inspire the basic approaches adopted by the existing network restoration algorithms. The survey article of Deo and Pang [11] classified the shortest path algorithms into four classes:

1) one (node) to all (nodes) [10],

2) one to one [20],

3) all to all [5], and

4) k-shortest path algorithms [23, 24, 25].

The k-shortest path algorithms try to find k paths between the source node and the sink node. Some of them find k shortest paths sequentially. Others find k paths where the total length of k paths are minimum. They can further be divided by the nature of the path found. Some find node disjoint paths. Others find edge disjoint paths. Dijkstra's shortest path algorithm [10], which is a labelling process initiated at the source node, has had great influence on the design of these k-shortest path algorithms and hence the design of existing network restoration algorithms. Mohr and Pasche [18] indicated that for finding the one-to-one shortest path, Nicholson's shortest path algorithm [20] is about 60% faster than Dijkstra's and Bovet's algorithm [4] performs far fewer iterations than Dijkstra's. The basic idea behind Nicholson's algorithm is a Two-Prong labelling process that is initiated simultaneously from both the source and the sink.

Routing algorithms proposed for computer networks [1, 3, 17, 19, 22] find the shortest paths from each node to all other nodes in a network. The approximate distributed Bellman-Ford Algorithms in [1] has polynomial message complexity and very fast response time. However, the goal of routing algorithms is quite different from that of the distributed network restoration, which finds the shortest paths between two disrupted nodes. The response time requirement of the network restoration algorithm is almost two orders of magnitude of that imposed on routing algorithms.

The network restoration problem can be formulated as a maximum flow problem [2, 6] where the goal is to find all the disjoint paths with the maximum flow between the two disrupted nodes. The efficient distributed maximum flow algorithm proposed by Goldberg and Tarjan [15] requires $O(n^2)$ waves of messages and $O(n^3)$ messages, where n is the number of nodes in the network. This algorithm, while finding the potential paths for restoration, can only be applied in small networks and is used in conjunction with a path selection algorithm to generate results as a benchmark to compare the efficiency of other network restoration algorithms.

In this paper we presents a distributed protocol called RREACT which can perform a complete restoration of fiber optic cable failures in under one second using conventional inter-nodal messages between the switching nodes such as the Digital Cross-connect Systems (DCSs). RREACT also supports restorations with multiple simultaneous link failures and provides efficient utilization of network spare channel resources. The protocol performs this function using a distributed algorithm which uses no prior network status or topography knowledge other than what is present locally at the switching node. In addition, enhancements to the basic algorithm are described which help to ensure near-optimal use of network spare channel resources and address such restoration difficulties as "Whalen's Dilemma" in [26], while still achieving the high levels of restoration performance.

RREACT outperforms the FITNESS network restoration algorithm [29] both in terms of restoration time and spare channel usage. It performs better than the Two-Prong [8] in some cases on a small test network and proves to be very reliable. When the network size increases, its performance degrades due to the large increases in message volume. Several heuristics are proposed and implemented to cut down the message volume. Note that the above distributed network restoration algorithms can also be applied to ATM-based networks for restoring virtual paths after a network failure.

The rest of this paper is organized as follows: First, the basic terms of the network model are defined. A brief review is then given for three existing distributed network restoration algorithms [8, 14, 29]. A network restoration simulation system called NETRESTORE is briefly described. NETRESTORE facilitates the design and comparison of network restoration algorithms. We then present RREACT network restoration algorithm and compare RREACT's performance with those of other three algorithms. The discussion on the enhancements of RREACT is followed by the concluding remarks.

DEFINITIONS AND NETWORK MODELS

To facilitate the presentation and the analysis of network restoration algorithms, let us define some of the basic terms used in this paper.

A network is an augmented undirected graph G(N, L) where N is the set of nodes and L is the set of links in the network. A link is associated with a certain bandwidth and this bandwidth is divided into channels. Each channel in the links of a network contains the same size bandwidth unit. A channel can only be in one of two states, working or spare. A channel which is in the working state is called a working channel. A channel which is in the spare state is called a spare channel. A link is labelled with a triple where the first label represent the number of working channels in the link, the second label represents the number of spare channels in the link, and the third label represents the transmission delay on the link. A link can be uniquely identified by the two end nodes of the link. The channels in a link are given a unique ID and can be uniquely identified in a network by concatenating the link ID with the channel ID.

A route is specified by the ordered set of concatenated link IDs. The hop-count of a route is the number of links in the route. A path is specified by an ordered set of concatenated channel IDs. We only consider paths which do not contain cycles. The hop-count of a path is the number of channels in the path. A working path is a path where all its concatenated channels are all working channels. A spare path is a path where all its concatenated channels are spare channels. A restoration path is a spare path being designated for restoring a disrupted working channel due to a network failure.

REVIEWS OF EXISTING DISTRIBUTED NETWORK RESTORATION ALGORITHMS

Grover's Self-Healing Network Algorithm

The first distributed network restoration algorithm for a DCS-based fiber network was proposed by Grover in [13] and detailed in his Ph.D. dissertation [14]. The protocol associated with the algorithm is called the Self-Healing Network SHN protocol. In the SHN protocol, one of the two DCS nodes on detecting the fiber cut becomes the Sender based on some arbitration rule such as larger DCS network ID and the other becomes the Chooser. Then, the request messages, called signatures, will be sent out along all the spare channels on all outgoing fibers. Each of these signatures will bear different indices. These signatures will be broadcasted to the intermediate nodes between the Sender and the Chooser. The Chooser on receiving a signature, will check the index. If it is the first time the Chooser has received a signature with this index number, then a reply signature will be sent back through the same request path. Each of the intermediate nodes on receiving a reply signature will generate a switch command to the DCS to connect the ports of the two spare channels. This is called reverse linking. When the Sender receives the reply signature, it will reconnect one of the disrupted channels to the new spare path and send the information of the restored channel ID through the spare path back to the Chooser. The Chooser, on receiving the information of the restored channel ID from the spare path, will reconnect the corresponding ports to restore the disrupted channel. This protocol basically requires three message transmissions between the Sender and the Chooser for each disrupted channel.

Note that one signature is sent out for each spare channel connected to Sender and is indexed for distinction. On the same route, these signatures compete for the computation resource in each DCS node for processing.

Bellcore's FITNESS Algorithm

Following Grover's publication, another distributed network restoration algorithm for DCS-based fiber networks was proposed by Yang and Hasegawa in [29]. This method is called FITNESS and also uses a Sender—Chooser relationship for the nodes adjacent to the fiber link cut. FITNESS reduces the potential large number of signatures that may be generated in SHN by requesting the aggregated maximum bandwidth that is allowed on a restoration route. The restoration process is initiated by the Sender which will broadcast restoration request messages, called help messages, on all links which contain spare channels. Each help message will contain the Sender address, Sender-Chooser pair ID, source of the message, destination of the message, requested bandwidth and hop count. Requested bandwidth will be the minimum of the working channels lost due to the fiber link cut and the spare capacity of the particular link over which the specific help message is being broadcasted.

Help messages will be selectively broadcasted by intermediate nodes. Each intermediate node will maintain a table of the help messages it has received. This table contains the source of the help message, requested bandwidth and hop-count of the path from the Sender. The first help message received will always be broadcasted. Successive help messages are broadcasted only if the requested bandwidth is greater than all previously received messages. When help messages with a requested bandwidth equal to earlier messages, but with a lower hop count are received, such messages will not be broadcasted. Instead, table hop count and source entries will be modified to reflect the discovery of the shorter path. Help messages with lower bandwidth than any table entries or equal bandwidth with higher hop count will be ignored.

The requested bandwidth in help messages broadcasted by intermediate nodes will be the minimum of the arriving message's requested bandwidth and the spare capacity of the link over which the help message is being sent.

On detection of fiber link failure, the node which becomes the Chooser will set a fixed time-out. The length of time for the time-out must be determined empirically, but appears optimal in the range of 250 to 350 msec. During the time-out, the Chooser will maintain a table of all help messages received. On the termination of the time-out, the Chooser will select the table entry corresponding to the largest requested bandwidth and send an acknowledgment message to the source of the selected help message.

On receipt of an acknowledgment message, each intermediate node will reply with a confirmation message and then will send the acknowledgment message to the next node along the path to the Sender. As this process continues, each node on receipt of a confirmation message will make cross-connection to restore lost working channels. If a single restoration path provides insufficient bandwidth to affect full restoration of all lost working channels, the Sender initiates a new wave of help messages and the process is repeated until all channels are restored, or no new paths can be found.

UCCS' Two-Prong: A fast distributed algorithm based on a Two-Prong approach

In [8] we propose a new fast distributed algorithm based on a Two-Prong approach. The Two Prong algorithm's name is derived from its fundamental approach to finding network restoration paths. Unlike other distributed approaches, the Two Prong algorithm does not use a *Sender-Chooser* relationship for the nodes adjacent to the fiber link cut to select one node to be the Sender and only to allow the Sender to initiate the restoration process. Instead, the two nodes perform nearly symmetrical roles throughout the execution of the Two Prong algorithm. On detecting a fiber link cut, one adjacent node will be arbitrarily designated the *Black-Origin* node. The other node will be designated the *Gray-Origin* node. Restoration will be initiated from both nodes with the Black-Origin node broadcasting *Black* restoration request messages and the Gray-Origin node broadcasting *Gray* restoration request messages. These messages are sent out on all links which contain spare channels. Each restoration request message (Black or Gray) will contain the pair ID of the Black-Origin and Gray-Origin nodes, the source of the message, the destination of the message and the requested bandwidth. Requested bandwidth will be the minimum of the working channels lost due to the fiber link cut and the spare capacity of the particular link over which the specific Black or Gray message is being broadcasted.

On receipt of a Black or Gray message, an intermediate node becomes designated as a Black or Gray node, respectively. Black intermediate nodes will selectively broadcast Black messages and Gray intermediate nodes will selectively broadcast Gray messages. A table is maintained of all messages which have been received, whether they are broadcasted or not. Selection for broadcasting is determined by the available spare channel capacity on links to neighboring nodes. If there are any available spare channels over a given link, the Black or Gray message will be broadcasted.

As Black and Gray messages propagate across the network, Black messages will begin arriving at Gray nodes and Gray messages will begin arriving at Black nodes. A node, upon receiving a different 'colored' request message, will make appropriate cross connections between the links over which the two different (Black and Gray) requests were received. Once the cross connection has been made, the request message will be forwarded over the newly connected link to the next node in the restoration path.

Backtracking will occur only when a request is forwarded and connections are not available to accommodate all or a portion of the requested bandwidth. The short fall will be reported in a Backtrack message. Upon receipt of a Backtrack message, the node will disconnect any channels to the false path and redirect the restoration path in another direction by generating a new request message of the same color as the Backtrack message. If no alternative path exists, the Backtrack message will be sent back along the path over which the original request message came. When a Backtrack message is received by a Black-Origin or Gray-Origin, disconnections will be made, as required.

As restoration paths become established between the Black-Origin and the Gray-Origin, Black messages will be converging on the Gray-Origin and Gray messages converging on the Black-Origin. On receiving a Gray message, the Black-Origin will insert an Ack message into the channels of the new restoration path. On receiving a Black message, the Gray-Origin will begin *listening* to the spare channels in that link for an Ack message from the Black-Origin. Once this message is received, the Gray-Origin will forward a *Confirm* message containing mapping information informing the Black Origin of which disrupted channels are to be connected to which restoration paths. Then the Gray-Origin will make final connections to restore its disrupted channels. The Black-Origin, upon receipt of the Confirm message from the Gray-Origin will make its final connections to restore its disrupted channels. Note that these Ack and Confirm messages are sent in-band and do not require message processing by intermediate nodes.

Once the Black-Origin has restored all lost channels, it will begin flooding *Cancel* messages to all neighbors. Cancel messages are used to terminate the algorithm and return the network to normal operation. Cancel messages also ensure that any cross-connections which are not used by paths in the final restoration solution are disconnected.

NETRESTORE NETWORK RESTORATION SIMULATION SYSTEM

An ongoing research effort at UCCS has developed a network restoration simulation tool to examine existing and proposed telecommunications restoration protocols. This tool, NETRESTORE, is based on a conventional discrete event simulation approach, runs on both PC and Unix workstation. The version on the UNIX workstation features an object-oriented X-windows user interface for specifying the network topology, control the simulation, and displaying the simulation results. NETRESTORE allows any network to be easily defined and then evaluated using user-definable restoration protocols.

With NETRESTORE, network processes are modelled at node level and passive links are modeled as aggregate spare and working channels. The protocol implementation changes are isolated In defined set of procedures and globals. The topology and status of a network and the failure scenarios are defined by simple text files and can be easily modified. Graphic tool are developed to facilitate the construction of network topology and status files. Command line interface are Implemented to specify parameters for simulation run. Pre- & post- run network states are included In NETRESTORE simulation log for further data analysis.

NETRESTORE has proved invaluable for examining the strengths and weaknesses of various restoration protocols, developing and debugging new protocols, and playing “what if” scenarios to determine protocol robustness, efficiency, and performance.

After analyzing various proposed protocols which address network restoration, we determined that significant improvements could be made in many areas. One new protocol we developed uses an unconventional Two-Prong restoration approach [8], while the other, RREACT, follows a more conventional methodology but offers substantial performance and network resource utilization improvements over similar protocols.

RREACT DISTRIBUTED NETWORK RESTORATION PROTOCOL

The RREACT protocol uses a distributed approach to network restoration. Upon detecting link failure, one node enters the SENDER state while the other node enters the CHOOSER state. RREACT utilizes the conventional inter-nodal message passing capabilities to identify all paths between the DCS nodes involved in the restoration effort with a single wave of SEEK messages. As paths are identified by the SEEK messages arriving at the CHOOSER node, a CONNECTION message retraces the original messages' path to set up the alternative connectivity to effect the restoration using whatever spare channel capacity is presently available in the network.

RREACT is a very reliable and fast distributed algorithm that follows the conventional Sender-Chooser network restoration strategy. It requests aggregated bandwidth. Its request messages include the path and bandwidth information and therefore are of variable length. With the additional path information, the Chooser is able to build "smart" picture of the current local network topology and status. This "smart" network information can be used in optimization options or enhancements and allows RREACT to find shorter paths and use the bandwidth of links wisely. RREACT establishes all restoration paths in one wave of request messages. It can support multiple link failures. The main disadvantage of RREACT is that its message volume is sensitive to the number of spans per node in highly connected meshes. Several heuristics are implemented to control the message volume during the flooding of restoration request messages.

The Basic RREACT Algorithm

RREACT operates as follows.

Initial Actions

Upon detecting a link failure with another switching node, the processor of a digital cross-connect switching node will determine whether it should change its state to either the SENDER or CHOOSER mode. This determination may be based on any convenient arbitration convention (e.g., node ID, link ID, etc.) such that each node directly involved with the restoration has a unique state. Both nodes will determine the number of channels needing restoration and save this information for future reference.

The SENDER node will flood out SEEK messages over all its links which are capable of supporting restoration channels. Each SEEK message contains fields of source, destination, message type, restoration ID, restoration start time, size, and two variable length vectors representing each node (ID) the message has passed through and the corresponding spare channel capacity for each node pair along the path it has followed.

The CHOOSER node will create a N by N network capacity table (where N is the number of nodes in the network) and initialize all entries to the value of the MAXIMUM spare channel capacity possible (e.g., MAXINT).

All intermediate (TANDEM) nodes between the SENDER and the CHOOSER nodes which receive the SEEK messages will flood out a modified copy of the original message over all their links (except the incoming link) which are capable of supporting restoration channels AND are not in identified in the incoming message's path description vector (i.e., no cycles in path). Message modifications will encompass increasing the message size field by one, adding the node's ID to the path description vector, and adding the spare channel capacity of the link over which it is to be sent to the spare channel capacity vector. As a simple heuristic to limit total message volume, if more than half the allowable time for the network restoration to occur has elapsed, the incoming message is killed and not flooded out.

CHOOSER node receives any SEEK message

Appropriate entries in the network capacity table will be updated with the spare channel capacity defined by the path vector and spare channel capacity vectors if each "new" value is less than the present value in the table. The updated network capacity table will then be used to update all fields in the spare channel capacity vector of the message currently being processed. The maximum channel capacity of the path described in the message is then determined by finding the minimum value among the elements in the updated spare channel capacity vector.

At this point, only non-zero capacity paths are processed further.

If the number of channels needing restoration equals zero, the new path and capacity vectors are saved in a list for possible future use. The list will be reexamined when a REJECT message is received.

The actual bandwidth of the new restoration path is determined by taking the minimum of the number of channels remaining to be restored, and the calculated maximum capacity of the new path. If this restoration bandwidth value is greater than zero, then a CONNECT message is created. The CONNECT message contains fields of source, destination, message type, the calculated maximum path capacity value, and a fixed length vector representing each node (ID) which the original message has passed through. Spare channels on the outgoing link of the CHOOSER for this restoration path are allocated based on the path capacity value and the corresponding DCS switch connect command is issued.

The total number of channels to be restored is decreased by the path capacity value of the new CONNECT message. The network capacity table is then updated to reflect the use of these spares.

The CONNECT message is sent to the neighboring node of the CHOOSER in the path description vector.

TANDEM node receives a CONNECT message

Spare channels on the incoming and outgoing links are allocated based on the path capacity value and the corresponding DCS switch connect command is issued.

If for any reason the channel capacity on the designated outgoing link to the previous node is less than the path capacity value included in the CONNECT message, special exception processing occurs. If only part of the path capacity value can be supported, the CONNECT message is changed to reflect this new limit and the CONNECT message is forwarded as originally intended. The number of channels which could not be supported by the changed link status are entered into a REJECT message which is sent back to the CHOOSER node along the original path using the CONNECT message's path vector. Examples of events which could cause changes in the channel capacity of a link could include additional link/channel failures or simultaneous link/channel restoration efforts originated elsewhere in the network. This feature allows for handling multiple link failures.

This CONNECT message is sent to the neighboring node (toward the SENDER side) of the current TANDEM node in the path description vector.

TANDEM node receives a REJECT message

The REJECT message is sent to the neighboring node (toward the CHOOSER side) of the current TANDEM node in the original path description vector. Spare channels on the incoming and outgoing links are deallocated based on the rejected path capacity value and the corresponding DCS switch disconnect command is issued.

CHOOSER node receives a REJECT message

Appropriate entries in the network capacity table will be updated with the rejected channel capacity. The total number of channels to be restored is increased by the rejected path capacity value of the REJECT message. If available, path and capacity vectors are retrieved from the saved list for originating new CONNECT message(s). As many of these saved paths are used to replace the rejected bandwidth value as needed; the original methodology used by the CHOOSER is employed to accomplish this.

SENDER node receives a CONNECT message

Spare channels on the incoming link are allocated based on the path capacity value and the corresponding DCS switch connect command is issued. The total number of channels to be restored is updated.

Network restoration is complete when the number of channels to be restored is reduced to zero and all associated CONNECT messages have successfully reached the SENDER node.

COMPARISON OF SHN, FITNESS, TWO-PRONG, AND RREACT

The following simple analysis gives us insight on the comparative performance of the above four distributed network restoration algorithms. Let the hop count of the restoration route be n and each link in the route be the same length, the message processing time in each node be p, and the message transportation time on each link be t, which includes the transmission and propagation time. The time for restoring traffic using Grover's SHN algorithm is $2(n+1)kp+3n$ where k is the additional queueing delay overhead caused by the competition for the DCS operation system with other signatures at each node. Let s be the number of spare channels on the link connected to the Sender. The Sender will send out s signatures in sequence and these s restoration requests will be processed concurrently along this route. For high speed fiber optical transmission systems, k, which is related to the number of spare channels, s can be pretty big. There will be $2(n+1)$ instances of message processing for each spare path along this route. The 3nt term is the message transportation time and reflects the three phase nature of the SHN protocol. The time for restoring traffic on this route using FITNESS is the time-out period set by the algorithm, which is not less than $(2n+1)p+2nt$. The time for restoring traffic on this route using Two-Prong algorithm is $(n+1)p+2nt$. The time for restoring traffic on this route using RREACT algorithm is $(2n+1)p+2nt+2t'$, where t' is the message transportation time due to the overhead of transmitting path information in the request message. Assume two bytes are used to encode a node in the path and T is the transmission speed of the network management channel, t' is equal to $\frac{2\times(1+2+3+\ldots+(n-1))}{T} = \frac{n\times(n-1)}{T}$. Note that we only have to specify the immediate nodes in the path information. The first intermediate node will send out a request message with its node ID in the path information field. The second intermediate node in the path will send out a request message with two node IDs. If we use 3 bytes of the DC fields in the SONET frame, T is 24Kbytes per seconds. For n=6, t' is only 1.25 mseconds. If the time-out of FITNESS is set as $(2n+1)p+2nt$, then FITNESS outperforms SHN. For p=10 mseconds, the rank of the restoration performance for the four algorithms from the fastest to the slowest is Two-Prong, FITNESS, RREACT, and SHN. The Two-Prong has the lowest restoration time among the four algorithms.

Note that after the first time-out, if the selected restoration route has less bandwidth than that of the disrupted working channels, then FITNESS will issue another wave of restoration request messages. Let w be the number of waves of restore request messages. The overall time for restoring traffic using FITNESS is $w \cdot \text{time-out}$. For RREACT, Two-Prong and SHN, the overall time for restoring traffic is the restoration time along the longest restoration path. In this situation the SHN can outperform FITNESS.

SIMULATION RESULTS

In the following, we present the simulation results of FITNESS, Two-Prong, and RREACT on two test networks: one is the New Jersey LATA test network defined in the FITNESS [29] paper and shown in Figure 1. The New Jersey LATA test network was designed to be fully restorable for any single link span failure.

Simulation results on the New Jersey LATA test network

The following assumptions were used in the testing of RREACT, FITNESS, and Two-Prong in order to get consistent comparison.

a. All messages have equal priority.
b. All messages are serviced by a node in the order they were received.
c. It requires 10 msec to process any incoming message.
d. It requires an additional 10 msec to generate each outgoing message.
e. Propagation speed of messages is 200000 km/sec.
f. Transmission delays are computed for variable length messages, using a 8kbytes/seconds transfer rate.

RREACT Simulation Results

The following tables provide some examples of the restoration result output of the RREACT protocol.

Table 1 - Sample of NJ Network RREACT Restoration Solutions

Path#	Break = N02 - N03 (53 working channels lost)		
1	(23 channels)	44% restored at	120 msec -> N03 N01 N02
2	(20 channels)	82% restored at	267 msec -> N03 N05 N06 N01 N02
3	(7 channels)	95% restored at	545 msec -> N03 N05 N06 N08 N04 N01 N02
4	(3 channels)	100% restored at	741 msec -> N03 N05 N11 N09 N04 N01 N02

Path#	Break = N03 - N01 (71 working channels lost)		
1	(53 channels)	75% restored at	119 msec -> N01 N02 N03
2	(18 channels)	100% restored at	195 msec -> N03 N06 N05 N03

Path#	Break = N05 - N08 (81 working channels lost)		
1	(16 channels)	20% restored at	140 msec -> N08 N07 N05
2	(27 channels)	54% restored at	161 msec -> N08 N06 N05
3	(17 channels)	75% restored at	182 msec -> N08 N11 N05
4	(1 channels)	76% restored at	318 msec -> N08 N06 N01 N03 N05
5	(17 channels)	97% restored at	396 msec -> N08 N04 N09 N11 N05
6	(3 channels)	100% restored at	402 msec -> N08 N04 N01 N03 N05

The following table summarizes the restoration performance of the RREACT protocol on NJ net for a set of selected link breaks.

Table 2. Sample NJ Network RREACT Restoration Performance

Break	Work	Restored	#Paths	Spares	# messages	Time(msec)
N01-N03	71	100%	2	160	73	195
N01-N05	53	100%	3	116	63	214
N02-N03	53	100%	4	189	149	741
N03-N05	16	100%	1	48	107	207
N04-N08	59	100%	6	222	108	594
N05-N08	81	100%	6	204	79	402

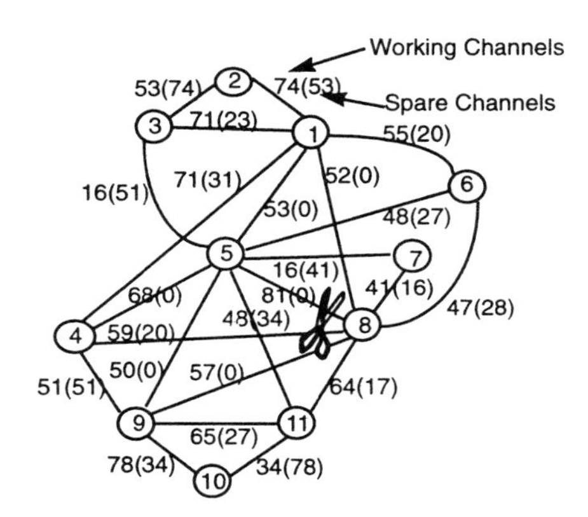

Figure 1. The New Jersey LATA Network.

FITNESS Simulation Results

The UCCS implementation of the FITNESS protocol was tested using NETRESTORE against the same New Jersey network. The following tables provide some examples of the restoration solution output of the FITNESS protocol. These examples correspond to the RREACT examples given above.

Table 3 - Sample NJ Network FITNESS Restoration Solutions

Path# Break = N02 - N03 (53 working channels lost)

1	(31 channels)	59% restored at	468 msec -> N02 N01 N04 N09 N10 N11 N05 N03
2	(22 channels)	100% restored at	724 msec -> N02 N01 N03

Path# Break = N03 - N01 (71 working channels lost)

1	(53 channels)	75% restored at	312 msec -> N01 N02 N03
2	(18 channels)	100% restored at	654 msec -> N03 N06 N05 N03

Path# Break = N05 - N08 (81 working channels lost)

1	(27 channels)	34% restored at	312 msec -> N05 N06 N08
2	(20 channels)	59% restored at	685 msec -> N05 N11 N09 N04 N08
3	(16 channels)	78% restored at	972 msec -> N05 N07 N08
4	(17 channels)	99% restored at	1439 msec -> N05 N03 N01 N04 N09 N10 N11 N08
5	(1 channels)	100% restored at	1756 msec -> N05 N03 N01 N06 N08

Table 4. Sample NJ Network FITNESS Restoration Performance

Break	Work	Restored	#Paths	Spares	# Messages	Time(msec)
N01-N03	71	100%	2	160	64	654
N01-N05	53	100%	2	157	61	645
N02-N03	53	100%	2	261	101	724
N03-N05	16	100%	1	48	30	343
N04-N08	59	100%	3	278	112	1127
N05-N08	81	100%	5	289	200	1756

Two-Prong Simulation Results

The Two-Prong protocol was tested using NETRESTORE against the same New Jersey network. The following tables provide some examples of the restoration solution output of the Two-Prong protocol. These examples correspond to the RREACT examples given above.

The following table summarizes the restoration performance of the FITNESS protocol on the NJ net for a set of selected link breaks.

Table 5 - Sample NJ Network Two-Prong Restoration Solutions

Path# Break = N02 - N03 (53 working channels lost)

1	(23 channels)	43% restored at	147 msec -> N03 N01 N02
2	(20 channels)	81% restored at	304 msec -> N03 N05 N06 N01 N02
3	(10 channels)	100% restored at	447 msec -> N03 N05 N11 N09 N04 N01 N02

Path# Break = N03 - N01 (71 working channels lost)

1	(53 channels)	75% restored at	107 msec -> N03 N02 N01
2	(18 channels)	100% restored at	200 msec -> N03 N05 N06 N01

Path# Break = N05 - N08 (81 working channels lost)

1	(16 channels)	20% restored at	115 msec -> N08 N07 N05
2	(27 channels)	53% restored at	158 msec -> N08 N06 N05
3	(17 channels)	74% restored at	231 msec -> N08 N11 N05
4	(1 channels)	75% restored at	304 msec -> N08 N06 N01 N03 N05
5	(3 channels)	79% restored at	325 msec -> N08 N04 N01 N03 N05
6	(11 channels)	93% restored at	345 msec -> N08 N04 N01 N03 N05
7	(6 channels)	100% restored at	386 msec -> N08 N04 N01 N02 N03 N05

The following table summarizes the restoration performance of the Two-Prong protocol on the NJ net for a set of the selected link breaks.

Table 6. Sample NJ Network Two-Prong Restoration Performance

Break	Work	Restored	#Paths	Spares	# Messages	Time(msec)
N01-N03	71	100%	2	160	106	120
N01-N05	53	100%	3	116	95	138
N02-N03	53	100%	3	186	97	321
N03-N05	16	100%	1	48	89	107
N04-N08	59	100%	7	222	157	595
N05-N08	81	100%	7	210	144	273

DISCUSSION OF SIMULATION RESULTS

The RREACT results show promise for improvement in overall network restoration performance in a robust manner. For the typical two second restoration requirement, RREACT's performance exceeded this requirement and our initial expectations by a wide margin. The RREACT solution versus those obtained from the FITNESS protocol shows its strength in a number of different areas:

Time to full (and intermediate stage) restoration.

The number of messages used to accomplish restoration were comparable.

The number of spare channels used to affect each restoration was considerably better.

The largest cause of these differences centers on FITNESS's use of multiple message waves and the penalty it exacts. The FITNESS algorithm chooses the largest identified restoration path during each wave of messages so that the number of discrete message waves required is minimized, and therefore the total time needed to complete the restoration. Unfortunately, this will (usually) result in inefficient use of the network's spare channel links for the restoration. Its performance directly scales with the number of discrete paths needed. Results for the simulation runs against the US network show slightly more favorable performance under most conditions, but when the number of needed restoration paths are above a certain threshold (approximately 5), FITNESS does not have enough time to complete the full restoration.

The simulation results on US Continental test network show that the number of messages generated by Two-Prong is about one half of that generated by RREACT and that the restoration time of Two-Prong is about half of that spent by RREACT. RREACT in nature will find all the paths between the disrupted nodes therefore as the network size increases the message volume increases dramatically. The Two-Prong on the other hand stops the flooding when the message reaches a node with a different color. However, in two cases, RREACT is able to find one additional path. The simulation results on New Jersey LATA test network show that the RREACT is quite robust and in some cases the restoration time and spare capacity usage are both better than the Two-Prong.

The overall excellent results of the basic RREACT protocol in initial simulation and evaluation prompted further investigation into use of available time to provide a more optimal solution to network restoration problem. The next section discusses enhancements to the basic protocol which greatly improve the optimality of the final restoration solution with minimal performance penalties.

RREACT ENHANCEMENTS

New versions of RREACT are being developed at UCCS to address the problem of optimal use of network resources (i.e., spare channels) during fast restorations.

The first version, called RREACT+, is based on a simple heuristic which addresses problems such as Whalen's Dilemma where selection of one restoration path early in the process has a negative impact on the protocol's ability to fully restore all working channels even when simple inspection reveals that the capacity exists for a full restoration and should be possible.

A second version uses the local topology information inherent in the incoming SEEK messages to derive an optimal allocation of spares (i.e., minimize the total number utilized for a given restoration) based on a dynamic programming solution. After this version of RREACT was implemented, we tried in on several networks and soon found that its large memory requirements made it impractical except for small networks.

A preliminary version of the RREACT+ algorithm has been implemented for the NETRESTORE system. For this version, a simulated hardware timer was used to determine when the CHOOSER should start making connections. RREACT+ was evaluated against the NJ network defined in Figure 1. As expected, this modification to the basic RREACT algorithm imposed a considerable penalty in restoration times. All restoration times for the NJ net increased into the 800 to 1300 msec range. The spare utilization efficiency of RREACT+ proved to be somewhat disappointing as almost half the link breaks resulted in identical spare usage to the basic RREACT algorithm, and the other half exhibited net increases in spare use (between 8 and 60 more). Three cases showed minor improvements in spare usage. The overall message count for each case remained virtually identical.

RREACT+ was designed to provide complete restoration solutions where other protocols would fail or otherwise perform very poorly. Further study and simulation is being performed to determine RREACT+'s performance, robustness and other characteristics under difficult network restoration topologies and various message processing delay scenarios.

CONCLUDING REMARKS

We have presented a distributed network restoration protocol called RREACT based on the Sender-Chooser paradigm. It tries to find all the paths between the two disrupted nodes by attaching path information in the broadcast request messages. The simulation results indicate that RREACT due to its simplicity is very robust and with single wave of request message is able to perform single link restoration under two seconds. Multiple simultaneous link failures are supported as well as other types of real-time changes in network configuration during restoration. A comparison with other important distributed restoration algorithms on a number of network configurations was also given.

ABOUT THE AUTHORS

C. Edward Chow (chow@quandary.uccs.edu) is an associate professor in the Department of Computer Science at the University of Colorado at Colorado Springs. He received his B.S.E.E. from National Taiwan University in 1977. He received his M.A. and doctorate from the University of Texas at Austin in 1982 and 1985, respectively. He is a member of IEEE and ACM. His research interests include network management, protocol engineering, distributed computing, and multimedia computing and communications.

Steve Craig McCaughey (scmccaug@sanluis.uccs.edu) is a senior scientist, working for the Science Applications International Corporation (SAIC) since 1987. Previously, he was an engineering project officer in the US Air Force. He received a Bachelor of Aerospace Engineering degree from the Georgia Institute of Technology in 1978. He received a M.S. in Computer Science from the University of Colorado at Colorado Springs in 1993. He is a member of the IEEE and ACM. His research interests include communications networks, simulation, scientific visualization, hyper-media, and orbital mechanics.

Sami Syed (0002912712@mcimail.com) is a Senior Manager working for MCI since 1984. He has over 20 years experience in Telecommunications industry at various technical and management levels. He received his Masters and Bachelors degrees in 1971 and 1969 respectively. He is a member of IEEE. His interests are in Intelligent Networks, Network Management, and Software Engineering.

REFERENCES

[1] B. Awerbuch, A. Bar-Noy, and M. Gopal. Approximate distributed bellman-ford algorithms. *Proceedings of IEEE INFOCOM*, pages 1206–1213, Bal Harbour, April 1991.

[2] J. E. Baker. Distributed Link Restoration With Robust Preplanning. *Proceedings of GlobalCom '91*, pp. 306-311, Dec. 1991.

[3] A. E. Baratz, J. P. Gray, P. E. Green, Jr., J. M. Jaffe, and D. P. Pozefsky. SNA networks of small systems. *IEEE Journal on Selected Areas in Communications*, SAC-3(3):416-426, May 1985.

[4] J. Bovet. Une amélioration de la methode de Dijkstra pour la recherche d'un plus court dans un réseau. *Discrete Applied Math.*, 13:93-96, 1986.

[5] B. A. Carré. An Algebra for network routing problems. *Journal Inst. Math. Appl.*, 7: 273-294, 1971.

[6] T. H. Cormen, C. E. Leiserson, R. L. Riovest. *Introduction to Algorithms*. The MIT Press, 1990.

[7] Coan, B, et al.. Using Distributed Topology Update and Preplanned Configurations to Achieve Trunk Network Survivability. *IEEE Trans. on Reliability*, 40(3):404-416, October 1991.

[8] C.-H. E. Chow, J. Bicknell, S. McCaughey, and S. Syed. A Fast Distributed Network Restoration Algorithm. *Proceeding of 12th International Phoenix Conference on Computers and Communications*, pages 261-267, Scottsdale, Arizona, March 24-26, 1993.

[9] D. K. Doherty, W. D. Hutcheson, and K. K. Raychaudhuri. High capacity digital network management and control. *Proceedings of GlobalCom '90*, pages 301.3.1–301.3.5, San Diego, December 1990.

[10] E. W. Dijkstra. A note on two problems in connexion with graphs. *Numer. Math.*, 1:269-271, 1959.

[11] N. Deo and C.-Y. Pang. Shortest-path algorithms: Taxonomy and annotation. *Networks*, 14:275–323, 1984.

[12] R. D. Doverspike and C. D. Pack. Using SONET for Dynamic Bandwidth Management in Local Exchange Network. *Proceedings of 5th International Network Planning Symposium*, Kobe, Japan, June 1992.

[13] W. D. Grover. The self-healing network: A fast distributed restoration technique for networks using digital cross-connect machines. *Proceedings of GlobalCom '87*, pages 28.2.1–28.2.6, 1987.

[14] W.D. Grover. *SELFHEALING NETWORKS: A Distributed Algorithm for k-shortest link-disjoint paths in a multi-graph with applications in real time network restoration.* in Doctoral Dissertation for the Department of Electrical Engineering, University of Alberta, Fall 1989.

[15] A. V. Goldberg and R. E. Tarjan. A New Approach to the Maximum-Flow problem. *Journal of the Association for Computing Machinery*, 35(4):921-940, October 1988.

[16] H. Komine, T. Chujo, T. Ogura, K. Miyazaki, and T. Soejima. A distributed restoration algorithm for multiple-link and node failures of transport networks. *Proceedings of GlobalCom '90*, pages 403.4.1–403.4.5, San Diego, December 1990.

[17] J.M. McQuillan, G. Falk and I. Richer. A review of the development and performance of the ARPANET routing algorithm. *IEEE Trans. on Communications*, COM-26(12): 1802-1811, December 1978.

[18] T. Mohr and C. Pasche. A parallel shortest path algorithm. *Computing*, 40:281-292, 1988.

[19] J.M. McQuillan, I. Richer and E.C. Rosen. The new routing algorithm for the ARPANET. *IEEE Trans. on Communications*, COM-28(5):711-719, May 1980.

[20] T. A. Nicholson. Finding the shortest route between two points in a network. *Computer J.*, 9:276-280, 1966.

[21] C. Palmer and F. Hummel, "Restoration in a partitioned multi-bandwidth cross-connect network," in Proceedings of GlobalCom '90, pages 301.7.1–301.7.5, San Diego, December 1990.

[22] M. Schwartz and T. E. Stern. Routing techniques used in computer communications networks. *IEEE Trans. on Communications*, COM-28(4):539-552, April 1980.

[23] J. W. Suurballe and R. E. Tarjan. A quick method for finding shortest pairs of disjoint paths. *Networks*, 14:325-336, 1984.

[24] J. W. Suurballe. Disjoint paths in a network. *Networks*, 4:125-145, 1974.

[25] D. M. Topkis. A k shortest path algorithm for adaptive routing in communications networks. *IEEE Trans. on Communications*, 36:855–859, July 1988.

[26] J. S. Whalen and J. Kenney. Finding maximal link disjoint paths in a multigraph. *Proceedings of GlobalCom '90*, pages 403.6.1–403.6.5, San Diego, December 1990.

[27] L. A. Wrobel. *Disaster Recovery Planning for Telecommunications*. Artech House, 1992.

[28] T.-H. Wu. *Fiber Network Service Survivability*. Artech House, 1992.

[29] Yang, C. H. and S. Hasegawa. FITNESS: Failure Immunization Technology for Network Service Survivability. *Proc. of GlobalCom '88,* pages 47.3.1-47.3.6, November 1988.

VI

DECISION SUPPORT SYSTEMS FOR PLANNING AND PERFORMANCE

INTRODUCTION AND OVERVIEW

There were seven talks in this session. The first discussed a network management platform and the others had themes of adaptive techniques, artificial intelligence, and fuzzy logic. The following is a brief synopsis of each talk.

Joachim Celestino discussed his experiences on the design and implementation of a computing platform for Telecommunications Management Networks (TMN). They used distributed processing software, e.g., ANSAware, to abstract away the complexity heterogeneity of the underlying host systems and communications protocols, and to provide distribution transparencies.

Azer Bestavros discussed communications techniques for systems which are time-constrained (e.g., multimedia). The message is that "quality of service", reliability, and security are all possible to ANY degree of confidence using surprisingly little redundancy if appropriate protocols are used (e.g., the AIDA protocol).

Robert Cahn proposed "modest" additions to network management to help solve network design problems. The theme is that unless enough hooks are in place in network management, it will be impossible to automate network design.

Teresa Rubinson discussed a fuzzy (logic) design procedure for network planning. The theme is that one needs to go beyond the confines of a particular design problem to see the broader implications of how the design evolves over time.

Lundy Lewis discussed a fuzzy logic representation of knowledge for detecting/correcting performance deficiencies and compared it to the rule-based reasoning approach. The theme is that the task of modeling/automating the expertise of a good network troubleshooter is hard; more advanced techniques are needed than are currently available.

Frank Lin discussed algorithms for "minimizing the total cost" for SMDS (Switched Multi-megabit Data Service) networks. Two combinatorial optimization problem formulations for the SMDS link set sizing problem are presented, and a near-optimal solution procedure is proposed.

Pierciulio Maryni discussed the definition and evaluation of an adaptive performance management strategy for a hybrid TDM network with multiple isochronous traffic classes.

Paul Prozeller
AT&T Bell Laboratories

EXPERIENCE ON DESIGN AND IMPLEMENTATION OF A COMPUTING PLATFORM FOR TELECOMMUNICATION MANAGEMENT NETWORKS

J. Celestino, Jr.,[1] J.N. de Souza,[1] V. Wade,[2] and J.-P. Claudé[1]

[1]University of Paris VI
MASI Laboratory
[2]Trinity College Dublin
Department of Computer Science

ABSTRACT

Telecommunications Management Networks (TMN) have been evolving and becoming more sophisticated. Different technologies from different vendors contribute to complicate telecommunication management so that the need for integration and interoperability is growing. A general TMN computing platform architecture was proposed by the TMN Computing Platform Special Interest ("CPSIG") Group, comprising representatives from RACE ("Research and Development in Advanced Communications in Europe") projects GUIDELINE, ADVANCE, AIM, NEMESYS, ROSA and SPECS whose main objective is to provide enough support for the contrasting TMN management applications. This paper discusses the distributed computing platform support developed in the ADVANCE project and its relationship with the general TMN computing platform architecture, as well as issues of prototyping experience in the related projects, particularly ADVANCE, AIM, NEMESYS and ROSA. The important services suported by both the implementation developed by the ADVANCE project and the more general TMN platform architecture are identified and discussed.

INTRODUCTION

The Integrated Broadband Communication Network (IBCN) is composed of a multitude of Network Elements with different technologies, from different manufacturers, which will be introduced at different rates across Europe. This complex environment needs to be managed in an efficient way, so that new applications and services are possible at a reasonable cost. The complexity increases when, beyond the technological aspects, one considers the regulatory and national environments, as well as the interconnection with existing and near future networks like ISDN.

Network Management and Control, Vol.2, Edited by
I.T. Frisch *et al.*, Plenum Press, New York, 1994

TMN management systems represent diverse telecommunication domains with wide ranging functionalities and requirements. On the one hand TMN management applications have different purposes, frequently employing different technologies, while on the other hand there is strong need for application integration and interoperability. In order to achieve this integration and realize open solutions an underlying TMN computing platform must offer the necessary support environment, connectivity and flexibility to allow management applications to achieve their goals[1].

The ADVANCE project aimed to produce recommendations for the implementation of the Integrated Broadband Communication (IBC) systems to ensure the effective administration of networks and services.

The goal of the ADVANCE project was to produce recommendations on how future Network and Customer Administration Systems (NCASs) should be designed and developed.

THE COMPUTING PLATFORM FOR TMNs

A Computing Platform is a set of facilities that provides a defined infrastructure upon which management applications can be built. The access point of these facilities and the underlying systems, is an interface provided by the platform, that management applications should use to intercommunicate and to access the network information.

A General TMN CP architecture, capable of supporting a wide variety of TMN management applications has been developed by the CPSIG group [1,2,3]. This General TMN CP architecture was the result of the prototyping experience of several independent RACE projects, namely ADVANCE, AIM, NEMESYS and ROSA. It consists of five layers, Figure 1. The goal of these layers is to deal with the complexity and heterogeneity of the underlying host systems and communications protocols. The layers also provide mechanisms for transparent for distribution, information access, and replication. These five layers are (i) CP Kernel, (ii) Distributed Processing Support, (iii) Computing Platform Interface, (iv) TMN Support Environment, and (v) User Generic Functions.

The CP-kernel consists of a number of components that mask heterogeneity, and provides a common view of the world.

The Distributed Processing Support layer essentially provides abstraction from the distribution of the applications.

The Computing Platform Interface provides the link between the programs of the application developer and the underlying processing support.

The TMN Support Environment provides the platform services that are required by the TMN Management Applications and the User Generic Functions to fulfill their respective roles in managing the telecommunication network.

User Generic Functions perform key tasks which are required across a range of TMN applications.

The ADVANCE project has developed its own Computing Platform for TMNs[4,5]. The design and implementation of this Computing Platform have taken two years through several phases. The final prototype developed in the last phase has been demonstrated in many European countries.

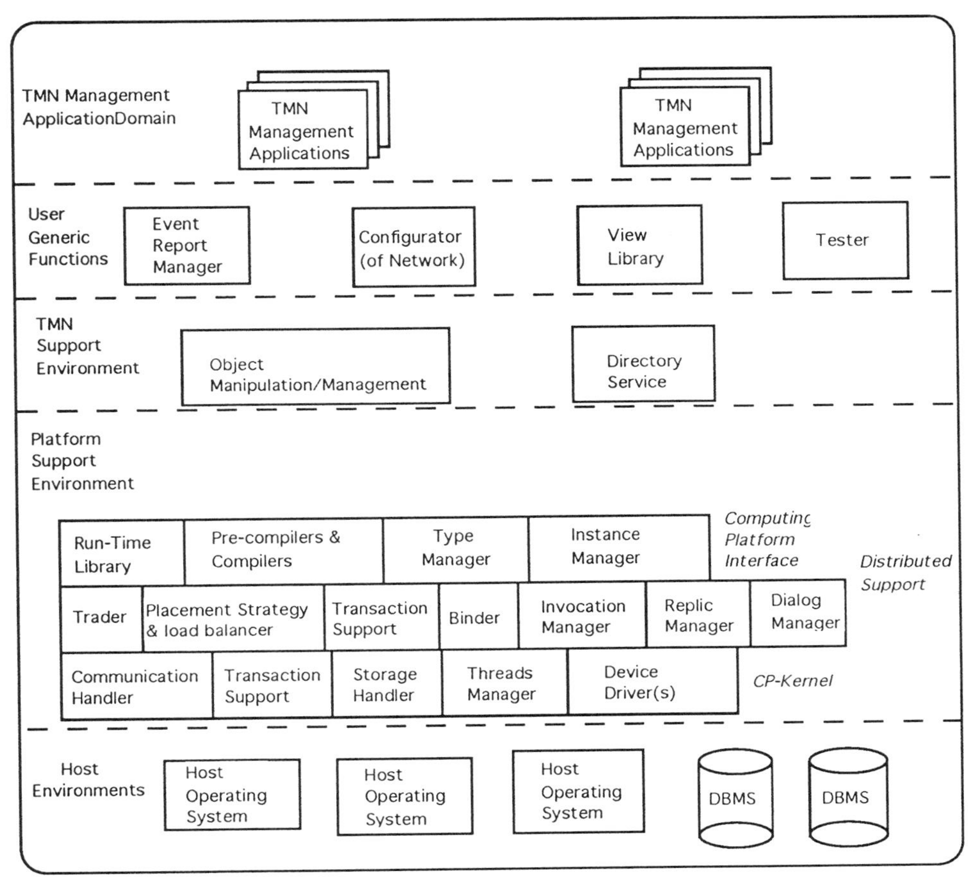

Figure 1. The General Computing Platform

The management system addressed by project ADVANCE is collectively know as NCAS (Network and Customer Administration System)[6]. To realize NCAS, there are a wide range of requirements which must be considered: openness, heterogeneity and distribution.

The NCAS is constructed from three types of components:

-Management Applications: are autonomous software components, which address a management task of NCAS;
-Delivery Mechanism: provides a means of communication between the components in a location transparent and access transparent manner (Analogous to the TMN Support Environment and CP Interface and Distributed Support Layers).
-Common Servers: Are autonomous software components which co-ordinate interaction with a number of heterogeneous and distributed shared resources (Analogous to the CP-Kernel).

THE ADVANCE COMPUTING PLATFORM

The NCAS computing Platform has been designed and developed to provide means of support for and the integration of the various components of the final integrated ADVANCE prototype[2].

The Computing Platform is composed of a number of individual components which either individually or collectively support one or more of the requirements. The Computing Platform components are shown in, Figure 2.

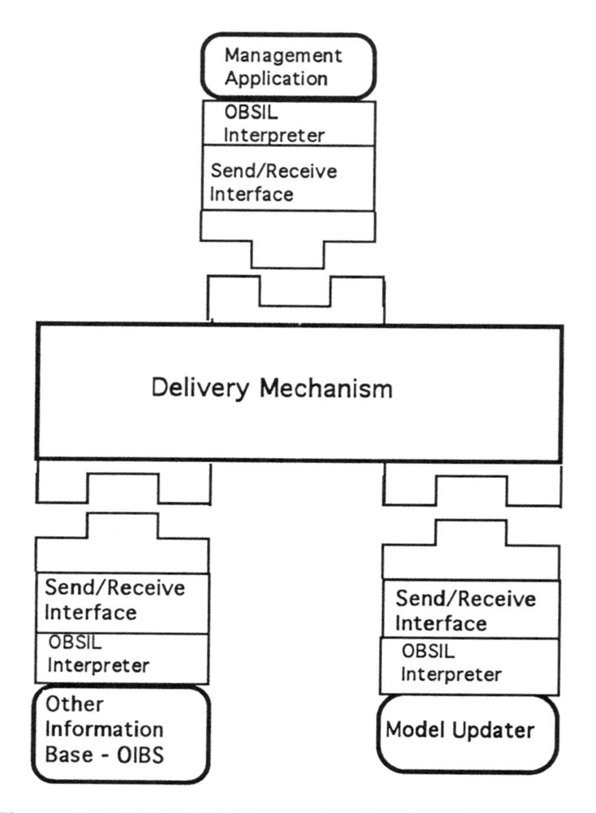

Figure 2. ADVANCE Computing Platform Components

Delivery Mechanism (DM). An obvious requirement for the realization of any architectural proposal is the ability of the component parts to inter-communicate. The DM provides the Computing Platform with means by which the inter-communication takes place. The Delivery Mechanism is at the heart of the Computing Platform. It is responsible for distributed processing support and the location transparency.

OBSIL("Object Based System Interaction") Language and Interpreters: The Computing Platform provides a number of interpreters which allow access to heterogeneous resources. These provide the application developer with access transparency through the use of OBSIL.

Send/Receive Interface: The Computing Platform provides application developers with a common access point to the platform via the Send/Receive Interface. This component can be seen as an appendix of the Delivery Mechanism component.

Model Updater: The Model Updater ensures consistency between the network resources and managed object representations. The Model Updater monitors the status of the resources and sends a notification of any changes to the appropriate managed object representation and visa-versa.

Other Information Base Server (OIBS): provides persistent storage of information. The main role of the OIBS is to provide a uniform view of stored information over a variety of multi-vendor relational databases.

DESCRIBING the ADVANCE COMPONENTS

The components showed above are defined in the ADVANCE architecture as Common Services. Another important component not depicted in this figure is called Common Information Model (CIM)[7].

The CIM provides a well-defined conceptual model of information. It is similar to, but broader in scope than, the OSI-SM MIB (Management Information Base). The CIM has extended the scope of the MIB to include representations of management information and functionality in addition to representations of the managed system.

OBJECT SYSTEM BASED INTERACTION LANGUAGE

The common interaction language facilitates interaction between the engineering components. By specifying a common interaction language, the engineering components have a common formalism for issuing and receiving requests. The main requirement for interaction is between the MAs and the CIM. The MA requests information from the object-oriented models of the CIM. The common interaction language needs to support queries on an object-oriented model. The ADVANCE project developed the basic function of the OBSIL language to do these interactions[8]. OBSIL enables MAs to remotely access and manipulate individual objects, relationships between objects, invoke management operations and create and delete object instances.

The OBSIL language provides for navigation. The world of discourse is presented to the user as an object oriented schema (i.e. CIM). The language provides mechanisms to specify paths by navigation through the model and qualifying/quantifying objects and/or relations of the model.

OBSIL is conceptually concise. It reduces the complexity of the query specification by providing extra constraint abstraction mechanisms built into the language. This approach is equivalent to the SQL language approach which reduces complexity of relational query through the provision of abstractional mechanisms within the query language.

For example a request to:

Retrieve customers who are older than 55 and subscribe to the 'tele-bingo' service

...is defined as:

```
GET Customer.name
  SUCH-THAT
    (age>55
      AND
        has_service->Service.name="tele_bingo")
```

OBSIL uses no artificial variables or explicit joins. It is argued that explicit joins are not needed as all the meaningful relations are represented in the object oriented model.

The effect of this is that queries are much easier to read as they are much closer to natural statements.

OBSIL language features

The main operations of the language are:

GET, ACTION, DELETE, CREATE, MODIFY and REPLY

The basic structure of the language is:

{<operation> <identification> <qualifier>}

The OBSIL language provides mechanisms to specify paths by navigating through the schema and qualifying objects and relations of the model. For example:

GET Customer.lives_at->Address.town

The above query refers to the attribute 'town' of the 'Address' object which is related to the 'Customer' through the relation 'lives_at->'. The result of this query will be a list of towns which customers live in.

THE OTHER INFORMATION BASE SERVER (OIBS)

The Other Information Base Server (OIBS) acts as a translation mechanism. It converts the object manipulation language of the Common Information Model (CIM) to a form required by the information base. The queries received by the OIBS are converted from the OBSIL languages to SQL. The prototype implemented is for database systems specifically, the selected database system is INFORMIX[4,5].

The OIBS is one of the Common Services of the NCAS Logical Architecture Infrastructure supporting the CIM requirements for transparent access to shared common resources. The services support the Management Application need for transparent access to data and knowledge. The OIBS component acts as an interface to heterogeneous resources. It provides common information for network management, as well as supporting Management Application access to their own private information which is not shared. By providing a consistent interface to underlying heterogeneous resources, the problem of interacting directly with the resources is hidden from developers of the CIM. The OIBS provides an interfacing mechanism to meet these requirements.

MODEL UPDATER

The Model Updater service is the point of the real objects. The CIM Module provides an abstract view of the real world to all the management applications. The objects defined in the CIM represent specific or abstract entities [9].

The Model Updater provides up-to-date information about the managed network elements accessed through the agent process. The Model Updater service must be able to answer requests from the CIM concerning the managed objects of its management system.

An other important functionality of the Model Updater is to notify the system of any change in the status of any of the real objects. This can be done by transmitting to the system the notification issued from a real object or by a mechanism which permits the Model Updater to be informed of any change of the real object's status.

THE SEND/RECEIVE INTERFACE

This Interface provides Send and Receive mechanisms for the two application development environments: C and SPOKE-C. It is currently intended only to support remote operation primitives as described below. It does not support transactions (where a number of remote operations are grouped together so that if one operation fails, the whole group fails and the target is restored to its original state) and each message is assumed to have one, and only one target[10,11].

Parser functions enable target interpreters to access the OBSIL messages, and decompose them into relevant parts, so that messages may be easily analyzed.

Protocols Primitives

Components communicating using OBSIL may interact in a number of ways:

CALL
A CALL is used to send a message to which a REPLY is expected, and the sending process waits for the response to the message it has sent. The sender of a CALL may receive a REPLY, a USER REJECT or a SERVICE REJECT in response.

CAST
A CAST is used to send a message to which no REPLY is expected by the sending entity. This means that the result of the requested processing need not be passed back to the sender. However, as a sending entity may wish to make sure that the message was at least delivered, there are three types of CAST, each with different levels of assurance about the delivery of the message:

> **WEAK CAST**
> There is never a response to a WEAK CAST, regardless of the what happens to the message. The sending process may continue processing immediately after sending a message.
>
> **CAST**
> If there is no response to a CAST, it has been successfully processed by the local component of the DM. If not, a SERVICE REJECT will be received in response to the message. The sending entity must wait until the message is accepted by the local component to the DM.
>
> **STRONG CAST**
> If there is no response to a STRONG CAST, the message has been successfully delivered to, and accepted by the target entity. If not, a SERVICE REJECT or USER REJECT will be received in response. The sending entity must wait until the message is accepted by the target entity.

REPLY
A REPLY contains the results returned from the normal termination of the remote operation.

USER REJECT
A USER REJECT indicates that the target entity has received the OBSIL message, but is unable or unwilling to respond to that message. There are a number of possible reasons for such a USER REJECT, such as: the sender is not authorized to request the specified operation; or OBSIL itself may not make sense to the target, even if it is syntactically correct.

SERVICE REJECT

A SERVICE REJECT indicates that the DM is unable or unwilling to deliver the OBSIL message. The most likely causes of this are communication (i.e. network) failure and failure to find the specified target.

THE DELIVERY MECHANISM

Applications should not be concerned with how information is stored. Transparency determines the extent that programmers need to be concerned with and have control over the distributed nature of the system. In a fully transparent system the application developer delegates all responsibility for distribution to the support environment provided (i. e Computing Platform).

The system is based on a client/server model. The Management Applications or clients use the service offered by other components, through the Delivery Mechanism. The Delivery Mechanism has the task of locating the component searched for by the Management Applications and must permit the communication between the two entities to be performed.

The DM is responsible for the transfer of information between the various engineering components, while hiding the underlying complexity of the communications network [12,13,14]. Therefore, it is a requirement on the DM to provide transparent distribution and to hide the heterogeneity.

The Computing Platform component spans a number of heterogeneous hosts. The heterogeneity transparency provided by the Computing Platform is achieved through OBSIL acting as the application interface of the CIM.

To support the interface the DM must provide functionality to:

- Allow components to make proper use of the interaction facilities, e.g. compilers, pre-compilers, run-time support, parser functions and functions, in order to facilitate the interpretation of interaction constructs;

- Realize the inter-communication required, e.g., implement and maintain a naming scheme, support synchronization and marshal messages, bind components, trade of interfaces, concurrence, replication management, etc.

The design of the Delivery Mechanism must provide support for the following transparencies:

. Access Transparency: provides identical invocation semantics for local and remote components.

. Location transparency: hide the exact location of a program component from any other.

. Failure transparency: applications must know or be informed of failed interactions so that they can take the necessary actions. However it is advantageous to provide transparent failure recovery and provide limited feedback to the applications.

Support for the other requirements identified by the architecture, such as implementing and maintaining a naming scheme, synchronization and marshalling of messages, binding of components, have been included in the design.

SOME DETAILS ABOUT THE IMPLEMENTATION

The two main tools used to implement the designs of the Delivery Mechanism are the ANSAware software®, (version 3.0)[15] and the UNIX®, System V Messages Queues. The DM is depicted in Figure 3.

ANSAware software is an implementation of the ANSA (Advanced Network Systems Architecture) architecture and is commercially available.

The Access to the ANSA based Delivery Mechanism is provided via capsules. Two basic types of capsules have been provided: a client and a server capsule. The application developer chooses the appropriate type depending on whether he intends to provide a service or use a service provided.

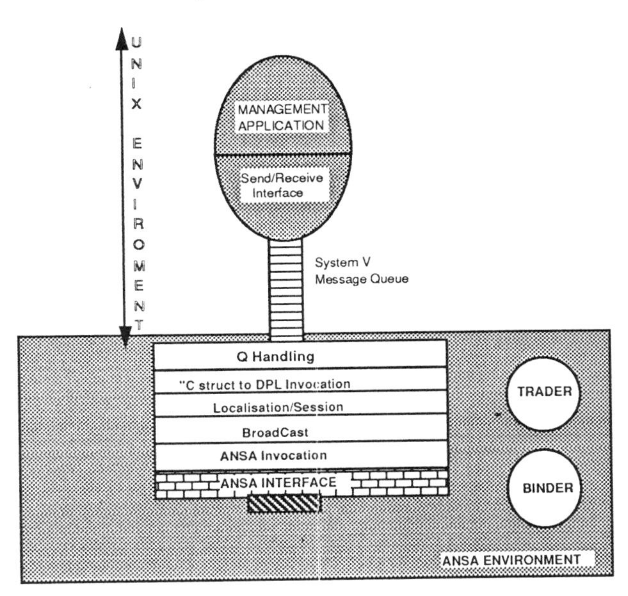

Figure 3. The Delivery Mechanism

The Delivery Mechanism has used the ANSAware software to provide components needed to implement distributed processing support (e.g., Trader, Binder, Threads, etc.).

ANSAware supports communication using sockets and TCP/IP. It does not provide access to any Database Systems or persistent storage mechanism other than those available using the native Operating System.

The Delivery Mechanism is encapsulated as a ANSA capsule and can communicate with all the processes that use the platform in two ways:
- The dialogue with the application that are not ANSA capsules is realized using the Kernel of the UNIX through the system V message queue.
- The dialogue with ANSA capsules is made directly using the interface references.

Reffering to figure 3, we can observe that:
- The communication between the Delivery Mechanism, the Management Applications and the OIBS is made using the Kernel of UNIX;

- The communication between the Delivery Mechanism and the Model Updater is made directly, because Model Updater is built using ANSAware software.

The location transparency offered by the Delivery Mechanism is made through the Trader and Binder provided by the ANSAware software. The Delivery Mechanism consults the Trader to obtain the interface reference where the server is attached and uses the binder to send and receive the OBSIL messages.

The ANSAware Trader is a form of directory service. The entries of the Directory relate to interface offers. An interface offer is exported to the trader by an object that supports a particular interface and may now be invoked. The instance that makes the offer is known as a server, and the invoker of that interface is known as client. A client looking for a particular service searches through the trader in order to find the particular one he wants. When the trader finds an offer that matches the request, it returns an interface reference to the client so that it may bind and invoke the server.

Once a client capsule possesses a reference to a server interface, it may invoke that interface (binding).

The role of the trader is to provide the correct destination for the Messages or Replies. This is achieved by mapping the context and target address of the message to its own directory service. Once the target server has been located, the Delivery Mechanism binds the client and server. This allows the transfer of the message across the platform and the return of a reply where appropriate. Binding between clients and services occurs on a message by message basis.

It is important to highlight the BroadCast mechanism that has been implemented in the Delivery Mechanism. If more than one interface reference has been provided to the Delivery Mechanism, by the Trader, the internal BroadCast mechanism is activated in order to send several copies of the same messages to each server that has been attached each interface reference. The BroadCast mechanism is also responsible for re-grouping all the responses and returning only one response to the client process.

NAMING SCHEMA

A system component is characterized by the type of its interface, the set of classes in the model and the existing relation between the classes defined in the model. Instances are not currently considered in the set of properties. The consequence of including instances could be unacceptable performance when matching a request with the properties exported by a system component[13], given the naming schema currently used within the system. Each system component is associated two properties: Classes and Relations, defined as follows:

Classes = {c | c is a class defined in the model}
Relations = {r | there exists the relation (ci r cj) in the model}.

AN EXAMPLE

The server makes use of the ANSAware software to offer services. This operation, called Export, is made using the Trader mechanism offered by the software. The following informationis included[1]:
-The interface type name;
-The context name for each server;

---

[1] The reader is invited to consult the ANSAware Manual if he is not familiar with this software.

-The names and values of properties (classes and relations).

This step is important because the Delivery Mechanism, (when acting as client and receiving a message) will build ANSAware queries to interrogate the trader so it knows the address of each server. The Query will be built providing the following information:
-The interface type name;
-The context name;
-An expression in terms of property names and values.

After this step, the Delivery Mechanism possesses the interface reference address that permits communication directly with the server(s).

CORRESPONDENCE OF ADVANCE COMPUTING PLATFORM TO THE GENERAL COMPUTING PLATFORM ARCHITECTURE

Mapping to the CP Kernel

The Communication Handler: the Advance CP platform supports communication over System V Messages Queues and TCP/IP networks.
Threads: The ADVANCE CP has used the multiple threads provided by ANSAware software.
Storage Handler: provides uniform interfaces to files and databases. This function is provided by the OIBS.

Mapping to the Distributed Processing Layer

Trader: The ADVANCE CP uses the mechanisms provided by ANSAware software.
Binder: The ADVANCE CP uses the mechanisms provided by ANSAware software.
Invocation Manager: is concerned with ensuring that interactions between system and application entities conform to the common computation model supported by the platform.

Mapping to the Computing Platform Interface Layer

Run-Time Library: messages are exchanged using a Run-Time Library, that contains the following components: the send/receive interface and the parser.
Pre-compilers: are provided by the ANSAware software for Interface Description Language and Distributed Process Language. They produce C code which is compiled with the appropriate libraries.

Mapping to the TMN Support Environment Layer

Directory: the Delivery Mechanism provides a naming schema.

Mapping to The User Generic Functions

Event Report Manager: event reports, of many types (eg. fault, state), will be received from the network elements. These must be relayed to the Management Applications that have registered interested either in the network element that generated the event report or in the type of event report. The Model Updater provides this function.

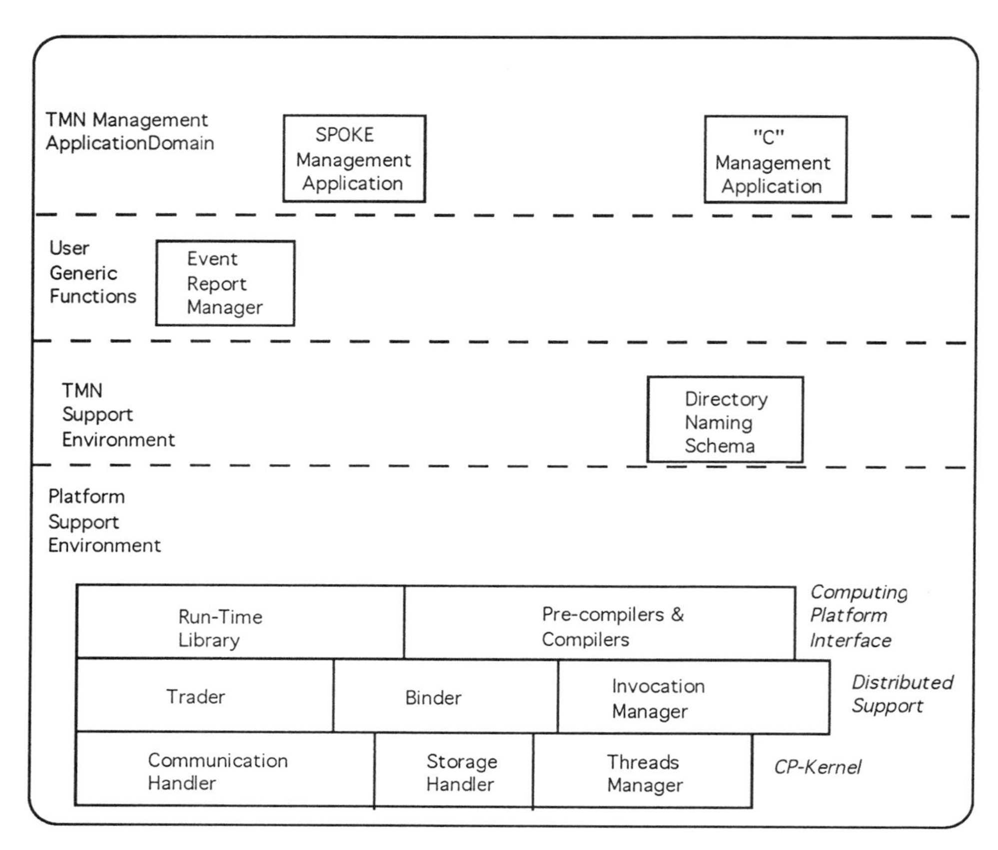

Figure 4. The ADVANCE Computing Platform

CONCLUSION

The ADVANCE project has produced recommendations for an architecture and implementation options for an integrated Network and Customer Administration System, a sub-set of a TMN system. The investigation studies the practical and theoretical aspects of a distributed computing platform and contributes to a general TMN CP architecture that is used as reference point by the other projects, like ICM (Integrated Communication Networks). This is a RACE II TMN project whose objective is to make use of components, choices and options from RACE I experiences to allow the development of other needed components in the TMN architecture.

ACKNOWLEDGEMENTS

The authors gratefully acknowledge Roke Manor Research Ltd. (UK), British Telecom (UK) and Broadcomm (IE) for their contributions. The two first authors have been supported by Banco do Nordeste do Brasil (BNB) and CAPES/CNPq.

ABOUT THE AUTHORS

Joaquim Celestino Junior (jc@masi.ibp.fr) is a Ph.D. candidate at the University of Pierre and Marie Curie (Paris VI). He holds a Msc from University of Paraiba-Campus of Campina Grande (Brazil), completed in 1990. His research interests are in Network

Management for broadband communications and distributed processing. He has worked in the ADVANCE project as part of the same programme but in its first phase (RACE I) and in the PEMMON project which was under the ESPRIT programme. He has produced several technical reports for the projects internal usage as well as technical papers published by international conferences.

José Neuman de Souza (jn@masi.ibp.fr) is a Ph.D. candidate at the University of Pierre and Marie Curie (Paris VI). He holds a Msc from University of Paraiba-Campus of Campina Grande (Brazil), completed in 1990. His research interests are in Network Management for broadband communications, with emphasis on the integration of heterogeneous networks in the Telecommunications Management Networks (TMN) environment. He is also interested in distributed computing and distributed information systems. For the past three years, he has been working in the area of TMN platform design and implementation. For the time being he is an effective technical member of the European Community project named ICM (Integrated Communications Management) which is under the RACE II (Research and Development in Advanced Communications in Europe) programme. He has also worked in the ADVANCE project as part of the same programme but in its first phase (RACE I) and in the PEMMON project which was under the ESPRIT programme. Has produced several technical reports for the projects internal usage as well as technical papers published by international conferences.

Vincent Wade (vwade@cs.tcd.ie) is a lecturer in the Computer Science Department, Trinity College, Dublin, Ireland. His current areas of research include Network Management for broadband communications, distributed computing and distributed information systems. For the past six years, he has worked in the area of communications platforms and support systems for network operations and maintenance. This work included several European Community funded projects under the RACE programme, principally GUIDELINE and ADVANCE. He holds a Msc from Trinity College Dublin and a Bsc from the University College Dublin and is also a member of IEEE.

Jean-Pierre Claudé (jpc@masi.ibp.fr) (Jean-Pierre.Claude@prism.uvsq.fr) is a doctor in computer science. He is lecturer at the UVSQ (University of Versailles Saint Quentin en Yvelines) University in the department of computer science. He leads a research team in Network Behaviour Control at the PRISM laboratory. He received his "doctorat d'Ètat" from the UPMC (University Pierre & Marie Curie) University in 1987. He is a member of IEEE. His research interests include IBCN network architecture, Multi-media network support, network management and performance evaluation.

REFERENCES

[1] Wade, V.et all.: "A Framework for TMN Computing Platforms", Proceedings of the 5th RACE TMN Conference, London Nov. 1991.

[2] Wade, V., Donnely, W., Roberts, S., Harness, D., Riley, K., Celestino, J., Shomaly, R.: " Experience Designing TMN Computing Platforms for Contrasting TMN Management Applications", published in "The Management of Telecommunications Networks", Ellis Horwood Limited, 1992, ISDN 0-13-015942-5.

[3] RACE Project R1009:"An Implementation Architecture for the Telecommunications Management Networks", Mar. 1991.

[4] ADVANCE - RACE Project R1009: "Deliverable 4: Recommendations and Conclusions", Document 09/BCM/RD3/DS/B/031/B1, Dec. 1992.

[5] ADVANCE - RACE Project R1009: "Deliverable 3: Report on Prototype Version 2", Document 09/MAR/MCS/DS/B/029/b1, Sept. 1992.

[6] ADVANCE - RACE Project R1009: "NCAS - Network Customer and Administration Systems", Document 09/BCM/RD3/DS/C/002/B1, Dec. 1988.

[7] Farley, P. Strang, C., Harkness, D. : "An Implementation Architecture for Network and Customer Administration Systems", proceedings of the 5th RACE TMN Conference, London Nov. 1991.

[8] Azmoodeh, M, Shomaly, R. : "OBSIL- An Object Based Query Language as a Basis for Telecommunication Management Systems Interactions", Proceedings of the 5th RACE TMN Conference, London, Nov. 1991.

[9] De Souza, J.N., Agoulmine, N., Claude,M. : "A System Architecture for Updating Management Information in Heterogeneous Networks", GLOBECOM'92, IEEE Global Telecommunications Conference, Dec. 1992.

[10]De Oliveira, M, De Souza, J.N., Celestino, J., Penna, M: "The Visible Interface for C Programmers", ADDN010, Apr.1991.

[11] Harkness, D. - "Proposed MA/CIM Interface Definition in SPOKE", ADPL161, Roke Manor Research, Feb.1991.

[12] Harkness, D: "The Delivery Mechanism Design", Internal report ADPL175, May 1991.

[13] Celestino, J., De Souza, J.N., De Oliveira, M., Penna, M: "The Locating Problem in the Delivery Mechanism", ADDN011, ADVANCE project, Apr.1991.

[14]Celestino, J - "A new approach for the GMS DM", ADDN032, Internal Report, Dec. 1991.

[15] ANSAware 3.0 Implementation Manual, Architecture Projects Management Ltd, Document RM.097.00, Jan. 1991.

AN ADAPTIVE INFORMATION DISPERSAL ALGORITHM FOR TIME-CRITICAL RELIABLE COMMUNICATION

Azer Bestavros

Computer Science Department
Boston University
Boston, MA 02215

ABSTRACT

AIDA – a novel elaboration on Michael O. Rabin's IDA [21] – is a communication protocol that uses redundancy to achieve *both* timeliness and reliability. In AIDA redundancy is used to tackle several crucial problems. In particular, redundancy is used to *tolerate* failures, to *increase* the likelihood of meeting tight time-constraints, and to *ration* (based on task priorities) the limited bandwidth in the system. AIDA is a *probabilistic* protocol in the sense that it does not guarantee the fulfillment of hard time constraints. Instead, it guarantees a lower bound on the *probability* of fulfilling such constraints. Such a bound could be lowered so as to satisfy any level of confidence in the timeliness and reliability of the system. In this paper we present AIDA and contrast it with traditional communication scheduling techniques used in conjunction with time-critical applications in general, and distributed multimedia systems in particular. The suitability of AIDA-based bandwidth allocation for a variety of time-critical applications is established and plans for future research experiments are mentioned.

INTRODUCTION

The successful execution of time-critical tasks running in a distributed environment often requires that a set of communication requests be successfully completed before some set deadlines. In many instances, such requests are periodic and require transfer rates *higher* than what a single client-server connection can provide. The large amount of data to be transferred, coupled with the possibility that it may not be available in full when the client-server connection is established (*e.g.* live transmission), limits the usefulness of buffering or data caching. Due to the extended period over which such client-server connections are expected to last, tolerance for failures and network traffic fluctuations becomes of utmost importance. We argue that for such time-critical systems, striping and redundancy must be employed to secure the required bandwidth and fault-tolerance. In this paper we propose a novel technique that allows the secure and fault-tolerant retrieval of information *striped* across a number of sinks in a distributed environment. While the sustained bandwidth that any single sink can confidently provide is low, the aggregate bandwidth achieved through striping and redundancy could be *demonstrably* much higher.

Network Management and Control, Vol.2, Edited by
I.T. Frisch *et al.*, Plenum Press, New York, 1994

Distributed multimedia represents an important class of time-critical applications with requirements similar to those mentioned above. A distributed multimedia presentation uses audio, video, text, and graphics from local and remote sources, some of which may be live. Multimedia data such as audio and video require special considerations when supported by a computer system. They have well-defined (*natural*) presentation timing constraints that must be satisfied by the system during presentation (or *playout*). Examples of natural timing constraints include the maximum tolerable delay in playing out a voice (or video) packet, beyond which dropping the packet might be necessary to avoid disturbing the continuity (smoothness) of the presentation. Other data types (*e.g.* text and graphics) do not have natural timing constraints, but can be subject to *synthetic* constraints, such as those arising from synchronization requirements (a piece of text or graphics may be required to appear when a particular sequence of video frames is displayed).

In order for any communication protocol to successfully schedule service requests, an accurate knowledge of the delays introduced by the communication network is often required. For communication scheduling purposes (hereinafter referred to as *bandwidth allocation*), such knowledge can be acquired either statically or dynamically. Using static techniques, worst-case delays are determined and accounted for *a priori* when scheduling the communication transactions. Alternatively, using dynamic techniques the average (or maximum) delays experienced through a communication network can be measured and used as an *estimate* for use with future communication transactions. Static communication scheduling (using a priori knowledge about the communication network delays) can be safely and efficiently used in systems with predetermined communication patterns (*e.g.* broadcasting) and systems with predetermined computation requirements (*e.g.* periodic tasks). For systems with unpredictable communication patterns or systems with sporadic computation requirements, dynamic communication scheduling becomes necessary.

Several techniques have been suggested in the literature for dynamic bandwidth allocation. Most of these techniques rely on the use of feedback from the communication network to establish a performance model that can be used in conjunction with a scheduling algorithm to allocate/reserve the communication bandwidth needed for the successful execution of time-critical tasks [15, 17].

While bandwidth allocation is an important consideration in the design of distributed time-critical systems (such as multimedia), it is not the only one. Issues of reliability, availability, and fault-tolerance are equally important. The most common technique used to tackle these issues is *replication*. For example, in distributed database applications [2, 7], several copies of a particular data object may be kept in a number of different sites so that the failure (whether intermittent or permanent [22]) of any proper subset of these sites would not render the data object unavailable.

For distributed applications operating under strict time constraints, replication alone may not be adequate. In particular, failures should not be allowed to increase the retrieval delay for data objects (at least not considerably). In this respect, techniques that rely on watchdog timers and/or retransmit protocols may not be adequate. Instead, techniques that use redundant communication (*e.g.* requesting the same data object from a set of failure-independent sites/paths) may be necessary.

Current fault-tolerant communication protocols do not attempt to use available redundancy (necessary for fault-tolerance) to improve other aspects of the system – timeliness, for example. In other words, the added redundancy serves a unique purpose: to boost the reliability of the system. It is our thesis that this approach is unnecessarily restrictive and, to some extent, inefficient. We argue that redundancy should be used to tackle many problems in an integrated manner.

In this paper we propose AIDA (Adaptive Information Dispersal Algorithm), a novel technique for dynamic bandwidth allocation, which makes use of minimal, controlled redundancy to guarantee timeliness and fault-tolerance up to any degree of confidence. Our technique is an elaboration on the Information Dispersal Algorithm of Michael O. Rabin [21], which we have previously shown to be a sound mechanism that considerably improves the performance of I/O systems and parallel/distributed storage devices [3, 6]. The use of IDA for efficient routing in parallel architectures has also been investigated [20].

REAL-TIME BANDWIDTH ALLOCATION: RELATED WORK

A real-time communication system must manage the communication of time-dependent data to provide timely and predictable data delivery; it provides performance guarantees for the delivery of data from source to destination. These guarantees can be absolute through deterministic scheduling and resource allocation, or probabilistic through the use of statistical approaches. In this section, we review a few of the representative techniques currently being used/investigated for real-time bandwidth allocation in multlimedia applications.

One way to schedule data transmission is to maintain statistics characterizing each of the communication resources (channels) in the system. Whenever the channel characteristics of the network change, the server responsible for delivering the time-critical data can adjust accordingly to maintain predictable service. This can be achieved by decreasing the demand on the network. For example, when a network becomes congested and the percentage of late data elements (missed deadlines) increases, dropping the demand on the network helps clear the congestion [12]. This effectively allows data elements scheduled for transmission to traverse the communication network and reach their destination *on time* rather than be lost due to lateness.

Another mechanism to deal with the adverse effect of network congestion is to distinguish between the various communication requirements. This was proposed in the Asynchronous Timesharing System (ATS) [15], in which data traffic is divided into four classes. A control class C has the highest priority; it delineates a class of communication where data loss or unpredictable communication delays cannot be tolerated. Class I is next on the priority scale; it delineates a class of communication where data loss cannot be tolerated, but a user-specified maximum end-to-end communication delay is allowed. Class II has a set maximum percent of lost packets and a maximum count of consecutive packets lost. Finally, class III has zero loss and no maximum end-to-end delay for communication that is not subject to time constraints.

A similar treatment of the different communication requirements imposed on a distributed system is under investigation at the University of California at Berkeley, where an experimental RAID-II network file server is being implemented. It treats requests with a low end-to-end delay requirement differently from requests with a high bandwidth requirement is being implemented [16].

The network protocol presented in [9, 8] handles performance requirements in a different manner. When a connection is requested, the user provides the network manager with maximum end-to-end delay, maximum packet size, maximum packet loss rate, minimum packet inter-arrival time, and maximum jitter, where jitter is defined as the difference in the delays experienced between two packets on the same connection. Three types of channels can be requested: *deterministic, statistical,* and *best-effort.* For deterministic channels, the communication delay is guaranteed to fall below a given time bound. For statistical channels, the probability that the delay is less than a given time

D is kept greater than or equal to a requested factor q. This can be thought of as establishing a *confidence interval* about the expected delay rather than a *deterministic bound* on that delay. Best-effort channels provide no guarantees for the percentage of messages reaching their destination on-time; they merely attempt to make the best use of the available bandwidth.[1]

Statistical approaches to overcoming delay and bandwidth limitations are attractive because they provide application programs with a flexible framework, in which a continuum of communication priorities can be easily expressed as confidence intervals. In particular, we argue that the distinction between deterministic, statistical, and best-effort channels in the protocol proposed in [9, 8] is artificial. Deterministic and best-effort channels can be thought of as *special* statistical channels, for which the confidence interval (determined by q) describing the communication delay is taken to its limits.[2] Therefore, in this paper (without loss of generality), we consider only statistical channels.

Current techniques for statistical bandwidth allocation [18] rely on choosing an end-to-end time delay T per packet that is larger than the delay expected to be experienced by a percentage P of the retrieved packets. This time T is used as an estimate for the time it will take to retrieve packets from a given source. Figure 1 illustrates a typical relationship between T and P. While such a delay function can accurately represent delay characteristics over a given period of time, network loading does change with time, possibly making the delay distribution (and thus the delay function) outdated.

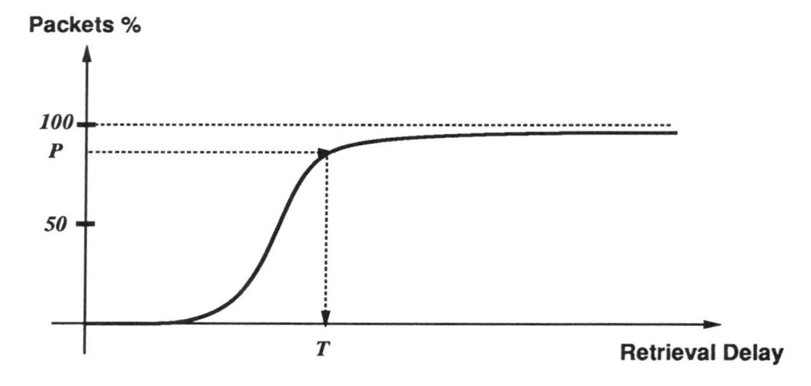

Figure 1. A typical end-to-end delay characteristic function.

One way to accommodate this dynamic behavior is to monitor the delays experienced by retrieved packets and adjust the delay function accordingly. In [10], a mechanism called *Limited A Priori* (LAP) scheduling is proposed, in which adjustments to the delay function are made either periodically or whenever sudden changes in network traffic are detected. Using second and higher order moments, linear and quadratic extrapolation of the network delay characteristics can be more accurately predicted. This research work, however, is yet to be pursued.[3]

All of the bandwidth allocation mechanisms described so far (with the exception of RAID-II) assume a single source of data for a given transaction. In a truly distributed environment, this is not likely to be the case; the storage of a single object may span

---

[1]Deterministic channels are necessary for computations with *hard* time constraints, whereas statistical channels are appropriate for computations with *soft* time constraints. Best-effort channels are adequate for computations with no time constraints.

[2]For deterministic channels, $q = 1$. For best-effort channels, $q = 0$.

[3]For more information, please contact the author.

a number of nodes to meet fault-tolerance requirements. For example, striping data for a video presentation over N nodes would increase the *availability* of the system by allowing a graceful degradation of the quality of the presentation by $1/N\%$, should any of the N nodes fail. Spreading the storage of a single object over a number of nodes may also be desired for caching purposes, especially for "hot" data objects. Finally, the distribution of a complex data object may be distributed as a result of placement and security constraints. Although it is possible to extend the aforementioned bandwidth allocation mechanisms to deal with data distributed over a number of nodes, the performance of these protocols will deteriorate significantly as a result of such distribution. The mechanism we are proposing in this paper is inherently distributed and, in that respect, is far more superior.

INFORMATION DISPERSAL AND RETRIEVAL

In this section we overview the original Information Dispersal Algorithm (IDA). We refer the reader to the original paper on IDA [21] for a more thorough presentation.

Let F represent the original data object (hereinafter referred to as the *file*) in question. Furthermore, let's assume that the storage of file F is to be distributed over N sites. Using the IDA algorithm, the file F will be processed to obtain N distinct pieces in such a way that recombining *any* m of these pieces, $m \leq N$, is sufficient to retrieve F. The process of processing F and distributing it over N sites is called the *dispersal* of F, whereas the process of retrieving F by collecting m of its pieces is called the *reconstruction* of F. Figure 2 illustrates the dispersal and reconstruction of an object using IDA. The dispersal and reconstruction operations are simple linear transformations using *irreducible polynomial arithmetic.*[4] Both the dispersal and reconstruction of information using IDA can be performed in real-time. This was demonstrated in [4], where we presented an architecture and a CMOS implementation of a VLSI chip[5] that implements IDA.

Let $|F|$ be the size of the file F. The IDA approach *inflates* F by a factor of $\frac{N}{m}$. In particular, the size of each one of the dispersed pieces of F would be $\frac{|F|}{m}$. This added redundancy makes the system capable of tolerating up to $N - m$ faults without any effect on timeliness. More importantly (as we will demonstrate shortly), this added redundancy will boost the performance of the information retrieval process *significantly.*

Several *redundancy-injecting* protocols have been suggested in the literature to deal with fault-tolerance issues. In most of these protocols, redundancy is injected in the form of (potentially distributed) parity blocks, which are only used for error detection and/or correction purposes [11]. The IDA approach is radically different in that redundancy is added *uniformly*; there is simply *no* distinction between data and parity. It is this feature that makes it possible for IDA to be used not only to boost communication fault-tolerance, but also to improve bandwidth allocation and utilization.

Unlike other redundancy-injecting protocols [23, 16], the amount of redundancy that IDA uses with a given object, or in a given session, does *not* have to be constant.

[4]For a concrete implementation and for examples, the reader is referred to our previous work on SETH [4] and IDA-based RAID I/O systems [5].

[5]The chip (called SETH) has been fabricated by MOSIS and tested in the VLSI lab of Harvard University, Cambridge, MA. The performance of the chip was measured to be about 1 megabyte per second. By using proper pipelining and more elaborate designs, this figure can be boosted significantly.

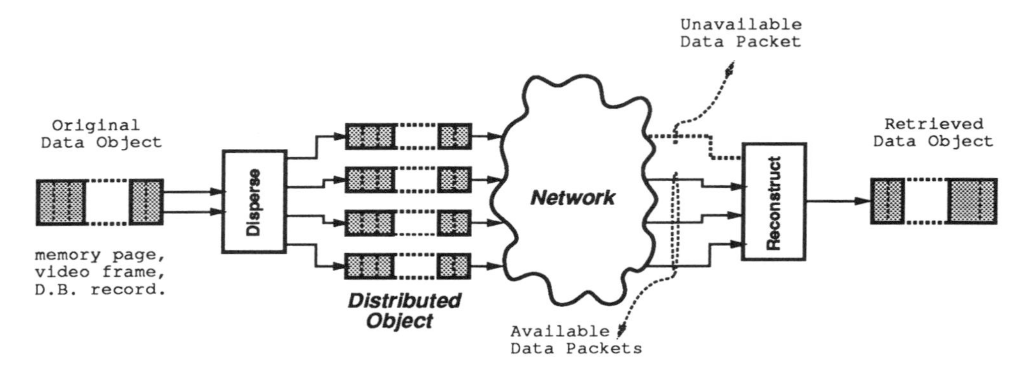

Figure 2. Dispersal and reconstruction of information using IDA.

In particular, as we will describe later, our AIDA-based bandwidth allocation strategy *controls* the amount of redundancy to be used with a particular object in a particular session so as to reflect the *priority* and/or the *urgency* of the transaction at hand. By increasing the redundancy allocated for a given communication session, the expected retrieval delay can be reduced, thus increasing the chances of meeting the possibly tight time constraint imposed on the transaction.

PERFORMANCE CHARACTERISTICS

Let X be a data object dispersed using IDA into N pieces, each residing in a different site. Let m be the minimum number of pieces needed to reconstruct X. Obviously, in order to retrieve X, at least m of the N sites must be consulted. It is possible, however, to consult more than m of these sites. Let n (where $m \leq n \leq N$) denote the total number of sites *consulted* for the retrieval of X. In this section, we derive an expression for the expected communication delay for accessing such an object. Later, we will use this result to establish the merits of our proposed AIDA-based bandwidth allocation protocol.

$$\text{Prob}(t \geq z) = \text{Prob}(\text{Response time of at least } (n-m+1) \text{ of the sites} \geq z) \tag{1}$$

$$= \sum_{r=n-m+1}^{n} \binom{n}{r} P^r (1-P)^{n-r} \tag{2}$$

where P is the probability that the response time of a single site will be z or more. P can be estimated using delay characteristic functions such as the one illustrated in figure 1.

Approximation Using a Uniform Distribution Delay Model

As a first and safe approximation, we will assume that the delays experienced through the communication network are *uniformly distributed* random variables with lower and upper bounds ($D_{\min}$ and $D_{\max}$) as illustrated in figure 3.

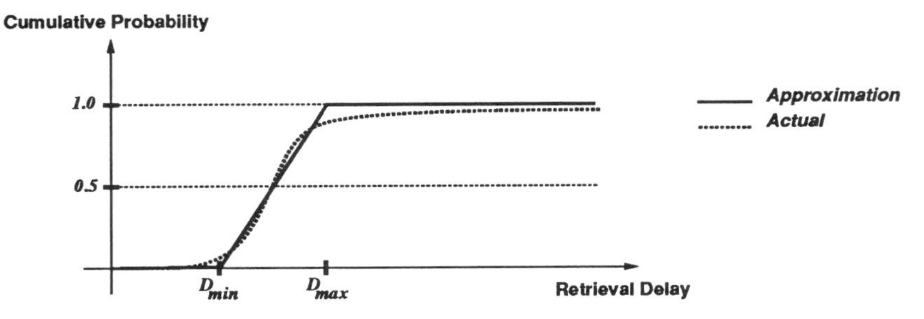

Figure 3. Delay characteristics under a uniform delay model.

We denote by P_u the value of P (in equation 2) under the uniform distribution assumption.

$$P_u = \begin{cases} 1 & \text{if } \quad 0 < z < D_{\min} \\ 1 - \frac{z - D_{\min}}{D_{\max} - D_{\min}} & \text{if } D_{\min} \leq z \leq D_{\max} \\ 0 & \text{if } D_{\max} < z < \infty \end{cases}$$

The random variable t (in equation 2) is simply the $(n-m+1)^{\text{th}}$ largest of these n uniformly distributed independent random variables. It can be shown that t follows the ***beta probability law*** and that the mean and standard deviation for t are given by:[6]

$$\tau_u = D_{\min} + \frac{m}{n+1}(D_{\max} - D_{\min}) \tag{3}$$

$$\sigma_u = \sqrt{\frac{m(n-m+1)}{(n+1)^2(n+2)}}\,(D_{\max} - D_{\min}) \tag{4}$$

Approximation Using an Exponential Distribution Delay Model

We denote by P_e the value of P (in equation 2) under the exponential distribution delay model.

$$P_e = \begin{cases} 1 & \text{if } \quad 0 < z < D_{\min} \\ e^{-\lambda(z - D_{\min})} & \text{if } D_{\min} < z < \infty \end{cases}$$

Let τ_e denote the average delay experienced using an IDA-based strategy under an exponential delay model with parameter λ (see figure 4). To compute τ_e, we proceed as follows:

$$\begin{aligned} \tau_e &= D_{\min} + E(t - D_{\min}) \\ &= D_{\min} + \int_{D_{\min}}^{\infty} \text{Prob}(t \geq z) \cdot dz \\ &= D_{\min} + \sum_{r=n-m+1}^{n} \binom{n}{r} \int_{D_{\min}}^{\infty} P^r (1-P)^{n-r} \cdot dz \end{aligned}$$

[6]Derivation is omitted for space limitations. For a reference, refer to [14].

$$
\begin{aligned}
\tau_e &= D_{\min} + \sum_{r=n-m+1}^{n} \binom{n}{r} \int_0^1 P^r (1-P)^{n-r} \cdot \frac{1}{\lambda P} \cdot dP \\
&= D_{\min} + \frac{1}{\lambda} \sum_{r=n-m+1}^{n} \binom{n}{r} \frac{\Gamma(r)\Gamma(n-r+1)}{\Gamma(n+1)} \\
&= D_{\min} + \frac{1}{\lambda} \sum_{r=n-m+1}^{n} \frac{\Gamma(n+1)}{\Gamma(r+1)\Gamma(n-r+1)} \frac{\Gamma(r)\Gamma(n-r+1)}{\Gamma(n+1)} \\
\tau_e &= D_{\min} + \frac{1}{\lambda} \sum_{r=n-m+1}^{n} \frac{1}{r} \qquad (5)
\end{aligned}
$$

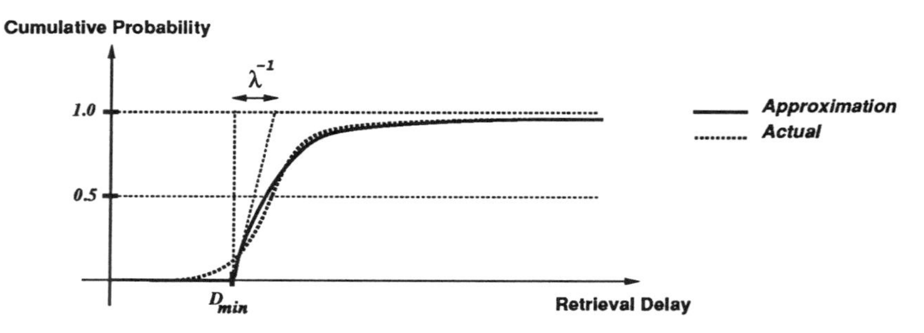

Figure 4. Delay characteristics under an exponential delay model.

The remainder of this paper assumes an exponential delay model.

Effect of Distribution and Redundancy on Delay Characteristics

There are a number of interesting observations to be made from the delay analysis of the previous section. By varying the values of n and m, the negative effect of data distribution and the positive effect of data redundancy on the delay characteristics can be demonstrated. The following cases can be readily examined:

a. $\underline{n = m = 1}$: This is the case when the object X is not distributed. The expected delay reduces to $\frac{1}{2}(D_{\min} + D_{\max})$ under the uniform delay model and reduces to $D_{\min} + \frac{1}{\lambda}$ under the exponential delay model. This corresponds to the average delay for one transmission.

b. $\underline{n > m = 1}$: This is the case when the object X is replicated over n sites. For $n \gg 1$, the expected delay approaches $D_{\min}$, which is the minimum delay for one transmission under both the uniform and exponential delay models.

c. $\underline{n = m > 1}$: This is the case when the object X is distributed without any added redundancy. For $n \gg 1$, the expected delay approaches $D_{\max}$, which is the maximum delay for one transmission under the uniform delay model. Under the exponential delay model, the expected delay approaches $D_{\min} + \frac{1}{\lambda}\ln(n)$, making the communication delay logarithmically proportional to the distribution level.

IDA-based communication attempts at striking a balance between the above three extreme setups. Figure 5 illustrates the improvement (speedup) in communication delay that can be achieved through the use of even remarkably small levels of redundancy.[7]

[7]These results were obtained under an exponential delay model, but can be easily reproduced for any other model.

For example, at a 20% redundancy level ($\frac{1}{5}$ of the communicated data is redundant), IDA cuts the expected delay through a communication network by almost 50% (a 2-fold speedup) for an object distributed over 8 sites. This gain is even larger for objects distributed over a larger number of sites. If the level of redundancy is increased further, the gain is substantial. For example, IDA can deliver a 5-fold speedup in communication with the redundancy level set at 50% for an object that is distributed over 32 sites.

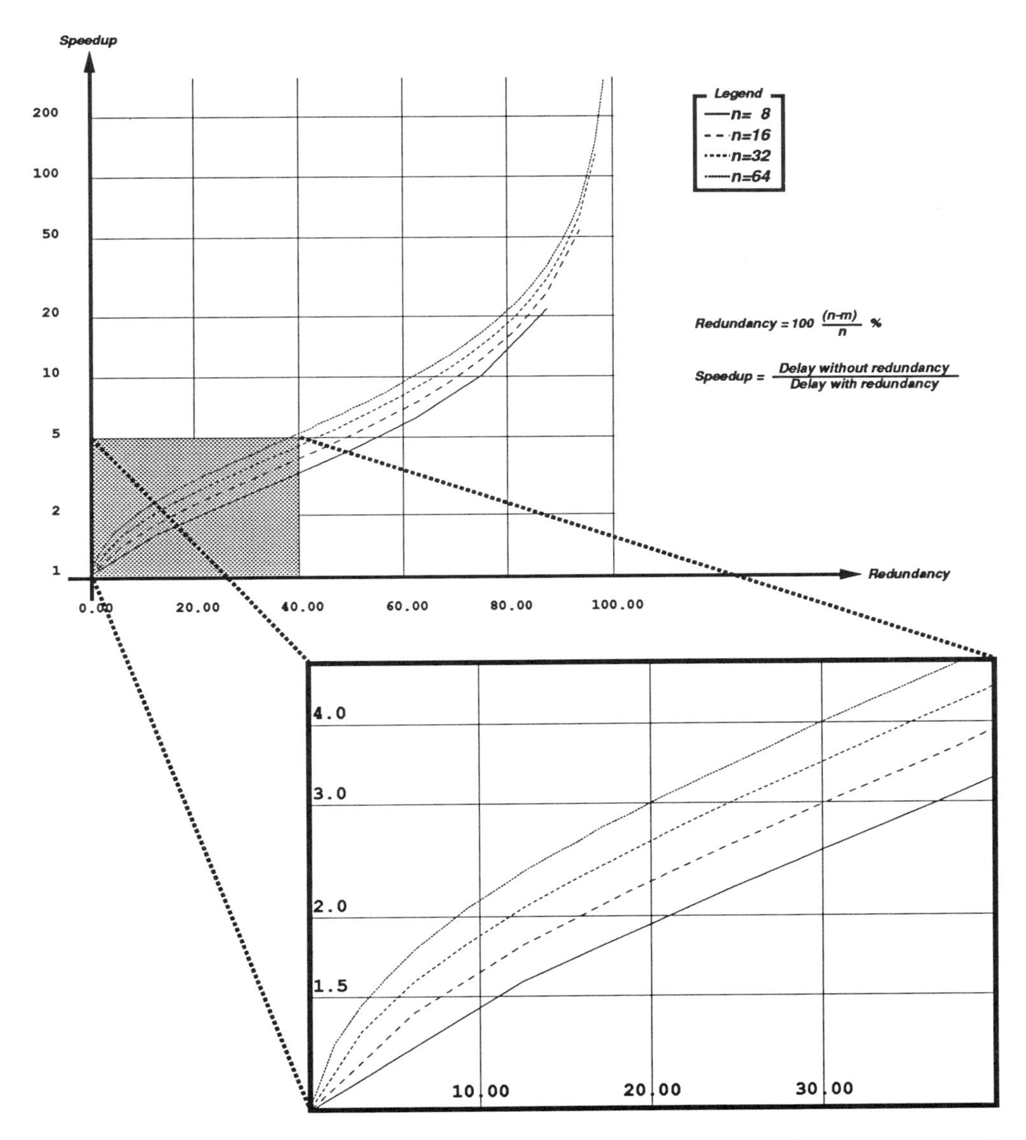

Figure 5. Expected AIDA speedups – Only the random part of the delay is considered.

For the same amount of redundancy, other protocols (such as replication) yield minuscule speedups compared to IDA. For example, if an object is replicated once ($\frac{1}{2}$ of the communicated data is redundant) and each of the two replica is distributed over 16 sites (for a total of 32-site distribution), then it can be shown that under the exponential delay model, the achievable speedup will be less than 1.1-fold. Under the same conditions, IDA delivers over 5-fold speedups.

AIDA-BASED BANDWIDTH ALLOCATION

In this section, we highlight the features of AIDA that enable it to deal effectively with deadline and priority issues in time-critical systems.

Using Redundancy to Control Communication Delays

Let the retrieval of an object X be subject to a soft time-constraint that requires X to be fetched within $T^X_{\mathbf{max}}$ units of time. According to equation 5 the expected delay in retrieving X decreases *predictably* as $n - m$ increases. Incorporating the time constraint in equation 5, we can solve for n as follows.

$$\begin{aligned} T^X_{\mathbf{max}} &\geq \tau_e \\ &\geq D_{\mathbf{min}} + \frac{1}{\lambda} \sum_{r=n-m+1}^{n} \frac{1}{r} \\ \lambda(T^X_{\mathbf{max}} - D_{\mathbf{min}}) &\geq \sum_{r=n-m+1}^{n} \frac{1}{r} \end{aligned}$$

Using the lower bound $[\ln(n) - \ln(n-m)]$ to approximate the value of $\sum_{r=n-m+1}^{n} \left(\frac{1}{r}\right)$, we get:

$$\lambda(T^X_{\mathbf{max}} - D_{\mathbf{min}}) > \ln(n) - \ln(n-m)$$

Solving the above inequality for a lower bound on n we get:

$$n > \frac{m}{1 - e^{-\lambda(T^X_{\mathbf{max}} - D_{\mathbf{min}})}} \tag{6}$$

In order to compute the appropriate value of n using equation 6, it is necessary to evaluate *dynamically* the values of $D_{\mathbf{min}}$ and λ. This can be done using statistical techniques similar to the those described in [10].

Priority-based rationing of redundant bandwidth

Equation 6 establishes a lower bound on n that guarantees an *expected* communication delay, not an *actual* communication delay. In other words, while it is very possible for the actual communication delay to be less than the desired expected delay (thus satisfying the imposed time constraint), it is very possible as well for the actual communication delay to exceed the desired expected delay (thus resulting in a violation of the imposed time constraint). This *randomness* factor can be accounted for and controlled by using second order moments (*e.g.* Standard Deviation) to build a *confidence interval* about the actual communication delay. One way of building such a confidence interval is to set the value of n so as to make $T^X_{\mathbf{max}}$, the available slack for completing the communication session, greater than or equal to $\tau_e + \alpha\sigma_e$ (rather than simply τ_e).

$$n > \frac{m}{1 - e^{-\lambda(T^X_{\mathbf{max}} - D_{\mathbf{min}} - \alpha\sigma_e)}} \tag{7}$$

The value of n in the above equation defines a confidence interval that corresponds to a specific probability of meeting the time constraint imposed on the communication session. This probability can be made arbitrarily high by increasing the value of α. This, however, is not without cost. In a distributed real-time system, the total communication bandwidth is *finite*, and increasing the amount of *redundant* information flowing in the system may adversely affect the end-to-end delay characteristics that we were aiming to improve in the first place!

One way of solving the aforementioned problem is to set the value of α in such a way that the total available bandwidth in the system is *rationed* among the different communication sessions in a way that reflects the *priority* assigned to these sessions. In other words, the value of α for a particular task is related to its priority *and* the priority of all the other tasks sharing the available bandwidth in the system.

It is important to notice that using AIDA, the priority of the transaction (how critical it is to the mission of the system) and the urgency of the transaction (how tight its time constraint is) are *both* taken into account when the value of n is determined. This stands in sharp contrast with protocols that deal only with either the priority of the transaction *or* its urgency, making it necessary for applications to express (artificially) one of these attributes using the other.

Fault-tolerance and Security Characteristics

The usual technique employed to deal with communication failures is to retransmit on errors (or timeouts). For time-critical applications, this detect-then-recover approach may not be feasible due to the time constraints imposed on the system. Instead, masking techniques are employed. In particular, error-correcting codes are used to tolerate communication failures, whereas replication and/or n-modular redundancy (NMR) techniques are used to protect against site failures [22]. The main drawback of these techniques is their excessive use of redundancy, which may adversely affect performance. For example, to mask one site failure an approach relying on replication will require that a particular object be retrieved from two different sites, thus doubling the network traffic. The blowup is even larger when error-correction for a relatively small number of communication-induced errors is taken into account.

The AIDA-based protocol we are proposing in this paper is a failure-masking protocol that is provably optimal in its use of redundancy. The main reason for AIDA's superiority is that it does not distinguish between communication failures and site failures, thus making the best use of allotted redundancy in the system. To tolerate up to r simultaneous failures, AIDA requires that the total number of sites from which the dispersed object X will be requested exceeds the minimum number of data pieces needed to reconstruct X by r. Thus, a total of $n = m + r$ sites is needed for every m pieces of data, a redundancy of $100(n - m)/m$ percent. For example, if an object X is to be dispersed over $n = 12$ sites and coverage for up to 3 failures is required, then using AIDA, the total redundancy would be 25% (a blowup of 33%). To provide the same coverage using replication, the total redundancy would soar to 75% (a blowup of 300%).

A common technique to insure communication security is to store and communicate information using some form of encryption, where only authorized users are enabled to decrypt the information through the use of appropriate *secret keys* [24]. The proven difficulty of decrypting the information without knowing the secret key guarantees a high level of *security*. The main disadvantage of this technique is that the information (although encrypted) is available in one site – whether stored in or communicated through that site – for long periods of time. This may make it possible for adversaries to break the secret key of the encryption.

The AIDA-based protocol we are proposing in this paper guarantees the security of the communicated information by making it unavailable as a whole in any one particular site. As a matter of fact, it is hard to get any clue about the original information unless at least m pieces from the dispersed file are collected. This makes the task of the adversaries more difficult, since they have to control m of the sites and not only one. Even if this happens, it is provably very difficult to reconstruct the original file unless the secret key is known.

Simulation Results for AIDA-based Communication

Predicting the effectiveness of AIDA using analytical techniques is limited to simplistic assumptions concerning the network delay characteristics, server response times, and communication capacity. Simulation becomes the only alternative, if realistic assumptions are to be adopted. We have conducted a number of simulations to further evaluate the promise of AIDA and check the accuracy of our analytical expectations. Figure 6 shows the simulation model used.

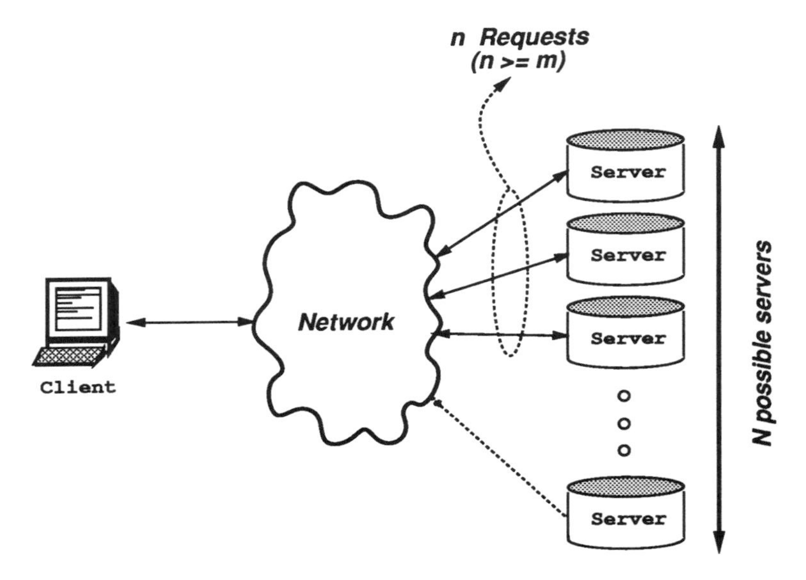

Figure 6. Simulation Model

Figure 7-a shows the speedups obtained by AIDA if the simplistic exponential delay model is replaced with a realistic multi-staged model, where each stage introduces an exponentially-distributed delay. The simulations show that AIDA still reduces the communication delay considerably.

Figure 7-b shows the speedups obtained by AIDA under a limited channel/server assumption. The simulations show that an expected *saturation* point exists. Obviously, increasing communication redundancy beyond that point is counter-productive. This confirms that a feedback redundancy control mechanism must be employed with any implementation of AIDA.

Two parameters determine the *achievable* access time for a given object (e.g. shared memory page, database record, or video frame sequence) in the system: distribution level m and dispersal level N. Figure 8-a shows the effect of varying these two parameters. This confirms the need to manage the distribution/dispersal level for the various objects in the system, based on the desired accessibility and timeliness constraints imposed on these objects. Needless to say, these constraints (and therefore the control mechanism employed to manage their distribution/dispersal levels) need to be dynamic to accommodate the changing priorities and modes of the system operation.

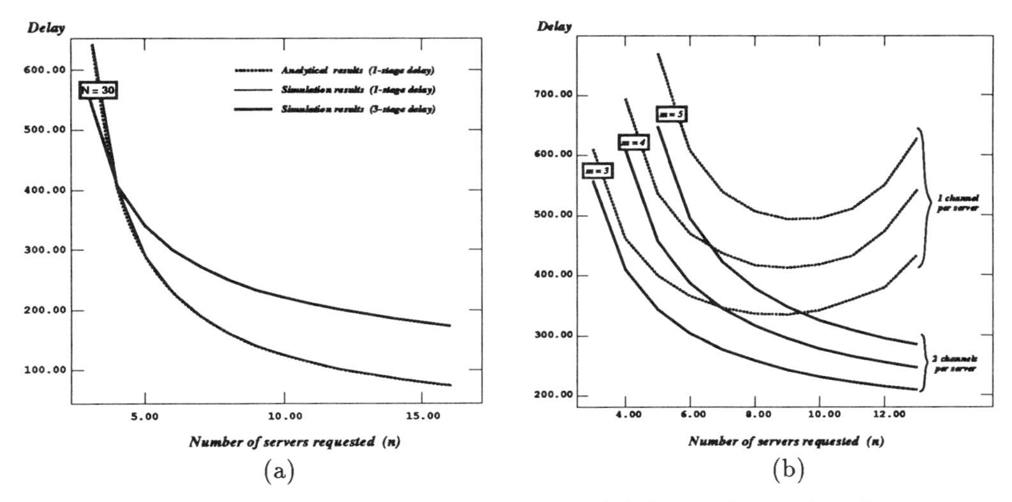

Figure 7. Simulation: (a) Staging effect (b) Channel capacity effect

Our analytical evaluation has assumed that all the servers in the system have identical responsiveness. If this assumption is not true, then the algorithm used to select the n out of N servers to be consulted for the retrieval of an object (where $m \leq n \leq N$) becomes crucial. Figure 8-b shows the performance of two such algorithms: *random selection* and *cyclic selection*. Both of these selection algorithms are blind in the sense that they do not account for the servers load or network traffic.

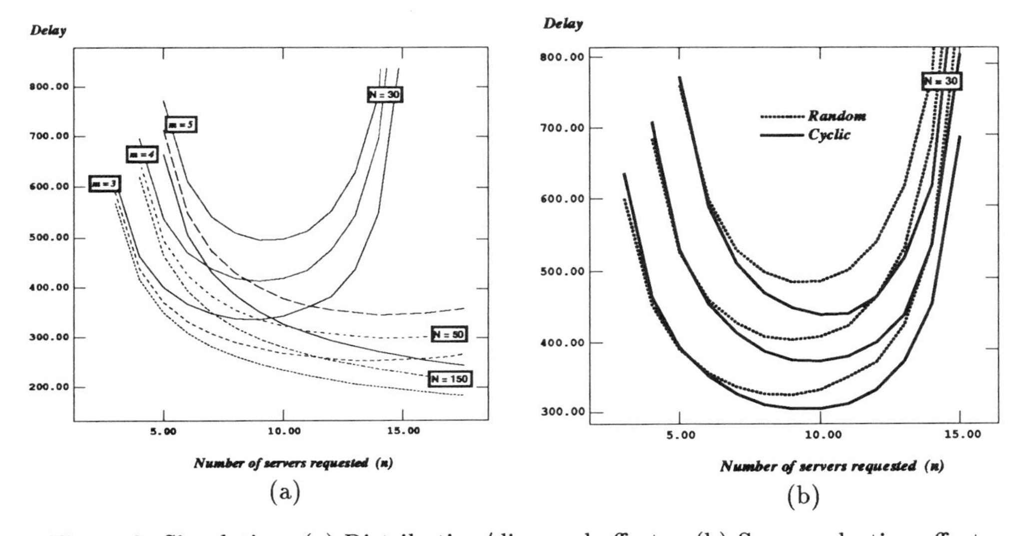

Figure 8. Simulation: (a) Distribution/dispersal effect (b) Server selection effect

We have performed a number of other experiments to study the effect of AIDA on data buffering/caching requirements (necessary for satisfying synchronization constraints in applications like Multimedia). One result was eminent: by reducing/controlling the uncertainty associated with communication delays, the buffering/caching requirements were greatly relaxed. This result suggests that AIDA should be used *in conjunction* with, and as an *integral part* of the I/O and memory management subsystems.

CONCLUSION

AIDA is a novel bandwidth allocation strategy suitable for distributed fault-tolerant time-critical systems. In AIDA redundancy is used to tackle several crucial problems; it is used to *tolerate* failures, to *increase* the likelihood of meeting tight time-constraints, and to *ration* (based on task priorities) the limited bandwidth in the system. Currently, we are in the process of building a prototype for an AIDA-based Network File Server (NFS), whose salient features are discussed below.

A request for data access made to the proposed AIDA-based NFS will specify both a priority level and a time constraint. The priority level will be used to establish the predictability requirement and thus the allotted redundancy (i.e. percentage of system bandwidth) to be granted for that request in order to meet its timing constraint. Using statistical data about the expected delays through the network, a subset of all servers that can grant such a request will be determined. The statistical techniques to be used will be similar to those suggested in [9, 15, 8, 18].

While the correctness and efficacy of AIDA are not dependent on the algorithm used to select the n out of N sites to be consulted for an object retrieval, its performance may benefit greatly. Performance gains may be achieved by classifying communication requests as was done in [15, 16]. This will likely reduce the uncertainty associated with communication delays, thus providing for a superior bandwidth allocation.

For a distributed real-time application, the NFS is only an added layer in the memory hierarchy. The addition of such a layer cannot be done in isolation from other system components (e.g. memory and I/O managers). In particular, the NFS might be able to improve the timeliness of data access considerably, if the access pattern (working set) for a given process can be reliability predicted so as to use local storage (e.g. memory or disk) to cache the NFS objects.

The reliability and accessibility requirements of various data objects in a distributed real-time application depend on the system *mode* of operation. For example, while the timely access of a data object (e.g. "location of nearby aircrafts") could be critical in a given mode of operation (e.g. "combat"), it might be less critical in some modes (e.g. "landing"), or even completely unimportant in others. The proposed AIDA-based NFS will be able to maintain different user-defined profiles of reliability and accessibility requirements for the various objects (files) in the system. This can be done dynamically (when a mode-change takes place) by controlling the levels of distribution and dispersal for the system objects.

In this paper, we have focused on using AIDA for information retrieval. Issues pertaining to information *update* were not tackled. These issues are particularly important in distributed time-critical systems to ensure data consistency and recency. In particular, it is of utmost importance to investigate the interaction between AIDA and other consistency-preserving protocols such as distributed shared memory protocols [25], caching protocols [1], and non-coherent memory protocols [13].

AIDA does not *guarantee* that time-constraints will be satisfied, rather it guarantees a lower bound on the *probability* of meeting these constraints. This probability can be made arbitrarily high if enough redundancy is secured. This, however, may not be feasible if the system is running close to capacity. One possible approach to deal with such a situation is to allow the *quality* of service to degrade gracefully. The integration of AIDA with *best-effort* techniques such as those presented in [8, 12] and *imprecise computation* techniques such as those suggested in [19] is an interesting research problem yet to be pursued.

ABOUT THE AUTHOR

Azer Bestavros (bestavros@cs.bu.edu) is an Assistant Professor in the Computer Science Department at Boston University, Boston, Massachusetts. He received his S.M. and Ph.D. degrees in Computer Science from Harvard University in 1989 and 1992, respectively. He is a member of several societies in the IEEE and the ACM. He is the moderator and maintainer of the electronic newsletter and archives of the IEEE-CS Technical Committee on Real-Time Systems. His research interests include reliable time-constrained communication and computation, embedded responsive systems, real-time databases, and parallel computing.

References

[1] J. Archibald and J-L. Baer. Cache coherence protocols: Evaluation using a multiprocessor simulation model. *ACM Transactions on Computer Systems*, 4(4):273–298, November 1986.

[2] A. Bernstein, A. Philip, V. Hadzilacos, and N. Goodman. *Concurrency Control And Recovery In Database Systems.* Addison-Wesley, 1987.

[3] Azer Bestavros. IDA-based disk arrays. Technical Memorandum 45312-890707-01TM, AT&T, Bell Laboratories, Department 45312, Holmdel, NJ, July 1989.

[4] Azer Bestavros. SETH: A VLSI chip for the real-time information dispersal and retrieval for security and fault-tolerance. In *Proceedings of ICPP'90, The 1990 International Conference on Parallel Processing*, Chicago, Illinois, August 1990.

[5] Azer Bestavros. IDA disk arrays. In *Proceedings of the First International Conference on Parallel and Distributed Information Systems*, Miami Beach, Florida, December 1991.

[6] Azer Bestavros, Danny Chen, and Wing Wong. The reliability and performance of parallel disks. Technical Memorandum 45312-891206-01TM, AT&T, Bell Laboratories, Department 45312, Holmdel, NJ, December 1989.

[7] Ramez Elmasri and Shamkant Navathe. *Fundamentals of Database Systems.* The Benjamin/Cummings Publishing Company Inc., 1989.

[8] D. Ferrari. Design and application of a delay jitter control scheme for packet-switching internetworks. In *Proceedings of the second International Conference on Network and Operating System Support for Digital Audio and Video*, Heidelberg, Germany, November 1991.

[9] D. Ferrari and D.C. Verma. A scheme for real-time channel establishment in wide-area networks. *IEEE Journal on Selected Areas in Communications*, 8(3):368–379, April 1990.

[10] John F. Gibbon, Azer Bestavros, and Tom Little. Limited a priori scheduling for distributed multimedia systems. Work in progress, November 1992.

[11] Garth Gibson, Lisa Hellerstein, Richard Karp, Randy Katz, and David Patterson. Coding techniques for handling failures in large disk arrays. Technical Report UCB/CSD 88/477, Computer Science Division, University of California, July 1988.

[12] M. Gilge and R. Gussella. Motion video coding for packet-switching networks – an integrated approach. In *Proceedings of the SPIE Conference on Visual Communications and Image Processing*, Boston, MA, September 1991.

[13] Abdelsalam Heddaya and Himanshu S. Sinha. An overview of MERMERA: a system and formalism for non-coherent distributed parallel memory. In *Proceedings of the 26th Hawaii International Conference on System Sciences*, January 1993.

[14] Harold Larson. *Probability theory and statistical inference, Third Edition.* John Wiley & Sons, 1982.

[15] Aurel A. Lazar, Adam Temple, and Rafael Gidron. An architecture for integrated networks that guarantees quality of service. *International Journal of Digital and Analog Cabled Systems*, 3(2), 1990.

[16] Edward Lee, Peter Chen, John Hartman, Ann Drapeau, Ethan Miller, Randy Katz, Garth Gibson, and David Patterson. RAID-II: a scalable storage architecture for high-bandwidth network file service. Technical Report CSD 92/672, University of California at Berkeley, Spring 1992.

[17] T.D.C. Little and A. Ghafoor. Scheduling of bandwidth-constrained multimedia traffic. *Computer Communications (Special Issue on Multimedia Communications)*, 15(6):381–387, July/August 1992.

[18] T.D.C. Little and J.F. Gibbon. Management of time-dependent multimedia data. In *Proceedings of the SPIE Symposium OE/FIBERS'92: Enabling technologies for multimedia, multiservice networks*, Boston, MA, September 1992.

[19] Jane Liu and Victor Lopez-Millan. A congestion control scheme for a real-time traffic switching element using the imprecise computations technique. In *Proceedings of the IEEE IPPS 1st Workshop on Parallel and Distributed Real-Time Systems*, pages 89–93, Newport Beach, CA, April 1993.

[20] Yuh-Dauh Lyuu. Fast fault-tolerant parallel communication and on-line maintenance using information dispersal. Technical Report TR-19-1989, Harvard University, Cambridge, Massachusetts, October 1989.

[21] Michael O. Rabin. Efficient dispersal of information for security, load balancing and fault tolerance. *Journal of the Association for Computing Machinery*, 36(2):335–348, April 1989.

[22] B. Randell, P. Lee, and P. Treleaven. Reliability issues in computing system design. *ACM Computing Surveys*, 10:84–98, June 1978.

[23] Martin Schulze, Garth Gibson, Randy Katz, and David Patterson. How reliable is a RAID? In *Proceedings of COMPCON-89, the Thirty-fourth IEEE Computer Society International Conference*, March 1989.

[24] A. Shamir. How to share a secret? *Communication of the ACM*, 22:612–613, November 1979.

[25] Avadis Tevanian and (et al). A Unix interface for shared memory and memory mapped files under Mach. Technical report, Carnegie-Mellon University, Department of Computer Science, February 1987.

NETWORK MANAGEMENT AND NETWORK DESIGN I

Robert S. Cahn[1] and Hong Liu[2]

[1]T.J. Watson Research Center
P.O. Box 704
Yorktown Heights, NY 10598
[2]Department of Computer and Information Science
University of Massachusetts Dartmouth
North Dartmouth, MA 02747

ABSTRACT

Network management and network design often work at opposite ends. Clearly, a network which uses shortest path routes and has components at 10% utilization will be easy to manage but will be horribly expensive. On the other hand, if the network is so tightly designed that it nudges the line separating optimality from unfeasibility then the network is unstable and a management nightmare. In order for neither situation to occur it is necessary that network design tools interoperate with network management tools. In this paper we describe a system which allows for the integration of network management and network design. We specifically discuss extensions to the network management needed for design tools to be tightly coupled in a network management system.

INTRODUCTION

Communication networks have evolved from voice telephony networks into computer networks. There is continuing evolution towards integrated networks operating at high speeds. As networks have become complex and ubiquitous, the need to manage heterogeneous networks has become apparent. Both the ISO and the communication industry are moving towards integrated network management through OSI. As these standards continue to develop, SNMP has emerged as a stop-gap which may become a lasting standard. The OSI network management standards will allow the devices of different vendors to interoperate with the following functions available through real time network management:

- Fault Management - detects and isolates abnormal network behavior then restores the system from outages or failures;
- Performance Management - gauges performance parameters such as throughput, delay, error rates, and etc., including historical information;
- Configuration Management - detects and controls the state of the network for both logical and physical configurations by providing the user with the ability to add/remove/rearrange nodes/links/routes and access/change device status;

- Accounting Management - collects and processes data related to resource consumption in the network;

- Security Management - controls access to network resources through the use of authentication techniques and authorization policies.

The critical elements of the ISO's OSI have been settled but remain under development:

- Common Management Information Service (CMIS) - provides a set of rules identifying the functions that an OSI interface performs between applications;

- Common Management Information Protocol (CMIP) - defines the rules for how information is exchanged between separate network management applications.

With CMIS and CMIP, the goal of integrated network management will be achieved since they together allow diverse network management tools from different vendors to communicate. Developments of CMIS and CMIP are major steps toward the goal of integrated network management since, together, they allow diverse network management tools from different vendors to communicate. However, the key issues of how management information is structured (Structure of Management Information - SMI) and defined (Management Information Base - MIB) are still fluid. At this stage, some interim methods have to be taken to resolve today's network problem.

The Internet community developed the Simple Network Management Protocol (SNMP) which has now been widely adopted by many vendors. The SNMP monitors and controls multivendor devices on TCP/IP-based internets. IBM is working towards integrating its diverse management products and those of others through NetView which is a host-based management system originally for the SNA environment. AT&T is also doing such integration through Accumaster Integrator which is a key element of AT&T's UNMA, a unified network management architecture controlling various resources via the public network. Many tentative systems have been proposed by universities and companies.

Currently, both the ambitious OSI standards and the applicable interim methods concentrate on monitoring, maintenance, and control. The ultimate goal of integrated network management should be to utilize the resources of computing and communication most efficiently and economically, not just for keeping networks running [1]. There is much anecdotal evidence that this goal is not being met. Networks are sometimes "over designed" because network managers wish to have a stable network over a long period of time. While underutilized voice trunks can be readily identified in networks, underutilized data links are not so easy to identify. Often changing traffic leads to unrecognized over capacity. It should be recognized that network design needs to cover the entire life-cycle of a network.

Integrated network management begins with network design - putting together a proper set of devices and facilities connected by the appropriate topology to support their communications with low cost and operational efficiency. Network design does not end with network implementation. It is an ongoing process. As the network is used, periodic reoptimization is necessary. Sometimes all that is required in a new routing pattern to deal with shifts in load. Eventually, when these measures can't deal with the new traffic, the network must be redesigned and extended. The lead time in provisioning the network makes it important to anticipate the need for redesign by a number of months. It is important that the network management system provide historical statistics for a network designer.

This paper articulates the relationship between network manageability and network design. We will discuss certain architectural changes that are required to network management so that the ideal environment we sketched above can be realized. We will then discuss the problem of redesign. Some of this work will be appear later in a second paper of the same name. A solution is given to take manageability into account for network design, with discussion on implementation feasibility. The same solution, used incrementally, can be applied to upgrade an existing system dictated by management information statistics, in addition to designing a network from scratch.

INITIAL DESIGN AND MANAGEABILITY

The manageability of an initial design can often be reduced to a short list of issues.

- Commonality - Can all the network elements talk with a single network management system? It is not uncommon to see networks with multiple management platforms. One monitors the multiplexers, another the modems, another the switches. Element Management Systems (EMS) should be chosen to provide real-time management information of the basic network elements on a network. An integrated platform, AT&T's Accumaster Integrator and MCI's INMS Workstation for instance [2], is needed to coordinate various EMSs. These management concerns can be put aside at the initial network design and left for later consideration. However, we will argue that ignoring manageability at the very beginning of network design, even at topology decision time, may result in an unmanageable network.

- Routing Diversity - A Cable cut is a major disaster in wide area networks (WAN). The damage becomes even more severe in broadband networks. To instantly detect and isolate the trouble spot then to restore the network operation manually or automatically, alternate paths to the affected nodes should be available for a network manager to collect diagnosis data and to reroute traffic for continued services. Thus, a network topology which is at least 2-connected is required for a network to be manageable, especially for fault management. Routing diversity can be an issue even in the local access. The ring-based local area network (LAN) was not popular in the United States until the mid 1980s. One of the principal reasons is the problem of poor survivability: one link break disables the entire network or at least obstructs all the upstream traffic. The task of fault management on a ring can be diminished by providing proper topology redundancy. A dual ring with two links, one for each of the two opposite circular directions, can self-heal from the failure of a link or a node [3].

  In addition to physical layer considerations discussed above, the logical layer has to be taken into account for fault management. Automatic cross-connects, that find available spare capacity in the network and reroute traffic around the failed spot, used to be the dominant method for fault management in WANs. Studies showed that this method is inferior due to high investment cost, slow restoration speed, additional operational expenses, difficult control requirements, and complex network planning [4]. Flanagan [5] presented a planning framework, consisting of three layers, that allows a network designer to build a survivable network against link failures at the physical layer, rather than depend on the upper logical layers to compensate.

- Spare Capacity - Routing diversity is not enough if there is not enough slack in the network to carry the load when a component fails. The reader can construct virtually any number of examples at his or her leisure. Simply take any network and routing pattern. If the least utilized link is over 75% utilized and the topology is spare enough that there are only one or two alternate routes for each requirement then deleting any link will produce a network which cannot carry all the requirements.

TRANSITION DESIGN AND MANAGEABILITY

The idea of transition design (model and algorithm) is first brought out by [5]. Given a multi-year plan representing evolution of communication requirements, a transition design finds a set of network topologies fitting the stages set by the plan while minimizing the total network PW cost. PW stands for present worth which takes into account the interest or discount rate on prices over time.

In practice, a company would spread capital investment over several years for both budgetary and technical reasons. Communications investments are only one of many areas where companies choose to invest. It is unusual for companies to build networks on the basis of a perceived need 5 years hence (service providers excepted.) Often it is unwise to do so. Many network owners find that it is almost impossible to estimate traffic requirements accurately before a network is implemented. Even after the network is in place estimating multi-year traffic from a 6 month sample is at best a guess.

A basic premise which we can state is the following. For networks to remain manageable it is necessary for network design tools to have the information in a timely fashion. Otherwise, networks will become unmanageable before the problem can be fixed.

Networks are often huge. Super tanker captains know that it can take miles to stop a huge ship or change the course. Similarly, network operators must have information months ahead of time if their networks are to be kept off the rocks. One of the most important things to accurately estimate is the amount of time left before the network becomes unmanageable versus the speed that the network can be changed. One of the most important gains from an integrated network management system will be realistic estimates of how close to the edge changing traffic has moved the network and how rapidly the network owner must move to control the situation.

Expandability needs to be considered at topology decision time. For instance, the self-healing ring mentioned in the last section has been widely accepted for fiber networks with survivability [5], but it is difficult to upgrade. While adding nodes is a simple matter changing the link speed is a global change to the entire ring.

Similarly to the fault management discussed in the last section, design decisions have to be made for performance management at the logical layer in addition to topological considerations. Filipiak [7] presented an analysis framework, consisting of three layers, that chooses state-dependent algorithms for automatic traffic management controls. At the logical layer, a traffic routing algorithm is run, off-line or on-line, to find a time-dependent sequence of routes in the order of the quality measured (e.g., end-to-end delay). At the link layer, the traffic is sent through the first choice link; alternative paths for overflow and on-line flow control to reject partial or all traffic will be activated, respectively, when the route quality or the link congestion exceeds a prespecified threshold. Actual calls are handled at the physical layer.

A SOLUTION OF THE TRANSITIONAL DESIGN PROBLEM

A Model for Manageable Network Design

The problem of designing manageable networks for multiple years can be stated as an optimization problem.

Given:

- Node locations with classification (e.g., terminals at the low level and gateways at the high level for a hierarchical topology) and clustering (e.g., nodes at a level are organized into clusters according to geographical nature or project assignment), set N
- Linkage choices with corresponding distance list, set L
- End-to-end multi-year traffic requirements, r_{sd}^t (the amount of traffic between ends s and d in the year t.)
- Mean link utilization constraint, U
- Mean end-to-end delay constraint, D
- Maximum number of links along a path from a source to a destination, i.e., maximum number of hops allowed, H
- Maximum number of paths allowed in a route for a source-destination pair, i.e., maximum number of bifurcations for a traffic requirement, B

- Candidate line speeds such as various SONET rates, T3, T1, or voice grade line, C_k
- Cost associated with each link, c_{ijk}^t.

Objective: Find a set of network topologies, i.e.,

- link capacities - n_{ijk}^t, number of lines between nodes i and j of type k during year t
- traffic routing - x_p, amount of flow routed on path p in the set P_{sd}^t

for a pre-determined multi-year period T such that the total network PW cost is minimized.

Subject to:

- End-to-end multi-year traffic requirements
- Mean link utilization constraint
- Mean end-to-end delay constraint
- Routing constraints such as maximum number of hops allowed in a path and maximum number of paths allowed in a route
- Topological constraints such as connectivity, topology candidates, and node classification and clustering.

The above qualitative problem model can be defined with a mathematical formula based on multi-commodity flow model in [8] as the following. It is a nonlinear and mixed-integer programming problem.

$$\text{Minimize } \Sigma_T \Sigma_L \; c_{ijk}^t \, n_{ijk}^t$$

subject to

$$\Sigma_{p \text{ in } P_{sd}^t} \; x_p = r_{sd}^t; \quad sd \text{ in } NxN, \; t \text{ in } T$$

$$\Sigma_{ij \text{ in } P} \; x_p \leq U \times (C_k \times n_{ijk}^t); \quad ij \text{ in } L$$

$$\Sigma_{p \text{ in } P_{sd}^t} \; x_p r_{sd}^t \times (\Sigma_{ij \text{ in } P} \; g_{ij}(.)) \leq D; \quad sd \text{ in } NxN, \; t \text{ in } T$$

$$0 < | x_p | \leq H; \quad x_p \text{ in } P_{sd}^t, \; sd \text{ in } NxN, \; t \text{ in } T$$

$$0 < | P_{sd}^t | \leq B; \quad sd \text{ in } NxN, \; t \text{ in } T$$

where the mean link delay $g_{ij}(.)$ (i.e., the link queuing and propagation delay) is computed with the method described in [9], assuming Markovian traffic flows.

This problem as stated exceeds the state of the art in network design. It contains a vast number of unknowns. One would like good heuristics to find near-optimal solutions. Many heuristic algorithms are available for various versions of this problem with different concentrations. To name a few, [6] is a multi-year planning algorithm determining when to expand each type of resource; [9] provided an algorithm adopting end-to-end mean delay constraint (as opposed to the commonly used network average delay) and considering various practical constraints on routing and topology; [10] is a globally-informed local optimization algorithm taking advantage of economics of scale. The key to solve the problem of designing manageable network is to integrate appropriate heuristics dealing with different parts into a systematic algorithm.

Estimating r_{sd}^t and Enabling Integrated Network Management

If we examine the inputs to the problem in the previous subsection we see that there are several types of data needed. Specifically we need:

- The set of sites;

- User set limits, i.e. delay, hops, utilization;

- The costs and capabilities of links;

- Some model of the capabilities of the nodes. This should include the delay, throughput and routing capabilities;

- The traffic, r_{sd}^t.

Each of these items except the last can be found by consulting an appropriate person or program. The first two items are given by the network operator. The link costs are usually supplied by tariff tools and the node capabilities are specified by suppliers or test departments. The only set of data which can't be supplied by people or programs is the traffic. Unfortunately, SNMP cannot observe these numbers. SNMP networks are not sufficiently instrumented to report the traffic to the network design tool. The rest of this section discusses changes in the MIB to remedy this situation.

Both MIB-I and MIB-II [11] do not contain enough information to even approximate the traffic between the nodes. Many practitioners try to deduce the traffic pattern from "sniffing" the LANs but this is generally ineffective.

If we are going to gather traffic the first thing to discuss is the granularity of the r_{sd}^t matrix. With individual workstations having IP addresses a full traffic matrix among all IP addresses would have thousands of rows and columns. Even with sparse matrix representations the granularity is too fine to be useful. One generally wants to know the traffic between sites and not between individuals. This produces several interesting problems.

The IDENTIFY Verb. We need to define what we mean by a site. 10 years ago sites correspond to a single network address. However, today, a site may consist of hundreds of IP addresses. A site in general will consist of a set of IP addresses. We can only hope that IP addresses have been assigned in a "sane" fashion. If, for example, a site, s, is represented by all IP addresses in the range from 129.35.20.0 to 129.35.26.255 we can ask the hardware to record packets to and from that range and deduce the total traffic into and out of a site. If another site, d, contains only IP addresses in the range from 120.22.15.0 to 120.22.16.255 then the traffic, r_{sd}^t, between the sites is derived by looking at all packets from an address in the first range destined to an address in the second range.

To deduce the traffic it is necessary to identify where it enters and leaves the network. This may not be obvious. When packets are routed using the RIP protocol the routes are quite distance insensitive. Therefore, if a local network is linked into an internet at multiple sites for reliability the traffic can enter the internet in quite unlikely places.

This motivates the definition of the IDENTIFY verb. This verb can be issued by a network design application, or the traffic collection process described below, against a network management system.

The syntax of the IDENTIFY verb is:

IDENTIFY IPMIN IPMAX PERIOD RETURN_IP_ADDRESS

This verb requests the location of where traffic enters or leaves the internet for a range of IP addresses during a time period. The result will be returned to the RETURN_IP_ADDRESS as a set of IP addresses of these entry and exit points. Alternatively, the result can be returned by the network management agents as a collection of datagrams.

The MONITOR Verb. Once the entry points for traffic into the network have been identified probes are needed to gather the traffic. The traffic is generally best probed at the entry point or points to the network but a probe might be activated at a single upstream location if this is more efficient.

The syntax of the MONITOR verb is:

MONITOR MONITOR_IP_ADDRESS SRC_IP_MIN SRC_IP_MAX
DST_IP_MIN DST_IP_MAX TWOWAY DURATION INTERVAL
RETURN_IP_ADDRESS

This verb will cause the network management system to activate a probe at location MONITOR_IP_ADDRESS. The probe will gather packets from source to destination at intervals and forward a summary of the activity to RETURN_IP_ADDRESS. The probing will last for a total period given by the DURATION field. If the TWOWAY variable indicates the probe will gather both traffic from the source to destination and in the reverse direction simultaneously.

The purpose of this verb is to allow the network design applications control of the instrumentation. By setting relatively short intervals the burstiness of the traffic can be gauged. By setting long intervals we can get a time-of-day profile. There are no standards for the quality of network design data and the data gathering must be sufficiently flexible to satisfy a variety of users. The exact amount of data returned by the probe will be the subject of the continuation of this paper. We will discuss in some detail the subject of, "How much and how accurate does traffic data need to be before we can reliably design a network?" there.

There is already some information available about the network status from RMON-MIB (Remote Network Monitoring MIB). RMON-MIB is a recent extension to the SNMP MIB [12] defining objects for managing remote network monitoring devices [13]. The information gathered is at the MAC rather than the network layer so it is useful for estimating link loads rather than the r_{sd}^{t} directly. However, these two verbs, together with the information available in RMON-MIB can be used to estimate network overhead.

IMPLEMENTATION CONSIDERATIONS

An integrated network management system for TCP/IP systems would then involve three principal pieces.

- A network management station and network management applications
- A network design tool
- A network traffic collection process

Figure 1 shows the architecture of a system for network management and planning. Sophisticated prototypes of the management and design components of the tool already exist. The network management station might be modeled after Netmate [14], a Network Management-Analysis-Testing Environment developed at Columbia University. The design tool component could be adapted from INTREPID [15], a Presentation Manager-based system developed at IBM Research, or Xnetdesign [16], an X-based network design system developed at Polytechnic

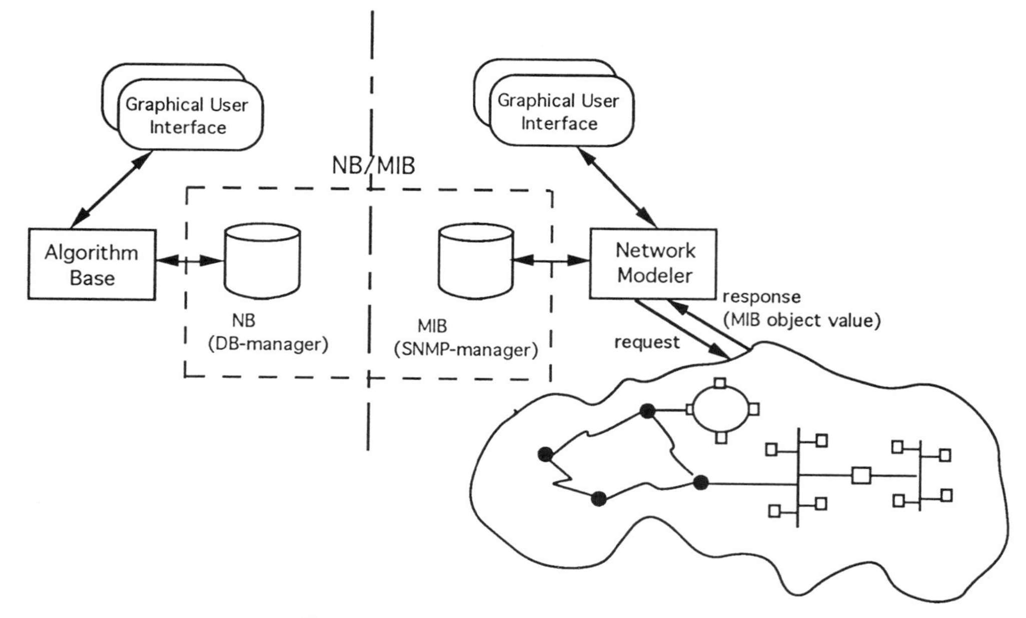

Figure 1. Integrated Network Management System

University. All of the tools cited have graphical front-ends and could be combined into a single, integrated network management tool with far less effort than that needed to write these modules anew.

The new and unique element is the traffic collection process (TCP). This component must gather network traffic without excessively loading the network. Further it must not make itself a locus of congestion. Consequently, it must limit the number of probes active in the network and limit the amount of information they return. Given this it may prove necessary to locate the traffic collection process in another site from the other components of an integrated network management system. It might further be necessary, in a very large network, to make this a distributed process. The design and analysis of this module will be discussed in our sequel paper.

The goal of this system is to automate as much as possible of the transition design. Ideally, the network design process would start the traffic collection process with one of several tasks. Tasks might be:

- Determine the total traffic between two sites,
- Determine the peak hour traffic between two sites,
- Locate the sites with the greatest traffic growth,
- Project the network traffic 6 months in the future,
- Survey the entire network traffic.

All of these represent the sort of data which a network designer, owner or manager might be interested in understanding. The TCP must decide which probes to set, issue requests to the network management subsystem, determine when the data from the probes is statistically significant, and determine how to use this data to deduce what is needed. The tasks are listed in increasing order of sophistication. Early implementations of the TCP will not be able to perform the most sophisticated tasks but will require considerable guidance from the user of the tool.

An integrated network management system will allow proactive as well as reactive network management. A complete system, such as we describe, can be run against the traffic gathered from the previous month or week. The network can be tuned by adjusting the routing and the estimates updated on how long the system will remain manageable.

CONCLUSIONS

Networks are not capable of being efficiently managed without being able to be efficiently designed. The current state of network management does not allow for network changes to be efficiently administered. The modest changes to existing management architectures proposed above will enable the integration of design and management in a single tool.

ACKNOWLEDGMENTS

The second author would like to express her appreciation to Yechiam Yemini for giving the opportunities and encouragement. She would also like to thank Alexander Dupuy and Soumitra Sengupa for their lectures on the Netmate system and their valuable suggestions on this work.

ABOUT THE AUTHORS

Robert S. Cahn (cahn@watson.ibm.com) received his S.B. from the University of Chicago in 1966 and his Ph.D. in Mathematics from Yale University in 1970. From 1970 until 1982, he was a member of the Department of Mathematics and Computer Science of the University of Miami and of the Mathematics Department of Lehman College from 1983-1985. In 1986 he joined the IBM Thomas J. Watson Research Center. At Watson he is a member of the Communications Department working on network design algorithms and network design tools. He is the author of over 20 research articles spanning his various interests.

Hong Liu (liu@cis.umassd.edu) received her B.S. with Honors and M.S. in Computer Science from Hefei Polytechnic University, China, in 1982 and 1984, respectively, and her Ph.D. in Computer Science from Polytechnic University, NY, in 1990. From 1987 until 1990, she was a member of the Computer Science Department at Polytechnic University. In 1990 she joined the Department of Computer and Information Science of the University of Massachusetts Dartmouth. Her research interests include computer network design, high-speed networking, programming languages and compilers, and computer graphics, for each of which she had publications.

REFERENCES

[1] K. Terplan, "Intergrated Network Management," *First IEEE Network Management and Control Workshop, Network Management and Control*, NY: Plenum Press, pp.31-58, 1989.

[2] M. Johnston-Turner, "Network Management Services from AT&T and MCI," *Business Communications Review*, pp.55-61, May 1991.

[3] T-H. Wu and M. Burrowes, "Feasibility Study of a High-Speed SONET Self-Healing Ring Architecture in Future Interoffice Fiber Networks," *IEEE Communications Magazine*, Vol.28, pp.33-42, November 1990.

[4] T-H. Wu, D.J. Kolar, and R.H. Cardwell, "Survivable Network Architectures for Broadband Fiber Optic Networks: Model and Performance Comparison," *IEEE Journal of Lightwave Technology*, Vol.6, pp.1698-1709, November 1988.

[5] T. Flanagan, "Fiber Network Survivability," *IEEE Communications Magazine*, Vol.28, pp.46-53, June 1990.

[6] T-H. Wu, R.H. Cardwell, and M. Boyden, "A Multi-period Design Model for Survivable Network Architecture Selection for SONET Interoffice Networks," *IEEE Transactions on Reliability*, Vol.40, pp.417-427, October 1991.

[7] J. Filipiak, "Analysis of Automatic Network Management Control," *IEEE Transactions on Communications*, Vol.39, No.12, pp.1776-1786, December 1991.

[8] H. Liu, "Models for T1 Problem," *Proceedings of 1992 International Conference on Microwaves and Communications*, Nanjing, China, pp. 10-15, June 1992.

[9] V.R. Saksena, "Topological Analysis of Packet Networks," *IEEE Journal on Selected Areas in Communications*, Vol.7, No.8, pp.1243-1252, October 1989.

[10] H. Liu, A. Kershenbaum, and R. Van Slyke, "Artificial Intelligence Applications to Communication Network Design with Bulk Facilities," *Proceedings of ACM 20th Annual Computer Science Conference*, pp.101-104, March 1992.

[11] K. McCloghrie and M. Rose, Management Information Base for Network Management of TCP/IP-based internets: MIB-II, Request for Comments: 1213, March 1991.

[12] J.D. Case, M.S. Fedor, M.L. Schoffstall, and J.R. Davin, A Simple Network Management Protocol, Request for Comments: 1157, May 1990.

[13] S. Waldbusser, Remote Network Monitoring Management Information Base, Internet Draft, August 1991.

[14] A. Dupuy, S. Sengupta, O. Wolfson, and Y. Yemini, "Design of the Netmate: Network Management System," *Proceedings of the Second IFIP Symposium on Integrated Network Management*, pp.101-107, April 1991.

[15] R.S. Cahn, P.C. Chang, P. Kermani, and A. Kershenbaum, "INTREPID: An Integrated Network Tool for Routing, Evaluation of Performance and Interactive Design." *IEEE Communications Magazine*, Vol.29, No.7, pp.40-47, July 1991.

[16] H. Liu and D. Hockney, "Visualization in Network Topology Optimization," *Proceedings of ACM 20th Annual Computer Science Conference*, pp.37-42, March 1992.

A *FUZZY* DESIGN PROCEDURE FOR LONG-TERM NETWORK PLANNING

Teresa Rubinson[1] and Aaron Kershenbaum[2]

[1]Pitney Bowes
Commercial Systems Applications
Shelton, CT
[2]IBM
Thomas J. Watson Research Center
Hawthorne, NY

ABSTRACT

The purpose of this paper is to present a heuristic network design and decision procedure (NEtwork Design and Decision) for long term network planning. NEDD incorporates a powerful design algorithm and a fuzzy dynamic programming procedure which allow the network designer to evaluate alternative design strategies based on multiple decision criteria. NEDD helps to identify robust topologies satisfying a range of requirements and design criteria, and points of transition when a topology must undergo major restructuring. NEDD is not dependent upon a particular form of technology. It is a general procedure that can be applied to many types of network design problems.

INTRODUCTION

By their very nature, design criteria are often competing and conflicting. For example, faster lines generally cost more than slower lines, and yet *both* speed and cost are important design considerations. Thus, a good network design reflects a balance of design trade-offs. However, traditional network design techniques usually optimize with respect to only one criteria - characteristically, cost - subject to constraints on the other design criteria. Inherently, when formulated this way, traditional network optimization techniques mask the effects of trade-offs in the design process.

Due to growth and the dynamic nature of network usage, the actual traffic requirements that must be supported by the network are usually not known with certainty. Because of this fact, a robust design is usually preferred to a more highly optimized design that gives superior performance over a narrower (and possibly unrealistic) range of conditions. Traditional network design techniques do not explicitly optimize with respect to "robustness", nor do they indicate how close a given solution is to optimal. Usually, iterative testing is performed on the design produced by traditional methods, to determine if improvements can be made. In the case of long term network planning, an initial best guess solution (See Figure 1) is selected, and then modified - by adding more links - to support more capacity. However, the basic topology and link type is preserved (See Figure 2). Ultimately, the quality of the solution over time is highly dependent upon the quality of the initial starting design.

Figure 1. Traditional Approaches for Selecting Initial Design Solutions

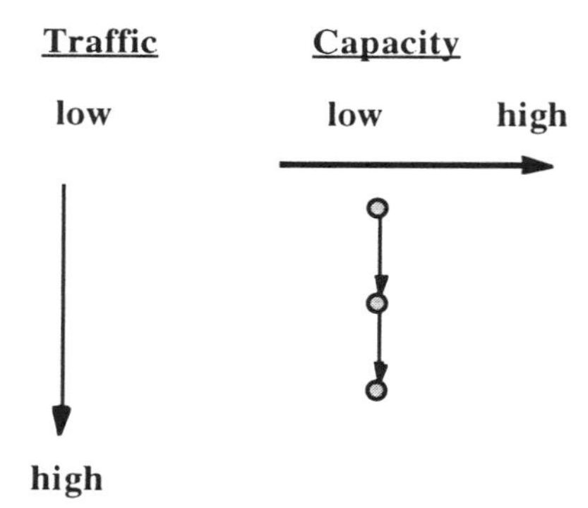

Figure 2. Traditional Long Term Network Planning

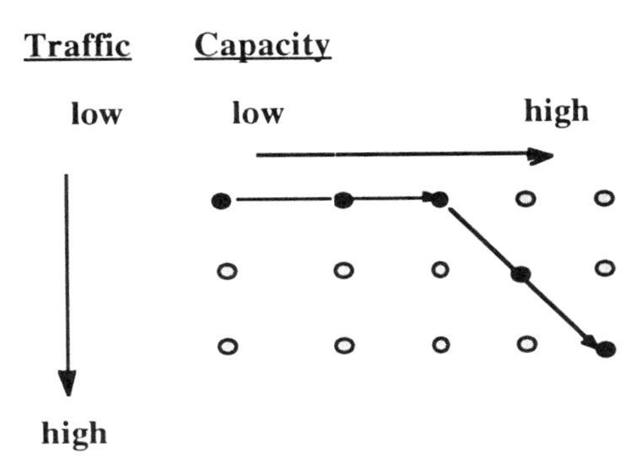

Figure 3. Long Term Network Planning Using NEDD

In contrast to traditional methods, NEDD allows a variety of network configurations and link types to be examined. In addition, NEDD indicates when a major restructuring is needed in the design to support growing requirements (See Figure 3).

OVERVIEW OF NEDD PROCEDURE

In contrast to traditional network planning approaches, NEDD is designed to explicitly explore of the effects and interdependencies of assumptions, criteria, and policies on network construction and selection. NEDD incorporates features from multiple disciplines, as outlined in Figure 4, to accomplish this task. One of the principal ways NEDD diverges from traditional methods is in its use of fuzzy decision functions to evaluate network designs.

Fuzzy decision evaluations are performed in NEDD in two phases. In the first phase (corresponding to steps two and three, as indicated in Figures 4 and 5), designs are evaluated and ranked solely on the basis of their static design qualities, and the best designs are culled for further evaluation. In a second phase (corresponding to step 4 of the NEDD procedure), a fuzzy dynamic programming algorithm is used to factor a time dependent criterion - the cost to change the networks over time - into the evaluation of the top ranking designs. At each review period, the fuzzy dynamic programming procedure examines the current network structure relative to new available topologies to determine if and when restructuring is needed. Final design selections are made based on how well "most" of the design criteria are satisfied, as determined from a fuzzy decision function incorporating an averaging OWA operator (See [3] for a formal definition of OWA operators). Automation of these evaluation and selection procedures gives NEDD the ability to rapidly assess many design alternatives from a variety of perspectives, thereby elucidating compromises in the choices at hand.

The four main procedures incorporated into NEDD are now reviewed. The first NEDD procedure uses the MENTOR algorithm to produce a variety of designs, based on traffic requirements and design input parameters. The design parameters are set to encourage production of a wide variety of tree-like, star-like, and hybrid topological structures for analysis. MENTOR is an expedient design tool for producing high quality topologies; however, except for practical considerations, there is no reason why other design algorithms could not be incorporated into NEDD. As is true for traditional methods, the quality of the solutions provided by NEDD is ultimately dependent upon the quality of the design templates available for its analysis. For more information on the MENTOR algorithm, the reader is referred to [1].

The second procedure invoked by NEDD computes fuzzy performance measures for each design candidate produced by MENTOR, by mapping actual network performance values into a fuzzy membership function. To this end, fuzzy membership functions are needed for each design criterion. The fuzzy membership functions used for performance analysis are derived from an automated, interactive procedure incorporated into NEDD. This interactive procedure solicits fuzzy ratings for ordered sets of performance values, for each

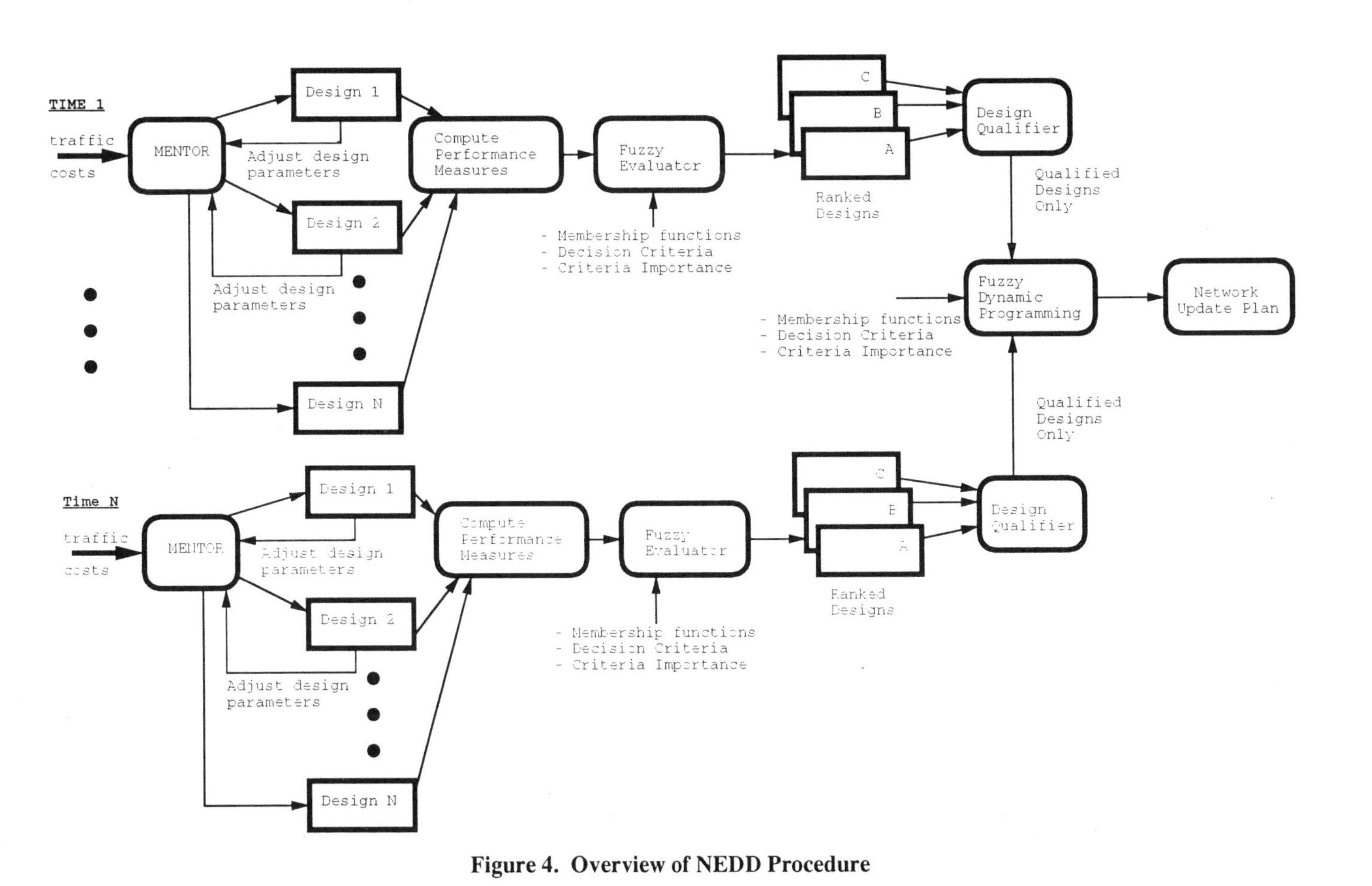

Figure 4. Overview of NEDD Procedure

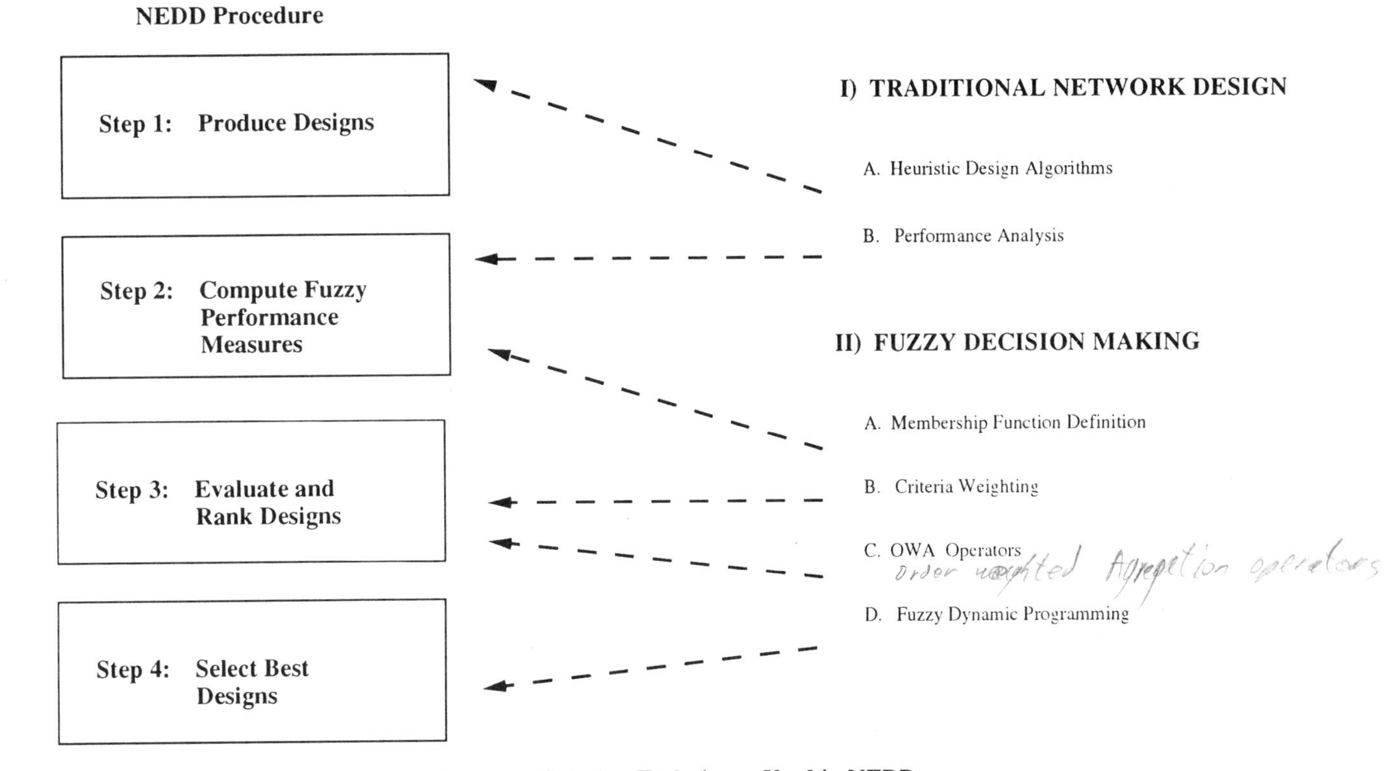

Figure 5. Solution Techniques Used in NEDD

design criterion of interest. These rating data are used in a regression analysis to compute a variety of equation forms. These equation forms are graphically displayed for an expert designer's review, and are re-formulated, as required, until a membership function is defined that is deemed to properly represent the intent of the designer. Examples of membership functions are provided in Figure 6.

Not surprisingly, the evaluations provided by NEDD are sensitive to the membership function shapes used for each design criterion. Thus, if the membership function shape is not represented properly, it is possible that ratings will be assigned that are too high or too low, relative to the designer's intent. However, in [2] practical guidelines are offered that to ensure that good membership functions are defined. In [2], it is also shown that imperfections in the membership function can be compensated for by systematic alterations to weights given to the design criteria. Assignment of decision priority weights - indicating the relative importance of each criterion in the evaluation process - provides, in effect, a means to alter the membership curves in controlled, prescribed directions, establishing a more direct relationship between each criterion and its importance in the evaluation process than could otherwise be determined from the substitution of different membership functions. As different sets of decision priority weights are used in the evaluation process, design clusters with significantly divergent performance characteristics emerge.

The third step in the NEDD procedure ranks design candidates, based on a ranking score assigned by a fuzzy decision function. There are three components to the fuzzy decision function: the design criteria membership functions, the relative priority weights of the design criteria, and an averaging OWA operator, which is used to compute a final aggregate score for each design (a formal definition of an Ordered Weighted Aggregation (OWA) operator is provided in [3]). The main feature of interest to the problem at hand is that the averaging OWA operator evaluates network performance across all the decision criteria, giving extra "credit" to designs that are well balanced. Given the necessity in network planning to balance satisfaction of the design criteria without any drastic compromises, the averaging OWA operator provides superior results to those provided by the more traditional AND and OR operators, which only look at the extremes of performance. The OWA operator defined by Yager is very versatile, and is theoretically capable of supporting an endless number of adaptations. Based on the ranking score computed in this step, poor quality designs are identified and removed from consideration. If no designs can be found to satisfy the design criteria, this too is reported so that the criteria can be re-specified if desired.

In the fourth and final step, NEDD uses a fuzzy dynamic programming algorithm - implemented as a fuzzy k-longest path procedure - to identify strategies for evolving promising design candidates over time to support anticipated future growth. Given designs at Period 0 and at Period N, and a series of intermediate designs at various planning periods stages, the network planning problem posed as a dynamic programming problem is to find the optimal transition path from Period 0 to Period N. More formally, this problem can be expressed:

Given:

o **Recursion equation:**

$$V_n(S,j) = \max [D_{(S,j)} + V_{n+1}(S,j)] \text{ for } n = 1, 2,...$$

$$\text{where: } D_{(S,j)} = [u_1 * u_2 *...* u_z] = \text{fuzzy decision evaluation}$$

o **Boundary condition:**

$$V_0(S,j) = 0 \quad \text{for } j = 1, 2, ...$$

Find:

$V_n(S,j)$ = maximizing decision in state S, <u>from</u> state j, at the nth planning stage,

where each state represents a design configuration

$A_n(S,j)$ = design alternative yielding $V_n(S,j)$

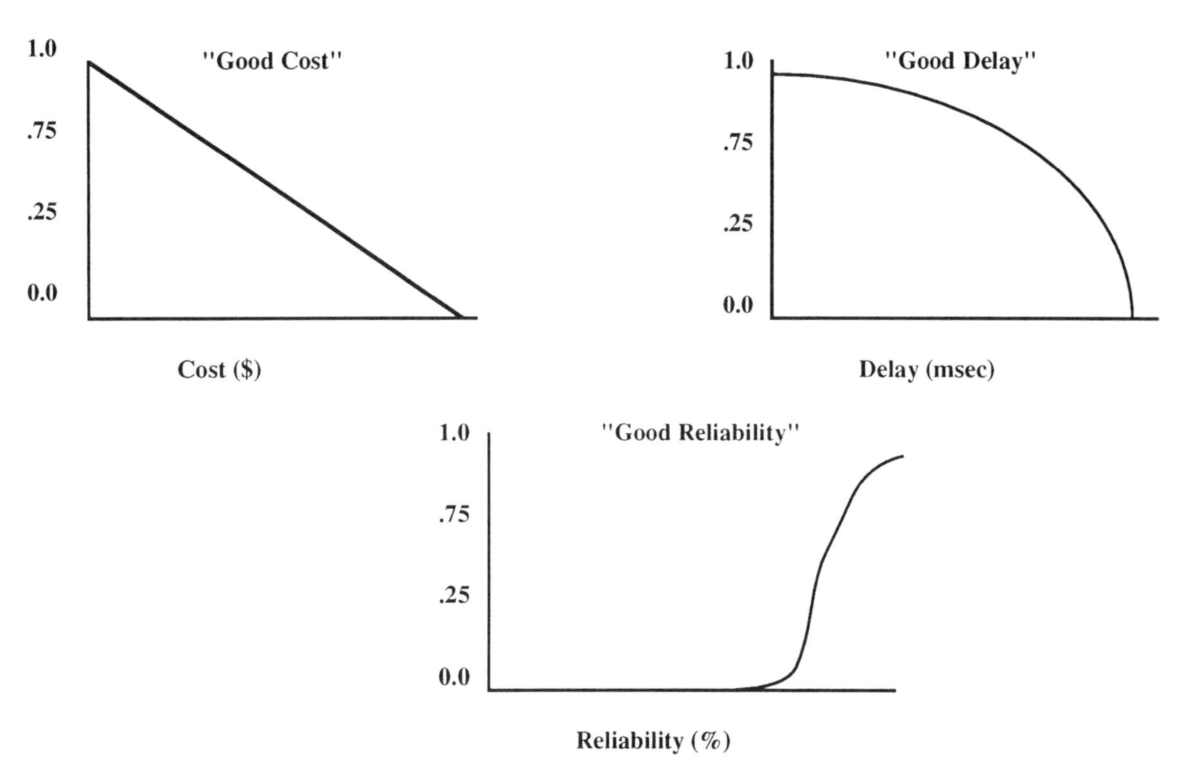

Figure 6. Sample Membership Functions

To summarize the solution procedure, the boundary condition is used to initiate the recursion equation calculations. A path, from each alternative at the initial stage to a final state at stage (S,j), is determined by recursively computing the value for $V_n(S,j)$. The best path is found by tracing a sequence of optimal solutions through time, and is the one for which $V_n(S,j)$ assumes its maximum value.

The fuzzy dynamic programming model used in NEDD is the same as a traditional dynamic programming model, except that the recursive optimization function is a fuzzy function. A fuzzy form of the recursive equation offers several advantages.

First, the fuzzy decision function allows both subjective and precise criteria to be combined in the same model.

Second, additional criteria can be included in evaluations, without increasing the dimensionality of the solution space. The fuzzy evaluations represented by the recursive optimization equation can be computed prior to initiating the dynamic programming procedure, and can encompass as many static criteria as can reasonably be dealt with. The only criteria affecting the dimensionality of the fuzzy dynamic programming procedure are those directly pertaining to time related interdependencies between design states.

If multiple time dependent criteria are included in the fuzzy dynamic programming model, computational complexity is increased, as each time dependent criterion has to be represented with an additional state variable. A major concern in implementing dynamic programming techniques is the potential for combinatorial explosion in the solution space, if the number of alternatives to be considered is too large. Steps two and three of the NEDD procedure are used to control the number of top ranking designs evaluated by the fuzzy dynamic programming procedure to a manageable size. However, if too few designs are considered, it is possible that designs facilitating the transition to dominant topologies will be weeded out because they do not meet the performance standards set in steps two and three of the NEDD procedure. Furthermore, if the designs available in each planning period are too dissimilar to provide good transitional opportunities, the fuzzy dynamic programming procedure used in NEDD tends to compensate by picking expensive designs with high capacity (that do not change much over the planning interval).

The ability to handle large numbers of design criteria and design alternatives in the evaluation process is highly dependent upon the computer resources available for implementing NEDD. The procedures used in NEDD are sufficiently flexible to allow more criteria and designs into the evaluation process, as the disk space, memory, and other computer resources permit. When faced with limited time and resources, it is reasonable to look at only the most promising alternatives; however, limiting the choices to only a few top ranking designs masks the effects of diversity and involves the establishment of somewhat arbitrary limits. In spite of the potential computational limitations of fuzzy dynamic programming techniques, they offer significant computational advantages over traditional dynamic programming approaches by reducing the state space required to represent the network planning problem.

An example is now provided to illustrate the approach. Assume that two design alternatives are available for each of two planning periods, as indicated in Figure 7. Associated with each design is a numeric performance rating, determined from a static fuzzy decision analysis. These ratings - referred to as $D_{(S,j)}$ in the previous discussion - provide the path weights used in the fuzzy dynamic programming calculations. In Figure 7, the cheapest design is found in the lower left cell. However, although this design is initially cheaper, in the long run it is not due to the costs required to convert this network to meet future requirements. NEDD is able to provide this type of information to the network designer. Additionally, NEDD can indicate that the design in the upper left cell is easily converted, at minimal cost, to a very high performance network (found in the lower right cell). If the design specifications are very demanding, NEDD would select the network in the lower right cell as the solution of choice for the *entire* the planning process.

A feature of the fuzzy dynamic programming procedure that deserves mention is the penalty function, which penalizes link additions and link removals, used by NEDD to quantify topological differences between networks. This penalty function can used to examine a variety of design problems. For example, under conditions of decreasing growth,

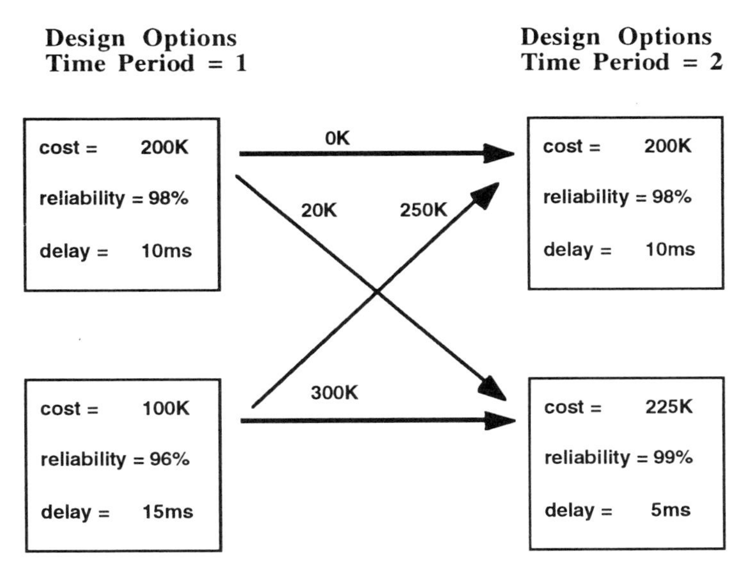

Figure 7. Example of NEDD Fuzzy Dynamic Programming

the penalty function can be set to favor designs with lower operating costs, while under conditions of increasing growth, the penalty function can be set to favor designs with lower conversion costs. The penalty function can also be used in an examination of the effects of different rates of traffic growth. Since the penalty function used in NEDD is based strictly on links comparisons, independent of the growth assumptions used to size the networks under comparison, the function can also be used to examine scenarios exhibiting fluctuating growth patterns. The penalty function is well suited to comparing networks of different link types, and was used to compare networks comprised of both pure and mixed link types. At the same time link comparisons are made, it may also be desirable to examine other topological features that may have a bearing on design decisions. For example, it may be useful to incorporate node types and costs into the design evaluations. The penalty function (and/or the fuzzy decision functions) used in NEDD can easily support extensions of this kind. For details, the reader is referred to [2].

The final output from NEDD is a time lapse representation of the network and its recommended evolution over time to satisfy anticipated growth requirements. An example of a final solution is given in Figure 8.

CONCLUSION

The quality of the solutions provided by NEDD were judged on the basis of their practicality and adherence to commonsense expectations and commonly accepted design practice. When evaluated in this context, the solutions provided by NEDD are consistent with specified design priorities, and, with few exceptions, good design practice and performance expectations. In the exceptional cases where the design goals could not be satisfied by the available design choices, there are two major options. First, the decision priority weights can be modified to provide greater emphasis to the criteria that have not been satisfied. Second, if the criteria that have not been satisfied have not been explicitly represented in the fuzzy decision models, they can be added to the fuzzy evaluation model. From the results reported in [2], NEDD is a robust procedure, providing good results with minimal specification.

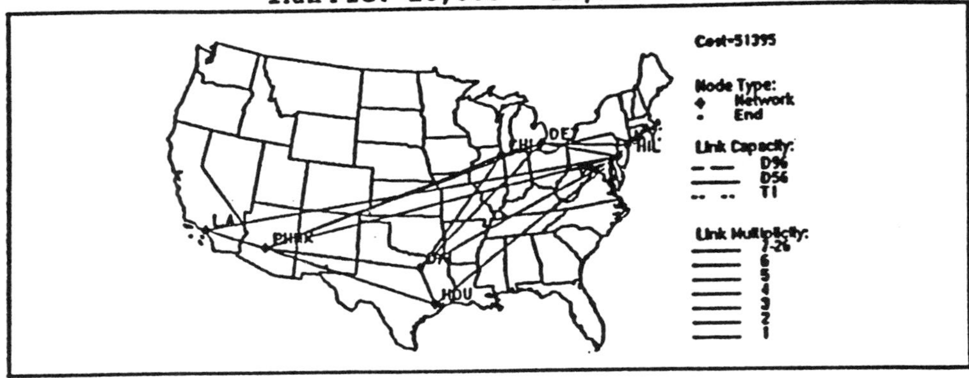

TRAFFIC: 12,100 - 13,310 BPS

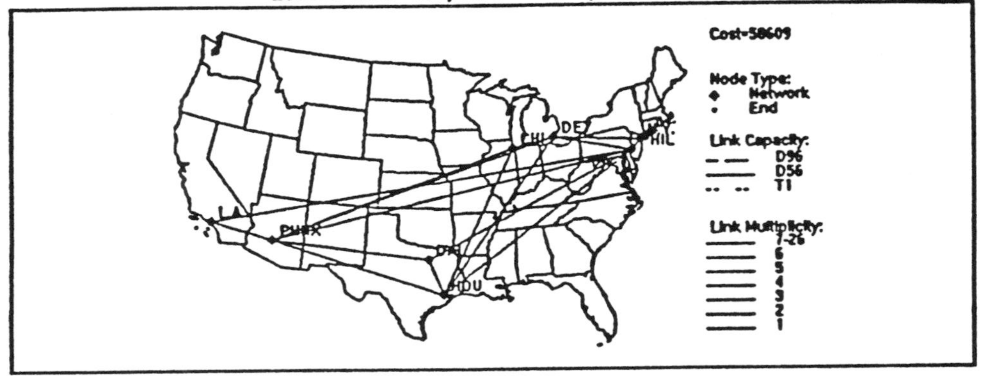

TRAFFIC: 13,310 - 30,000 BPS

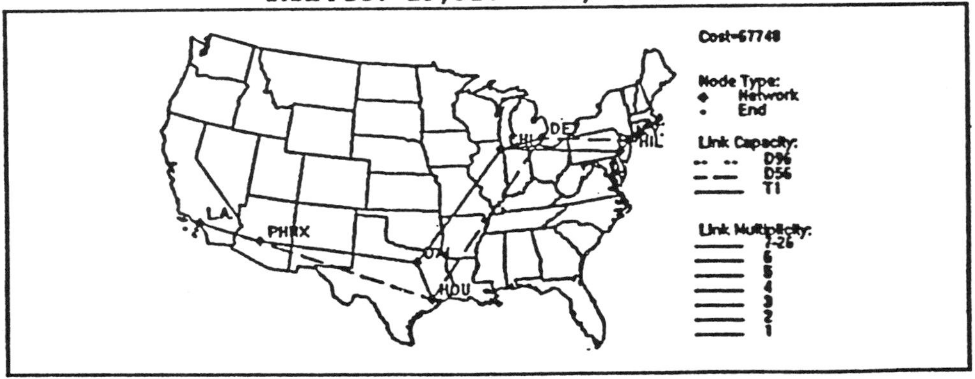

Figure 8. Example of NEDD Network Evolution

In summary, NEDD is a four step procedure developed to assess the implications of varied design policies and strategies. Unlike many existing design techniques, NEDD provides a means to explore many competing and interacting effects in the design process, and is a powerful tool for visualizing possible scenarios for network evolution. NEDD is a highly extensible procedure that can easily be modified to accommodate a variety of designs and design criteria. Ultimately, assessment of the quantitative and qualitative implications of diverse design policies and strategies promotes more informed and rational long term planning decisions.

ABOUT THE AUTHORS

Teresa Rubinson works for Pitney Bowes in the Commerical Systems Applications group as a consultant in the areas of operations research, systems analysis and planning, simulation and modeling, and artificial intelligence technology. Prior to Pitney Bowes, she held management/technical positions at Pepsico, IT&T, and Boehringer Ingelheim, working in both MIS and R&D environments to design state of the art systems incorporating database and artificial intelligence technologies. She has lectured and published extensively, and has long standing research interests in data base design, data segmentation and classification techniques, neural nets, fuzzy logic, expert systems, and network design. Dr. Rubinson has a Ph.D. in Operations Research from Polytechnic University, an MBA from Iona College, and a BS in Biology from the University of Illinois. She is a member of IEEE, and ORSA/TIMS. She is also an Adjunct Associate Professor at Polytechnic, where she teaches network design.

Aaron Kershenbaum has been a member of the Network Design and Analysis Tools Group at IBM T.J. Watson Research Center in Hawthorne, New York since 1989. Prior to that, he was at Polytechnic University from 1978 to 1989 where he was Professor of Computer Science and Director of the Network Design Center (which is part of the New York State Center for Advanced Technology in Telecommunications) at Polytechnic. From 1969 to 1978, he was the with Network Analysis Corporation, where he become Vice President of Software Development. He has played an active role in the design of many of the largest voice and data communications networks in the United States, including the ARPANET and AT&T network. He has also created software tools for the design and on-going management of networks which are widely used in industry. He has supervised over 20 Ph.D. theses in the area of network design and is the author of over 50 publications in journals and technical proceedings. He is the author of the text "Telecommunications Network Design Algorithms" and he co-edited the volume "Network Management." Dr. Kershembaum is a Fellow of the IEEE.

REFERENCES

[1] Kershenbaum, A., Kermani, P. and Grover, G.,"MENTOR: An Algorithm for Mesh Network Topological Optimization and Routing," *IEEE Transactions on Communications*, Volume 39, No. 4, (1991).

[2] Rubinson, T. , "A Fuzzy Multiple Attribute Design and Decision Procedure for Long Term Network Planning," Ph.D. dissertation, (1992).

[3] Yager, R., "On Ordered Weighted Averaging Aggregation Operators in Multicriteria Decisionmaking," *IEEE Transactions on Systems, Man, and Cybernetics*, Volume 18, No. 1., January/February (1988).

A FUZZY LOGIC REPRESENTATION OF KNOWLEDGE FOR DETECTING/CORRECTING NETWORK PERFORMANCE DEFICIENCIES

Lundy Lewis

Cabletron Systems R&D Center
9 Executive Park Drive
Merrimack, NH 03054 USA

ABSTRACT

A viable approach to the early detection and correction of network performance problems is to construct algorithms that translate streams of numeric readings of network monitors into meaningful symbols, and to provide an inference mechanism over the symbols that captures the knowledge of the best experts in the field. Current implementations of this sort of solution represent the requisite knowledge in a rule-based framework. Unfortunately, rule-based systems can suffer the problems of brittleness and knowledge acquisition bottleneck. In this paper we describe an alternative implementation in which knowledge is expressed in a fuzzy logic framework.

INTRODUCTION

Current network monitoring tools are very good at reporting values of network parameters such as network load (%), packet collision rate, packet transmission rate (%), packet deferment rate (%), channel acquisition time (mS), and file transfer throughput (KBytes/s). Daemons can be attached to these parameters so that values that exceed a given threshold result in an alarm. In addition, there are good graphics tools that can display this information in the form of bargraphs, X-Y plots, histograms, and scatterplots. However, except for clear faults, there are few capable experts that can interpret these values and alarms in common sense terms and point to reasons for performance deficiencies such as network entropy, hanging/resetting, and slow file transfer throughput [1] [2]. Reasons for these deficiencies might include an overloaded network link, a router with an insufficient CPU, or an incorrectly adjusted timer for a transmit buffer. In addition, the task of detecting/correcting performance problems is becoming harder with the advent of increasingly large and heterogeneous networks.

Network Management and Control, Vol.2, Edited by
I.T. Frisch *et al.*, Plenum Press, New York, 1994

Performance problems are expensive, difficult to detect, and hard to correct -- especially in real or close to real time. A 1989 study of Fortune 1000 companies found an annual average of $3.5 million in lost productivity from downtime stemming from these problems, causing a loss of about $800,000 in profits [3]. An approach towards solving these problems is to simulate a network with a mathematical model. One can then predict the nature of performance deficiencies by running the model with simulated network conditions. Unfortunately, most networks do not lend themselves to mathematical modeling, either because they are too complex and dynamic to be modeled or because the computational expense of running the model is prohibitive.

A second approach is to simulate the expertise of a good network trouble-shooter. The usual way to do this is to construct algorithms that translate streams of numeric readings of network monitors into meaningful symbols, and to provide an inference mechanism over the symbols that captures the knowledge of the best experts in the field. Current implementations represent the requisite knowledge in a rule-based framework. However, as explained below, this sort of solution has shortcomings. In this paper we describe an alternative implementation in which knowledge is expressed in a fuzzy logic framework.

A RULE-BASED APPROACH TO THE PROBLEM

Most research and commercial systems that try to represent network troubleshooting knowledge first need a human expert to translate raw network data into meaningful symbols. They then use a rule-based reasoning (RBR) system for drawing inferences from the symbols [4]. A good review of network applications of RBR (a.k.a. expert systems, production systems, or blackboard systems) is [5]. Figure 1 shows the basic RBR paradigm (from [5]). A RBR system consists of a working memory (WM), a knowledge-base consisting of rules, and an inference engine. For a network application, the WM typically contains a representation of characteristics of the network, including topological and state information. The knowledge-base contains rules that indicate the operations to perform when the network malfunctions. If the network enters an undesirable state, the inference engine selects those rules that are applicable to the current situation. Of the rules that are applicable, a pre-determined control strategy selects a rule that is actually executed. A rule can perform tests on a network, query a database, provide directives to a network configuration manager, or invoke another expert system. With these results, the system updates the WM by asserting, modifying, or retracting WM elements. This cycle continues until a desirable state in WM is achieved.

There exist several variations on the basic RBR paradigm. For example, the control procedure can be enhanced with a belief revision capability. The control procedure keeps a list of rules selected on each cycle and may backtrack to a previous cycle to select an alternative rule if progress is not being made towards a desirable state (assuming that no operation has been performed that cannot be undone). In addition, the rule-base can be functionally distributed and a meta-control strategy can be provided that selects the component RBR system that should be executed for specific kinds of tasks (e.g. [6] [7] [8]). Alternatively, the rule-base can be distributed component-wise such that there exist specialized rule-bases for each component in the network (e.g. [9]).

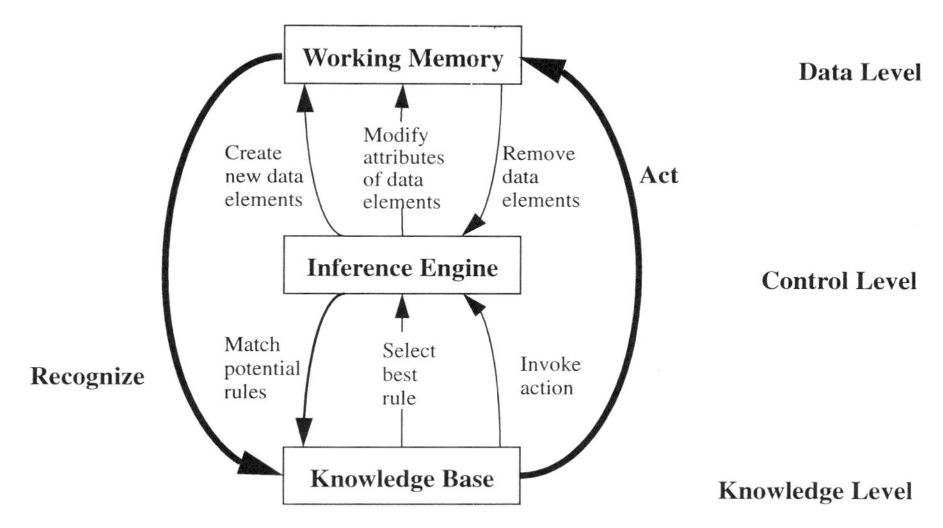

Figure 1. The basic rule-based reasoning paradigm

PROBLEMS WITH THE RBR APPROACH

The usual procedure for constructing a RBR fault resolution system is (i) to define a description language that represents the problem domain, (ii) to extract expertise from multiple domain experts and/or trouble-shooting documents, and (iii) to represent the expertise in the RBR format. The procedure can require several iterations of an interview/implement/test cycle in order to achieve a correct system. If the domain and the problems encountered remain relatively constant, a correct system needs little maintenance (e.g. see the discussions in [10] [11] [12]). However, if the system is used to resolve faults in unpredictable or rapidly changing domains, two problems inevitably occur: (i) the system suffers the problem of brittleness and (ii) the development process suffers the problem of knowledge acquisition bottleneck. Brittleness means that the system will fail when it is presented with a novel problem. The counterpart of the brittleness problem is the system's lack of ability to adapt existing knowledge to a novel situation. A knowledge acquisition bottleneck can result when a knowledge engineer tries to incorporate additional knowledge into the system that will cover unforeseen situations. When this happens, the system typically becomes unwieldy, unpredictable, and unmaintainable. With rapidly changing domains, the system can become obsolete quickly. The alternatives at this stage are to limit the coverage of the RBR system or to search for other approaches.

For example, equation (1) shows a simple function that describes a set of rules for issuing notices about the load of a network:

$$\text{notice} = \begin{cases} \text{alarm} & \text{if network\_load} <= 10\% \\ \text{alert} & \text{if } 10\% < \text{network\_load} <= 20\% \\ \text{ok} & \text{if } 20\% < \text{network\_load} <= 30\% \\ \text{alert} & \text{if } 30\% < \text{network\_load} <= 40\% \\ \text{alarm} & \text{if network\_load} > 40\% \end{cases} \qquad (1)$$

In this example there is a WM element, *network_load*, that is updated by a network monitor. The value of *network_load* is compared to the rules at pre-specified time increments, and one rule fires to issue a notice. In some cases, the reading of a value along an interval of length .02 could make a big difference, whereas in other cases the reading of a value along an interval of length 9.98 makes no difference. For example, a value of *network_load* = 9.99 issues an alarm and a value of 10.01 issues an alert, whereas the values 10.01 and 19.99 both issue an alert. Of course, this is so because the rule set describes a function which is discontinuous, as shown in Figure 2.

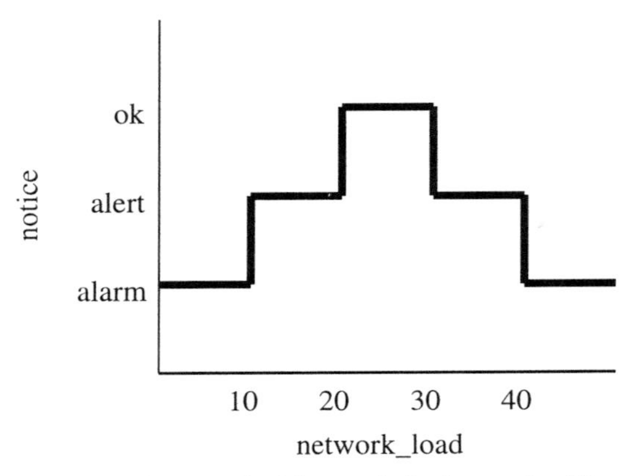

Figure 2. A graph of the set of rules in equation (1)

For issuing alerts and alarms, perhaps this is acceptable. However, the lack of continuity of a rule set becomes problematic for other useful variables. Suppose we are interested in a variable *reroute%* that tells us the percentage of traffic to re-route in order to maintain a notice of *ok*. A possible implementation of this function is equation (2):

$$\text{reroute\%} = \begin{cases} 5\% & \text{if } 30\% < \text{network\_load} <= 40\% \\ 15\% & \text{if network\_load} > 40\% \\ 0\% & \text{otherwise} \end{cases} \tag{2}$$

This function is unsatisfactory; the primary reason is that the rules are "brittle". The antecedent (the "if" part) of a rule must be either true of false, the output (*reroute%*) is returned in total, and only one rule can fire at any one time. One approach to getting around the brittleness problem is to add more rules. However, this approach is likely to result in a proliferation of rules, and thus introduces the knowledge acquisition bottleneck problem.

These problems are grounded in what we call the "lean semantics" of the RBR approach. We would like to describe the load of a network in terms like "heavy", "very heavy", "slightly heavy", etc. We would like to look at a load measurement of say 29% and say that it is "not heavy but not ok either" or simply "slightly heavy". Further, a rule of thumb sometimes can tell us something useful even though its antecedent is not perfectly true. The fuzzy logic framework described in the rest of this paper allows us to interpret network behavior in such common sense terms with a rigorous mathematical underpinning and affords us a richer semantics than the RBR framework. Further, the fuzzy logic approach shows promise to alleviate the brittleness and knowledge acquisition bottleneck problems inherent in RBR systems.

A FUZZY LOGIC REPRESENTATION OF KNOWLEDGE

We are working on an alternative problem-solving paradigm for network fault management, the details of which are described in [13]. An important component of the approach is a fuzzy logic representation of knowledge for detecting/correcting network performance deficiencies. The seminal paper on fuzzy logic is [14]. [15] provides a thorough analysis of fuzzy logic and [16] provides an accessible introduction.

With the fuzzy logic approach, we construe network parameters as reported by monitors (e.g. network load, collision rate, etc.) as linguistic variables and provide membership functions that translate the parameters' numeric values into degrees of membership in a fuzzy set. A linguistic variable is simply a variable that takes linguistic values rather than numeric values. The variable *network_load* might take values "light", "ok", and "heavy". For each of these values, we describe a function that maps a numeric value into a degree of membership into a linguistic fuzzy representation. This allows us to represent an interpretation of *network_load* such as "ok but very slightly heavy".

A simple comparative illustration is as follows. Let the variable *network_load* have a universe U over the interval [0,100%]. Figure 3a shows the concept "heavy" in crisp logic (i.e. non-fuzzy, two-valued logic). With the fuzzy logic framework, we could define a fuzzy set over U that describes the common sense term "heavy" with equation (3):

$$\text{heavy} = \begin{cases} 0.0 & \text{for } 0 < y \text{ in } U <= 25 \\ (1 + ((y - 25)/5)^{-2})^{-1} & \text{otherwise} \end{cases} \tag{3}$$

Figure 3b illustrates the fuzzy concept "heavy". A numeric value of *network_load* less than 25 would have 0.0 grade of membership in the concept "heavy", a value of 30 would have .5 grade of membership, and a value of 40 would have a .9 grade of membership. In similar fashion, we define fuzzy sets for the concepts "ok" and "light". Figure 4 shows what these functions might look like. Note that a value of 25 would have a 100% grade of membership in the concept "ok" but would have 0.0 in "heavy" and "light". Also note that now a value of 30 would participate to degree .5 in "heavy" and say .8 in the concept "ok".

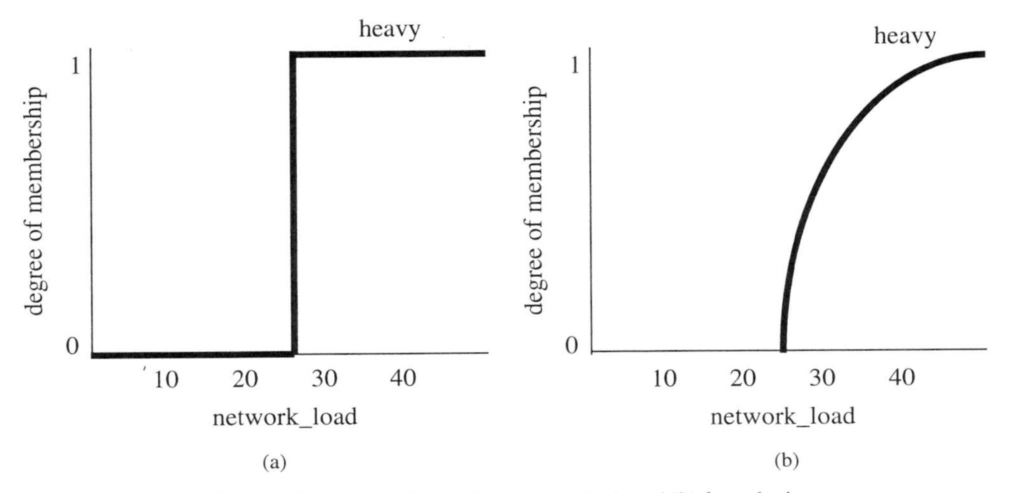

Figure 3. The concept "heavy" in (a) crisp logic and (b) fuzzy logic

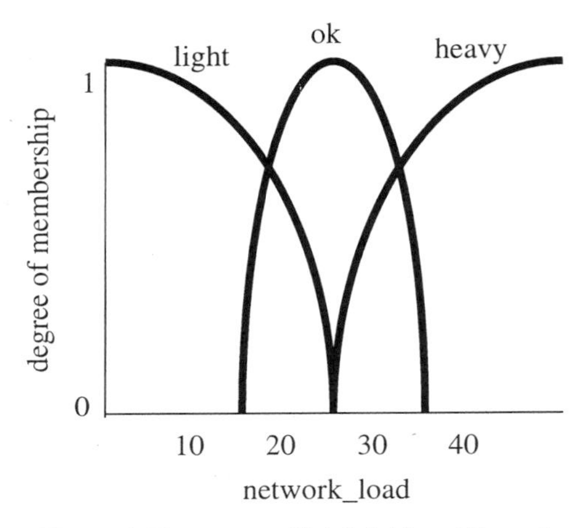

Figure 4. The concepts "light", "ok", and "heavy"

In this manner, we define concepts representing (i) input variables from network monitors, e.g. packet collision rate, packet transmission rate, packet deferment rate, channel acquisition time, etc., (ii) other input variables that represent network behavior as perceived by users, e.g. slow file transfer throughput and command execution response time, and (iii) output variables, e.g. notices, network load adjustment, and transmit buffer timer adjustment. Then we build a grammar over the concepts that defines well-formed expressions in the language. Next, we allow experts to define resolution strategies (i.e. fuzzy rules) that connect input variables and output variables. Figure 5 shows the general engineering framework for building and fine-tuning a fuzzy logic system.

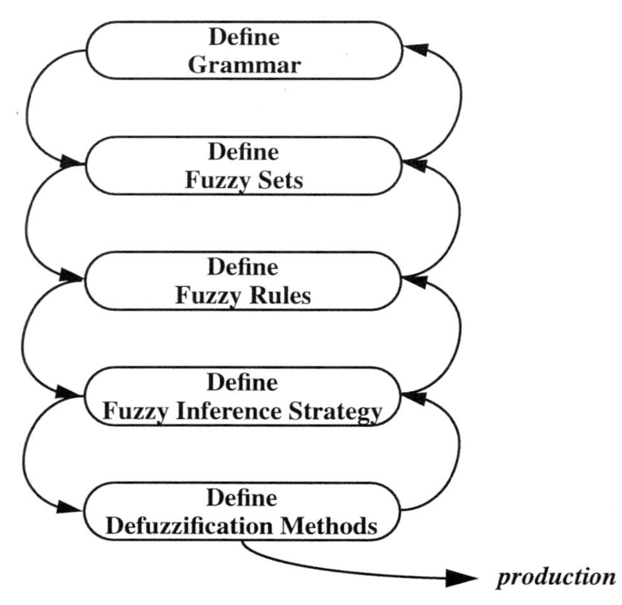

Figure 5. The engineering framework for building fuzzy logic systems

Example fuzzy rules include the following:

If *network_load* is heavy and *file_transfer_throughput* is slow then *bandwidth_adjustment* is small increase

If *network_load* is not heavy and *packet_collision_rate* is high then *transmit_buffer_timer_-adjustment* is small increase

If *network_load* is very heavy then *notice* is strong alert and *reroute%* is medium decrease

If *network_load* is medium and *rate_of_load_change* is high increase then *notice* is alert and *reroute%* is small decrease

The detection/correction of performance problems operates by mapping numeric data into common sense terms, and then matching the common sense terms with the fuzzy rules. The operation is shown in Figure 6. Unlike the RBR approach, however, all fuzzy rules that participate in the "truth" of the input data will fire and thus contribute to the overall solution. Further, a rule might not be an exact match with the input data, although enough in the ball-park to be applicable. Essentially, the output variables of a rule are tweaked relative to the degree of match between the input variables of the rule and the current readings returned by a network monitor (or user). The inference mechanism that does this is the compositional rule of inference. Here we do not describe the underlying calculations. See [14] or [15] for good discussions of standard formulae for generic fuzzy concepts and standard inference mechanisms.

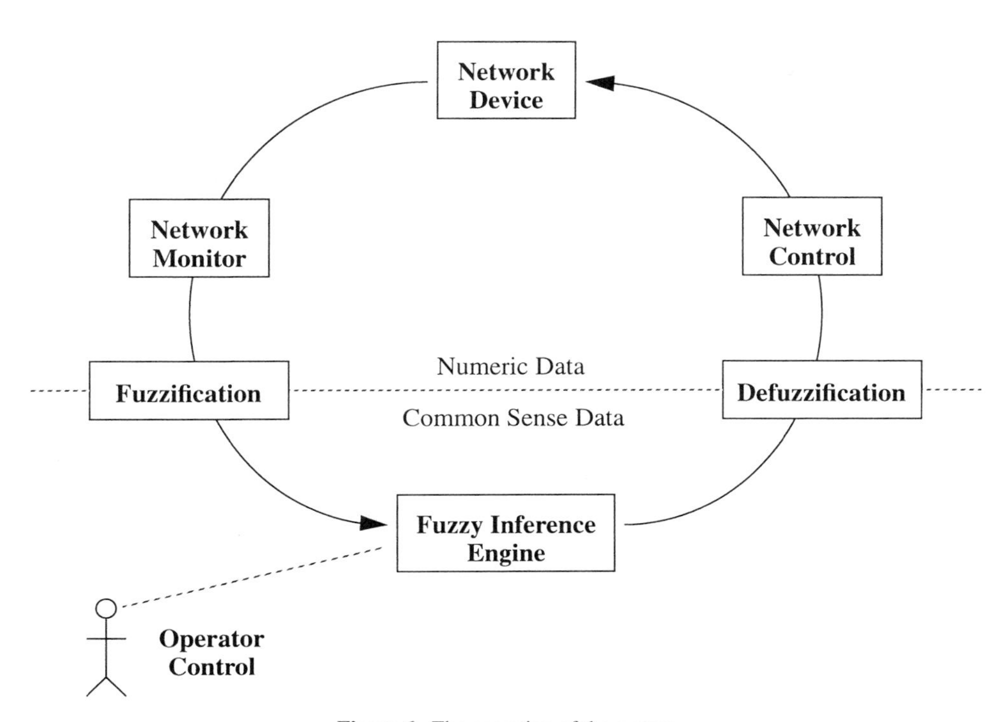

Figure 6. The operation of the system

CONCLUSION AND CURRENT STATUS

Performance deficiencies are difficult to detect and correct, and there are few well-trained experts in this area. An approach towards alleviating the problem is to collect the knowledge of the best experts in a RBR system. Unfortunately, RBR systems have shortcomings. They can become brittle and unmaintainable as networks evolve. In the paper we described an alternative approach in which knowledge is represented in a fuzzy logic framework. The advantages of the fuzzy logic approach are that (i) numeric network data is represented as understandable common sense terms, (ii) a fuzzy rule does not have to be a perfect match with the input data in order to contribute to a solution, (iii) the implementation of knowledge is intuitive and straightforward, and (iv) the approach has a sound mathematical underpinning.

Currently, we are defining a grammar for representing expertise involving management of performance deficiencies, defining membership functions that map network parameters to well-formed phrases in the grammar, and experimenting with alternative fuzzy inference strategies and defuzzification techniques such as those described in [15]. For our proof-of-concept, we are using the netsnoop utility of Silicon Graphics' NetVisualizer to monitor and return raw network data. Later we intend to use Cabletron System's SpectroWATCH tool, currently under development. At that time we will be in a better position to report the pros and cons of the approach.

ACKNOWLEDGEMENTS

The author is grateful to members of the Network Management Applications Group at Cabletron Systems for helpful discussions and valuable comments on this work.

ABOUT THE AUTHOR

Lundy Lewis received the BS in Mathematics and BA in Philosophy from the University of South Carolina, the MS in Computer Science from Rensselaer Polytechnic Institute, and the PhD in Philosophy from the University of Georgia in the USA. He is an engineer and researcher in the Network Management Applications Group at Cabletron Systems R&D Center and an Adjunct Lecturer of Computer Science at Rivier College in Nashua, New Hampshire USA.

REFERENCES

[1] S. Francis, V. Frost, and D. Soldan. Measured Ethernet Performance for Multiple Large File Transfers. *Proceedings of the 14th Conference on Local Computer Networks*. Minneapolis, 1989. pp. 323-327.

[2] B. Lance. LAN Analyzers Move to AI. *BYTE Magazine*. March 1992. pp. 287-290.

[3] S. Wilkinson and T. Capen. Remote Control. *Corporate Computing*. October 1992. pp. 101-117.

[4] L. Lewis. A Review of Rule-Based Approaches to Network Management. Technical Note ctron-lml-92-02. Cabletron Systems Research and Development Center. 1992. 7 pages.

[5] R. Cronk, P. Callahan, and L. Berstein. Rule-Based Expert Systems for Network Management and Operations: An Introduction. *IEEE Network Magazine*. Vol. 5, No. 4, September 1988. pp. 7-21.

[6] M. Frontini, J. Griffin, and S. Towers. A Knowledge-Based System for Fault Localization in Wide Area Networks. *Integrated Network Management II* (editors: I. Krishnan and W. Zimmer). Elsevier Science Publishers (North Holland). 1991. pp. 519-530.

[7] B. Hitson. Knowledge-Based Monitoring and Control: An Approach to Understanding the Behavior of TCP/IP Network Protocols. *Proceedings of the ACM SIGCOMM '88*. Stanford, CA. pp. 210-221.

[8] M. Sutter and P. Zeldin. Designing Expert Systems for Real-Time Diagnosis of Self-Correcting Networks. *IEEE Network Magazine*. Vol . 5, No. 4, September 1988. pp. 43-51.

[9] J. Schroder and W. Schodl. A Modular Knowledge Base for Local Area Networ k Diagnosis. *Integrated Network Management II* (editors: I . Krishnan and W. Zimmer). Elsevier Science Publishers (North Holland). 1991. pp. 493-503.

[10] G. Vesonder, S. Stolfo, J. Zielinski, F. Miller, and D. Copp. ACE: An Expert System for Telephone Cable Maintenance. *Proceedings of the Eighth International Joint Conference on Artificial Intelligence*. Karlsruhe, West Germany. August 8-12, 1983. pp. 116-121.

[11] T. Williams, P. Orgren, and C. Smith. Diagnosis of Multiple Faults in a Nationwide Communications Network. *Proceedings of the Eighth International Joint Conference on Artificial Intelligence*. Karlsruhe, West Germany. August 8-12, 1983. pp. 179-181.

[12] J. Wright, J. Zielinski, and E. Horton. Expert Systems Development: The ACE System. *Expert System Applications to Telecommunications* (editor: J. Liebowitz). John Wiley and Sons, New York. 1988. pp. 45-72.

[13] L. Lewis. A Case-Based Reasoning Approach to the Resolution of Faults in Communications Networks. *Integrated Network Management* (editors: H.-G. Hegering and Y. Yemini). Elsevier Science Publishers (North Holland), pp. 671-682. Also published in the *Proceedings of IEEE INFOCOM '93*, IEEE Computer Society Press, pp. 1422-1429.

[14] L. Zadeh. Outline of a New Approach to the Analysis of Complex Systems and Decision Processes. *IEEE Transactions on Systems, Man, and Cybernetics*, SMC-3. 1973. pp. 28-44.

[15] C. Lee. Fuzzy Logic in Control Systems: Fuzzy Logic Controller (Parts I and II). *IEEE Transactions on Systems, Man, and Cybernetics*. Vol. 20, No. 2, March/April 1990. pp. 404-435.

[16] E. Cox. Fuzzy Fundamentals. *IEEE Spectrum*. October 1992. pp. 58-61.

LINK SET SIZING FOR NETWORKS SUPPORTING SMDS

Frank Y.S. Lin

Bell Communications Research
Piscataway, New Jersey 08854

ABSTRACT

To appropriately size networks that support Switched Multi-megabit Data Service (SMDS), we must determine how much additional capacity is needed and where it is needed so as to minimize the total capacity augmentation cost. We consider two combinatorial optimization problem formulations. These two formulations are compared for their relative applicability and complexity.

A solution procedure based upon Lagrangean relaxation is proposed for one of the formulations. In computational experiments, the proposed algorithm determines solutions that are within a few percent of an optimal solution in minutes of CPU time for networks with 10 to 26 nodes. In addition, the proposed algorithm is compared with a Most Congested First (MCF) heuristic. For the test networks, the proposed algorithm achieves up to 152% improvement in the total cost over the MCF heuristic.

PROBLEM DESCRIPTION

Switched Multi-megabit Data Service (SMDS) is a high-speed, connectionless, public, packet switching service that will extend Local Area Network (LAN)-like performance beyond the subscriber's premises, across a metropolitan or wide area [1] [2]. To ensure the performance objectives, a backbone network supporting the SMDS service (referred to as an SMDS network) must be carefully managed. The INPLANS™ system is developed by Bell Communications Research (Bellcore) to provide a *single* environment to support Bellcore Client Company (BCC) network planning and traffic engineering across different networking technologies instead of building individual systems for each type of networks[3] [4]. The INPLANS integrated network monitoring capability supports studies that monitor the ability of in-place networks to meet performance objectives. When performance exceptions are identified, corrective actions are needed to reduce the degree of congestion[3]. One possible action is to adjust the routing assignments. In [5], a responsive routing algorithm is proposed to balance the network load. Usually, routing is a cost-effective solution to network congestion caused by short-term traffic fluctuation. However, when the network load exceeds the network capacity and routing adjustment can no longer relieve the network congestion, additional capacity is needed. The process of determining the minimum amount of additional capacity needed for an exhausted network and where to add the capacity is referred to as sizing. The sizing approach usually involves ordering/installing equipments and therefore is not intended to be adopted on a (near) real-time basis.

INPLANS is a trademark of Bellcore.

Network Management and Control, Vol.2, Edited by
I.T. Frisch *et al.*, Plenum Press, New York, 1994

In this paper, a sizing approach to reducing network congestion is described. The proposed link set sizing algorithm can be used as one of the initial functionalities in the INPLANS integrated network servicing capability to support SMDS networks, which will take corrective actions when performance exceptions are identified by the integrated monitoring capability. To size the link sets, the routing strategy must be specified since routing and link set capacity assignment are closely related. The default routing algorithm for SMDS networks is specified in [6]. A brief review of the default Inter-Switching System Interface (ISSI) routing algorithm is given below.

The routing algorithm used for SMDS networks is referred to as ISSI Routing Management Protocol (RMP). The RMP is derived from the Open Shortest Path First (OSPF) specification Version 2[7]. The main features of the RMP are as follows:

- All routers have identical routing databases where a router is defined to be a Routing Management Entity (RME);
- Each router's database describes the complete topology of the router's domain;
- Each router uses its database and the Shortest Path First (SPF) algorithm to derive the set of shortest paths to all destinations from which it builds its routing table.

Each link set is assigned a positive number in the RMP called the link set metric. The default link set metric of each link set is inversely proportional to the aggregate link set capacity. One can apply standard shortest path algorithms, e.g. Dijkstra's algorithm[8] to calculate a shortest path spanning tree for every origin. Ties are broken by choosing the switch with the lowest router ID number.

Two types of traffic are supported by SMDS -- individually addressed message and multicast (group addressed) message. The individually addressed message is transmitted from the origin to the destination over the unique path in the shortest path spanning tree. The multicast message is destined for more than one destination (may not be for all destinations, which is referred to as broadcasting). However, one copy of the multicast message will be transmitted over every link in the shortest path spanning tree. A multicast message will be discarded by a leaf (termination) switch in the shortest path spanning tree if the message is not for any user connected to the switch.

The sizing problem for SMDS networks is difficult when the default routing algorithm and link set metrics are adopted. From the mathematical formulations shown in the next section, the difficulty is attributed to (i) the nonlinear arc weights with respect to the link set capacities (the default link set metric is inversely proportional to the aggregate link set capacity) and (ii) usually a discrete set of available link set capacities (e.g. in units of DS3 lines). If the probably most commonly used greedy heuristic, i.e. to place additional capacity on the most congested link set in each iteration, is applied, we predict and show by an example in Figure 1 that it is possible that a link set will be even more congested after its capacity is augmented (more traffic than the added capacity is rerouted over this link set). For illustration and comparison purposes, we develop a Most Congested First (MCF) heuristic.

The MCF Heuristic

1. If no overflow is found, stop; otherwise, go to Step 2.
2. Find the link set with the most overflow where ties are broken arbitrarily.
3. Add one link on the link set identified in Step 2.
4. Calculate the new link set metrics and reroute the traffic.
5. Go to Step 1.

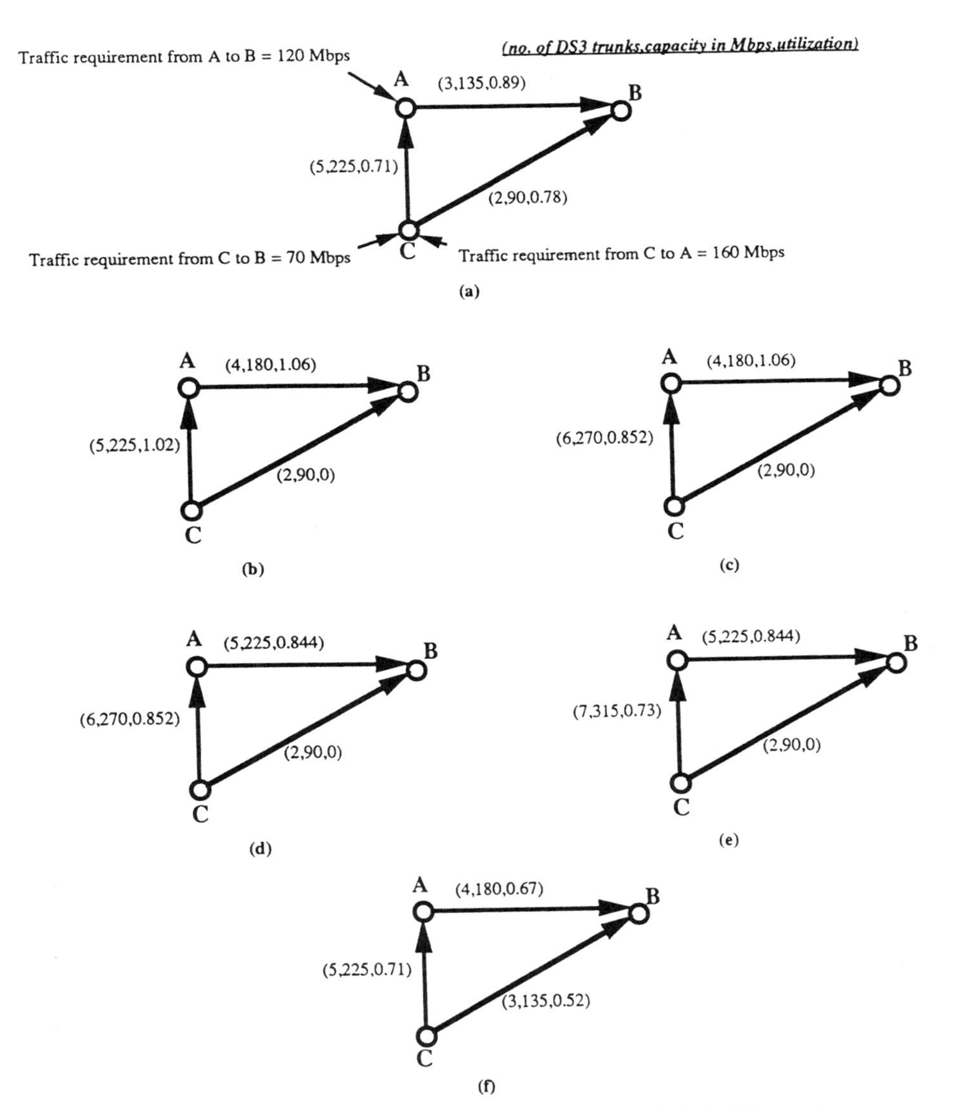

Figure 1. An example comparing the proposed algorithm with the MCF heuristic

Below, we go through an example where the MCF heuristic is applied.

In Figure 1(a), a network with three nodes (switches) and three link sets is shown. Assume that the default routing scheme (the OSPF protocol) and link set metrics (inversely proportional to the link set capacities) are applied and that the utilization threshold for each link set is 0.85 (the engineering thresholds can be calculated using the methods proposed in [9]). In Figure 1(a), link set (A,B) is overloaded (with 0.89 utilization). Assume that the MCF heuristic is used to solve the sizing problem where one DS3 line (45 Mbps) is added to the most congested link set (in terms of the amount of overflow) in each iteration. The routing assignments are adjusted accordingly and the aggregate link set flows are recalculated. If any of the link set is still overloaded, then this process is repeated until the utilization of each link set is no greater than 0.85. Figure 1(b) show the network status after 1 DS3 line is added to link set (A,B). This additional link changes the link set metric of (A,B) and therefore the routing assignments. After the traffic is rerouted according to the new set of link set metrics, it is observed that the utilization of link set (A,B) becomes 1.06 which is even larger than the value before a DS3 line is added, i.e. 0.89.

This seemingly counterintuitive phenomenon is attributed to the static nature of the OSPF routing with the default link set metrics, which does not react to the network load (and does not consider the capacity constraints). Also shown in Figure 1(b) is that (C,A) becomes overloaded while (C,B) is completely unused. Figures 1(c) to 1(e) depict the intermediate steps and the final result when the MCF heuristic is further applied. Consequently, 4 DS3 lines are added.

In contrast, Figure 1(f) shows the optimal solution where the number of additional lines required to satisfy the capacity/utilization constraints is minimized. (For this example problem, the proposed algorithm to be introduced in a later section finds the optimal solution.) The optimality can be verified easily. In the optimal solution, 2 additional DS3 lines are needed. Compare the optimal solution with the solution obtained by the MCF heuristic, it is observed that the MCF heuristic is not effective in this case (the proposed algorithm achieves a 100% improvement in the total capacity augmentation cost over the MCF heuristic). Another observation from the optimal solution is that network planners/administrators may need to put additional capacity on a link set with normal load originally (e.g. link set (C,B) in this example). It is thus clearly demonstrated that once a congested area of the network is identified, one needs to study the whole network rather than an isolated area to determine how much additional capacity is needed and where it is needed.

In this paper, we present two integer programming formulations for the SMDS link set sizing problem. In the first formulation, the objective is to minimize the total routing cost (to enforce the OSPF routing with the default link set metrics) subject to a budget constraint. In the second formulation, we minimize the total capacity augmentation cost subject to a set of shortest-path-routing constraints. Comparisons are made between the two formulations and an efficient near-optimal solution procedure is developed for the first formulation.

Three networks (8 test cases) with up to 26 nodes were tested in the computational experiments. The proposed algorithm determines solutions that are within a few percent of an optimal solution within minutes of CPU time. Compared with the MCF heuristic, the proposed algorithm achieved 7.1% to 152% improvement in the total capacity augmentation cost.

This work has the following significance. First, the problem is formulated as mathematical programs, which facilitates optimization-based solution approaches. Second, the proposed near-optimal sizing algorithm can help the BCCs expand SMDS network capacities in an economical way. Third, the formulations and algorithm developed can easily be generalized to consider the joint link set and node sizing problem for SMDS networks (by a different interpretation to the graph model). Last, by letting the existing node and link set capacities for potential locations be zero, this work can be used to solve the topological design and capacity assignment problem for SMDS networks.

The remainder of this paper is organized as follows. In the second section, two formulations of the SMDS link set sizing problem are given and compared. In the third section, a solution procedure based upon Lagrangean relaxation is proposed for the first formulation. In the fourth section, generalization and additional constraints on the problem formulations are considered. In the fifth section, computational results are reported. The last section summarizes this paper.

PROBLEM FORMULATIONS

An SMDS network is modeled as a graph $G(V, L)$ where the switches are represented by nodes and the link sets are represented by links. Let $V = \{1,2,...,N\}$ be the set of nodes and L be the set of links in the graph (network)[1]. Let W be the set of all origin-destination (O-D) pairs (single destination) in the network. According to the ISSI routing scheme, all traffic of an O-D pair is transmitted over exactly one (shortest) path. Furthermore, multicast traffic from an origin is transmitted over a shortest path spanning tree, which is the union of the shortest paths from the origin to each destination. As explained earlier, the multicast traffic from one origin to each of its associated multicast groups is *broadcast* to all the other Switching Systems (SSs) over the same shortest path spanning tree. For each O-D pair $(o,d) \in W$, the mean arrival rate of new individually addressed traffic is γ_{od} (packets/sec), while the aggregate (sum over all associated multicast groups) mean arrival rate of multicast traffic originated at origin o is α_o (packets/sec). Let P_{od} be the set of all possible simple directed paths from the origin to the destination for an O-D pair (o,d). The overall traffic for O-D pair (o,d) is transmitted over one path in the set P_{od}. Let P be the set of all simple directed paths in the network. Let T_o be the set of all spanning trees rooted at o. The multicast traffic originated at o is transmitted over one spanning tree in the set T_o. Let T be the set of all spanning trees in the network. For each link $l \in L$, the existing capacity is C_l packets/sec and the added capacity is A_l packets/sec (a decision variable).

For each O-D pair $(o,d) \in W$, let x_p be 1 if path $p \in P_{od}$ is used to transmit the individually addressed packets for O-D pair (o,d) and 0 otherwise. In an SMDS network, all of the packets of an O-D pair are transmitted over one path from the origin to the destination. Thus $\sum_{p \in P_{od}} x_p = 1$. For each path $p \in P$ and link $l \in L$, let δ_{pl} be 1 if link l is on path p and 0 otherwise.

For each origin o, let y_t be 1 if spanning tree $t \in T_o$ is used to transmit the multicast message for origin o and 0 otherwise. SMDS switches have the capability of duplicating packets for multiple downstream branches of a spanning tree used to carry the multicast traffic. When a packet is multicast from the root to the destinations using tree t, exactly one copy of the packet is transmitted over each link in the tree. Similar to the single-destination case, $\sum_{t \in T_o} y_t = 1$ for every origin o. For each tree $t \in T$ and link $l \in L$, let σ_{tl} be 1 if link l is on tree t and 0 otherwise.

Let $\Phi_l(A_l)$ be the cost to add capacity A_l to link l. This cost can include a fixed charge to change the capacity. Usually A_l is chosen from a discrete set K_l, e.g., in units of DS3 lines. A_l can be negative when existing capacities are allowed to be removed form the network. Let ρ_l be a prespecified threshold on the utilization factor of link l. The end-to-end delay objectives for SMDS networks will be satisfied if those utilization thresholds are not exceeded. These thresholds can be calculated using the schemes proposed in a recent work on allocating end-to-end delay objectives to individual network elements[9]. The SMDS link set sizing problem can be formulated as the following two combinatorial optimization problems.

1. As will be shown in the fourth section, if the node (switch) sizing problem is considered jointly with the link sizing problem, the links then represent either the switches or the link sets, while the nodes represent junctions between link sets and switches.

Formulation 1

Let B be the total budget available for capacity augmentation.

$$Z_{\overline{IP1}} = \min \sum_{l \in L} \sum_{(o,d) \in W} \sum_{p \in P_{od}} \frac{x_p \, \delta_{pl} \, \gamma_{od}}{C_l + A_l} \qquad (\overline{\text{IP1}})$$

subject to:

$$\sum_{p \in P_{od}} x_p = 1 \qquad \forall (o,d) \in W \qquad (1)$$

$$x_p = 0 \text{ or } 1 \qquad \forall p \in P_{od}, (o,d) \in W \qquad (2)$$

$$\sum_{(o,d) \in W} \sum_{p \in P_{od}} x_p \, \delta_{pl} \, \gamma_{od} + \sum_{o \in V} \sum_{t \in T_o} y_t \, \alpha_o \, \sigma_{tl} \le (C_l + A_l) \, \bar{\rho}_l \qquad \forall l \in L \qquad (3)$$

$$\sum_{d \in V - \{o\}} \sum_{p \in P_{od}} x_p \, \delta_{pl} \le (N-1) \sum_{t \in T_o} y_t \, \sigma_{tl} \qquad \forall l \in L, \; o \in V \qquad (4)$$

$$\sum_{t \in T_o} y_t = 1 \qquad \forall o \in V \qquad (5)$$

$$y_t = 0 \text{ or } 1 \qquad \forall t \in T_o, \; o \in V \qquad (6)$$

$$A_l \in K_l \qquad \forall l \in L \qquad (7)$$

$$\sum_{l \in L} \Phi_l(A_l) \le B. \qquad (8)$$

The objective function and Constraints (1) and (2) ensure that the individually addressed traffic for every O-D pair be routed over exactly one shortest path where each arc weight is inversely proportional to the corresponding link set capacity. The left hand side of Constraint (3) denotes the aggregate flow (including individually addressed and multicast traffic) over link l. Constraint (3) requires that the utilization factor of each link not exceed a prespecified value (to guarantee the end-to-end delay objectives). Constraints (5) and (6) require that all of the multicast traffic from one origin be transmitted over exactly one spanning tree. The left hand side of Constraint (4) (together with (1) and (2)) is the number of selected paths (for individually addressed traffic) rooted at origin o and passing through link l, while the right hand side of Constraint (4) (together with (5) and (6)) equals $N-1$ if link l is used in the spanning tree for root o to multicast messages and 0 otherwise. Recall that $N-1$ is the maximum number of selected paths originated at node o and passing through link l. Therefore, Constraint (4) requires that the union of selected paths from one origin to all the destinations for individually addressed traffic be the same spanning tree rooted at the origin to carry multicast traffic. (Note that this constraint implies that the selected paths from one origin to carry individually addressed traffic form a spanning tree.) Constraint (7) requires that the capacity added to each link be allowable. Constraint (8) requires that the total capacity expansion cost not exceed the given budget B.

It would be interesting to investigate the hardness of the above problem. If A_l is a constant and only Constraints (1) and (2) are considered, the problem is a well known shortest path problem. However, with the consideration of the capacity constraint (3), the problem becomes NP-complete and no existing polynomial time algorithm is available to solve the problem optimally. Next, the arc weight $(C_l + A_l)^{-1}$ is a nonlinear function of the discrete decision variable A_l. Moreover, The knapsack type of constraint (8), the integrality constraint (6) and the routing constraint (4) for y_t add another degree of hardness to the problem.

An equivalent formulation of $\overline{\text{IP1}}$ is

$$Z_{IP1} = \min \sum_{l \in L} \frac{f_l}{C_l + A_l} \qquad (\text{IP1})$$

subject to:

$$(1)-(8)$$

$$\sum_{(o,d)\in W}\sum_{p\in P_{od}} x_p\,\delta_{pl}\,\gamma_{od} \le f_l \qquad \forall l \in L \tag{9}$$

$$0 \le f_l \le (C_l + A_l)\,\bar{\rho}_l \qquad \forall l \in L. \tag{10}$$

For each link l, an auxiliary variable f_l is introduced. We interpret those variables to be aggregate flows attributed to individually addressed traffic. Since the objective function is strictly increasing with f_l and (IP1) is a minimization problem, equality of (9) will hold in an optimal solution. As the reader will see in the next section, the introduction of f_l decouples the problem into three independent subproblems in the Lagrangean Relaxation. Constraint (10) gives the range of f_l.

Formulation 2

$$Z_{IP2} = \min \sum_{l\in L} \Phi_l(A_l) \tag{IP2}$$

subject to:

$$\sum_{p\in P_{od}} x_p = 1 \qquad \forall (o,d) \in W \tag{11}$$

$$x_p = 0 \text{ or } 1 \qquad \forall p \in P_{od},\ (o,d) \in W \tag{12}$$

$$\sum_{(o,d)\in W}\sum_{p\in P_{od}} x_p\,\delta_{pl}\,\gamma_{od} + \sum_{o\in V}\sum_{t\in T_o} y_t\,\alpha_o\,\sigma_{tl} \le (C_l + A_l)\,\bar{\rho}_l \qquad \forall l \in L \tag{13}$$

$$\sum_{d\in V-\{o\}}\sum_{p\in P_{od}} x_p\,\delta_{pl} \le (N-1)\sum_{t\in T_o} y_t\,\sigma_{tl} \qquad \forall l \in L,\ o \in V \tag{14}$$

$$\sum_{t\in T_o} y_t = 1 \qquad \forall o \in V \tag{15}$$

$$y_t = 0 \text{ or } 1 \qquad \forall t \in T_o,\ o \in V \tag{16}$$

$$A_l \in K_l \qquad \forall l \in L \tag{17}$$

$$\sum_{l\in L}\sum_{q\in P_{od}} \frac{x_q\,\delta_{ql}}{C_l + A_l} \le \sum_{l\in L} \frac{\delta_{pl}}{C_l + A_l} \qquad \forall p \in P_{od},\ (o,d) \in W. \tag{18}$$

The objective function is to minimize the total cost of capacity augmentation. Constraints (11)-(17) are the same as (1)-(7). The left hand side of (18) (together with (11) and (12)) is the routing cost for O-D pair (o,d) (for one unit of flow on the selected path). The right hand side of (18) is the cost of path $p \in P_{od}$. Constraint (18) requires that for each O-D pair a shortest path be used to carry the individually addressed traffic.

A Comparison between Formulations 1 and 2

It would be useful to make a comparison between Formulations 1 and 2 for their relative applicability and complexity. An apparent difference between (IP1) and (IP2) is the objective functions and the last constraints. However, there is a dual relation between these two formulations. The objective function of Formulation 1 (together with constraints (1) and (2)) enforces the shortest path routing strategy, while the last constraint of Formulation 2 explicitly serves this purpose. The objective function of Formulation 2 is to minimize the total capacity augmentation cost, while the last constraint of Formulation 1 imposes an upper limit on the total capacity augmentation cost.

One potential drawback of Formulation 1 is that the shortest path routing strategy is enforced by the objective function but not constraints. The constraint set of (IP1) allows an O-D pair to choose an alternative route when the true shortest path with respect to the default link set metrics is congested (under the capacity constraint (3)). It is therefore possible that (IP1) is feasible with respect to the constraint set but is infeasible with respect to the shortest path routing strategy. One can increase the given budget when no desired solution is found. However, it is undesirable to assign too much budget, which will make link sets overengineered. Consequently, it may take several iterations to adjust the given budget when one wants to determine the minimum budget required. Whereas, the optimal objective function value of (IP2) is the minimum budget needed.

In view of the number of constraints, Formulation 1 is better than Formulation 2 since (18) is potentially comprised of a huge number of constraints (equals the number of simple paths in the network). It is difficult, if not intractable, to consider the numerous constraints in a solution procedure. Although (2) and (6) ((12) and (16) as well) have the same nature, in the proposed solution procedure, these two sets of constraints are considered implicitly in a shortest path problem and a minimal cost spanning tree problem, respectively. As a result, no aforementioned complexity problem will be incurred.

A SOLUTION PROCEDURE

Due to the difficulty of developing a solution procedure to Formulation 2 as mentioned in the previous section, we attempt to solve only Formulation 1 in this paper. A solution procedure to Formulation 2 will be developed and presented in a forthcoming paper so that the relative computational complexity and solution quality trade-off of Formulations 1 and 2 can be compared.

The basic approach to the development of a solution procedure to Formulation 1 is Lagrangean relaxation. Lagrangean relaxation is a method for obtaining lower bounds (for minimization problems) as well as good primal solutions in integer programming problems. A Lagrangean relaxation (LR) is obtained by identifying in the primal problem a set of complicating constraints whose removal will simplify the solution of the primal problem. Each of the complicating constraints is multiplied by a multiplier and added to the objective function. This mechanism is referred to as dualizing the complicating constraints.

For Formulation 1 (Problem (IP1)), we dualize constraints (3), (4), (8) and (9) to obtain the following relaxation

$$Z_{D1}(v,s,\beta,u) = \min \sum_{l\in L} \frac{f_l}{C_l + A_l} + \sum_{l\in L} v_l \Big[\sum_{(o,d)\in W} \sum_{p\in P_{od}} x_p\, \delta_{pl}\, \gamma_{od} + \sum_{o\in V} \sum_{t\in T_o} y_t\, \alpha_o\, \sigma_{tl} - (C_l + A_l)\, \bar{\rho}_l \Big]$$
$$+ \sum_{l\in L} \sum_{o\in V} s_{ol} \Big[\sum_{d\in V-\{o\}} \sum_{p\in P_{od}} x_p\, \delta_{pl} - (N-1) \sum_{t\in T_o} y_t\, \sigma_{tl} \Big] + \beta \Big[\sum_{l\in L} \Phi_l(A_l) - B \Big]$$
$$+ \sum_{l\in L} u_l \Big[\sum_{(o,d)\in W} \sum_{p\in P_{od}} x_p\, \delta_{pl}\, \gamma_{od} - f_l \Big] \qquad (LR\,1)$$

subject to:

$$\sum_{p\in P_{od}} x_p = 1 \qquad \forall (o,d)\in W \qquad (19)$$
$$x_p = 0 \text{ or } 1 \qquad \forall p \in P_{od},\ (o,d)\in W \qquad (20)$$
$$\sum_{t\in T_o} y_t = 1 \qquad \forall o\in V \qquad (21)$$
$$y_t = 0 \text{ or } 1 \qquad \forall t\in T_o,\ o\in V \qquad (22)$$
$$A_l \in K_l \qquad \forall l\in L \qquad (23)$$
$$0 \le f_l \le (C_l + A_l)\,\bar{\rho}_l \qquad \forall l\in L \qquad (24)$$

where v, s, and u are the vectors of $\{v_l\}$, $\{s_{ol}\}$ and $\{u_l\}$, respectively. Note that the constraints are dualized in such a way that the corresponding multiplers are nonnegative.

(LR1) can be decomposed into three independent subproblems. Note that the constant terms, e.g. βB, were omitted in the objective function in the subproblems.

Subproblem 1:

$$Z_{D1}^{1}(v,u,\beta) = \min \sum_{l \in L} \left\{ \frac{f_l}{C_l + A_l} - u_l f_l - v_l \bar{\rho}_l A_l + \beta\, \Phi(A_l) \right\}$$

subject to:

$$A_l \in K_l \qquad \forall l \in L \tag{25}$$

$$0 \le f_l \le (C_l + A_l)\bar{\rho}_l \qquad \forall l \in L \tag{26}$$

Subproblem 2:

$$Z_{D1}^{2}(v,s,u) = \min \sum_{l \in L} \sum_{(o,d) \in W} \sum_{p \in P_{od}} [(u_l + v_l)\gamma_{od} + s_{ol}]\, x_p\, \delta_{pl}$$

subject to:

$$\sum_{p \in P_{od}} x_p = 1 \qquad \forall (o,d) \in W \tag{27}$$

$$x_p = 0 \text{ or } 1 \qquad \forall p \in P_{od},\ (o,d) \in W \tag{28}$$

and

Subproblem 3:

$$Z_{D1}^{3}(v,s) = \min \sum_{l \in L} \sum_{o \in V} \sum_{t \in T_o} (v_l\, \alpha_o - (N-1)\, s_{ol})\, y_t\, \sigma_{tl}$$

subject to:

$$\sum_{t \in T_o} y_t = 1 \qquad \forall o \in V \tag{29}$$

$$y_t = 0 \text{ or } 1 \qquad \forall t \in T_o,\ o \in V. \tag{30}$$

Subproblem 1 is composed of $|L|$ (one for each link) problems. Since A_l is discrete and bounded, the problem can be solved by successively fixing A_l to all possible values that satisfy (25). The following observation may greatly improve the efficiency of the solution procedure. For a fixed A_l, the objective function becomes minimizing a linear function of f_l over a simple interval of f_l specified in (26). The minimum objective function value can be found at one boundary point of the simple interval. As a result, to solve the problem for each link, only $2\,|K_l|$ points need to be evaluated. If $\Phi(A_l)$ possesses a certain property, e.g. convexity or concavity, the computational load can be further reduced. For example, if $\Phi(A_l)$ is linear, only the 4 extreme points of the convex hull of (25) and (26) need to be evaluated.

Subproblem 2 consists of $|W|$ (one for each O-D pair) shortest path problems where $(u_l + v_l)\,\gamma_{od} + s_{ol}$ is the arc weight for link l and O-D pair (o,d). Dijkstra's algorithm can be applied to solve the shortest path problems. It is worth mentioning that Constraint (4) can also be written as

$$\sum_{t \in T_o} y_t\, \sigma_{tl} \le \sum_{d \in V-\{o\}} \sum_{p \in P_{od}} x_p\, \delta_{pl} \qquad \forall l \in L,\ o \in V. \tag{31}$$

However, this may cause the difficulty of negative cycles in the shortest path problems.

Subproblem 3 consists of $|V|$ (one for each root) minimum cost spanning tree problems where $v_l\, \alpha_o - (N-1)\, s_{ol}$ is the arc weight for link l and root o. One can apply Prim's or Kruskal's algorithm[10] to solve the problem.

For any $(v,s,\beta,u) \ge 0$, by the weak Lagrangean duality theorem, the optimal objective function value of (LR1), $Z_{D1}(v,s,\beta,u)$, is a lower bound on Z_{IP1} [11]. The dual problem (D1) is

$$Z_{D1} = \max_{(v,s,\beta,u) \ge 0} Z_{D1}(v,s,\beta,u). \tag{D1}$$

To find the greatest lower bound, we solve (D1). Another common approach to finding a lower bound on the optimal objective function value of a minimization integer programming problem is to use linear programming relaxation (the integrality constraints are relaxed). However, the objective function of (IP1) is in general nonconvex with respect to f_l and A_l (examining the Hessian of $f_l / (C_l + A_l)$). No standard procedure can be applied to solve the linear programming relaxation optimally to obtain a legitimate lower bound.

There are several methods for solving the dual problem (D1). One of the most popular methods is the subgradient method. Let a $(2+|V|)\,|L|+1$ vector b be a subgradient of $Z_{D1}(v,s,\beta,u)$. In iteration k of the subgradient optimization procedure, the multiplier vector $m^k = (v^k, s^k, \beta^k, u^k)$ is updated by

$$m^{k+1} = m^k + t^k b^k.$$

The step size t^k is determined by

$$t^k = \delta \, \frac{Z_{IP1}^h - Z_{D1}(m^k)}{\|b^k\|^2} \qquad (32)$$

where Z_{IP1}^h is an objective function value for a heuristic solution (upper bound on Z_{IP1}) and δ is a constant, $0 < \delta \leq 2$. To solve (D1), the subgradient method is used.

The above procedure is for solving the dual problem and obtaining good lower bounds on the optimal primal objective function value. We next describe a procedure for finding good primal feasible solutions. In each iteration of solving the dual problem (where an (LR1) is solved), one can calculate the aggregate link set flows using the routing assignments form the solution to the (LR1). From these aggregate link set flows, the minimum link set capacities required to satisfy the capacity/utilization constraints can be calculated. We then use these minimum link set capacities to calculate a new set of link set metrics and to reroute the traffic accordingly. If any of the capacity/utilization constraints is violated, we may apply the MCF heuristic to place additional capacity. If the total cost is less than the given budget, then a primal feasible solution is found. Another alternative is to apply the following All Congested First (ACF) heuristic to find primal feasible solutions.

The ACF Heuristic

1. Set the counter limit K to be a prespecified value.
2. If $K = 0$, return; otherwise, decrease K by 1.
3. Find the link sets where the capacity (utilization) constraints are violated.
4. Add the minimum number of links on each link set identified in Step 3 to satisfy the current flows.
5. Calculate the new link set metrics and reroute the traffic.
6. If no overflow is found, return; otherwise, keep the routing assignments, remove the links added in Step 4 and go to Step 2.

To find the tightest budget constraint (the lowest cost), one may apply the concept of bisecting search in adjusting the given budget B. However, this may require solving a significant number of (IP1)'s. In addition, for a given budget B, it may be difficult to determine whether the problem is feasible (an integer programming problem). The following implementation attempts to achieve better efficiency.

The Overall Algorithm

1. Apply the MCF heuristic to calculate an initial value of the given budget.
2. Solve the current (IP1).

3. Record the lowest capacity augmentation cost for the feasible capacity augmentation plans in solving (IP1).
4. If the lowest cost from Step 3 is smaller than the current given budget, construct a new (IP1) by replacing the given budget with the lower value and go to Step 2; otherwise, stop.

GENERALIZATION AND OTHER CONSTRAINTS

In this section, a generalization of Formulations 1 and 2 to jointly consider node (switch) and link (link set) sizing for SMDS networks is considered. A new graph model is presented and a number of associated modifications on Formulations 1 and 2 are described. In addition, two types of additional constraints, i.e. switch termination constraints and symmetric link set capacity constraints, are considered. The switch termination constraints require that the number of Subscriber Network Interface (SNI) and ISSI terminations to a switch not exceed a given number, depending upon the capacity requirement of each SNI and ISSI connection. The symmetric link set capacity constraints require that for two adjacent switches the link set capacities be the same in both directions.

Node Sizing

To jointly consider node and link set sizing, a new graph model is first introduced. An SMDS backbone network is modeled as a graph where each link in the graph corresponds to a delay element (e.g., a link set or a switch). Let L_T be the set of links associated with link sets. Let L_S be the set of links associated with switches. Let L be the set of links in the graph ($L_T \cup L_S$). The nodes in the graph represent junctions among the delay elements. Figure 2 depicts a node splitting technique to model a switch, say switch i, as a link. The procedure is described as follows.

For each node i in the old graph model, do the following:

1. Split node i in the old graph model into two nodes, i and i', in the new graph model.
2. Connect all the inbound links to switch i in the old graph model to node i in the new graph model.
3. Connect all the outbound links from switch i in the old graph model to node i' in the new graph model.
4. Add a new link (i,i'), which now represents switch i.

After this node splitting procedure is applied, it is clear that all the incoming traffic to switch i must flow through link (i,i') and thus the aggregate nodal flow can be considered. In the event that the architecture of a switch involves more than one major delay elements, the node splitting procedure can be further applied in a similar fashion to capture the aggregate flows on those delay elements.

Since in the default OSPF routing algorithm only the link set metrics (but not the switch metrics) are considered, the arc weight associated with a link in L_S is set to 0. The current formulations can be modified as follows. For Formulation 1, the set L in the objective function is replaced by L_T. For Formulation 2, the set L in the last constraint (18) is replaced by L_T. Another modification that needs to be made is (i) to separate the set of nodes V in the new graph into 2 sets, V_I and V_O, where V_I is the set of originating nodes for links in L_S and V_O is the set of terminating nodes for links in L_S and (ii) to replace the node set V in both formulations by V_I. It is clear that in the generalized formulations, an origin o must be in V_I and a destination d must be in V_O. One last modification is to replace L in Constraints (9) and (10) by L_T since there is no need to introduce an auxiliary variable f for each link associated with a switch.

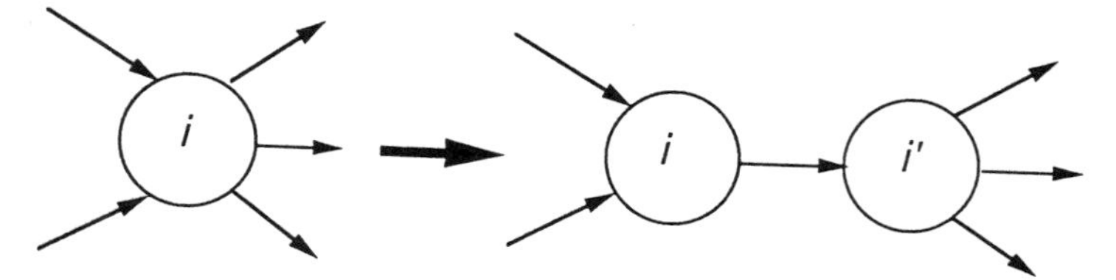

Figure 2. Node splitting technique

Switch Termination Constraints

A switch may have a given number of terminations (ports) where each termination can handle a fixed number of SNI and/or ISSI connections. Given this architecture, one may need to explicitly consider the switch termination constraints, which can possibly become the limiting factors than the switch capacity (utilization) constraints. Let R_i be the number of free terminations for switch i (represented by link i in the new graph model), taking into account the existing SNI and ISSI connections. For switch (link) i, let I_i and O_i be the sets of inbound and outbound links, respectively. Also let $\Omega_i(A_l)$ be the number of ports needed for switch i to connect an incident link set l of capacity A_l. Then the following switch termination constraint needs to be considered in the formulations.

$$\sum_{l \in I_i \cup O_i} \Omega_i(A_l) \le R_i \qquad \forall i \in L_S.$$

This constraint requires that for each switch the total number of terminations needed for installing additional links be no greater than the number of free terminations.

This constraint couples elements in $\{A_l\}$ and is dualized in the Lagrangean relaxation. Consequently, two more terms need to be considered in Subproblem 1 for each link. However, the solution procedure to Subproblem 1 remains the same.

Symmetric Link Set Capacity Constraints

For a switch, it may be required that the link sets be installed in pairs with the same capacity in both directions. In this case, a symmetric capacity constraint needs to be considered. To express this constraint, it is neater to denote a link by its originating node i and terminating node j as (i,j) than by an index number as l. Then the symmetric capacity constraint is given below.

$$\begin{cases} A_{(i,j)} = A_{(j,i)} & \forall (i,j),\ (j,i) \in L \quad \text{(for the old graph model)} \\ A_{(i',j)} = A_{(j',i)} & \forall (i,i'),\ (j,j') \in L_S,\ (i',j),\ (j',i) \in L_T \quad \text{(for the new graph model).} \end{cases}$$

This additional constraint has an impact on the decomposition of Subproblem 1 of the Lagrangean relaxation. Without the symmetric capacity constraint, Subproblem 1 can be further decomposed into $|L|$ independent problems. Whereas, with the constraint, those decomposed and independent problems need to be solved in pairs (problems corresponding to two link sets in opposite directions are solved jointly). Fortunately, this does not greatly complicate the solution procedure.

COMPUTATIONAL RESULTS

Two sets of computational experiments are performed. In the first set of experiments, we test the proposed sizing algorithm with respect to its (i) computational efficiency and (ii) effectiveness in determining good solutions. In the second set of experiments, we quantify how much the total capacity augmentation cost can be reduced by the sizing algorithm compared with the Most Congested First (MCF) heuristic described in the first section.

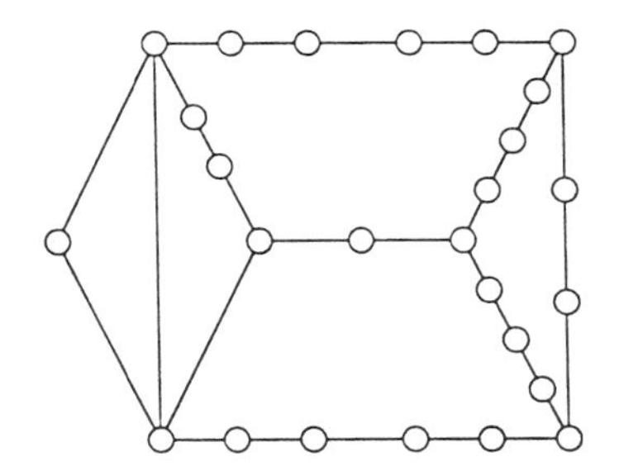

Figure 3. 26-node 60-link OCT Network

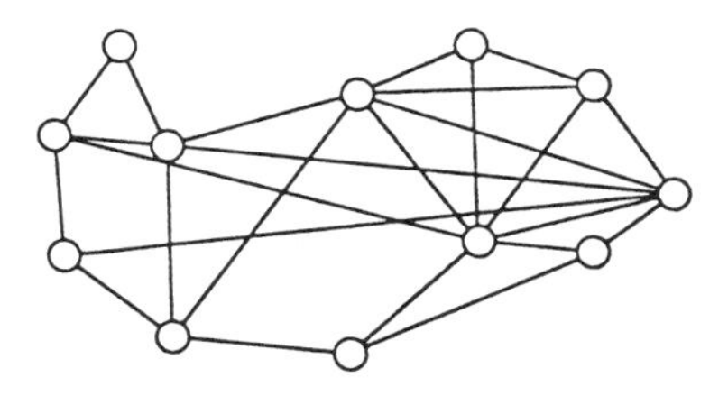

Figure 4. 12-node 50-link GTE Network

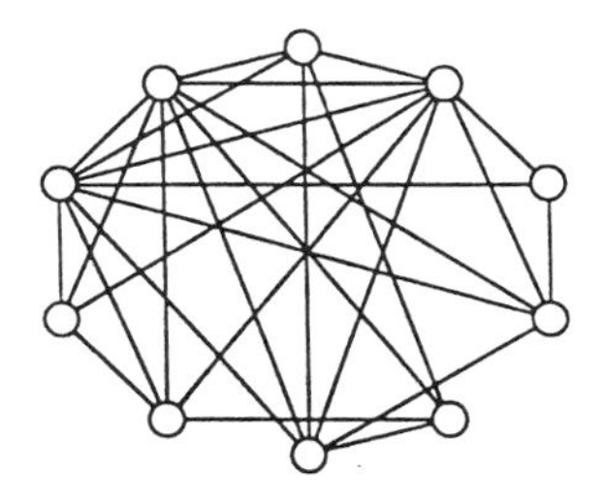

Figure 5. 10-node 56-link SITA Network

The link set sizing algorithm for SMDS networks described in the third section was coded in FORTRAN 77 and run on a SUN SPARC file server[2]. The algorithm was tested on three networks: OCT (26 nodes), GTE (12 nodes), and SITA (10 nodes) whose topologies are shown in Figures 3, 4 and 5, respectively. For each of the three networks, it was assumed that for each O-D pair the total individually addressed traffic rate at which packets are generated is the same (uniform traffic demand). Since the amount of group addressed traffic is expected to be small

2. Bellcore does not recommend or endorse products or vendors. Any mention of a product or vendor in this paper is to indicate the computing environment for the computational experiments discussed or to provide an example of technology for illustrative purposes; it is not intended to be a recommendation or endorsement of any product or vendor. Neither the inclusion of a product or a vendor in a computing environment or in this paper, nor the omission of a product or vendor, should be interpreted as indicating a position or opinion of that product or vendor on the part of the authors or Bellcore.

compared with the individually addressed traffic, the group addressed traffic is not considered in the computational experiments. It is also assumed that only one type of links (DS3 lines) are available and that the cost for each additional link is 1.

The given budget B is initially calculated by applying the MCF heuristic. The ACF heuristic is applied to to find primal feasible solutions. As mentioned in the third section (the overall algorithm), the lowest feasible capacity augmentation cost found in solving the current (IP1) is recorded and used as the new (tighter) budget in a new (IP1). This process is repeated until no tighter budget is found.

To solve (D1), the subgradient method described in the third section was applied. In our implementation, Z^h_{IP1} was initially chosen as $\sum_{l \in L} \overline{\rho_l}$ (an upper bound on the total link set utilization factors if (IP1) is feasible) and updated to the best upper bound found so far in each iteration. In Equation (32), δ was initially set to 2 and halved whenever the objective function value did not improve in 30 iterations. The initial values of u_l, v_l and β were chosen to be $1/C_l$, 0 and $\sum_{l \in L} \overline{\rho_l} / \sum_{l \in L} C_l$, respectively.

We first show that the proposed sizing algorithm performs well under tight budget constraints. The given budget is chosen so that in the course of solving (IP1) all the feasible capacity augmentation plans found use up the given budget. Table 1 summarizes the results of the computational experiments with the proposed sizing algorithm. The second column gives the traffic requirement for each O-D pair (normalized by the DS3 line capacity) . The third column specifies the existing capacity of each link in each network (also normalized by the DS3 line capacity). The fourth column specifies the value of counter limit K used in the ACF heuristic to find primal feasible solutions. The fifth column gives the number of iterations (the number of (LR1)'s solved) executed to solve (IP1)/(D1). The sixth column provides the given budget. The seventh column is the largest lower bound on the optimal objective function value found in the number of iterations specified in the fifth column. Recall that this is the best objective function value of the dual problem. The eighth column gives the best objective function value for (IP1) in the number of iterations specified in the fifth column. The ninth column reports the percentage difference ([upper-bound – lower-bound] × 100 / lower-bound) which is an upper bound on how far the best feasible solution found is from an optimal solution. The tenth column provides the CPU times which include the time to input the problem parameters.

Table 1 shows that the sizing algorithm is efficient and effective in finding near-optimal solutions given tight budget constraints. For every test problem (networks with 10 to 26 nodes), the algorithm determines a solution that is within 4 percent of an optimal solution in less than 2 minutes of CPU time on a SUN SPARC file server.

Another set of experiments was performed to compare the sizing algorithm with the MCF heuristic. For each test problem, a number of (IP1)'s are solved and the given budget is updated (reduced) until no more improvement is achieved (see the Overall Algorithm in the third section). A comparison of the performance of the sizing algorithm with the performance of the MCF is reported in Table 2. The first five columns in Table 2 provide the same information as the first five columns in Table 1. The sixth column gives the number of (IP1)'s solved before the algorithm terminates. The seventh column reports B^{MCF}, the cost obtained by applying the MCF heuristic. This value is used as the initial given budget. The eighth column reports B^h, the best primal objective function value found by the proposed sizing algorithm. The ninth column gives the percentage improvement of the sizing algorithm over the MCF heuristic $[100 \times (B^{MCF} - B^h)/B^h]$.

The results in Table 2 show that using the proposed sizing algorithm results in an 7% to 152% (79% on the average) improvement in the total capacity augmentation cost over the MCF heuristic. In addition, the number of (IP1)'s solved for each test case is at most 3.

Table 1. Summary of computational results of the proposed sizing algorithm

Net ID	Req.	Cap.	K	Ite.	B	Lower Bounds	Upper Bounds	Perc. Diff.	Time (sec)
OCT	0.10	5	4	800	100	38.64	40.14	3.889	95.0
OCT	0.15	10	4	800	56	36.50	36.84	0.943	77.2
GTE	0.60	4	5	1000	28	28.00	28.33	1.161	20.2
GTE	0.45	3	5	1000	20	28.27	28.58	1.091	20.2
GTE	0.30	2	5	1000	16	27.20	27.91	2.610	19.8
SITA	0.20	1	5	1000	5	21.20	22.00	3.774	15.4
SITA	1.00	5	5	1000	16	22.64	22.91	1.190	14.4
SITA	1.50	5	5	1000	50	29.80	30.30	1.678	14.6

Table 2. Comparison of the proposed sizing algorithm with the MCF heuristic

Net ID	Req.	Cap.	K	Ite.	# of (IP1)'s solved	B^{MCF}	B^h	Perc. Impru.	Time (sec)
OCT	0.10	5	4	800	3	224	100	124.0	258.8
OCT	0.15	10	4	800	2	60	56	7.1	153.1
GTE	0.60	4	5	1000	2	45	28	60.7	40.1
GTE	0.45	3	5	1000	2	30	20	50.0	40.0
GTE	0.30	2	5	1000	2	27	16	68.8	40.3
SITA	0.20	1	5	1000	2	11	5	120.0	29.1
SITA	1.00	5	5	1000	2	24	16	50.0	27.1
SITA	1.50	5	5	1000	5	126	50	152.0	29.2

SUMMARY

Switched Multi-megabit Data Service (SMDS) is a high-speed, connectionless, public, packet switching service that will extend Local Area Network (LAN)-like performance beyond the subscriber's premises, across a metropolitan or wide area. The SMDS service is considered as the first step towards the BISDN service and is thus strategically important for the BCCs.

To satisfy the performance objectives and, on the other hand, to avoid excessive overengineering, it is essential that the capacity of SMDS networks be carefully managed. When performance exceptions are identified by a monitoring process, one may either reroute the traffic or resize the network to reduce the degree of network congestion. However, when the load exceeds the network capacity, routing alone cannot resolve the congestion problem and additional capacity is needed.

In this paper, a sizing approach to reducing network congestion is described. The objective is to determine the minimum amount of additional capacity needed for an exhausted network and where to add the capacity. As demonstrated by a simple example, a commonly used greedy heuristic failed to provide satisfactory solutions. An optimization-based approach is then taken to attack the problem. We consider two combinatorial optimization problem formulations. In the first formulation, the objective is to minimize the total routing cost (to enforce the default routing protocol in SMDS networks) subject to a budget constraint. In the second formulation, we minimize the total capacity augmentation cost subject to a set of shortest-path-routing constraints. These two formulations are compared for their relative applicability and complexity.

A solution procedure based upon Lagrangean relaxation is proposed for the first formulation. In computational experiments, the proposed algorithm determines solutions that are within a few percent of an optimal solution in minutes of CPU time of a SUN SPARC file server for networks with 10 to 26 nodes. In addition, the proposed algorithm is compared with a Most Congested First (MCF) heuristic. For the test networks, the proposed algorithm achieves 7% to 152% (79% on the average) improvement in the total capacity augmentation cost over the MCF heuristic.

This work has the following significance. First, the problem is formally formulated as mathematical programs, which clearly demonstrates the difficulty of the problem and facilitates optimization-based solution approaches. Second, the proposed sizing algorithm has been computationally shown to be efficient and effective. The algorithm can thus help the BCCs expand SMDS network capacities in an economical way. Third, the formulations and algorithm developed can easily be generalized to consider the joint link set and node sizing problem for SMDS networks (by a different interpretation to the graph model). Last, by letting the existing node and link set capacities for potential locations be zero, this work can be used to solve the topological design and capacity assignment problem for SMDS networks.

ABOUT THE AUTHOR

Frank Y.S. Lin received his B.S. degree in Electrical Engineering from National Taiwan University in 1983 and his Ph.D. degree in Electrical Engineering from University of Southern California in 1991. Since 1991, he has been with Bell Communications Research where he is involved in developing network capacity management algorithms for advanced technologies and services. His current research interests include computer network optimization, high-speed networks, distributed algorithms and performance modeling.

REFERENCES

1. Bell Communications Research, "Generic System Requirements in Support of SMDS," TR-TSV-000772, Issue 1, May 1991.
2. T. Cox and F. Dix, "SMDS: The Beginning of WAN Superhighways," *Data Communications,* April 1991.
3. J.L. Wang and E.A. White, "An Integrated Methodology for Supporting Packet Network Engineering," *Proc. IEEE Globecom,* December 1990.
4. J.L. Wang, "An Integrated Methodology for Supporting Network Planning and Traffic Engineering with Considerations to SMDS Service," *Proc. IEEE Globecom,* December 1991.
5. F. Y.S. Lin and J.L. Wang, "Minimax Open Shortest Path First Routing Algorithms in Networks Supporting the SMDS Service," *Proc. ICC,* May 1993.
6. Bell Communications Research, "Generic Requirements for SMDS Networking," TA-TSV-001059, Issue 2, August 1992.
7. OSPFIGP Working Group of the Internet Engineering Task Force, "The OSPF Specification, Version 2," *Network Working Group Request for Comments (RFC 1131),* June 1990.
8. E.W. Dijkstra, "A Note on Two Problems in Connection with Graphs," *Numerical Mathematics,* Vol. 1, pp. 269-271, October 1959.
9. F. Y.S. Lin, "Allocation of End-to-end Delay Objectives for Networks Supporting SMDS," this book.
10. A.S. Tanenbaum, *Computer Networks,* Prentice-Hall, 1988.
11. A.M. Geoffrion, "Lagrangean Relaxation and Its Uses in Integer Programming," *Math. Programming Study,* 2:82-114, 1974.

DEFINITION AND EVALUATION OF AN ADAPTIVE PERFORMANCE MANAGEMENT STRATEGY FOR A HYBRID TDM NETWORK WITH MULTIPLE ISOCHRONOUS TRAFFIC CLASSES

R. Bolla, F. Davoli, P. Maryni, G. Nobile, G. Pitzalis, and A. Ricchebuono

Department of Communications, Computer and Systems Science (DIST)
University of Genoa
Via Opera Pia, 11/A - 16145 Genova, Italy

ABSTRACT

The fair sharing of a common bandwidth among a user population is considered, in a Metropolitan Area Network based on hybrid TDM frames, dedicated to carrying both isochronous and asynchronous traffic types. The model can be applied, for instance, to a DQDB network, where asynchronous packets are handled as being part of a common distributed queue. Different classes of isochronous traffic are considered, according to their speed and originating user site. The slots in each frame are dynamically shared among all traffic types (isochronous and asynchronous), by controlling the admission of isochronous connection requests, and by periodically adjusting upper bounds on the number of isochronous slots that may be assigned to each traffic class. The upper bounds are updated on the basis of long-term averaged as well as instantaneous information, by attempting to minimize a cost function related to call blocking and packet delay. Simulation results show a fair sharing of the whole bandwidth and the achievement of a desired grade of service.

INTRODUCTION

Telecommunication networks are increasingly characterized by the integration of several types of services, which involve the transport of traffic classes with very different

Network Management and Control, Vol.2, Edited by
I.T. Frisch *et al.*, Plenum Press, New York, 1994

statistical behaviour and performance requirements. In this environment, it is necessary to exert control actions, often at various management levels, in order to guarantee a fair allocation of resources (like bandwidth and buffer space) and prevent network congestion. Typically, controls of this kind may consist in access strategies at the call level for connection-oriented services, whereby a specific connection is accepted only if the required quality of service can be maintained. Integrated local area networks (ILAN's) or metropolitan area networks (MAN's) present the additional problem of dynamic bandwidth sharing among a sparse user population; in this context, both multiaccess strategies for asynchronous packet traffic (as in the QPSX (DQDB) distributed queue algorithm[1]) and admission strategies for isochronous circuit-switched connections have been developed[1,2].

Obviously, the specific mode adopted for transport within the network has a great influence on the amount and nature of control; in particular, if flexibility is required in the allocation of bandwidth, care must be taken in ensuring fairness and in avoiding congestion. In this respect, transfer modes that are somehow more "structured" (e.g., those based on Time Division Multiplexing (TDM) frames, as opposed to the "unstructured" Asynchronous Transfer Mode (ATM)) are more easily managed. However, even in this case, the presence of different traffic and service types motivates the problem to decide the sharing of the resource (bandwidth) in order to maintain a good performance level (measured, for instance, in terms of blocking probability for isochronous, circuit-switched, traffic and average delay or packet loss rate for packet traffic), while ensuring a fair allocation. The bandwidth allocation problem in this context has been treated, for instance, in [2-10].

In the case of a single user with two traffic types, a circuit-switched one and a packet switched one, some attempts have been made at finding optimal dynamic (i.e., based on the instantaneous system's state) allocation strategies[7,8]. However, the extension of this approach to a larger population of several users accessing an integrated network seems to pose nontrivial computational difficulties.

A possible way of treating an integrated management problem of this kind has been considered in Aicardi[9] and Bolla[10] within the framework of a TDM drop/insert network with hybrid frames carrying the two above mentioned traffic types, and later extended to a DQDB network[11,12]. The method is based on parametric optimization, i.e., it solves an essentially static mathematical programming problem, whose solution is periodically recomputed at specified time instants, by taking into account real time information as well as long term time averages of parameters characterizing traffic statistics. More specifically, the slots in each hybrid frame are partitioned between the two traffic types, by controlling the admission of isochronous connection requests, and by assigning upper bounds on the number of isochronous slots per frame that each network interface unit is allowed to use. These assignments are periodically updated, by attempting to minimize a suitable cost function.

In this paper, we treat the access control and bandwidth allocation problem in a network of this kind, by considering also the presence of isochronous traffic at various different speeds. We suppose that these flows may be assigned a slot every n frames, and thus actually “see” frames of different duration. After introducing the mathematical model, the control system’s structure and the admission control rules in the next Section, we consider the optimization problem for bandwidth reallocation in Section III. Simulation results are reported and discussed in Section IV, and the conclusions are drawn in Section V.

CONTROL SYSTEM'S ARCHITECTURE

We consider a MAN, where M user stations, through their respective Network Interface Units (NIU), see a basic frame of b seconds duration, consisting in a total of C isochronous and asynchronous slots. Each NIU is able to receive isochronous traffic at H different speeds, which will be indicated by $v^{(1)}$, ..., $v^{(H)}$ (in slots/frame), in order of ascending speed. We suppose $v^{(H)}=1$, i.e., the maximum speed achievable by assigning no more than one slot/frame. We also restrict each $v^{(h)}$, h<H, to be of the form $v^{(h)}=1/n^{(h)}$, where $n^{(h)}$ corresponds to the number of basic frames between two successive slots of a connection at speed $v^{(h)}$. Thus, a slot is always completely filled by the bits corresponding to an isochronous connection; however, a slot needs not be available at every frame for a connection belonging to a speed class that is less than the maximum allowable one. In this respect, an isochronous flow at speed $v^{(h)}$, h<H, effectively “sees” a frame (which we may call a “virtual” frame of speed class h) that is $n^{(h)}$ times longer than the basic frame. For simplicity reasons that will be apparent in the following, $n^{(1)}$ (corresponding to the lowest speed traffic) is further supposed to be a multiple of all $n^{(h)}$, $h\neq 1$.

In our subsequent model, time is discrete and, as a matter of fact, each isochronous traffic class (whose dynamics need to be described, in terms of call arrivals and duration), has a different time scale with respect to the basic frame, determined by $n^{(h)}$. For this reason, we will denote the time variable of an isochronous traffic flow of speed class h by $t^{(h)}$, and consider the whole block of $n^{(h)}$ basic frames as its discrete time unit.

For notational simplicity, it is convenient to express $t^{(h)}$ in terms of $t^{(1)}$ as follows. Let

$$f^{(h)} = \frac{n^{(1)}}{n^{(h)}}$$

Then, $t^{(h)} = t^{(1)} \cdot f^{(h)}$, and we can define $t^{(1)}=t$ and use the longer time unit as a base reference. Fig. 1 depicts the relationship among the various time units.

Access to the asynchronous slots (that are those unused by the isochronous traffic), is regulated by a distributed access protocol (e.g., the distributed queue mechanism of

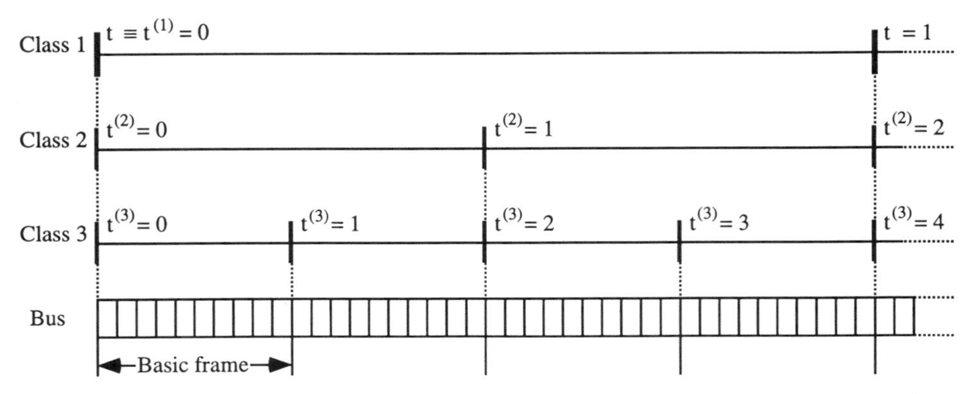

Figure 1. Example of the relation among the time scales for C=10, H=3, $n^{(1)}$=4, $n^{(2)}$=2 (obviously, $n^{(3)}$=1).

DQDB). We will not need to take into account the details of this access algorithm; for our purposes, it suffices to consider asynchronous slots as part of a common pool.

Summing up, the basic frame, seen by all users, is essentially divided into two parts, dedicated to carrying isochronous and asynchronous traffic, respectively; the isochronous part is itself divided among the isochronous traffic classes that are distinguished by their originating NIU and their speed. Obviously, these subdivisions are not fixed; we actually want to dynamically control the bandwidth shares, by acting on the acceptance of isochronous calls and on the distribution of isochronous slots among the different classes. The control system's architecture described in the following is depicted in Fig. 2.

At fixed decision instants, each NIU_i, i=1, ..., M, is assigned capacities $c_{t_k}^{(i,h)}$, h=1, ..., H, in slots/frame, that can be used for the isochronous traffic. We denote the decision instants by t_k, k=0,1,... , and suppose that each assignment lasts over the time units (in terms of the time variable t as defined above) $[t_k, t_k+1, ..., t_{k+1}-1]$. Any part of this capacity unused by the isochronous traffic can be utilized to carry packets of the asynchronous one, which access it according to the multiaccess protocol, and are therefore regarded as a whole undistinguished stream.

Variables $c_{t_k}^{(i,h)}$ will be considered as continuous; the actual maximum achievable capacity for class h of user i will therefore be $\lfloor c_{t_k}^{(i,h)} \rfloor$, where $\lfloor x \rfloor$ denotes the largest integer less than or equal to x, and we have defined the virtual capacity $\tilde{c}_{t_k}^{(i,h)} = c_{t_k}^{(i,h)} \cdot n^{(h)}$.

We suppose that, in general, each NIU_i may be offered H+1 traffic types, namely, H circuit-switched isochronous traffic flows with average call arrival rate $\lambda_1^{(i,h)}$ calls/s and average holding time $1/\mu^{(i,h)}$ s, and a packet switched traffic with average arrival rate $\lambda_2^{(i)}$ packets/s, with fixed length packets (one slot). Circuit-switched traffic requires continuation of service until the end of the call.

We do not assume any specific distribution for packet arrivals, as our scheme will not be based on the dynamics of the packet queue; actually, the only information that will be needed will regard the overall average packet arrival rate

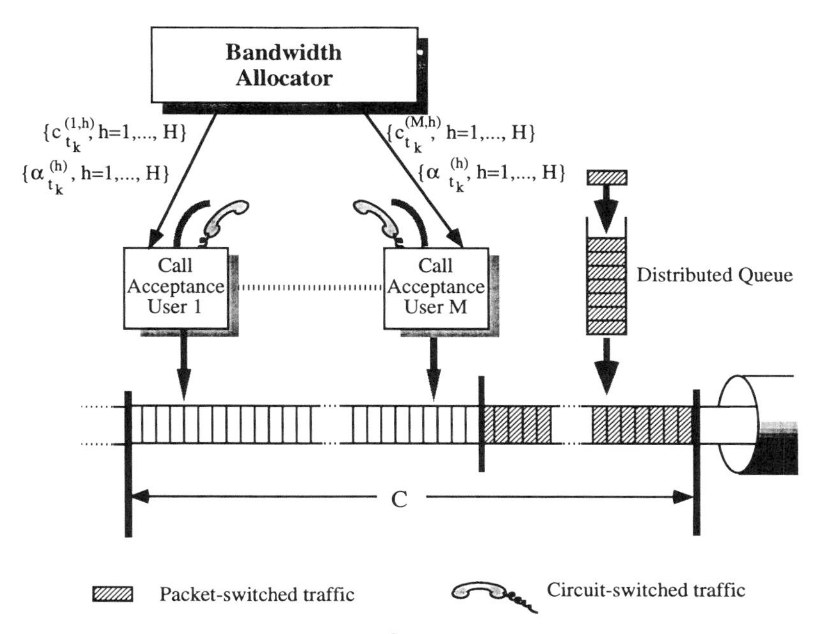

Figure 2. System structure.

$$\lambda_2 \equiv \sum_{i=1}^{M} \lambda_2^{(i)} \tag{1}$$

On the other hand, circuit-switched connections of speed class h are supposed to be described by a birth-death process with respect to their corresponding time unit $n^{(h)} \cdot b$ (i.e., the probability of more than a call of class h per user arriving or ending within a time unit of class h is considered to be infinitesimal; this implicitly implies $\lambda_1^{(i,h)} n^{(h)} \cdot b<1$ and $\mu^{(i,h)} n^{(h)} \cdot b\tilde{c}_{t_k}^{(i,h)}<1$). Moreover, each NIU_i has a local access controller that acts according to a fixed randomized strategy. More specifically, at the starting instant $t^{(h)}$ of each time unit of speed class h, NIU_i either accepts an incoming call with probability $\beta_{t^{(h)}}^{(i,h)}$ or blocks it with probability $1-\beta_{t^{(h)}}^{(i,h)}$. Let $r_{t^{(h)}}^{(i,h)}$ be the number of calls that require continuation of service in the next "virtual" frame (i.e., $n^{(h)}$ basic frames later). As done in [9,10] in a different context, we have chosen $\beta_{t^{(h)}}^{(i,h)}$ to be a function of $r_{t^{(h)}}^{(i,h)}$, whose shape is determined by $\tilde{c}_{t_k}^{(i,h)}$ and by a further parameter $\alpha_{t_k}^{(h)}$, as follows

$$\beta_{t^{(h)}}^{(i,h)} = \begin{cases} 1-\left[\dfrac{3\left(r_{t^{(h)}}^{(i,h)}\right)^2}{\left(\tilde{c}_{t_k}^{(i,h)}-1\right)^2} - \dfrac{2\left(r_{t^{(h)}}^{(i,h)}\right)^3}{\left(\tilde{c}_{t_k}^{(i,h)}-1\right)^3}\right]^{\alpha_{t_k}^{(i,h)}} & \text{if } \tilde{c}_{t_k}^{(i,h)} > r_{t^{(h)}}^{(i,h)}+1 \\ 0 & \text{otherwise} \end{cases}$$

$$i=1, \ldots, M;\ h=1, \ldots, H;\ t^{(h)}=t_k, t_k+1, \ldots, t_{k+1}-1 \tag{2}$$

It is not difficult to verify that $\beta^{(i,h)}_{t^{(h)}} \in [0,1]$ decreases with a reduction of the "available space" $\tilde{c}^{(i,h)}_{t_k} - r^{(i,h)}_{t^{(h)}}$; moreover, if considered as a function of $\tilde{c}^{(i,h)}_{t_k}$, it tends to a step function for $\alpha^{(h)}_{t_k} \to \infty$, whereas it flattens on the horizontal axis for $\alpha^{(h)}_{t_k} \to 0$.

It should be clear by now that the structure of our control system closely resembles a two-level hierarchical one[13]. The "local" acceptance rules that have been just defined represent a first control layer, that acts on the basis of local decentralized information; the higher layer will be constituted by a central controller, to be described in the next Section, which will act on the parameters $\alpha^{(h)}_{t_k}$ and $c^{(i,h)}_{t_k}$, i=1,...,M, h=1, ..., H. However, our local controllers are very simple ones. Actually, randomized strategies appear in the literature for several optimal access control problems in the presence of multiple traffic classes (see, e.g., Ross[5] and Maglaris[7]). Zukerman[2,3] also determines access strategies for isochronous traffic in a DQDB network; however, they are not NIU-selective (which would not allow to establish a fairness criterion for the isochronous traffic), and exhibit a threshold behaviour that would make them non differentiable with respect to $c^{(i,h)}_{t_k}$. For these reasons, we prefer the use of local heuristic strategies, which will be somehow compensated by the intervention of the central agent. An alternative to our limited "hierarchical" scheme would be a totally dynamic scheduling on the part of the central agent, which may become hardly tractable, as a Markov chain optimization problem, due to the possibly large dimension of the state space.

With our choice of $\beta^{(i,h)}_{t^{(h)}}$, each $r^{(i,h)}_{t^{(h)}}$, i = 1, ... , M, h=1, ... , H, is thus a controlled Markov chain with transition probabilities (due to the birth-death assumption)

$$p^{(i,h)}_{jk} \equiv P(r^{(i,h)}_{t^{(h)}+1} = k \mid r^{(i,h)}_{t^{(h)}} = j) = \begin{cases} \mu^{(i,h)} b^{(h)} j & k = j-1 \\ 1 - \mu^{(i,h)} b^{(h)} j - \lambda^{(i,h)}_1 b^{(h)} \beta^{(i,h)}_{t^{(h)}} & k = j \\ \lambda^{(i,h)}_1 b^{(h)} \beta^{(i,h)}_{t^{(h)}} & k = j+1 \\ 0 & \text{otherwise} \end{cases} \quad (3)$$

where we have defined $b^{(h)} = n^{(h)} \cdot b$. Let

$$\pi^{(i,h)}_{t^{(h)}}(s) = \Pr\left\{ r^{(i,h)}_{t^{(h)}} = s \right\} \quad (4)$$

Clearly $0 \le r^{(i,h)}_{t^{(h)}} \le \left\lfloor \tilde{c}^{(i,h)}_{t_k} \right\rfloor, \forall t^{(h)} \in \left[t^{(h)}_k, t^{(h)}_{k+1} \right]$. Then, we shall take $\left\{ 0,1,\ldots,\left\lfloor \tilde{c}^{(i,h)}_{t_k} \right\rfloor \right\}$ as the state space, and define

$$\pi^{(i,h)}_{t^{(h)}} \equiv \left[\pi^{(i,h)}_{t^{(h)}}(0), \ldots, \pi^{(i,h)}_{t^{(h)}}\left(\left\lfloor \tilde{c}^{(i,h)}_{t_k} \right\rfloor \right) \right]^T \quad (5)$$

where T denotes transpose.

The timing structure with respect to the decision instants is illustrated in Fig. 3.

Let L_t be the total number of packets in the distributed packet queue at time t; this quantity is obviously part of the overall system's state, which may be written as $X_t \equiv \mathrm{col}[\, r^{(i,h)}_{t^{(h)}}, t^{(h)} = t \cdot f^{(h)}$, i=1, ..., M, h=1, ..., H; $L_t]$. However, the description of the dynamics of L_t depending on the multiaccess protocol, would be far more complicated than that of the isochronous traffic. Fortunately, we will see in the following that we can avoid a direct dependence of the cost function on the statistics of L_t, by making it only a function of the value L_{t_k} at time t_k.

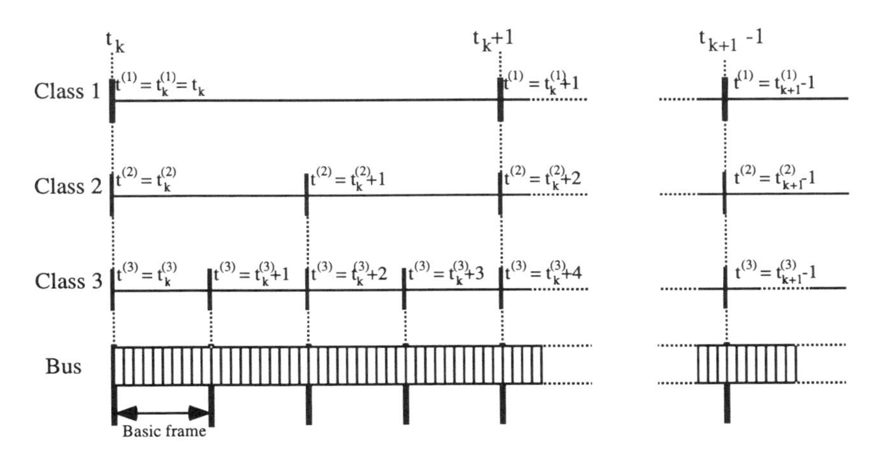

Figure 3. Timing structure and decision instant t_k.

OPTIMIZATION PROBLEM

We may turn now to the definition of a suitable optimization problem for the central agent that may physically reside in the Network Management Unit, and whose purpose is that of periodically reassigning the parameters $\alpha_{t_k}^{(h)}$ and $c_{t_k}^{(i,h)}$, i=1, ..., M, h=1, ..., H. To this aim, it is necessary to devise a cost function capable of reflecting the performance requirements of the different traffic classes; these will typically be expressed in terms of blocking probability for the isochronous, circuit-switched traffic, and of average delay for packets. We have chosen to minimize a weighted sum of terms related to these quantities, evaluated over a finite time horizon. This is simpler in our case than enforcing a minimum blocking probability as a constraint; however, through an appropriate choice of the weight to be attributed to one of the quantities with respect to the other, it should be possible to keep a desired bound very closely.

Let us define $\beta^{(i,h)}(s) = \beta_{t^{(h)}}^{(i,h)}(s)$ for $r_{t^{(h)}}^{(i,h)} = s$, and $G \equiv t_{k+1} - t_k \equiv T/n^{(1)}b$ (i.e., G and T represent the duration of the interval $t_{k+1} - t_k$ in time units of the slower speed class and in seconds, respectively). At each instant t_k, k = 0, 1, ..., the central agent, basing on the knowledge of all "a priori" information (constituted by $\lambda_1^{(i,h)}$, i=1,...,M, h=1, ..., H, and λ_2), and of X_{t_k}, wants to assign the maximum capacities $c_{t_k}^{(1,1)},...,c_{t_k}^{(1,H)},...,c_{t_k}^{(M,1)},...,c_{t_k}^{(M,H)}$ and the coefficients $\alpha_{t_k}^{(1)},...,\alpha_{t_k}^{(H)}$, in order to minimize

$$J_{t_k}(X_{t_k}) = \sum_{t=t_k}^{t_{k+1}-1} \left\{ \frac{1}{M \cdot H} \sum_{i=1}^{M} \sum_{h=1}^{H} \sum_{z=(t-1)f^{(h)}+1}^{t \cdot f^{(h)}} \left[\frac{1}{f^{(h)}} \sum_{s=0}^{\lfloor \tilde{c}_{t_k}^{(i,h)} \rfloor} \left(1-\beta^{(i,h)}(s)\right)\pi_z^{(i,h)}(s) \right] + \right.$$

$$\left. +\sigma \cdot \max{}^2 \left[0;1-\frac{\sum_{i=1}^{M}\sum_{h=1}^{H}\sum_{z=(t-1)f^{(h)}+1}^{t \cdot f^{(h)}} \sum_{s=0}^{\lfloor \tilde{c}_{t_k}^{(i,h)} \rfloor} \left(\tilde{c}_{t_k}^{(i,h)} - s\right)\pi_z^{(i,h)}(s)}{\frac{L_{t_k}}{G} + \lambda_2 n^{(1)} b} \right] \right\} \tag{6}$$

where σ is a weighting coefficient.

The first term in (6) represents the percentage number of blocked calls (connection requests) of the isochronous traffic per "frame" at user i, for all speed classes. In the second term, the ratio in the nonzero argument of max[0 ; .] represents an approximation of the ratio between the average output flow and the average input flow of the distributed packet queue, during a time unit $n^{(1)}b$ of the slowest isochronous class. Actually, we consider the number of packets in the queue at time t_k, L_{t_k}, as uniformly spread over the decision interval $[t_k, t_{k+1}-1]$, and approximate the output flow with the average space available for packets.

In the capacity assignment the following constraints must be taken into account:

$$\sum_{i=1}^{M}\sum_{h=1}^{H} c_{t_k}^{(i,h)} = C, \qquad k = 0,1, \ldots \tag{7}$$

$$\tilde{c}_{t_k}^{(i,h)} \geq r_{t_k^{(h)}}^{(i,h)}, \qquad \begin{array}{l} i = 1, \ldots, M \\ h = 1, \ldots, H \\ k = 0,1, \ldots \end{array} \tag{8}$$

where (8) imposes continuation of the outstanding calls. Moreover, as the minimization has to be extended also over the $\alpha_{t_k}^{(h)}$'s, we must consider the additional constraints

$$\alpha_{t_k}^{(h)} > 0, \qquad \begin{array}{l} h = 1, \ldots, H \\ k = 0,1, \ldots \end{array} \tag{9}$$

We may say that the $c_{t_k}^{(i,h)}$'s set the partition of the total capacity C among the users and speed classes in relation to the isochronous traffic, and the $\alpha_{t_k}^{(h)}$'s tend to influence the subdivision of the whole available space between the asynchronous and the isochronous traffic.

It is worth noting at this point that, at each intervention time, the actual numbers of active calls $r_{t_k^{(h)}}^{(i,h)}$, i=1,...,M, h = 1, ... ,H, and the number of packets in the distributed queue L_{t_k} should be known by the centralized decision maker: the first quantities are needed both to correctly initialize the Markov chains and to write constraints (8), whereas L_{t_k} explicitly appears in the cost function and creates a feedback effect. Actually, the values of $r_{t_k^{(h)}}^{(i,h)}$,.

$i=1,...,M$, $h = 1, ... ,H$, should not create a problem, as calls may be thought of as being assigned by a management function that resides in the same head station as the central controller.

On the other hand, L_{t_k} might be known, with a certain delay, by direct communication on the part of each NIU_i by means of dedicated signaling slots, or might be estimated, with the estimate being used in lieu of the unknown true value.

A final remark must be made with regard to the knowledge of values of the average intensities $\lambda_1^{(i,h)}$, $i=1, ..., M$, $h=1, ..., H$, and λ_2, which is also required to evaluate (6). Actually, local estimates can be costructed at NIU_i of $\lambda_1^{(i,h)}$, λ_2, by observing the arrival processes over a time window. These local estimates, in turn, can be communicated to the Network Manager and updated avery time that the difference between the old and the new measured values falls above a certain threshold.

Finally, as regards the optimization problem, once computed the gradient, the optimization can be carried out by applying, for example, the gradient projection technique in the same way as in Aicardi[9], with the only difference of having added a new set of optimization variables, namely, $\alpha_{t_k}^{(h)}$, $i = 1,...,M$, $h = 1, ... ,H$. In fact, the single equality constraint (7) gives the opportunity to obtain very simple gradient projections, that closely follow the procedure used in Bertsekas[15] for optimal routing.

SIMULATION RESULTS

We report and comment in the following several simulation results whose purpose is to evaluate the effectiveness of the proposed management and control scheme. To fix ideas, we refer to a DQDB network. The simulation is a discrete event one and represents a DQDB network according to the standard[15].

Network model

All results have been obtained with a network made up of five user stations and three different classes of isochronous traffic for each user station. Since the DQDB network in real life situation presents delays between user stations and the access protocol may present unfairness as regards asynchronous traffic, a bandwidth balance technique (BWB) has been utilized.

Now, let us specify the characteristics of the network in full details. Only one bus is considered; the situation is symmetric for the reverse bus. It is important to remember that one of the five NIU's, which corresponds to the head station of the reverse bus, has no traffic on the bus in question and is not counted in M.

The frame duration has been set to a standard value of 125 μs. Given a total capacity of 600 Mbit/s and a number of bits per slot of 424 (53 bytes), 177 slots/frame are obtained. These slots will be equally subdivided among the different classes of every user at every decision instant in accordance with their isochronous traffic. After fixing a decision

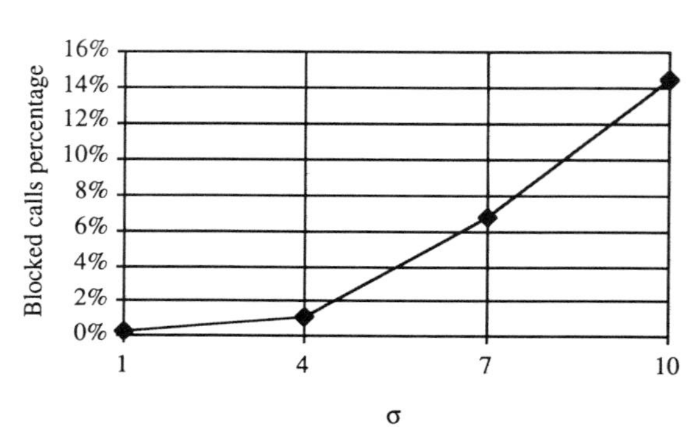

Figure 4. Blocked calls percentage vs. σ with load = 1.1.

interval duration of 0.4 s, the number of frames between decision instant results to be 3200. Each simulation whose results are illustrated below has a duration of 8 s, so as to get 20 decision intervals.

As regards especially the average duration of connections, the values chosen are particularly short, in order to limit the length of the simulation runs necessary to obtain a significant number of events. However, one of the main purposes of the hierarchical scheme we have been using is that of coping, to a certain extent, with dynamic variations in the traffic characteristics; in our cases, with relatively short connections, this is achieved by keeping the reallocation interval also relatively short. In case of longer connections (with the same traffic intensity), the situation would not be substantially different from this point of view.

In the following, the notation (i,h) indicates node i and class h. A Poisson distribution has been assumed for packet arrivals in all simulations.

Behaviour with different σ

In all the following simulations we have used a traffic pattern equally subdivided between isochronous and asynchronous. The isochronous traffic is the same for each class of every user station. The following data generates a total traffic flow which exploits the network potential to the utmost. We refer to this traffic flow as an offered load 1; an offered load "x" corresponds to the same data except that $\lambda_1^{(i,h)}$, i=1,2,3,4, h=1,2,3 and λ_2 are multiplied by x.

The network characteristics (corresponding to offered load 1) are:

$M = 4;\quad H = 3;\quad n^{(1)} = 4\,,\ n^{(2)} = 2\,,\ n^{(3)} = 1$

$\lambda_1^{(i,h)} = 41$ calls/s $\forall\, i, \forall\, h$

$1/\mu^{(i,h)} = 0.18$ s $\forall\, i, \forall\, h$

$\lambda_2 = 708{,}000$ packets/s; BWB = 8

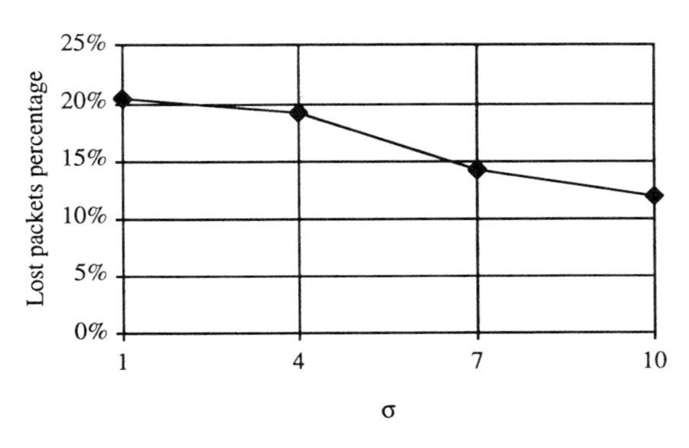

Figure 5. Lost packets percentage vs. σ with load = 1.1.

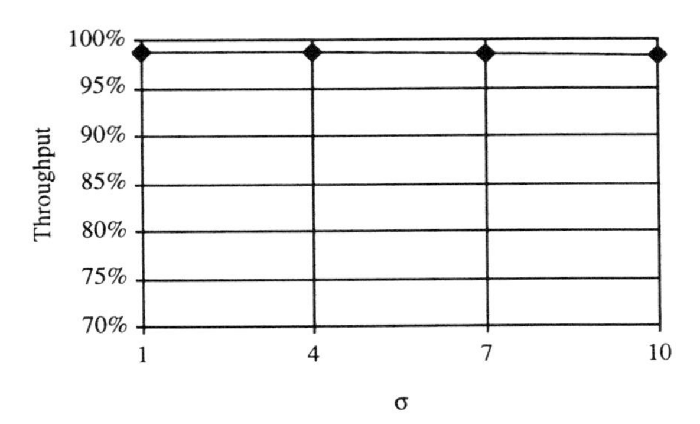

Figure 6. Throughput vs. σ with load = 1.1.

Propagation delay between adjacent nodes = 5 μs

Figs. 4, 5 and 6 represent the system behaviour with increasing values of σ and a fixed offered load of 1.1. Moreover they show the blocked calls percentage, the lost packets percentage and the overall throughput of the network, respectively.

The effectiveness of σ in controlling the share of the capacity between asynchronous and isochronous traffic without influencing the global performance becomes evident from the plots under examination. In fact, for increasing values of σ the blocked calls percentages increase and lost packets percentages decrease, while the throughput does not change.

Behaviour with increasing load

The following plots show the behaviour of the network with increasing values of

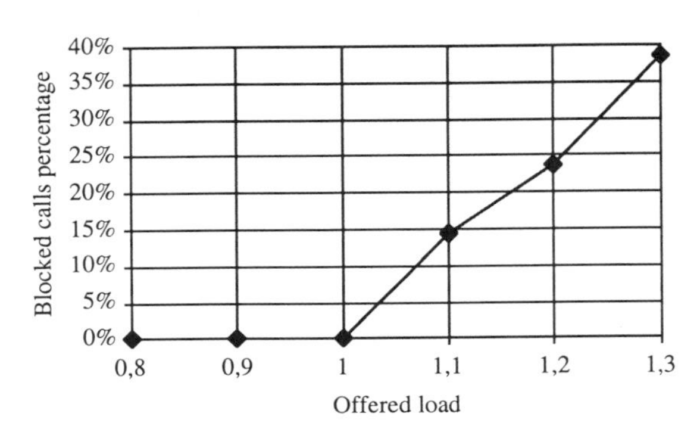

Figure 7. Blocked calls percentage with increasing load and σ =10.

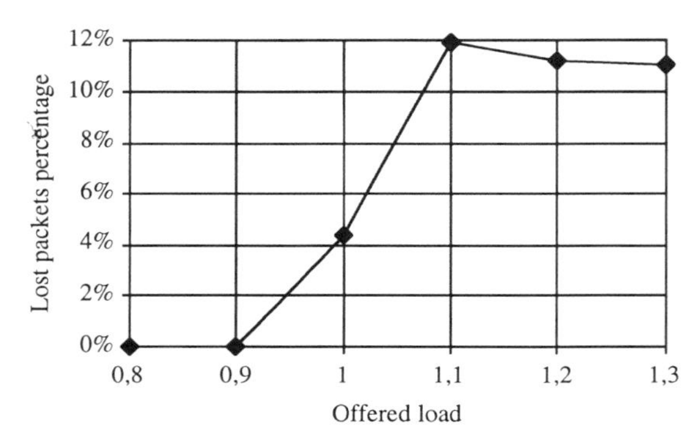

Figure 8. Lost packets percentage with increasing load and σ =10.

offered load. The characteristics of the network and the reference offered load value are the same as introduced in the previous situation. We have chosen σ = 10, which appears to give almost similar performaces between isochronous and asynchronous traffic.

Figs. 7, 8 and 9 show the blocked calls percentage, the lost packets percentage and the overall throughput of the network, respectively. It can be seen that the percentage of refused calls is near zero up to an offered load of 1 and that the percentage of lost packets starts to become significant from an offered load of 1. For saturation values of the offered load, the degradation in performance for packets is faster than for the isochronous traffic; this is possibly due to the fact that the call acceptance is the only directly controlled action in the system. Moreover, for saturation values of the offered traffic, the throughput behaviour (Fig. 9), has values near the physically achievable maximum (100%), which shows that the overall performance of the network is rather satisfactory.

In Fig. 10, the percentage of blocked calls is represented for a value of the offered

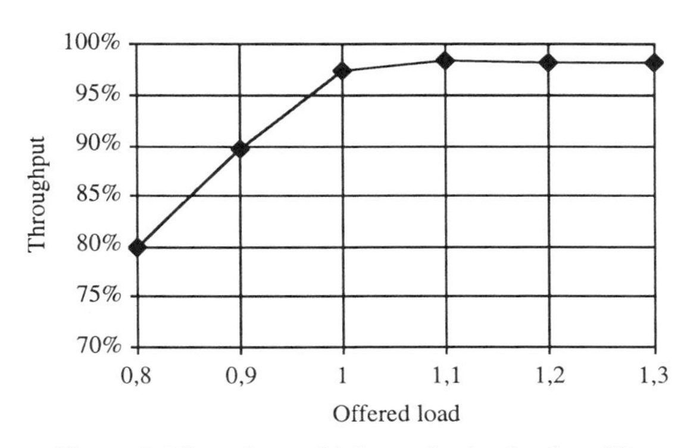

Figure 9. Throughput with increasing load and σ =10.

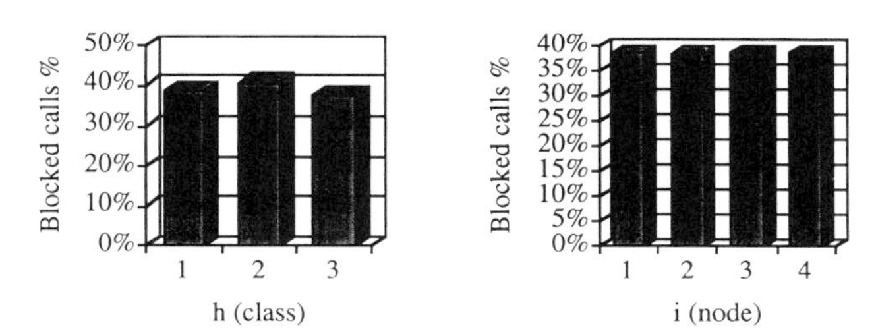

Figure 10. Blocked and generated calls with load = 1.3 and σ =10.

load that significantly exceeds the saturation value. The two plots refer to the overall loss per isochronous class (over all nodes) and per node (over all isochronous classes), respectively. The high losses (owing to the high load) are nevertheless shared quite fairly.

CONCLUSIONS

A dynamic management and control scheme for resource allocation in a MAN structure based on hybrid (i.e., integrating isochronous and asynchronous traffic) frames has been described in the paper. The discrete time analytical model takes into account the presence of multiple isochronous traffic classes, differentiated by user station and transmission speed, whereas packets are supposed to be handled by a distributed multiaccess protocol, whose details must not be necessarily known to the network manager.

The scheme has been tested by simulation. The results obtained show a good behaviour for what regards the optimization of network bandwidth utilization, the ability of control the balancing between isochronous and asynchronous traffic by using different values of the wheight σ and the fair servicing of the different users and of the different classes inside each user.

ACKNOWLEDGEMENT

This work was supported by the Italian Ministry of Education and Scientific Research.

ABOUT THE AUTHORS

Raffaele Bolla was born in Savona, Italy, in 1963. He took the "laurea" degree in Electronic Engineering from the University of Genoa in 1989. Since 1990 he has been a Ph.D. student in Telecommunication Networks at the Department of Communications, Computer and Systems Science (DIST) of the University of Genoa. His research interests are in management and control of STM and ATM networks, multimedia communications, and dynamic routing and reconfiguration in optical networks.

Franco Davoli was born in Genoa, Italy, in 1949. He received the "laurea" degree in Electronic Engineering from the University of Genoa in 1975. Since 1985 he has been Associate Professor and since 1990 Professor of Telecommunication Networks at the University of Genoa, where he is with the Department of Communications, Computer and Systems Science (DIST). From 1989 to 1991 he was also with the University of Parma, Italy, where he held a course in Telecommunication Networks. His current research interests are in bandwidth allocation, admission control and routing in STM and ATM networks, multimedia communications and services, and integrated mobile packet radio networks.

Piergiulio Maryni was born in Genoa, Italy, in 1965. He took the "laurea" degree in Electronic Engineering from the University of Genoa in 1990. Since 1990 he has been a Ph.D. student in Telecommunication Networks at DIST-University of Genoa. In 1993, he spent one year at the Center for Telecommunications Research (CTR) of Columbia University in New York. His research interests are in management and control of STM and ATM networks and multimedia communications .

Giorgio Nobile was born in Savona, Italy, in 1968. He was a student at the Faculty of Engineering of the University of Genoa, where he took the "laurea" degree in Electronic Engineering in 1993, with a thesis on resource allocation and access control in a DQDB network. He is currently studying for a master degree in Business Administration.

Giuseppe Pitzalis was born in Savona, Italy, in 1968. He was a student at the Faculty of Engineering of the University of Genoa, where he took the "laurea" degree in Electronic Engineering in 1993, with a thesis on resource allocation and access control in a DQDB network. He is currently doing the military service in the Italian army.

Alberto Ricchebuono was born in Savona, Italy, in 1968. He was a student at the Faculty of Engineering of the University of Genoa, where he took the "laurea" degree in Electronic Engineering in 1993, with a thesis on resource allocation and access control in a DQDB network. He is currently studying for a master degree in Computers and Telecommunications.

REFERENCES

1. R. M. Newman, Z. L. Budrikis, J. L. Hullet, "The QPSX Man", *IEEE Commun. Mag.*, vol. 26, pp. 20-28, April 1988.
2. M. Zukerman, "Circuit allocation and overload control in a hybrid switching system", *Comput. Networks and ISDN Syst.*, vol. 16, pp. 281-298, 1989.
3. M. Zukerman, "Bandwidth allocation for bursty isochronous traffic in a hybrid switching system", *IEEE Trans. Commun.*, vol. COM-37, pp. 1367-1371, Dec. 1989.
4. B. Kraimeche, M. Schwartz, "Analysis of traffic access control strategies in integrated service networks", *IEEE Trans. Commun.*, vol. COM-33, pp. 1085-1093, Oct. 1985.
5. K.W. Ross, B. Chen, "Optimal scheduling of interactive and non interactive traffic in telecommunication systems", *IEEE Trans. Automat. Contr.*, vol. 33, pp. 261-267, March 1988.
6. K.W. Ross, D.H.K. Tsang, "Optimal circuit access policies in an ISDN environment: a Markov decision approach", *IEEE Trans. Commun.*, vol. COM-37, pp. 934-939, Sept. 1989.
7. B.S. Maglaris, M. Schwartz, "Optimal fixed frame multiplexing in integrated line- and packet-switched communication networks", *IEEE Trans. Inform. Theory*, vol. IT-28, pp. 283-273, March 1982.
8. I. Viniotis, A. Ephremides, "Optimal switching of voice and data at a network node", *Proc. 26th IEEE Conf. on Decision and Contr.*, Los Angeles, CA, Dec. 1987, pp. 1504-1507.
9. M. Aicardi, R. Bolla, F. Davoli, R. Minciardi, "A parametric optimization approach to admission control and bandwidth assignment in hybrid TDM networks", *Internat. J. of Digital and Analog Commun. Syst.*, vol. 6, pp. 15-27, Jan 1993.
10. R. Bolla, F. Davoli, "Dynamic hierarchical control of resource allocation in an integrated services broadband network", *Computer Networks and ISDN Syst.*, vol. 25, no. 10, pp. 1079-1087, May 1993.
11. R. Bolla, F. Davoli, "A traffic control strategy for a DQDB-type MAN", *Proc. 16th IEEE Annual Conf. on Local Comput. Networks*, Minneapolis, MS, Oct. 1991, pp. 195-203.
12. R. Bolla, F. Davoli, "Connection and bandwidth management in a DQDB network", *Proc. Internat. Workshop on Advanced Commun. and Appl. for High Speed Networks*, Munich, March 1992, pp. 327-334.
13. K. Malinowski, "Hierarchical control under uncertainty: formal and practical aspects", *Proc. 5th IFAC Symp. on Large Scale Systems*, Berlin, Germany, Aug. 1989.
14. D. Bertsekas, R. Gallager, "Data networks", Prentice-Hall, Englewood Cliffs,1987.
15. IEEE 802.6 Distributed Queue Dual Bus - Metropolitan Area Network - Draft Standard - Version D.15.

AUTHOR INDEX

SUBJECT INDEX